# Constructive Modeling of Structural Turbulence and Hydrodynamic Instabilities

# Constructive Modeling of Structural Turbulence and Hydrodynamic Instabilities

O. M. Belotserkovskii

Russian Academy of Sciences, Russia

World Scientific

NEW JERSEY • LONDON • SINGAPORE • BEIJING • SHANGHAI • HONG KONG • TAIPEI • CHENNAI

*Published by*

World Scientific Publishing Co. Pte. Ltd.
5 Toh Tuck Link, Singapore 596224
*USA office:* 27 Warren Street, Suite 401-402, Hackensack, NJ 07601
*UK office:* 57 Shelton Street, Covent Garden, London WC2H 9HE

**British Library Cataloguing-in-Publication Data**
A catalogue record for this book is available from the British Library.

**CONSTRUCTIVE MODELING OF STRUCTURAL TURBULENCE AND HYDRODYNAMIC INSTABILITIES**



ISBN-13 978-981-283-301-3
ISBN-10 981-283-301-3

Printed by FuIsland Offset Printing (S) Pte Ltd, Singapore

# Preface

The book provides an original approach in the research of structural analysis of free developed shear compressible turbulence at high Reynolds number on the base of direct numerical simulation (DNS) and instability evolution for ideal medium (integral conservation laws) with approximate mechanism of dissipation (FLUX dissipative monotone "upwind" difference schemes) and *does not use any explicit sub-grid approximation and semi-empirical models of turbulence.* Convective mixing is considered as a principal part of conservation law. Appropriate hydrodynamic instabilities (free developed shear turbulence) are investigated from unique point of view. It is based on the concept of large ordered structures with stochastic core of small scale developed turbulence ("turbulent spot"). Decay of "turbulent spot" are simulated by Monte Carlo method. Proposed approach is based on two hypotheses "statistical independence of the characteristic of large ordered structures (LOS) and small-scale turbulence (ST)" and "weak influence of molecular viscosity (or more generally, dissipative mechanism)" on properties of large ordered structures.

Two versions of instabilities, due to Rayleigh–Taylor and Richtmyer–Meshkov are studied in detail by the three-dimensional calculations, extended to the large temporal intervals, up to turbulent stage and investigation turbulent mixing zone (TMZ).

The book covers both the fundamental and practical aspects of turbulence and instability and summarizes the result of numerical experiments conducted over 30 years period with direct participation of the author.

**Appreciation of contribution Academician O. M. Belotserkovskii to the world science are given by Academician A. S. Monin (Russia), Prof. Y. Nakamura (Japan, Nagoya University) and Prof. F. Harlow, (USA, Los-Alamos).**

## *OPINION by Academician A. S. Monin (Russia)*

Reproduction of coherent structures in turbulent flows by numerical calculation of the hydrodynamic equations — is very complicated problem, especially at high Reynolds number, when the spectrum of turbulent structures with turbulent cores is very wide and it is practically impossible to provide their full resolution from max to min scale even with the help of the modern super computers.

That is why the modeling of the large vortices is needed to be conducted separately from numerical description of small-scale turbulence taking into account molecular viscosity (which at present time is accessible only at $Re < 1000$). Original approach to the task of estimation LOS is developed in the works of Academician O. M. Belotserkovskii and his school. It consists in *the Direct Numerical Simulation* (DNS) of the hydrodynamic ideal flow. For this task non-stationary Euler equations, evaluating proper conservation laws with *approximate* account of energy dissipation created by small-scale turbulence are used. Turbulent flow is represented as relatively ordered slow movement of coherent structures — weakly unstable large structures, which transfer flow with developed turbulence from one flow area to another. This approach can be used also for description of "residual" coherent structures of bifurcation origin (dynamic structures).

Academician RAS A. S. Monin

["About coherent structures in turbulent flows" in the book. "Sketch about Turbulence" (Moscow, Nauka, 1994) pp. 7–17.]

See also: Investigation of stochastic turbulence, A. S. Monin, A. M. Yaglom, Statistical fluid mechanics, eds. J. Lumiey (MIT Press, Cambridge, MA; Vol. 1, 1971; Vol. 2, 1975).

## *OPINION by Professor Yoshiaki Nakamura (Japan)*

Although the theoretical studies of turbulence have a long history, physical principles of turbulence are still not well clear (for example, what is the source of the energy of chaotic motion). Currently, some experimental data evidently point at the existence of Large-Scale Ordered ("coherent") Structures (LOS), especially in the case of fully developed turbulence, where the basic part of the transferred energy is disposed. The Karman's Lecture of 1976 delivered by Professor Oleg M. Belotserkovskii was one of the turning points in the theoretical study of turbulence (see Appendix A to this book).

Since 1976 the LOS as well as the related hydrodynamic instabilities have been intensively investigated by Russian researchers headed by Professor O. M. Belotserkovskii. "Constructive Modeling of Structural Turbulence and Hydrodynamic Instability" is one of the most influential books on the subject ever to appear. It is devoted to the studies of multi-dimensional developed turbulence and hydrodynamic instability on the basis of structural analysis and numerical calculations with the usage of high performance computers. This monograph puts together the most important works of Professor O. M. Belotserkovskii and his school for more than 30 years. It demonstrates constructive approach to calculation of Large-Scale Ordered Structures (LOS), which transfer energy. And together with the stochastic models such an approach allows to study all phenomena as a whole. One of the most interesting results is that the inner structure of "large vortices" has stochastic character of "Small Turbulence" (ST). Calculations of LOS were made on the base of Integral Conservation Laws (ICL) without taking into consideration for molecular viscosity effect, of which influence on the characteristics of big structures is quite insignificant. Such approach allowed to decrease demands to possibilities of Computers (power and operational memory) and conduct researches of a number of important tasks such as: wakes after moving objects, shear layers, jets and other phenomena. It is important to stress that for LOS calculations one should not use semi-empirical models of turbulence and sub grid approximations. Investigations of ST were carried out on the basis of rational approaches to spectral methods (Navier–Stokes equations) by F. Harlow, Y. Kaneda or at the kinetic level by the Monte Carlo approach (O. Belotserkovskii, Y. Chlopkov, V. Vanitskii).

The revised approach to the studies of structural turbulence problems and hydrodynamic instability was presented by the author at the lectures and workshops in Belgium at Karman's Lecture, in the USA on lecture in Mathematical Institute of R. Courant, at the work-shops in Los Alamos organized by Dr. F. Harlow and also at numerous seminars in Germany, Poland, Japan, India and other countries. Russian–Japanese cooperation in this area appeared very successful, which found acknowledgement in numerous workshops and joint project "Turbulence and Instability" between Nagoya University, Japan and Institute for Computer-Aided Design, Russian Academy of Sciences.

We can assert that the monograph by Academician O. M. Belotserkovskii is based on the structural analyses of the nature of the investigated phenomena and the achieved results demonstrate it.

In a whole the book is outstandingly interesting and can be useful from both theoretical and practical point of view.

Professor Yoshiaki Nakamura 16 May 2008
Department of Aerospace Engineering
Nagoya University

See also: O. Belotserkovskii, Y. Kaneda, and I. Menshov (eds.), *Investigation of Hydrodynamical Instability and Turbulence in Fundamental and Technological Problems by means of Mathematical Modeling with Supercomputers* (Nagoya-Press, Japan, 2007).

## *OPINION by Professor F. Harlow and Dr. Hopson (USA, Los Alamos)*

### Memorandum of understanding

For assumed cooperation between the Institute for Computer Aided Design, Russian Academy of Sciences (ICAD RAS), and Group T-3, Los Alamos National Laboratory (LANL).

Academician Oleg Belotserkovskii, Director of the ICAD RAS, visited the Los Alamos National Laboratory, Group T-3, in a consultant position from September 29, 1994, to January 11, 1995. During this time he established various scientific contacts and recognized different scientific and applied directions that represent mutual interest for common activity between ICAD RAS and Group T-3 and with other Divisions of LANL. Academician Oleg Belotserkovskii presented 14–15 official seminars and lectures representing topics of significant interest to Los Alamos. The main directions of the scientific activity of Oleg Belotserkovskii were the following:

- direct modeling of turbulence (large ordered structures, turbulent background, transition to chaos);
- instabilities studies: Rayleigh–Taylor, Richtmyer–Meshkov instabilities (multidimensional and multi-mode interactions, nonlinear late-time stages, a sequential transition to turbulence, etc.);
- development and adapting of new algorithms for the calculations of unsteady multi-dimensional (2D-3D) problems (FLUX-method, grid-characteristic approach, and others);

- Monte-Carlo technique;
- difference schemes with positive approximation on unstructures grids;
- practical applications (computational fluid dynamics, solid mechanics plasma physics);
- application of mathematical methods and computers in medicine;
- demonstration diskettes using a color visualization and others.

For the above-mentioned activities Academician Oleg Belotserkovskii was very highly qualified. Two reports are being prepared for publication by LANL on the mentioned subjects.

The visit of Acad. Belotserkovskii developed much friendship and was a great pleasure for those of us who had the opportunity for extensive discussions.

For Group T-3, LANL
Dr. John Hopson, Group Leader of T-3
Dr. Frank Harlow, Group T-3
January 6, 1995

# Introduction

The problems of research on turbulent and hydrodynamic instabilities occupy the scientific minds for many years. The creation of highly efficient computational technique and its usage gave a new impulse to the study of such complicated phenomenon, which are described by "multidimensional nonlinear equations in partial derivatives with complicated internal structure." Many ideas developed in this monograph have been published earlier in Russia and elsewhere, but this book is the most complete one, in a certain way it synthesizes the ideology of computational turbulence, suggested by the author and his followers.

The history of the turbulence theory begins from the classical researches of Reynolds and Richardson. In the middle of 1920s, Keller and Fridman proposed the idea of stochastic description of turbulent processes that had invaluable principal significance. A number of important achievements were made (for instance, Kolmogorov–Obukhov spectrum). Nevertheless, the physical principles of turbulence development remain unclear (for instance, what is the source of energy of chaotic motion). Currently, the experimental data point clearly to the existence of the large-scale "coherent" structures (especially, for fully developed turbulence), where the main part of energy transferred is disposed. Let us note that, *for high Reynolds number* (Re), *the energetic part of spectrum is far from the dissipative one in terms of wave numbers.*

The last decades have been marked by a new approach[1,2] in the study of turbulence, namely, by the direct numerical simulation of the processes of hydrodynamic flow on the basis of the solution of hydrodynamic and kinetic equations. The results of such an approach demonstrate an important difference from the traditional statistical methods. *For instance, the significant role of large-scale "ordered" structures in the turbulence development was shown.*

The advent of the highly efficient computational technique made it possible to realize the simulation of the quite complicated phenomena and processes in mechanics, physics, and even in ... medicine. Generally speaking, here we have to do with multi-dimensional (spatial) nonstationary problem, whose original mathematical formulation is rather confusing. The solution to such problems is carried out, as a rule, within the frame of a **computational experiment**, when the systematical computations might be realized with the variations of the original problem setting, so that the eventual results correspond both to the criteria prescribed and (within certain limits) to the initial physical process. Moreover, if one takes into account the need to the development of the effective "rational" numerical methods (which would permit to bring about their realization using the "admissible" computer time), then it becomes quite clear that the organization of computational cycle on the whole demands an utmost skill.

Quite evidently, such a class of problems belong to those of a numerical simulation of the phenomena of turbulence and of the instabilities' development. Taking into account the *structural* character of turbulence (large-scale "coherent" structures, statistical background, laminar–turbulent transition, etc.), which has, generally speaking, various mechanisms of interaction, it seems necessary to consider also various original formulations (sets of equations), which would be the most adequate to the processes under investigation. Only by taking into account all the factors indicated it might prove possible to accomplish effectively the numerical realization of the wide spectrum of the problems of turbulence.

Treated in Chapter 1 is one of the complicated types of fully developed shear turbulent flows, namely, the phenomena related to the high Reynolds number within the wake (both the close-range and long-range ones) behind the moving body, oceanic flows, etc. In this case apart from the three dimensionality and unsteadiness, one takes into consideration the medium's compressibility and the effect of viscosity (where the prevailing role is played by the molecular mechanism of interaction) as well. Thus, all the observations, as a whole, form the basis for the "rational" approach, which is not quite ordinary by the numerical modeling of structural turbulence.

## Main Ideology

1. To put it briefly, the main features of our approach might be described as follows. For the wide class of phenomena of that type, by the high

Reynolds numbers within the low-frequency and inertial intervals of turbulent motion, the effect of molecular viscosity and of the small flow elements in the largest part of perturbation domain is not practically essential neither for the general characteristics of macroscopic structures of the flow developed, nor for the flow pattern in general. This makes it possible to disregard the effects of molecular viscosity when studying the dynamics of large vortices, and to implement their study on the basis of the models of the ideal gas — "discrete" Euler equations for compressible gas (using the methods of "rational" averaging, but without applying any explicit subgrid approximation and semiempirical models of turbulence). The main ideology of "rational" approach for direct numerical simulation (DNS) of the characteristics of the fully developed shear turbulence by high Re — the investigation of the large-ordered structure (LOS) and small-scale stochastic turbulence (ST) is based on the two hypotheses: the "independence of LOS and ST" and the "weak influence of molecular viscosity (or, more generally, the mechanism of dissipation) on properties of LOS".[1,2] Eventually, Kolmogorov has received the form of a spectrum in an inertial interval assuming nothing in general about a kind of dissipative mechanism. Among the problems to be studied here are those of the jet-type flow in the wake behind the body, the motions of ship frames with stern shearing, the formation of anterior stalling zones by the flow about blunted bodies with jets or needles directed to meet the flow, etc.

2. At the same time, the properties of flows within the boundary layers and within the thin layers of mixing, at the viscous interval of turbulence, as well as those of flows by the moderate Reynolds numbers and in the domain of laminar–turbulent transition, are primarily determined by the molecular diffusion, and for these flows it is necessary to consider Navier–Stokes models.
3. The pulsational motions in turbulence are of chaotic type and have an unstable, irregular character, thus constituting a stochastic process. In this respect, one can speak here only about obtaining the mean characteristics of that type of motion (like the moments of various orders) by means of the statistical processing of the results using kinetic Monte Carlo approaches.

The studies of various kinds of instabilities (like, for example, **Richtmyer–Meshkov, Rayleigh-Taylor, Kelvin–Helmholtz instabilities**) are of high interest, especially for the calculations extended to the large temporal intervals, up to the turbulent stage. The main difficulty, which one encounters in considering the above-indicated class of problems, consists in the development of the **general concept** of the creation of constructive numerical models of turbulence, of the instabilities' development, and of the transition to chaos.

When studying complicated physical phenomena that take place either on the galactic scale, for example, in astrophysical investigations of the substance-mixing processes in supernovas, or on the microscale inherent to nuclear physics, for example, in realizing the inertial confinement fusion, there arises the necessity to analyze physical mechanisms and their adequate description for various hydrodynamic instabilities. These are the Rayleigh–Taylor, Kelvin–Helmholtz, and Richtmyer–Meshkov instabilities. As numerous experimental and theoretical investigations show, the change of dimensional characteristics of phenomena, namely, the transition from two-dimensional to three-dimensional flows is accompanied by the appearance of new physical effects. In lower dimension problems, these effects are either absent or manifest themselves to the degree inaccessible for observation. Among the hydrodynamic instabilities, the three-dimensional flows formed when developing the Richtmyer–Meshkov instability are the least studied because, aiming to realize their numerical simulation, it is necessary, not only to have an extended difference grid, but also higher-quality algorithms for considering strong discontinuities, in particular, shock waves and their interactions. It is this fact that substantially complicates experimental diagnostic investigations. In the numerical simulation of Rayleigh–Taylor and Kelvin–Helmholtz instabilities, it was established that, for the identical initial amplitudes of perturbations and wavelengths, the growth rate for perturbations is higher in the three-dimensional case as compared to the two-dimensional one, while the process of formation of mushroom-shaped structures proceeds more slowly. Similar results for the Richtmyer–Meshkov instability were obtained by other experiments. In this connection, two types of questions arise. They are associated, first, with studying the physical mechanisms that lead to the observable phenomena, and, second, with establishing the relationship between the growth rates for perturbation amplitude and a number of geometrical and physical quantities. These are the amplitude and the duration of the perturbation, its shape, the Mach and Atwood numbers, the thermodynamic properties of substances etc. which require systematic investigations (both theoretical and experimental).

Presented in Chapter 2 are the results of comparative calculation studies of the development of Richtmyer–Meshkov instability (RMI) for the two-dimensional Cartesian case and for the corresponding axially symmetrical and three-dimensional versions.

The investigation of Rayleigh–Taylor instability (RTI) is of considerable practical interest at present. A numerical experiment is an effective means of investigation. The solutions to the problem of a nonlinear phase of spatially three-dimensional perturbations have been obtained through it, and the spatially two-dimensional RTI problem has been investigated in detail. There is an urgent need to study RTI because it must be taken into account in analyzing problems of practical importance, concerning the instability of the boundary between detonation products and a gas in an explosion, the compression of spherical laser targets, etc.

In the analysis of the actual RTI processes, it is necessary to investigate not the individual perturbations with a fixed wavelength but the development of a collection of perturbations with different wavelengths and the interaction of these perturbations.

Chapter 3 represents the results of the numerical experiment carried out by the various methods. A complete vortical system of Euler equations of the compressible medium motion is solved. We consider the influence of the interaction of two or more waves on the evolution of the contact surface in an RTI for the two- and three-dimensional cases.

Finally, Chapter 4 is devoted to the investigation of the turbulence problems, using the statistical Monte Carlo method. The approach of that kind proved to be highly effective when considering the problems of rarefied gas-dynamics. However, as concerns the turbulent motions, the possibility of the application of Monte-Carlo methods is still evaluated on a comparatively small scale.

There are also five appendices in the book.

Appendix A is a reprint of my paper[1] dated 1976, where the main principles of our ideology were published at first. I also included in the book two papers — "Formation of large-scale structures in the gap between rotating cylinders: the Rayleigh–Zeldovich problem" (Appendix B) and "Universal technology of parallel computations for the problems described by systems of the equations of hyperbolic type: a step to supersolver" (Appendix C) published in the last two years. Appendix D presents itself my scientific report "Mathematical modeling using supercomputers with parallel architecture" given on October 15, 2003, at the meeting of Presidium of Russian Academy of Sciences.

And at last in Appendix E, "On nuts and bolts of structural turbulence and hydrodynamic instabilities", are formulated the main positions for the analyses of so complicated and multifunctional problems, that it appeared as the result of many discussions (Academicians O. M. Belotserkovskii, Dr. A. M. Oparin, Dr. O. V. Troshkin, and Dr. V. M. Chechetkin). This part can be used as a program for the further researches, which is opened for talks over.

The classical fundamental results of research in the area of turbulence and instabilities are well known. One might cite the monographs and papers by G. I. Taylor, G. K. Batchelor, H. Lamb, A. N. Kolmogorov, A. M. Obukhov, F. Harlow, J. O. Hinze, W. Heisenberg, L. D. Landau and E. M. Lifshitz, A. S. Monin and A. M. Yaglom, A. Einstein, C. C. Lin, A. A. Townsend, V. M. Ievlev, and many others. Actually, following these works, the elaboration of the general numerical methods for solution of nonlinear problems of aerodynamics, turbulence, and plasma physics was carried out.

The approaches which are mainly discussed further on, are those which use (for the description of the free-developed turbulent flow for extended temporal intervals) the **complete** (and **closed**) set of the dynamical equations for true values of velocities and pressure, as well as the statistical methods. The combined application of both these approaches (based on the use of hydrodynamic equations and on the statistical Monte Carlo method) permits to understand in more detail the structures of turbulence, and to determine the rational ways of constructing the corresponding mathematical models. The series research in this field was started from Karman's Lecture[1] given by O. M. Belotserkovskii in March 15–19, 1976. This new approach to the study of turbulence was published (in rather detailed form) in a paper.[2]

One should admit that some postulates of this approach have a heuristic and intuitive character but the results obtained speak for themselves. The overviews of these approaches were presented in different periods at the seminars of Academicians P. L. Kapitza, N. N. Yanenko, V. V. Struminskii, A. M. Obukhov, A. S. Monin, and, also, G. Batchelor, H. Daiguji *et al.* The detailed discussion was held in 1994–1995 in Los Alamos (USA) in collaboration with Dr. F. Harlow and his colleagues, where the author spent about 4 months. As a rule, the concepts proposed were apprehended quite adequately. One might think that a sufficiently minute account on the approach developed would be of an indubitable interest both from the purely academic point of view and for the practical experts in the corresponding field.

I thank V. V. Demchenko for his help in writing Chapter 2 and V. E. Yanitskii for providing the materials for Chapter 4.

I specially thank my colleagues Dr A. M. Oparin, Dr V. M. Chechetkin, and Dr O. V. Troshkin, who worked together for many years and who pleasantly gave material for this book.

I also thank my assistants I. Tarkhanova, N. Nosova, and Dr A. I. Lobanov for helping me to prepare the manuscript for publication.

I express my gratitude to the Director of the Abdus Salam International Centre for Theoretical Physics (ICPT) — Professor Katepalli R. Sreenivasan for suggesting me to the publication of the book *Constructive Modeling of Structural Turbulence and Hydrodynamic Instabilities* in "World Scientific".

I will also be grateful to the readers for any remarks and suggestions, which I take into account in the next work.

I invite readers to the mutual cooperation!

O. M. Belotserkovskii
Academician of the Russian Academy of Sciences
Institute for Computer-Aided Design
President of Russian-Indian Centre for
Advanced Computing Research

19/18 2-nd Brestskaya st.
Moscow 123056, Russia

E-mail : o.bel@icad.org.ru

December 2007

## References

1. O. M. Belotserkovskii, Computational experiment: direct numerical simulation of complex gas-dynamics flows on the basis of Euler, Navier–Stokes, and Boltzmann models, Karman's Lecture, von Karman Institute for Fluid Dynamics, 15–19 March 1976, in *Numerical Methods in Fluid Dynamics*, eds. H. J. Wirz and J. J. Smolderen (Hemisphere, Washington, London, 1978), pp. 339–387.
2. O. M. Belotserkovskii, Direct numerical modeling of free induced turbulence, *Zh. Vychisl. Mat. Mat. Fiz.* (in Russian) **25**(12), 1857–1883 (1985) [translated in *J. Comput. Math. Math. Phys.* **25**(6), 166–183 (1985)].

# Contents

# Chapter 1

# Constructive Modeling of Free Developed Turbulence — Coherent Structures, Laminar–Turbulent Transition, Chaos

Approaches to numerical simulation of the free developed shear turbulence are discussed. A discrete dissipation model, obtained from the nonstationary averaged Euler equations (integral conservation laws) is proposed for the direct simulation of coherent ("ordered") structures in the low frequency and inertial intervals of turbulence. The dissipation mechanism is generated by corresponding "rational" averaging (asymmetric compact difference schemes of a high order of accuracy — the smoothing filter). The stochastic component of turbulence is modeled at the kinetic level by a statistical Monte Carlo method. The laminar–turbulent transfer and the transition to chaos are analyzed on the basis of the full Navier–Stokes equations. The efficiency of the proposed algorithms is demonstrated for the problem of aseparated flow around bodies, valuation of the turbulent wake, gas-dynamics of clean rooms, oceanic circulation, stratified flows, Rayleigh–Taylor's instability and simulation of a "forest-fire" problems for two- and three-dimensional cases.

## 1.1. Introduction

It is already for several decades that the problem of study of the hydrodynamic turbulence occupies the heads of scientists, engineers, designers and even of politicians. Thus, in a rather short paper on turbulence written by the French scientists,[a] the authors have chosen as an epigraph the words by the famous French politician G. Pompidoux: "Everything will disappear into the turbulence!"

It is very interesting and at the same time very hard to speak on turbulence. It is reported that in 1961, at the inauguration ceremony of the

[a]Brissaud A., Frish V., Leorat I. *et al.* Catastrophe energetiqué at nature de la turbulence // *Ann. geophys.*, **29**(4), 539–546 (1973).

Institute of Statistical Methods of Turbulence in Marseille, where among the persons present were von Kármán, Kolmogoroff and Taylor. Professor Kármán in his introductory speech has mentioned that when he, at last, will appear before Creator, then the first thing he will ask God would disclosing the secret of turbulence.

There are a great number of different turbulent flows observed in nature (see, e.g., Ref. 1). The most of fluids, which one could find on the Earth, are in the turbulent state. However, prior to speaking on the essence of these phenomena, one should formulate the definition of hydrodynamic turbulence. And at this point we found ourselves practically powerless, for up to the present moment there is no stable and unique definition of turbulence.

The phenomenon of turbulence is rather complicated and multilateral one. The essential part is played here by a great diversity of spatial structures of flow. Quite frequently, the initiation of nonstationary turbulent pulsations is preceded by the appearance of extremely complicated quasi-stationary formations, and so on.

The aim of hydrodynamics consists in the description and forecast of the motion of fluids induced by certain forces. As concerns the Newtonian incompressible fluid, the magnitude of these forces in many situations is described by dimensionless Reynolds number ($Re$). When this number is small, the dynamics of fluids is rather simple due to the existence of unique correspondence between the prescribed boundary and volumetric influences and the motion forecasted. As the Reynolds number is increased, the uniqueness of a solution disappears (steady flows have extremely complicated structure), and from the multitude of solutions one should single out the stable ones and those, which might be observed. The stationary motions might be replaced by periodical, quasiperiodical, or aperiodical ones (in respect to time); the gradual merging of turbulent domains takes place and, eventually, the whole flow proves to be involved in this process.

The first serious scientific study of the transition to turbulence was carried out by Reynolds as early as in 1883. Reynolds noticed that the transition from laminar flow to the "wavy" one occurs by certain critical value $Re_{\mathrm{cr}}$. He wrote: "The general cause of the stationary flow to be changed for a vortical one, as it was noted in 1843 by Professor Stokes, lies in the fact that at certain conditions this stationary flow becomes to be unstable..." Such an instability essentially depends on the nonlinearity. It might be interesting to note that in the already cited paper (see p. 7) was expressed a somewhat different suggestion concerning the nature of appearance of a completely developed uniform and isotropic turbulent flow: the turbulence is not a result of the instability's development occurring due to the increase of the Reynolds number of a viscous flow, but appears after the finite time $t_0$ due to the energetic catastrophe in an inviscid flow (prior to $t_0$ the spatial covariation keeps to be analytical, the spectrum possesses the finite moments of any order, and the energy is conserved; after $t_0$ the covariation is not already analytical, the spectrum with high wave numbers is submitted to the Kolmogoroff's law, and the finite dissipation of energy takes place with zero viscosity).

It is usually assumed that the hydrodynamic turbulence is a purely chaotic (sometimes it is said — nonpredictable) motion of the fluid. While during the first years of the study of turbulence these phenomena were treated as completely stochastic processes, at the present period there occurred, from our standpoint, quite a radical turn toward the understanding of these phenomena. The physical experiments[1] and theoretical investigations have shown (see, e.g., Refs. 2–11 and others) that the wide class of turbulent flows is characterized by the presence of the nonstationary "organized" motion of the large-scale formations having "stochastic" structure.[6] The question of the relationship between the deterministic basis and the chaotic one is presently subjected to an active study.

Somewhat different aspect of the nonlinear mechanics, which is also related to the turbulence and was discovered quite recently, consists in the phenomenon of "stochastization" of the dynamical dissipative systems. As a matter of fact, even some very simple dynamical systems (describing the temporal evolution) reveal the irregular, chaotic behavior reminding the turbulence. The key to understanding this behavior lies in the "perceptible dependence on the initial conditions" (the term introduced by Lerau for the description of systems of similar type[8]). It is quite natural that here arises a lot of questions (concerning the mechanism of transition to chaos, its properties, the systematization of various types of the chaotic dynamics, etc.), but the actual transition to the irregular behavior is realized just by several definite ways. The study of such phenomena (especially for sufficiently "real" dynamical systems) is of an unquestionable interest.

Based on the above considerations one is able to propose the following ("operative") definition of the turbulence. The hydrodynamic turbulence is just a property of vortical flows (of liquid or of a gas) to take up a stochastic character. The problems connected with a study of turbulent flows gave a powerful impetus to the development of the whole branches of mathematics, mechanics and physics. Among those are such areas of science as hydrodynamic instability, theory of bifurcations, strange attractors, fractality, and some others.[7,8,10,11]

The hydrodynamic instability and turbulence are studied intensely for more than a century. However, by these studies the problems of two types were mainly considered: the maximally developed turbulence (with very large Reynolds number), and manifestation of the first instability (bifurcation) within supercritical flow. The situation has drastically changed when the investigators began to use modern computational technique and the advanced mathematical methods. There arose the possibility "to work" with sufficiently complete spatial, nonstationary, nonlinear models, which permitted to enlarge essentially the class of problems under investigation — one is able to consider the properties of the ordered structures, the phenomena of laminar–turbulent transition, to study various "scenarios" of the transition to chaos, etc.

The approaches based on the computational experiment consider within the frame of a single complex the mathematical formulation of a problem, the method of its solution, and the process of its realization on a computer. Such a way of investigation gives a possibility during the calculations either to modify the initial problem's formulation or to make it more precise, depending on the "reality" of the results obtained. To be sure, all these tricks with "a feedback" demand a great artistry to be revealed and proper substantiations to be made. Only the combined analytical, computational, and experimental approaches might lead to a success in the study of such a complicated phenomenon as turbulence.

At this stage arises quite a natural question: on the basis of which models — that of an ideal (inviscid) medium, that of Navier–Stokes equations, or that of a kinetic description should one realize the construction of initial imitational systems for the study of turbulent flows? Restraining ourselves from the detailed consideration of these problems, we are going only to observe that it seems to be unreal to create a universal model of turbulence. At the same time, when judging in accordance to the properties of the structural turbulence, prevailing by the different regimes of motion are, generally speaking, the different types of the perturbations' interaction mechanism. Thus, the undulatory (dynamical) processes play the decisive role by the consideration of large-scale formations for large *Re* numbers, the molecular (viscous) effects reveal themselves within the narrow zones of merging — the boundary layers, the domains of laminar–turbulent transition, the range of dissipation; finally, the "random" stochastic (pulsational) processes should be taken into account by the investigation of a "stochastic" component of turbulent motions. From there follows a necessity in the "construction" of various adequate mathematical models. It seems to be possible to claim that in the nearest future the success of inculcation of the methods of computational experiment into the practice of the applied and fundamental research will depend not only (and not decisively) on the computer's power, but on the quality of working out and realization of the numerical models (brain-ware).

The contents of the present book is based on the original papers of some of the authors who, by the way of action in different directions (theoretically, numerically or experimentally), are studying the wide class of the problems of hydrodynamic turbulence and instabilities with the help of various approaches.

At the present time, one of the actual problems of continuum mechanics and plasma physics is that of modeling of the complicated transitional and turbulent motions (here included are the multi-dimensional problems, those with consideration of the medium's compressibility, for various regimes of motion, in the wide range of variation of the flow's parameters, and so on) with the help of a modern computational technique, of the algorithms and approaches of the applied mathematics. The commonly encountered in practice, so-called transitional flows, are characterized by the clearly pronounced nonstationarity and nonlinearity of the processes involved, by the presence of the large medium's displacements, by the diverse and complicated mechanism of interaction, and by the energy's dissipation. Among the problems are, for example, such as gas-dynamic ones, transonic diffraction problem of the flow within a wake behind the moving body, of the interaction of a jet injected with a main stream, etc. In the mechanics of a solid deformable body, the transition phenomena take place, for example, by study of elasto-visco-plastic states. In the plasma physics, the processes of similar type are observed by the numerical simulation of the problems of interaction of a powerful laser radiation with a matter, etc. In some cases (flow within a wake, that with "blow-in"; processes accompanying the laser compression of shells, and so on) the phenomena involved are of a *turbulent character*.

The main problem of the theory of turbulence consists in the study of general dynamics of turbulence and the nature of its development, that is, in the study of evolution of large-scale formations and of the statistical representation of a turbulent motion in the core as the time increases. As it is usually assumed, the spatial and temporal variation of the velocity distribution is determined by Navier–Stokes equations. In principle, the analysis of development of turbulence and of its degeneration might be conducted by one of two ways[b]:

(i) the first one consists in a general solution of Navier–Stokes equations with arbitrary initial distribution of velocities in the space;
(ii) the other way is reduced to a derivation of the sole dynamic equation, which would describe the temporal variation of the total probability distribution (a problem of presentation of statistical distribution in a functional space).

It is quite natural that the main characteristics of turbulent motions — the "structural" character of turbulence, the nonstationarity and nonlinearity of the processes involved, the possibility of a transfer of molecular "groups" (*molar transfer*), the existence of a continuous flow of energy over the cascade of vortices, the different mechanisms of interaction for different scales of motion, the phenomenon of viscous dissipation, etc. — should be, as much as possible, reflected during the numerical simulation of turbulence. Such a simulation might pursue, generally speaking, the following three goals:

(i) determination of the main characteristics of large-scale turbulence without the detailed study of its structure at the limiting regimes of motion (when $Re \to \infty$);
(ii) calculation of the area of transition of laminar flow into turbulent one ("disintegration" of a stable laminar flow by $Re_{\mathrm{cr}}$);
(iii) analysis of the statistical turbulent characteristics (including the effects of fluctuations) and of the transition to chaos.

The advent of computers and the development of efficient numerical methods for solving nonlinear problems in mathematical physics and mechanics have created the basis for the *direct* numerical simulation of complex fluid flows, turbulence included. The use of a *numerical experiment* substituting a physical one opens new vistas for understanding natural phenomena; it also permits us to estimate relative significance of different factors and to determine the limits of the applicability of abstract schemes and mathematical models. The numerical experiment encompasses the mathematical formulation of the problem, the optimal method of solving it

[b]See, for example, Refs. 2, 4 and 11.

and the process for realizing it. Actually, it analyzes a model system, which provides an adequate (in a certain sense) description of the most important characteristics of the phenomenon under study. The initial problem formulation is refined in the course of the calculations, and the presence of feedback ensures results that are quite reliable with a moderate cost in computer time.

Below we consider a direct numerical analysis of a modern problem in aerodynamics based on *comprehensive models without employing semiempirical theories.* We focus on the study of separated (generally, turbulent) flows in the wake behind a body for the critical flow modes corresponding to high Reynolds numbers, $Re$[12] and within the region of the laminar–turbulent transition. The major objectives of this paper are the construction of discrete nonequilibrium dissipation models (corresponding to this class of flows) and the development of intelligent numerical algorithms that are realizable on modern computers. This chapter mainly deals with the general concept of the *direct numerical simulation of turbulence*, a subject pursued by Academician Belotserkovskii for many years (see Refs. 12–14) and implemented by his pupils and colleagues at the Institute of Computer Aided Design and the Computer Center of the Russian Academy of Sciences. A number of original papers on this problem are published under this cover.

The study of separated flows in the wake behind a body is closely related to the analysis of free developed shear turbulence. In analyzing this class of flows, the study of averaged large-scale macrostructures and statistical characteristics of turbulence in itself seems to be both feasible theoretically and justified from the physical point of view. To our mind, the pivotal question is: "On the basis of which models, viz., that of an ideal medium, that of the Navier–Stokes equations, or those on a kinetic level, should schemes be constructed for studying the separated and turbulent flows corresponding to various types of motion?" Our approach is based on the following considerations.

For large Reynolds numbers in the low-frequency and inertial-scale ranges of turbulent motion, the effect of the molecular viscosity and small flow elements in the larger portion of a perturbed-flow region on the general characteristics of developed-flow macrostructures and the flow pattern as a whole is practically insignificant for a broad class of flows of this type, and thus allows us to ignore the molecular viscosity effects while studying the dynamics of large vortices and to analyze them on the basis of ideal-medium models (using a "rational" averaging but without employing semiempirical models of turbulence). This refers to jet flow problems of the wake behind a body, the motion of a ship with a truncated stern, the formation of bow separated flows past blunt bodies with jets or needles, etc. At the same time, the properties of flows in boundary layers, thin mixing layers, within the viscous-turbulence range, and for moderate Reynolds numbers within the range of the laminar–turbulent transition are primarily defined by molecular diffusion and must be considered on the basis of the Navier–Stokes model. Fluctuating motions in turbulence are unstable and irregular: in fact, this is a random process.

Therefore, these motions may only be described by averaged characteristics (moments of different order) obtained by an appropriate statistical reduction of the calculated results, using, for instance, kinetic approaches.

The major difficulty we confront while analyzing these problems is the development of a *general concept* for the construction of rational numerical models of turbulence.

Below we concentrate on approaches based on the use of a complete (and closed) *system of dynamic equations* for determination of the true values of the velocity and pressure, as well as of statistical methods for describing the free developed turbulent motions over extensive time intervals. A combined use of approaches (based on the consideration of the hydrodynamic equations and statistical Monte Carlo techniques) yields a better understanding of the structure of turbulence and helps to find rational routes to the construction of adequate mathematical models.

Generally speaking, the numerical realization of the general concepts of the theory of turbulence is proceeding in various directions (see Refs. 3–7, 11–18) and other publications):

(i) integration of full non stationary Navier–Stokes equations without any additional assumptions concerning the nature of transfer (finite-difference approaches, spectral Fourier methods and so on, which are used for comparatively simple problems of convection and diffusion, to simulate the "disintegration" of laminar flow, etc.);
(ii) calculation of the same models on a coarser difference grid using semiempirical theories and variable transfer coefficients, such as the effective viscosity;
(iii) solution of the Reynolds or Boussinesq equations for averaged flow elements of motion and determination of the Reynolds stresses simultaneously with solution of the approximate transfer equations;
(iv) the use of differential equations for moments of various orders and with different types of closure.

The major difficulties encountered while analyzing developed turbulent flows in wakes and jets arise because the flows are characterized by the perturbed-motion scales ranging over several orders of magnitude. The spectrum of wave numbers in real turbulent separated flows stretches by four to five orders, so that the *construction of a comprehensive model of turbulence for large Reynolds numbers seems to be unrealistic, at least in foreseeable future.* Prediction[15] is that a practical analysis of topics such as the numerical simulation of large vortices and semiempirical models of turbulence for realistic aircraft configurations may only be possible by the end of the century and will require computers with $10^{10}$–$10^{11}$ words of memory and speeds of $10^5$–$10^6$ Mflops. Figure 1.1 presents the estimated computer resources needed for calculating an airfoil, a fuselage at an angle of attack, etc. for the following tasks: (1) solution of the nonlinear equations of inviscid flows, (2) that of the Reynolds-averaged Navier–Stokes equations, and (3) numerical simulation of large vortices taking into account turbulence and using semiempirical models.

One approach to circumvent the difficulties reduces the analysis to *a straightforward* study of a three-dimensional unsteady turbulent flow within the range of scales exceeding a certain fixed dimension $h*$ (where

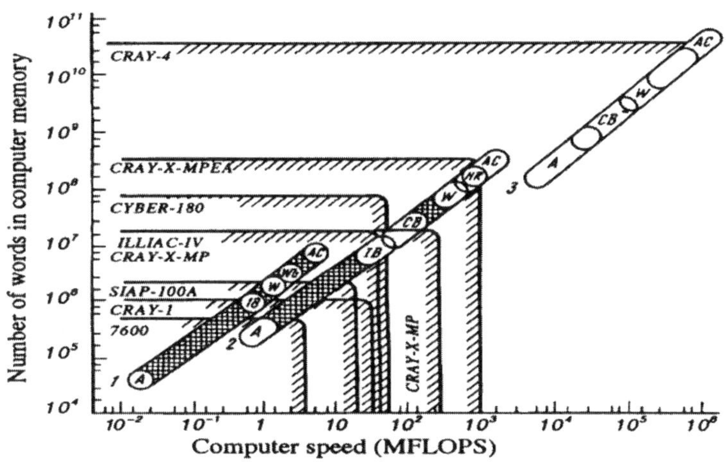


**Fig. 1.1.** Computer resources needed to solve various problems in computational aerodynamics: (1) nonlinear inviscid equations; (2) Navier–Stokes equations averaged with respect to Reynolds number; (3) numerical simulation of large vortices taking into account turbulence and using semi-empirical models. AC, an aircraft; W, a wing; HR, a helicopter rotor; CB, turbine/compressor blades; IB, a fuselage at an angle of attack; A, an airfoil.

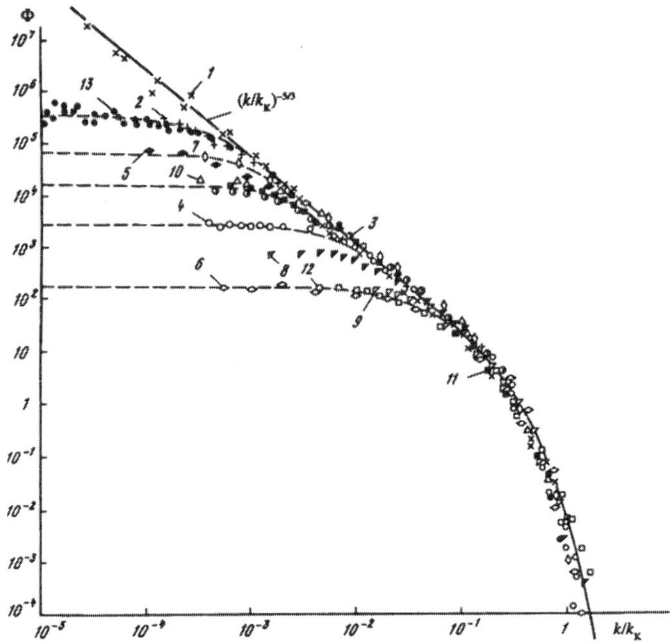

**Fig. 1.2.** Spectra of energy density $E_1(k)$ of longitudinal velocity component fluctuations for various flows. $\Phi = E_1(\varepsilon v^3)^{-1/4}$, $k$ is a wave number. (1) $Re_\lambda = 2000$, a tide tank, $Re_D \approx 10^8$; (2) $Re_\lambda = 780$, a round jet; (3) $Re_\lambda = 170$, a tube flow, $Re_D \approx 5 \times 10^5$; (4) $Re_\lambda = 130$, constant-shear flow; (5) $Re_\lambda = 380$, the wake behind a cylinder; (6) $Re_\lambda = 23$, the wake behind a cylinder; (7) $Re_\lambda = 540$, turbulence downstream of a honeycomb; (8) $Re_\lambda = 72$, turbulence downstream of a honeycomb; (9) $Re_\lambda = 37$, turbulence downstream of a honeycomb; (10) $Re_\lambda = 401$, a boundary-layer flow, $y/\delta = 0.5$, $Re_\delta = 3.1 \times 10^5$; (11) $Re_\lambda = 282$, a boundary-layer flow, $y/\delta = 0.22$, $Re_\delta = 5.6 \times 10^5$; (12) $Re_\lambda = 23$, a boundary-layer flow, $y/\delta = 1.2, Re_\delta = 3.5 \times 10^5$; (13) $Re_\lambda = 850$, a boundary-layer flow over water surface, $y/\delta = 0.6, Re_\delta = 4.0 \times 10^5$.

$h^* \sim k/k_K \sim 10^{-2}$ is, for example, the numerical grid spacing, see Fig. 1.2). The turbulence scales for which direct resolution is impossible are modeled by subgrid turbulence using turbulence viscosity or some other rational approximation of the transfer processes. This approach is based on the assumption that the small-scale structure of turbulence is *nearly universal* for various problems (in particular, for large wave numbers $k$ when $k/k_K > 10^{-2}$, where $k_K$ are the Kolmogorov-scale wave numbers), and hence the exact solution is unnecessary.[12,c] This is illustrated by Fig. 1.2

[c]For instance, this means that the effect of high-frequency spectrum on large structures is negligible or very small for large Reynolds numbers.

which is borrowed from Ref. 15 and shows spectra of the energy density $E_1(k)$ of fluctuations in the longitudinal velocity components for various turbulent flows

$$\int_0^\infty E_1(k)dk = \langle u'^2 \rangle.$$

$\Phi = E_1(\varepsilon v^3)^{-1/4}$ is the non-dimensional spectral density, $\varepsilon$ is the local energy dissipation rate per unit mass, $v$ is the kinematic viscosity and $k_K = E_1(\varepsilon v^{-3})^{1/4}$ is the Kolmogorov wave number. The structure of large vortices incorporating turbulence, which varies markedly from flow to flow or from one group of conditions to another, is analyzed in a straightforward manner (in essence, this is the concept we develop below). Another approach to the direct study uses the "vortex dynamics" (or "discrete vortices"), in which a turbulent zone is simulated by a set of multiple discrete inviscid vortices evolving in time.[17–19]

Next, we have to consider how to develop *constructive models* and the corresponding numerical algorithms. Techniques for studying free developed turbulence based on the Navier–Stokes equations are closely related to the analysis of solutions of the equations for small molecular viscosity[d] (looking, in particular, for the appearance of randomness). Usually, the following *Reynolds hypotheses* are used:

(i) for $Re > Re_{\rm cr}$ laminar flows are unsteady;
(ii) for $Re > Re_{\rm cr}$ chaotic motions must be described by the Navier–Stokes equations;
(iii) the system of equations or its solutions must be averaged using some unknown parameters with the help of a function which is also unknown;
(iv) since the averaged system of equations is *unclosed*, a model of closure must be employed.

In this formulation the major difficulty in simulating large-scale turbulence is associated with the construction (for high supercritical Reynolds numbers) of an unstable (with respect to the averaged characteristics) numerical solution, which is *coincident* with that of Navier–Stokes equations. In order to ensure the conditions for the approximation and stability of the numerical solution, the grid spacing must be chosen so as to ensure that errors in the approximation of convective terms are much smaller than difference representations of viscous terms. Estimates made for model

[d]Neither the validity nor the possibility of this formulation is discussed here (see Refs. 8 and 20).

equations[21] show that a real solution may be approximated for $Re_h \ll 1$. This means that if flows with a molecular mechanism of dissipation effective for large (turbulent) Reynolds numbers can at all be calculated, this may be done using extremely fine difference grids whose dimensions are comparable, generally speaking, with those of the fluctuation modes.

It should be stressed that molecular viscosity may be ignored as compared with the effective turbulent viscosity due to turbulent transfer (usually, $v_{\mathrm{t}}/v_{\mathrm{m}} \approx 10^4 - 10^6$). In addition, to analyze turbulence, we should consider three-dimensional (in space variables) Navier–Stokes equations. For large Reynolds numbers the problems may have multiple solutions (though the Navier–Stokes equations have been proved to possess invariably unique solutions for both unstable one-dimensional and two-dimensional problems over the entire time interval $t > 0$). Generally speaking, the conditions for the appearance of turbulence must correspond to the conditions for the nonuniqueness of solutions to the three-dimensional nonstationary Navier–Stokes[e] equations[20] and so on. Hence, for large Reynolds numbers a detailed direct analysis of developed large-scale structures on the basis of the Navier–Stokes equations is barely achievable even using the most powerful modern computers.[f]

## 1.2. Rational averaging of large vortex structures

Lately, the following two approaches to the numerical analysis of turbulence have been seriously developed. Orszag[3] called them models of small-scale motions and direct numerical simulation.

The approach initiated by Deardorff[24] and Ferziger[25] approximates turbulent transfer only for those scales of motion, which cannot be explicitly resolved by a numerical approximation of the Navier–Stokes equations, while smaller scales are considered with the help of a statistical approximation during a detailed analysis of larger scales. The effect of unresolvable

---

[e]Following Struminskii[22,23], it is natural to suppose that "the major form of motion of matter, turbulent one, must be described by the laws of mechanics and physics without employing any additional hypotheses and assumptions... The equations of mechanics and statistics were used to obtain Boltzmann equation from which the Navier–Stokes equations follow, providing a straightforward description of laminar flows only. Evidently, the main class of turbulent motion has been lost somewhere, and more rigorous analysis is required to find it."

[f]As in the case of direct simulation, the large scales are calculated in straightforward manner.

small scales on resolvable larger scales is accounted for by turbulent-viscosity coefficients with semiempirical constants. The most popular is the Smagorinsky approach[26] in which the turbulent viscosity $v_{\mathrm{t}}$ substitutes for the molecular viscosity $v_{\mathrm{m}}$ and is represented in the forms

$$v_{\mathrm{t}} = (c\Delta x)^2 \left| \left(\frac{\partial v_i}{\partial x_j}\right)^2 + \left(\frac{\partial v_j}{\partial x_i}\right)^2 \right|^{1/2},$$

$$v_t = (c'\Delta x)^3 \left| \frac{\partial}{\partial x_i} (\nabla \times \mathbf{v}) \right|$$

for three-dimensional and two-dimensional flows, respectively. Here $\Delta x$ is the spacing of a grid, and $c$ and $c'$ are constants equal, approximately, to 0.1 and 0.2. The formulae are only valid when the spatial derivatives are approximated to the first or second order of accuracy.[3,26]

At the first glance, this approach imposes no restrictions on the Reynolds number; in other words, it accounts for the effect of small-scale turbulence (this is in fact its advantage over direct numerical simulation). However, the reality is far from this. The closure scheme for the small-scale component is satisfactorily accurate only when the *partition of the flow into small- and large-scale components does not have any significant effect on the evolution of large-scale structures.* Thus, the efficiency with which the invariance of the relative division of scales is attained remains an unanswered question.

In addition, the methods of accounting for small-scale effects have a number of drawbacks, such as the use of arbitrary models of transfer for small-scale structures (with semi-empirical coefficients) and ignoring all the stochastic effects of fluctuations of the characteristics of this component on the generation of fluctuations in large-scale formations.

In the case of the *direct numerical simulation of turbulence* based on the Navier–Stokes equations with large Reynolds numbers it is proposed to decrease the value of *Re* artificially until the flow can be modeled with sufficient accuracy on present-day computers. In fact, it is assumed that *large-scale motions remain invariable*, though it is clear that the motion corresponding to all the scales of a model flow cannot stay invariable. Thus, the large-scale characteristics of turbulent flow are independent of the Reynolds number if both the boundary and initial conditions are also independent thereof.[27] Large-scale turbulent motion displays a strong tendency to self-regulation resulting in *large-scale structures being independent of the details of the dissipation mechanism.*

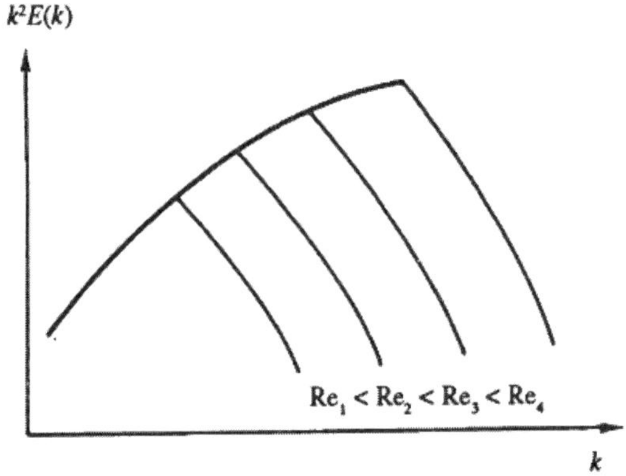


**Fig. 1.3.** Variation in the energy dissipation spectrum due to increase in the Reynolds number. In accordance with the hypothesis concerning independence of the Reynolds number $Re$, the scales of motion with wave numbers below the maximum value in the spectrum $k^2E(k)$ (i.e., for $k < 1/\lambda_0$) are practically independent of the Reynolds number.

As stated in Ref. 3, the hypothesis that macrostructures are independent of the Reynolds number means that not only statistical characteristics but also the detailed structure of motions with wave numbers $k < 1/\lambda_0$ are independent of the Reynolds number.[g] (Here $\lambda_0 = (v^3/\varepsilon)^{1/4} \approx \ell/Re^{3/4}$ as $Re \to \infty$, and $\lambda_0$ and $l$ are the scales of small and large vortices, respectively.) In fact, as $Re$ increases, the small-scale three-dimensional turbulence restructures itself to ensure the required rate of dissipation. This is accompanied by an extension of the spectrum $k^2E(k)$ of the mean-square-root vorticity into increasing $k$, while its behavior at small $k$ (macrostructures) remains practically invariable (Fig. 1.3).

Thus, one may assume[2,28] that turbulent transfer proceeding simultaneously on many scale levels results in *turbulent flows being self-similar with respect to the Reynolds number*. Hence, the averages determined by large-scale velocity variations are independent of the Reynolds number if $Re \to \infty$ (*the hypothesis that there is statistical independence of large- and small-scale motions*). This has been confirmed experimentally.[28] The statistical independence hypothesis has a wide range of applications since it allows us to *split* many of correlations between the macro- and micro-properties of turbulence for $Re \to \infty$.

[g]This applies to statistical characteristics only (see Sec. 1.5).

Note that a description of large-scale velocity variations alone cannot be closed since their evolution is determined, among other things, by viscous dissipation. However, if we consider the fluctuations over the entire spectrum of scales (taking into account the effects of molecular viscosity), then, generally speaking, we need to cope with superfluous information (since the major characteristics of turbulence are independent of the Reynolds number). Another approach is based on the search for universal relationships between the characteristics of small- and large-scale fluctuations. The Kolmogorov and Obukhov theory[9,10] is that such relationships actually exist if the scales of characteristic fluctuations, which determine the turbulent energy and its dissipation, differ significantly. These relationships are also consequences of the self-similarity of turbulence with respect to the Reynolds number for large $Re$.[28]

The separation of scales may be also interpreted in the following way: let us consider the spectral distribution of energy of oscillations $E(k)$ in the velocity field of an isotropic uniform turbulent flow as an example. According to Kolmogorov's assumption,[9] for high Reynolds numbers there must be an *intermediate* range of wave numbers (the *inertial interval*) within which energy is neither produced nor dissipated but only advanced to larger wave numbers.[h] From dimensional analysis it may be deduced that for the inertial interval[i] (Fig. 1.4).

$$E(k) \sim \varepsilon^{2/3} k^{-5/3}.$$

Thus, it can be stated that *for large Reynolds numbers* $Re_\lambda = \lambda \langle u'^2 \rangle^{1/2} / v$ *the spectral energy interval* (I) *and the dissipation interval* (II) *are widely separated in frequency* (see Fig. 1.4), thus confirming the reality of the statistical independence hypotheses. Let us consider this in detail (see Ref. 11).

The spectral Kármán–Howarth equation has the form

$$\frac{\partial E(k,t)}{\partial t} = T(k,t) - 2vk^2 E(k,t), \tag{1.1}$$

where $T(k,t)$ is the function determining energy redistribution over the spectrum.

---

[h]Kármán's assumption is that a similar process takes place in boundary layers where an *intermediate equilibrium layer* exists between the large-scale vortex motion (where energy is produced) and the viscous-dissipation region. Dimensional analysis provides the logarithmic equation for boundary-layer flow.

[i]The issue of universality has yet to *be explored.*

Function $T(k,t)$ describes the variation in the energy of a spectral component of turbulence with wave number $k$ due to the *nonlinear inertial terms* of the hydrodynamic equations. It should be noted that the variation reduces to the redistribution of energy between spectral components *without* changing the total energy of turbulent motion.

The last term on the right-hand side of (1.1) describes *energy dissipation* due to viscous forces. We see that viscosity decreases in the kinetic energy of perturbations with an increase in the wave number $k$, which is directly proportional to the strength of the perturbations multiplied by $2vk^2$. Thus, the energy of long-wave perturbations (corresponding to small $k$) decreases due to viscosity much more slowly than that of short-wave perturbations, in conformity with the law that the frictional forces are directly proportional to the velocity gradient.

Figure 1.4 shows schematically the energy spectrum $E(k)$, the energy dissipation spectrum $2vk^2E(k)$ and the function $T(k)$. The behavior of the function $T(k)$ (which is negative for small $k$ and positive for large $k$) is consistent with the conclusion that turbulent mixing must result in the partition of turbulent perturbations, i.e. the transfer of energy from large-scale components of motion to small-scale ones, whose energy is directly spent to overcome viscous friction. Since viscosity is important for relatively small-scale components of motion (characterized by large local velocity gradients), the maximum of the energy dissipation spectrum $2vk^2(k)$ shown in Fig. 1.4 is displaced along the wave number axis to the right part of the energy spectrum $E(k)$.[11]

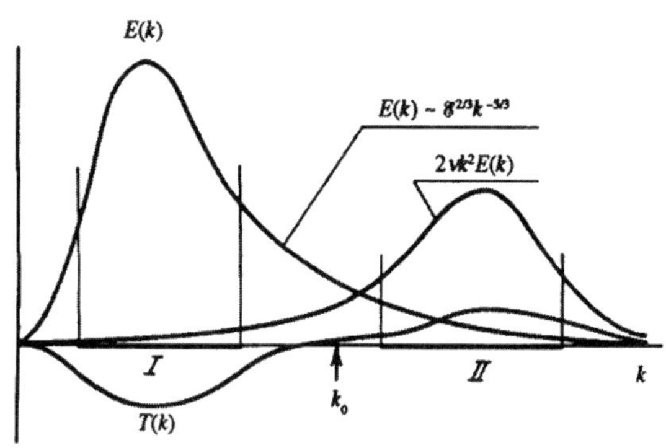


**Fig. 1.4.** Energy spectrum $E(k)$, the energy dissipation spectrum $2\nu k^2E(k)$ and the function $T(k)$ for large Reynolds numbers: (I) the energy interval; (II) the dissipation interval; in between is the inertial interval.

Most of the total fluctuation energy

$$\langle u'^2 \rangle = \int_0^\infty E(k)dk \cong \int_0^{k_0} E(k)dk \tag{1.2}$$

(say, 80% to 90%) is concentrated within the *energy interval* (I), while most of the total dissipation

$$\varepsilon = 2v \int_0^\infty k^2 E(k,t)dk \cong 2v \int_{k_0}^\infty k^2 E(k,t)dk \tag{1.3}$$

is concentrated within the *dissipation interval* (II). Here $k_0$ is an intermediate wave number lying within the inertial range (which lies between the energy and dissipation intervals).

Equation (1.1) may be integrated over the whole frequency interval

$$\frac{\partial \langle u'^2 \rangle}{\partial t} = -2v \int_0^\infty k^2 E(k,t)dk \equiv -\varepsilon,$$

since for isotropic turbulence

$$\int_0^\infty T(k,t)dk = 0. \tag{1.4}$$

However, if (1.1) is integrated over the intervals $(0, k_0)$ and $(k_0, \infty)$, then taking into account (1.2) and (1.3), we get

$$\frac{\partial \langle u'^2 \rangle}{\partial t} \cong \int_0^{k_0} T(k,t)dk, \quad 0 \cong \int_0^\infty T(k,t)dk - \varepsilon,$$

and if Eq. (1.4) is taken into account, then

$$\frac{\partial \langle u'^2 \rangle}{\partial t} \cong \int_0^{k_0} T(k,t)dk = -\int_{k_0}^\infty T(k,t)dk = -\varepsilon. \tag{1.5}$$

We see that the rate of energy dissipation $\varepsilon$ is also defined within the energy interval (which fact was used to construct the semiempirical Prandtl theory).

Next, we determine the *order of magnitude* of energy dissipation in a turbulent flow.[10] This energy is drawn from the large-scale motion and is transmitted to ever decreasing scales until it dissipates within the viscous interval. Although dissipation is generated by viscosity, the order of magnitude of $\varepsilon$ may be estimated solely by quantities characteristical for large-scale motions, such as density $\rho$, length $l$, and velocity $\Delta u$ (the variation in the mean velocity corresponding to the characteristical dimension of large-scale vortices $l$). There exists a unique combination of these quantities with the dimensionality of the average energy dissipation (per unit time per unit mass), viz.,

$$\varepsilon \sim \frac{(\Delta u)^3}{l}. \tag{1.6}$$

By describing the properties of a turbulent flow by turbulent viscosity $\nu_{\mathrm{t}} \sim \Delta u l$, we get

$$\varepsilon \sim v_{\mathrm{t}} \left(\frac{\Delta u}{l}\right)^2. \tag{1.6$'$}$$

It should be noted that the molecular viscosity $v_{\mathrm{m}}$ defines energy dissipation via the derivatives of the true velocity with respect to coordinates, while $v_{\mathrm{t}}$ relates dissipation to the gradient of the mean flow velocity ($\sim \Delta u/l$).[10] The relationships in (1.6) and (1.6′) may be used to construct numerical models.

Thus, *for large Reynolds numbers Re the major features of turbulent flow are determined by the energy (inviscid) interval, i.e. within the region of long-wave perturbations (corresponding to small k).*[j] *Note that this is the region of integral scales* $\Lambda$.

Let us estimate the position of the energy interval within the scale of turbulence (following Onufriev). The energy integral $\langle u'^2 \rangle$ taken over the spectrum is mostly concentrated within the interval $0.1 < k\Lambda < 10$. Using the generalized Kármán model (the EVK model) for a uniform isotropic turbulence, Driscall and Kennedy[29] derived approximations for the integral scale $\Lambda$ and Taylor (transverse) microscale $\lambda$

$$\frac{\Lambda}{\eta} = Re_\lambda^{1/2}[2.47 + 0.081 Re_\lambda^{1/2}(Re_\lambda^{1/2} - 1)] \to 0.81 Re_\lambda^{3/2},$$

$$\frac{\lambda}{\eta} = (15^{1/4}), \tag{1.7}$$

where $\eta = (v^3/\varepsilon)^{1/4}$ is the Kolmogorov microscale, $Re = \Lambda U/v$ and $Re_\lambda = -\langle u'^2 \rangle^{1/2}/v$. From (1.7), we get: $\Lambda/\lambda \sim 0.1 Re_\lambda$. Thus, we obtain the estimate

$$\frac{Re}{Re_\lambda} \sim \frac{U\Lambda}{\lambda \langle u'^2 \rangle^{1/2}} \sim \frac{U}{\lambda \langle u'^2 \rangle^{1/2}} \frac{\Lambda}{\lambda} \sim 100 \times 0.1 Re \sim 10 Re_\lambda,$$

and hence $Re \sim 10 Re_\lambda^2$ and for

$$Re = 10^7, \quad Re_\lambda = 10^3, \quad \Lambda/\eta = 10^3 - 10^4, \quad \Lambda/\lambda = 10^2, \tag{1.8}$$

[j] Usually, turbulence is studied within the viscous interval. In the numerical analysis of large-scale structures this creates formidable (and *fundamental*) difficulties.

i.e. the longitudinal integral scale $\Lambda$ is two (or more) orders of magnitude larger than the Taylor microscale $\lambda$, and three to four orders of magnitude larger than the Kolmogorov microscale $\eta$. Thus, it suffices for us to consider the low-frequency and initial zones of the inertial range of the spectrum. In order to accumulate the energy $\langle u'^2 \rangle$ needed to carry out the calculations, we have to use scales that are smaller than $\Lambda$ by a factor of 10–20. The estimates in (1.8) indicate that calculations on "real" grids may hopefully yield reliable results.

Let us discuss an approximate method for determining the *longitudinal integral turbulence scale*. Onufriev and coworkers[30,31] experimentally studied the behavior of the energy spectrum in a turbulent flow through a rotating (with respect to the longitudinal axis) pipe. They discovered a rather universal non-dimensional spectral distribution, which proved to be similar to the generalized Kármán model (the EVK model) for uniform isotropic turbulence.

This statement (let us call it *Onufriev's rule*) may be formulated as follows:

*For the semiempirical isotropic EVK model of a spectrum at large Reynolds numbers, the maximum dimensionless quantity $\varphi$ for a one-dimensional spectrum* ($\max\varphi = \max[kE_1(k)/\langle u_x'^2 \rangle]$) *corresponds to* $k\Lambda \cong 1$. For $Re_\lambda > 100-200$, the maximum varies by less than 10%.

Figures 1.5 and 1.6 present $\varphi$ for the EVK model against the experimental data[30,31] for a developed turbulent flow in a tube. While approaching the wall, the flow becomes strongly nonuniform and the spectral representation loses its universality (the systematic deviation in the spectrum may be accounted for and ignored for the larger portion of the spectrum). The above estimation, according to which $\max\varphi = \max[kE_1(k)/\langle u_x'^2 \rangle]$ for $k\Lambda = 1$, holds for anisotropic flows too. Hence, if the statement is assumed to be the rule for determining the integral scale $\Lambda$, then $u_x'^2(t)$ and $\langle u_x'^2 \rangle$ may be calculated from a rather narrow interval of scales at low frequencies, i.e. we have a one-dimensional spectrum $E_1(k)$ within the interval and may determine $\Lambda$.

Using the approximation of the Kármán spectral distribution for both the low frequency and inertial ranges of the spectrum, we find that the energy dissipation rate is

$$\varepsilon = \frac{(2/3)^{3/2}[\Gamma(5/6)/\Gamma(1/3)]^{5/2}}{(9/55)^{3/2}\pi^{1/4}(c_1)^{3/2}} \frac{E_1^{3/2}}{\Lambda} \approx \frac{0.71}{(c_1)^{3/2}} \frac{E_1^{3/2}}{\Lambda},$$

$$\text{or} \quad \varepsilon \sim \langle u'^2 \rangle^{3/2}/\Lambda \tag{1.9}$$

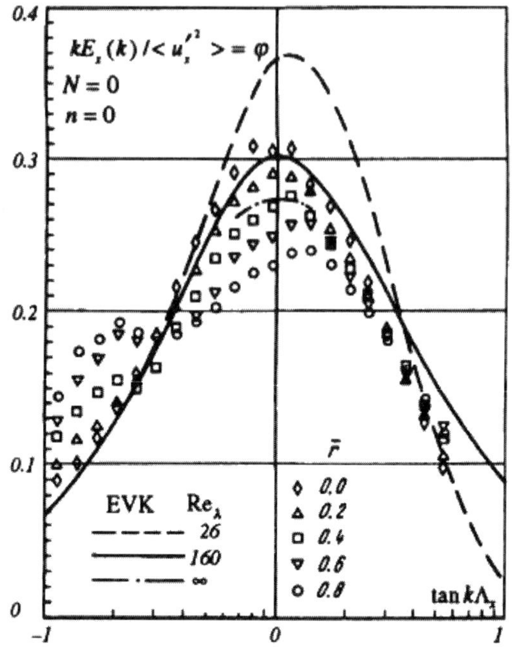


**Fig. 1.5.** Spectral distribution in the system of coordinates $kE_x(k)/ < u_x'^2 >= \varphi$ and $k\Lambda_x$ in an irrotational flow: $N = 0$; °, $\Delta$, etc, are experimental data for various radial positions; the line corresponds to the EVK model.

(see Ref. 4). Here $c_1 = 1.4$ is a constant of the three-dimensional spectrum for the inertial range[k]: $\varepsilon(k) = c_1\varepsilon^{2/3}k^{-5/3}$ (see Fig. 1.4). Next, the relationships for a uniform isotropic flow may be used to determine the Taylor ($\lambda$) and Kolmogorov ($\eta$) microscales and $Re_\lambda$ (see Refs. 4 and 11), viz.,

$$\lambda^2 = \frac{10\nu\langle u_x'^2\rangle}{\varepsilon}, \quad \eta = \left(\frac{\nu^3}{\varepsilon}\right)^{1/4}, \quad Re_\lambda = \frac{\langle u_x'^2\rangle^{1/2}\lambda}{\nu}. \tag{1.10}$$

Of course, we must account for flow anisotropy and nonuniformity affecting the integral scale and the amplitude of the spectral distribution. However, in the first approximation these effects may be ignored.[l] Thus, low-frequency (long-wave) structures dominate turbulence for large Reynolds numbers, and these large-scale structures must not be significantly distorted during calculation.

---

[k]In Ref. 32, the variational approach was used to calculate the constant $c_1 = 1.2$.

[l]The author is thankful to Prof. A. T. Onufriev for material pertaining to this problem and his useful discussions.

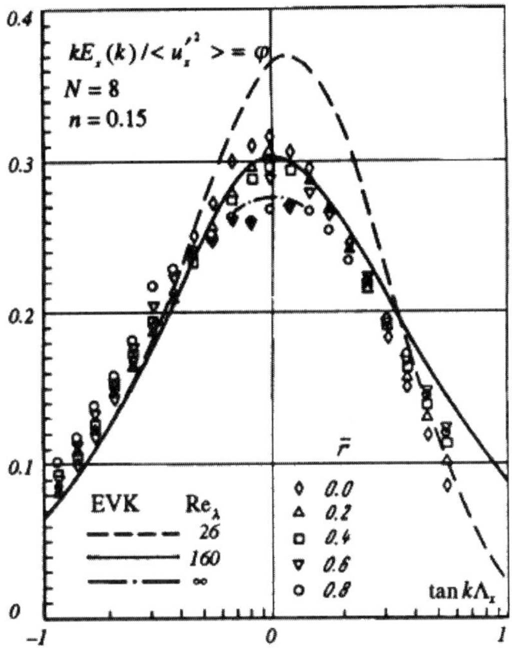


**Fig. 1.6.** Spectral distribution in the system of coordinates $kE_x(k)/ < u_x'^2 >= \varphi$ and $k\Lambda_x$ in a swirling flow. The rotation rate of the exit channel section $N = 8$ rps. For notation, see Fig. 1.5.

Using the above statements and hypotheses (large-scale formations practically independent of the character of dissipation, and about the relative positions of the energy and dissipation intervals), we can readily calculate the dynamics of large-scale turbulence, employing rather approximate averaging techniques for the subgrid turbulence (in a certain sense this is the pivotal point of the present approach).

Taking into account the above, below we shall consider techniques for rational numerical averaging while calculating large-scale turbulent structures and based on *integral laws of conservation for an ideal medium.*

Recently, it has become increasingly clear that both the presence and the dynamics of turbulent structures are extremely important for studying turbulence. The presence of ordered structures in chaotic motion results in the formation of regions where vorticity is intensely concentrated, namely, vortex tubes and vorticity layers.

In what follows, the dynamics of large-scale turbulence is considered to be crucial to the study of large-scale formations in fully developed turbulence. It is natural to divide a flow into large- (supergrid) and low-scale (subgrid) turbulence. We shall try to construct an approach *without* introducing semi-empirical models of closure for the subgrid scales. We start by considering the two ideas at the basis of our methodology.[12–14]

In Ref. 10, it is stated that "...for large Reynolds numbers $Re$, the Reynolds numbers of large-scale motions, $Re_\lambda$, are also large ($Re_\lambda \sim U_\lambda \lambda / v$, where $U_\lambda$ is the order of velocity corresponding to the scale $\lambda$). However, large Reynolds numbers are equivalent to low viscosity... Hence, for the large-scale motion that dominates any turbulent flow, the fluid viscosity is insignificant and may be ignored, making it possible to describe the motion by the Euler equations.[m] In particular, this means that no considerable energy dissipation takes place in large-scale motion."

Above we have shown that for the low-frequency energy interval (where the energy is generated) the effect of viscosity is negligibly small. Analogously, the process of kinetic energy transmission down the cascade of vortices, proceeding with the rate $\varepsilon$, dominates the inertial range of motion (corresponding to the vortex scales $l > \lambda \gg \lambda_0$, where $\lambda_0$ is the internal scale of turbulence; $Re_{\lambda_0} \approx 1$). In the process, vortices are fragmented by inertial forces, while viscosity is insignificant (the second similarity hypothesis formulated by A. N. Kolmogorov, which treats energy flow conservation). These motions occur at large Reynolds numbers, and vortex scales $\lambda \gg \lambda_0$ should be calculated using the local turbulence relationship.[4,9–11] In essence, the simulation of free turbulence should start from *local* turbulence scales, while the dynamics of large-scale turbulence (ordered structures and large-scale motions) may be calculated on the basis of the full dynamic equations for an ideal medium, i.e. ignoring the effects of molecular viscosity.[n]

Let us finally consider how to formulate a "*rational*" *averaging of subgrid scales*. Usually, two approaches are used. The first (see above) is to average the equations of motion over elementary grid volumes (see Refs. 24 and 33). In this case the effect of turbulence with the scales $\lambda < h$ manifests itself in the form of Reynolds stresses, and semi-empirical models of closure must be used.

However, an alternative approach is possible and is based on the theory of filtration.[25,34] In this case, first a *smoothing operator* is introduced, which helps to single out small-scale motion. The application of this filter operator

---

[m]The applicability of these equations to turbulent flows is determined by distances of the order $\lambda_0$.[10]

[n]Naturally, viscosity cannot be ignored in the vicinity of a body.

to the equations of motion allows us to determine the effect of stresses due to small-scale turbulence, via volume-averaged quantities. The approximation preserved after filtering nonlinear terms of the form $\overline{\bar{u}_i\bar{u}_j} \neq \bar{u}_i\bar{u}_j$ results in a *nonzero* interaction of velocities or Leonard stresses. Then the resulting equations are solved numerically.

Following Ref. 33, "ideally the filter would exclude all the contributions of small-scale motion above a certain wave number *without* changing the modes in the range of wave numbers smaller than the limiting one. Until now such filters have not been used in finite-difference approximations."[o]

This ability (viz., *to "extinguish" small-scale fluctuations without practically affecting large-scale structures*) proves to be characteristical for dissipative finite-difference schemes based on *asymmetric approximations.* In this case it is highly desirable to replace differential equations of motion by balance relationships (in the form of integral laws of conservation) for which an elementary volume coincides with a cell of the calculation grid. This *method of preliminary "integral smoothing"* of fine vortices seems to be most natural for calculating large-scale turbulence numerically. However, the "splitting" scheme must ensure the approximation of turbulent transfer.

In connection with the above, we would like to present here some general considerations concerning the *"rational" numerical averaging* used in the approach proposed (here we use results of Tolstykh[35]).

Numerical simulation of large vortex structures of a turbulent flow based on the use of difference schemes is always accompanied by the appearance of errors which may be conveniently divided into *amplitude* and *phase errors.* Since both exact and difference solutions may be represented as superpositions of harmonics with different frequencies and amplitudes (in the form of a Fourier series or integral), it would be natural that comparatively large structures correspond to long-wave and medium-wave harmonics, while small structures correspond to short-wave harmonics.

Instead of the wave exponent $\exp(ikx)$ (where $k$ is the wave number and $\Lambda = 2\pi/k$ is the wavelength), a difference solution incorporates the representation $\exp(ikx_n) = \exp(ikhn)$ (where $h$ is the grid spacing and $n$ is the node number). By periodicity, we have $0 \leq kh \leq 2\pi$. The short waves correspond to $a = kh \approx n$ (since for $kh = \pi$ we have $\Lambda = 2\pi/k = 2h$, i.e. the shortest waves allowed by the grid). The values of $kh \ll 1$ correspond to long waves, which should be described accurately. Let us consider the

[o]According to Ref. 33, they can only be used in spectral approaches.

simplest analog of the equation of vorticity transfer

$$u_t + cu_x = 0, \quad c = \text{const}. \tag{1.11}$$

The amplitude and phase errors may be represented in the form $a_{\text{dif}}/a_{\text{exact}} = 1 - d(a), C_{\text{dif}}/C_{\text{exact}} = 1 - l(\alpha), \alpha = kh$, where $a$ is an amplitude, $C$ is the phase velocity, $d$ is dissipation and $l$ is a function characterizing dispersion.

As a rule, all schemes yield small errors for long waves ($kh < 1$) and large errors for short waves ($kh \approx \pi$). Usually, the dependence characterizing the difference between $C_{\text{dif}}$ and $C_{\text{exact}}$ has the form of the solid line shown in Fig. 1.7(a). As far as the description of large vortices (the long- and medium-waves for which $\alpha = kh$ is relatively small) is concerned, the errors in the representation of "small" structures (short waves with $\alpha \approx 1$) cannot be ignored because these errors (small-scale perturbations) propagate with a group velocity $C_{\text{dif}}^{\text{gr}}$ which differs from the exact velocity of the perturbation propagation $C$ (see the dashed line in Fig. 1.7(a)) even more significantly than $C_{\text{dif}}$ does. For $kh \approx \pi, C_{\text{dif}}^{\text{gr}}$ may become negative and the perturbations propagating upstream may sometimes distort the structure of large vortices considerably.

In the nonlinear case (for instance, the Euler equations) the situation is even worse since the energy of small-scale perturbations may be transmitted to larger scales resulting in the nonlinear instability of the scheme.

The situation may be improved (or even saved) by a dissipation $d$. Let us consider both nondissipation ($d = 0$) and dissipation ($d \neq 0$) schemes.

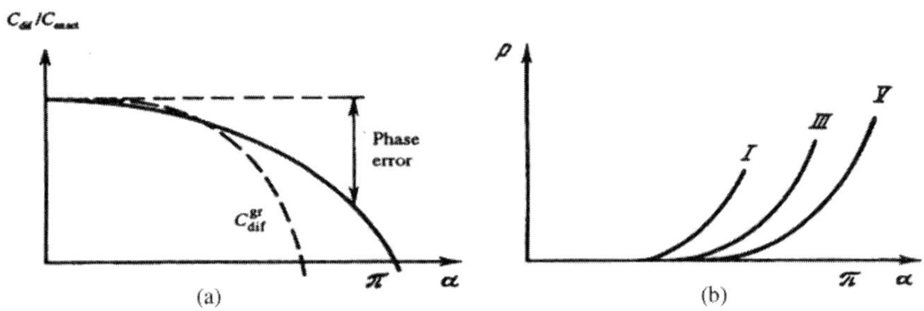


**Fig. 1.7.** Analysis of the model equation: (a) $C_{\text{dif}}/C_{\text{exact}}$ curves characterizing the departure of the difference phase velocity from the exact one (the solid curve) and an approximate plot of the difference group velocity $C_{\text{dif}}^{\text{gr}}$ versus the development of small-scale perturbations (the dashed line); (b) a typical plot of the function $p(\alpha)$ characterizing the dissipation in difference schemes of different orders of accuracy ($\alpha = kh$, $k$ is the wave number and $h$ is the grid spacing).

Figure 1.7(b) shows typical curves for dissipation schemes with different degrees of accuracy ($p(\alpha)$ is a function characterizing dissipation). We see that the dissipation schemes provide comparatively low dissipation in the large-scale region ($kh < 1$) and a rather high one in the small-scale region ($kh \approx \pi$). Since dissipation characterizes the attenuation of amplitude perturbations, *the dissipation-stable schemes describe large structures (such as large vortices)well and attenuate (or extinguish)unneeded small-scale perturbations.*

In schemes with symmetric approximations dissipation is absent as well as the mechanism for suppressing short-wave errors which obstruct the description of large vortices (the only way out is filtering, but this is equivalent to the introduction of dissipation). However, for a given scale $l$ of a large vortex the dissipation may be so high (for not small an $h/l$) that the vortex will extinguish itself (will be smeared by the scheme). Therefore, special care must be taken while choosing the grids (actually, the ratio $h/l$) for high-dissipation schemes (such as schemes with first order accuracy).

Let us consider these issues in more detail. We noted above that the role of the dissipation mechanism incorporated in difference schemes used for solving convective problems might be most conveniently analyzed on the example of the transfer equation (1.11). Let us substitute the $cu_x$, in this equation, by the difference approximation $c\delta_x$, leaving $u_t$ unchanged (in order to exclude the effect of time discretization). Like any other operator, $c\delta_x$ may be represented as a sum of *self-conjugate*, $c\delta_x^{(0)}$ and *antisymmetric*, $c\delta_x^{(1)}$ *components*

$$c\delta_x = c\delta_x^{(0)} + c\delta_x^{(1)},$$

where $(\delta_x^{(0)}u, v) = (u, \delta_x^{(0)}v), (\delta_x^{(1)}u, v) = -(u, \delta_x^{(1)})$ and (,) denotes the scalar product of the grid functions.

At the first glance it may seem that since $c(\partial/\partial x)$ is an antisymmetric operator, $c\delta_x$ operator must be an antisymmetric too (i.e. $c\delta_x^{(0)} = 0$). For instance, $\delta_x$ may be set equal to the usual central difference $\delta_x u_j = (u_{j+1} - u_{j-1})/2h$. Such approximations were used at the earlier stages of the study (in the 1960s) for solving comparatively simple problems. However, the presence of strong-gradient regions often makes these "natural" approximations difficult to use since the accompanying parasitic (scheme-generated) oscillations either prevent a stable solution or strongly distort one, if obtained.

The difficulty can be overcome by introducing additional, artificial viscosity terms that act, in particular, as smoothing filters. However, the filter parameters must be chosen for each concrete problem (or a class of problems). Sometimes this cannot be done efficiently enough (especially in three-dimensional and unstable problems).

Another approach is based on *the utilization of schemes in which the dissipation mechanism (the smoothing filter) is incorporated into the operator $\delta_x$ itself, which then approximates convective transfer.* In this case, the self-conjugate component of operator $c\delta_x^{(0)}$ must be positive (i.e. $(c\delta_x^{(0)}u, u) > 0$). In fact, upon discretization with respect to $x$, Eq. (1.11) may be presented in the form

$$\frac{du}{dt} + c\delta_x u = 0. \tag{1.12}$$

By multiplying the latter equation by $u$, one gets

$$\frac{d}{dt}(u, u) = \frac{d}{dt}\|u\|^2 = -(c\delta_x u, u).$$

Since $\delta_x^{(1)}$ is an antisymmetric operator, i.e. $(\delta_x^{(1)}u, u) = 0$, we have

$$\frac{d}{dt}\|u\|^2 = -(c\delta_x^{(0)}u, u) < 0.$$

Hence, *the norm of the solution decreases and dissipation takes place.* However, this statement does not indicate which components of a function (or its harmonics) are suppressed more quickly. To answer the question, we assume that the initial data for (1.12) are specified in the form

$$u(x, 0) = \exp(-ikx).$$

In this case, the exact solution of Eq. (1.11) has the form

$$u(x, t) = \exp(-ik(x - ct)).$$

Let us look for the solution of Eq. (1.12) in the form

$$u(x, t) = v(t)\exp(ikx), \quad x = jh, \;\; j = 0, \pm 1, \pm 2, \ldots,$$

where the real values of $p(k)$ and $q(k)$ satisfy the relationships

$$\delta_x^{(0)}\exp(ikx) = p\exp(ikx), \quad c\delta_x^{(1)}\exp(ikx) = iq\exp(ikx).$$

If $\delta_x^{(0)} > 0$, then $p > 0$.

Hence, we have

$$v(t) = \exp(-p(k)t) \cdot \exp(-iq(k)t).$$

The multiplier $\exp(-iq(k)t)$ is the oscillatory component of the difference solution, which defines dispersion. The other component, $\exp(-p(k)t)$, is the attenuation coefficient for harmonics with the wave number $k$ during the time $t$ (it may be considered as characteristic of dissipation). The larger the value of $p(k)$, the sooner a harmonic dies out. Figure 1.7(b) shows schematically $p(k)$ curves for dissipation schemes of different orders of accuracy. We see that in dissipative schemes short waves (corresponding to large $kh$) die out first. In particular, this is true for nonphysical scheme-generated oscillations ($kh \approx \pi$). The higher the order of the scheme the larger the region of $kh$ within which harmonic amplitudes do not vary (as in the case of the exact solution). Thus, higher-order dissipative schemes incorporate a natural filter which suppresses small-scale or nonphysical saw-toothed oscillations, but does not suppress harmonics which describe the physical process (see Fig. 1.7(b)).

Dissipative schemes (with positive operators) use oriented, i.e. directed (uncentered) differences. *By reversing their direction (depending on the sign of $c$ in (1.11)), the operator $c\delta_x$ may be kept positive for any sign of $c$.* Besides the effect of filtering, the positiveness of the operators makes the schemes more stable — a property which is especially important for solving stable problems and analyzing the dynamics of large vortices in developed free turbulence.

However, we must keep in mind that in unstable problems, such as weather forecasts, $\exp(-pt)$ may extinguish, for small $k$, physical harmonics as well. From this point of view the higher-order schemes have an important advantage because for them the above situation occurs much later than for low order schemes (see Fig. 1.7(b)).

Presently, we use only dissipation-stable schemes, such as the methods of large particles and fluxes, the approximation of viscous equations, compact schemes of a high order of accuracy, etc. (see Refs. 13, 14, 35, and 36). In these cases, dissipation is achieved by orienting differences along the flow (asymmetric schemes) without introducing explicit terms with artificial viscosity. The above techniques constructed based on the splitting schemes allow us to calculate complicated three-dimensional flows on standard computers.

Here, we propose a radically new (as compared with Refs. 3, 24 and 25) approach to rational averaging, viz., one based on the construction of *dissipation-stable and divergent-conservative difference schemes.* Let us list in brief the major ideas of our approach as applied to the calculation of free, shear-generated fully developed turbulence and transition phenomena:

(i) large-scale coherent structures are studied using balance relationships, i.e. the integral laws of conservation for an ideal medium;
(ii) asymmetric difference approximations are used for averaging convection terms over the volume of an elementary cell (this results in the formation of a dissipation mechanism in the difference equations which ensures the stability of solutions and, generally speaking, should account for the effect of small-scale subgrid vortices[p]);
(iii) the energy integral is built up by directly calculating macrostructures on a variety of approximation grids for as long as necessary until the results stabilize, and the effective value of the dissipation mechanism may be determined (as numerous calculations have shown) using the principle of a stable solution for a required degree of resolution; this approach allows us to satisfy the requirement of the "invariance of splitting" into large- and small-scales and to obtain a stable (equilibrium) solution while determining the nonstationary macrostructures of a flow;
(iv) no semiempirical models of subgrid closure are used to calculate the dynamics of large structures (approximate models of the effective viscosity should only be introduced in local turbulence zones or for estimation of the random component forming the turbulent background);
(v) the algorithms are naturally extended for three-dimensional problems (the dissipation mechanism forms automatically due to the above process of averaging, and there is no need to consider, as was necessary in Ref. 26, different kinds of the turbulent-viscosity coefficients for two- and three-dimensional flows);
(vi) the technique of direct numerical simulation allows us to use smoothed equations to determine space-time structures in the dynamics of large-scale turbulence and (by statistically processing the fluctuating, unsteady flow mode) to find their averaged "fluctuational" characteristics, namely, the Reynolds stresses and the turbulent energy (the correctness of the process is discussed below);

[p]Calculations have shown that the characteristics of large-scale flows are practically independent of the structure of dissipation mechanism since in the calculations of the coherent structures small-scale high-frequency turbulence may be either extinguished or accounted for roughly if the universal portion of the spectrum is mostly on the subgrid scale (see Figs. 1.8(e) and 1.8(f)).

(vii) the laminar–turbulent transition is analyzed on the basis of the full Navier–Stokes equations;

(viii) the transition to chaos (numerical experiment) usually requires full dynamic models, but may be implemented solely in the presence of external perturbations which enhance the natural nonlinear inertial mechanism of a system.

Finally, depending on the problem it seems worthwhile to use schemes of a high order of accuracy. In this case an ever-growing proportion of long-wave harmonics may be resolved, which are not extinguished by dissipation. To realize this approach, much less computer capacity is required. This is the essence of the general methodology presented here and its *principal distinction* from the alternative approaches to the direct numerical simulation of turbulence (see Refs. 3, 16, 19, 24–26, 33, and 34). Below the approach is discussed in more detail (see, Refs. 12–14).

## 1.3. Some experimental and theoretical investigations

At the dawn of the study of turbulence, the phenomenon was treated as a *stochastic process* (determined by random distributions of fluctuating quantities). However, we believe that a truly revolutionary change has recently occurred in our understanding of the phenomenon: turbulence was found to incorporate *ordered motions* of almost coherent structures, and the relationship between the deterministic and chaotic ones is presently subjected to scrutiny (see Refs. 5, 6, 37, and 38).

Different experimental and theoretical studies have shown (see Refs. 1–11, and 15) that a wide class of turbulent flows with transverse shear is characterized by the unsteady ordered motion of large-scale formations ("large vortices") which oscillate weakly (*ordered motion of stochastic structures*[q]) and possess a stable and typical (for a fixed problem) space–time structure (see Fig. 1.8). In the case of jet problems the inner flow zone is turbulent (stochastic) in nature and is formed by disordered small-scale fairly intense fluctuations, although its structure is approximately uniform[r] (see Refs. 1, 4–6). Figure 1.8 presents photographs (borrowed from Ref. 1)

[q]Concerning ordered (organized, coherent, collective, etc.) structures, it should be said that the above class of motions is stochastic in both time and space.[6,38]

[r]Cantwell[6] thinks it is possible that in a number of cases (such as plane mixing layers) "even the smallest-scale motions will prove to be highly ordered."

(a)

(b)

(c)

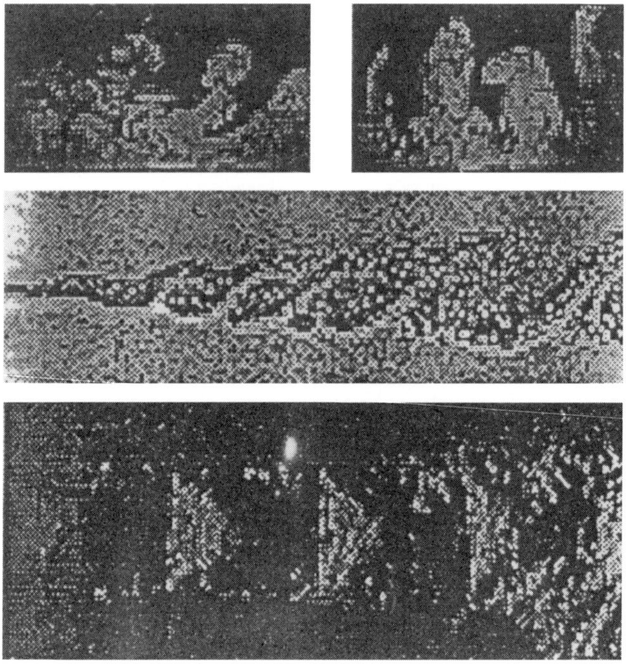

**Fig. 1.8.** Forms of ordered turbulent structures[1]: (a) a turbulent boundary layer; (b) a mixing layer between nitrogen (above) and a helium–argon mixture; (c) a supersonic jet; (d) the wake behind a circular cylinder: $(d_1) M_\infty = 0.64$, $Re = 1.35 \times 10^6$; $(d_2) M_\infty = 0.80$; $(d_3) M_\infty = 0.90$; $(d_4) M_\infty = 0.95$; $(d_5) M_\infty = 0.98$; (e) the wake behind an inclined flat plate ($\alpha = 45°$, $Re = 4300$); (f) oil flowing from a damaged tanker ($\alpha = 45°$, $Re = 10^7$). In the latter two cases the wake structures are remarkably similar.[1]

demonstrating the types of ordered structures, characteristic of different kinds of turbulent flow.[s]

Thus, in free shear flows (such as wakes and jets) a double ("intermittent") turbulence structure is observed.[5] The major zone occupied by the turbulent fluid is comparatively small and uniform. However, an external system of steady "slow" (nonturbulent) large vortices is superposed onto it. This transfers the turbulent fluid from one part of the flow to another (Fig. 1.9). Thus, ordered structures are typical and form the basis of shear turbulent motion.[t] Following Ref. 38, we shall classify turbulent

[s]Figures 1.8(e) and 1.8(f) illustrate the negligible influence of molecular effects on the large-scale wake structure (which is approximately the same in both cases).

[t]Recent Russian research in the theory of structural turbulence is presented in Ref. 38.

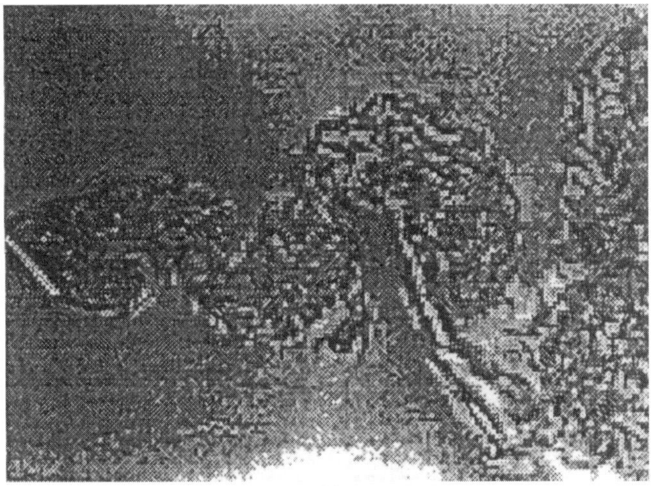

(e)

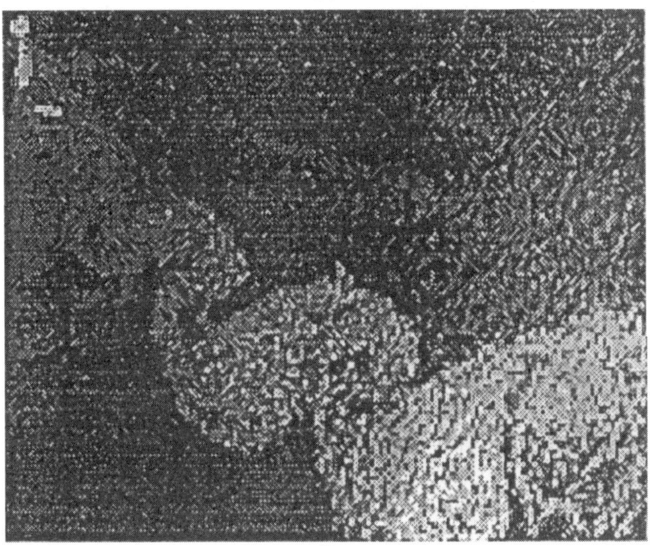

(f)

**Fig. 1.8.** (*Continued*)

structures as:

(i) *dynamic structures of bifurcational origin*, which exist near the laminar–turbulent transition point (structural stochasticity: *chaos born of order*);

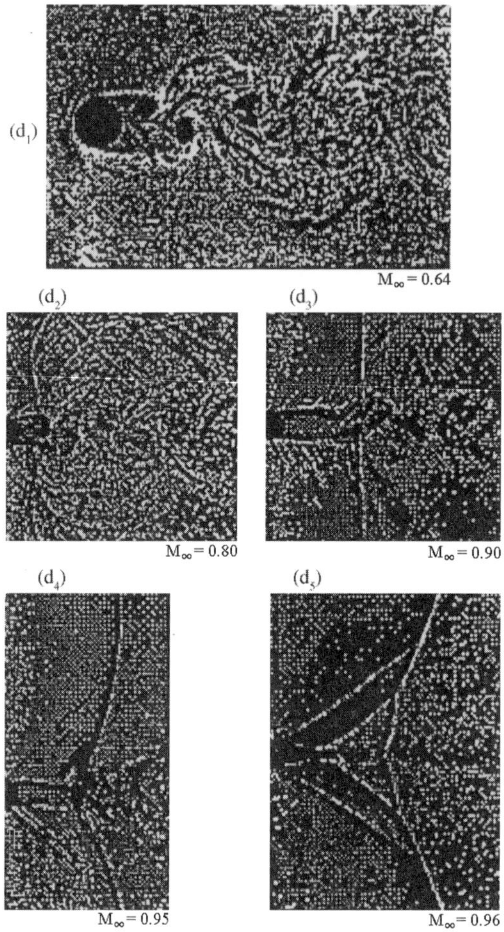


**Fig. 1.8.** (*Continued*)

(ii) *quasi-equilibrium structures* formed within the regions of developed turbulence where chaotic motion is so highly developed that a system is close to thermodynamic equilibrium (the ordered structure of such a stochastic motion: *order born of chaos*);

(iii) *flow structures* occupying a position halfway between the two enumerated above.

In free shear turbulent flows, such as jets, wakes, and mixing layers, ordered large-scale formations are observed for *arbitrarily large Reynolds*

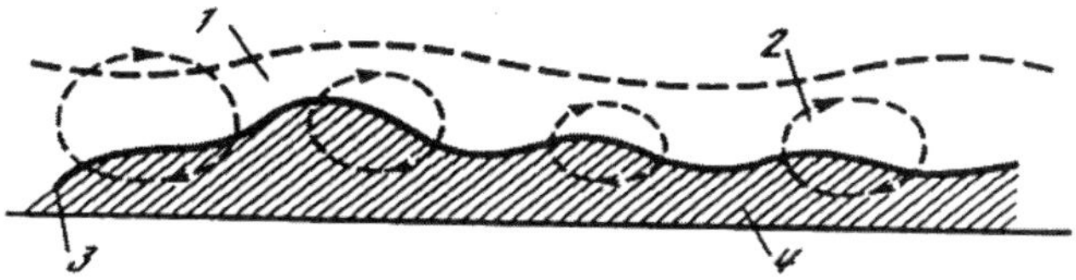


**Fig. 1.9.** Schematic flow in free developed turbulence (Townsend[5]): (1)unturbulized flow; (2) the motion of large vortices; (3) the boundary of turbulized fluid; (4) uniform turbulence.

*numbers.* Often these structures are two-dimensional (or quasi-two-dimensional).[38] The strength, scale, and shape of this low-frequency ordered motion are quasi-deterministic (i.e. individual) for a given flow and should be described by the hydrodynamic equations (not statistically). The dimensions of large vortices are comparable with the characteristic flow dimension and are much larger than the scales of the vortices forming the turbulent motion.[5,6] The turbulence within the central portion of a wake is characterized by a high *degree of local isotropy* (with respect to the strengths of velocity pulsations in various directions, and hence with respect to the energy characteristics[4]).

In numerical analysis, it is important to simulate transfer processes correctly. There are three types of energy-related phenomena characteristic of real turbulent motion,[39] viz.,

(i) generation of large-scale vortices, which depends on the specific properties of the flow as a whole;
(ii) fragmentation of the vortices into smaller-scale ones caused by nonlinearity, and the transmission of energy down the spectrum without considerable loss (the Kolmogorov cascade process);
(iii) viscous energy dissipation for the smallest scales.

Until recently only types (ii) and (iii) have been subjected to analysis by statistical methods and the theory was developed ignoring the vortex-generation mechanism (the necessary fluctuation energy was added from outside by introducing a random external force). However, the vortex-generation aspect of a large-scale turbulence (which is mostly deterministic) is very important since it incorporates the source of the turbulence as well as the mechanism for sustaining it.

Evidently, large-scale transfer of a turbulent fluid is primarily implemented by an ordered motion of groups of large vortices that distort the boundaries of a turbulent flow field and transfer the turbulent fluid in the transverse direction.[5,6] Hence, the motion of ordered and large-scale turbulent structures is the major factor determining the dynamic, kinematic, and energy characteristics of a flow as a whole; also, it determines the properties of the depth turbulence responsible for energy dissipation.

However, for free developed shear turbulent flows with high Reynolds numbers the backward influence of small-scale turbulence (as well as of the molecular diffusion) on the major characteristics of the large-scale flow is obviously insignificant since these are effects of different orders of magnitude; small-scale structures are "almost" universal (see Fig. 1.2), and the process of energy transfer itself is one-sided.

Listed below are some important, on our opinion, theoretical results of the investigation of coherent structures, which are exposed following the fundamental work by Monin.[40]

It is the intent of this section:

(i) to substantiate the thermodynamic necessity of turbulent fluid flow's spatial structuralization;
(ii) to give a dynamic definition of coherent structures;
(iii) to present an outline of the first step toward construction of the coherent structures' dynamic theory.

A fluid with a shear and/or temperature gradient is a thermodynamically nonequilibrium system (see Landau and Lifshitz[10]). Although the density $\rho$, the internal energy $\varepsilon$, and the velocity $\mathbf{V}$ are then defined in the usual manner ($\mathbf{V}$ being defined as the fluid momentum per unit mass), the remaining thermodynamic quantities, such as entropy and pressure, are defined as the same functions of $\rho$ and $\varepsilon$, as they are at the thermal equilibrium, but retain their exact physical meaning only up to the terms that are small compared with the shear and/or temperature gradient. Developed locally isotropic turbulence in the inertial range, which does not exhibit a uniform distribution of kinetic energy over the wave number space (i.e. an energy distribution $F(k) = \text{const.}$), but rather the strongly nonuniform Kolmogorov spectrum $F(k) \sim k^{-11/3}$, and doesn't behave as a system in thermodynamic equilibrium. In other words, turbulent disorder (or chaos) is quite different from the complete chaos that is characteristic of thermodynamic equilibrium, in that it exhibits a certain degree of order.

That this order exists also implies that at any given time some spatial structure or structures will exist in the flow. While each type of laminar flow is characterized by a specific spatial structure, in turbulent flows, ordered structures may be generated at random points and random times.

These structures are randomly oriented and exist for random periods of time (which, however, are long enough to be observable against a rapidly varying disordered background). As far as we know, they were first observed by Velikanov in channel flows in rivers and were called "macrostructural elements". At present in the new discipline of "synergetics" concerning self-organization processes, which includes the creation of "order from chaos" (or "the order in chaos"), these macrostructural elements are called "coherent structures". The first attempt to describe these structures theoretically seems to have been undertaken by Brown and Roshko[41] in the case of a planar mixing layer.

Given the progress in the study of stochastization processes of functions of time (represented by solutions of ordinary differential equations), there is now a good basis for an analysis of processes involving the stochastization of functions of several variables, represented by solutions of partial differential equations, in particular, the equations of fluid mechanics. In other words, now there exists a basis for a systematical study of coherent structures. This can be done by visualization of the flows in laboratory experiments (see Offen and Kline,[42] who discovered and measured oriented horseshoe vortices in the boundary layers near walls), in numerical experiments (see paper of Hussain[43]) and by constructing and solving model equations.

Some enthusiasts of the study of coherent structures have taken the extreme position of demanding the cessation of financial support for theoretical research based on Reynolds averaging (see Ref. 44). However, they forget that

(i) quite often there is a need for data on averaged turbulent flow fields;
(ii) coherent structures are randomly distributed over space and time, and the study of their individual forms, properties, and behavior cannot provide a comprehensive description of turbulence.

This means that a statistical description of, say, spatial and temporal distributions of various forms of coherent structures is needed, and hence the theory of turbulence must remain statistical.

Moreover, to identify coherent structures and analyze their individual properties, one has to use certain specific techniques for a conditional averaging of ensembles of visualized flows, say, by relating them to the singular points of the topology of streamlines (sometimes accompanied by calibration of the distances between the points) as well as by selecting the elements of an ensemble from the maximum of the histogram of the time intervals between them.

Goldshtik[38] has proposed a classification of coherent structures in residual flows (the "successors" of ordered subcritical stationary or quasi-periodic forms often encountered in turbulent flows), nonequilibrium flows (which in the case of a developed turbulence that is far from thermodynamic equilibrium, such as Kolmogorov turbulence, may possess universal self-similarity properties) and the theoretically imaginable, quasi-equilibrium flows (synergetically born from chaos close to the thermodynamic equilibrium, for instance, due to the ordering effect of certain conservation laws). Examples of the latter are given by the formation of zonal flows out of a two-dimensional turbulence in an ideal fluid at the surface of a rotating sphere or a beta-plane where almost all the energy passes over to the zonal flows, and almost all the entropy — to the residual chaos. However, different authors so far define the notion of coherent structures in different ways.

Thermodynamic nonequilibrium in, say, incompressible fluid is generated by the velocity shear. For the sake of simplicity, we start by considering plane-parallel flows and the boundary-layer flows on a flat plate in a uniform oncoming flow; or, as another example, we shall consider flows with constant (in space) velocity shear. In all these cases, the most natural coherent structures are rolls whose axes are perpendicular to the plane of shear.

Owing to secondary instabilities, vortex tubes may meander and suffer varicose expansions and contractions (in an ideal compressible fluid this results in variations in density, and, in the presence of gravity, in the portions of the tubes with lower density floating up due to buoyancy forces). Similar structures, for example, toroidal rolls, may appear in the flow regions with a more complex geometry. It a viscous fluid, neighboring locally antiparallel vortex tubes may reconnect, forming pairs of oppositely oriented horseshoe vortices with residual "bridges" between them.

Sometimes these vortex structures may form quasi-periodic one-dimensional chains or two- and three-dimensional lattices. Defects may arise within the latter, i.e. defects "frozen" into the lattice or involved into a "Brownian motion" on it; if the number of defects is large, then a "turbulence of defects" capable of evolving into fully developed turbulence may merge (e.g., see Refs. 10 and 38).

Taking into account that the topology of coherent structures is best represented by the vorticity vector lines, we propose to define coherent structures as *preferable nonlinear superpositions of large-scale spatial components of the turbulent vorticity field which are long-lived or, in other words, are the least unstable with respect to perturbations existing in the turbulent flow.*

To our mind, turbulent-flow coherent structures should be distinguished from the quite different laminar-flow spatial structures. Examples of the latter are given by axisymmetrical toroidal vortices in three-dimensional space (such as "smoke rings") and by similar two-dimensional dipole vortices in a plane described by some general self-similar solutions of the nonlinear vorticity equation with initial singularities. Therefore, we doubt that Barenblatt *et al.*[45] correctly define a coherent structure as a laminar flow modeled by a self-similar specific solution of the two-dimensional *linear* diffusion equation for vorticity. Naturally, the linear diffusion equation describes the final stages of viscous attenuation of vorticity, but the singular solution[45] presented in the form of the sum of a vortex dipole and a potential point source described by the stationary stream function $\psi = Q\theta/2\pi$ (where, in addition, the strength $Q$ of a source is artificially put proportional to the fluid viscosity $\nu$) seems unnatural since it does not satisfy the *nonlinear* vorticity equation. Latter one of the authors of Ref. 45 replaced the solution by another one which does not contain an unnatural source of mass.

Note also that the attempt in Ref. 45 to compare solutions of the two-dimensional vorticity equation and the results of two-dimensional laboratory experiments with the dipole vortices sometimes observed at satellite photographs of the ocean surface seems to be incorrect, since the motions at the latter cannot be described by two-dimensional hydrodynamic equations and are actually induced by three-dimensional motion within the ocean. Unlike two-dimensional hydrodynamics, the cyclonic and anticyclonic vortices composing an ocean dipole are essentially different: within the former water flows up, while within the latter it flows down. In addition, if an ocean vortex is not small (according to observations, its dimensions may be of order of tens of kilometers), then it may be affected by the so-called beta-effect, which creates favorable conditions for the survival of anticyclones in which, unlike cyclones, the non linearity may compensate for the dispersive spreading. An analysis of the anticyclone–cyclone asymmetry in ocean dipoles may be an interesting and as yet unformulated problem of the quantitative study of available satellite photographs. In this respect, the model presented in Ref. 45 may be misleading.

Although the laminar flows mentioned above cannot be classified as coherent structures, identification of real coherent structures in the ocean surface layer (as well as the reflections in it of coherent structures existing in the atmospheric boundary layer) is very important. The structures may appear as successors to both primary and secondary instabilities of the Ekman boundary layers in the ocean and atmosphere.

The reproduction of coherent structures within the turbulent flows by the numerical solution of hydrodynamic equations is very difficult, especially for high Reynolds numbers, when the spectrum of turbulence scales is very wide and cannot be fully resolved from the maximum to the minimum scales even using the fastest modern computers. Therefore, the large eddies simulation (LES) must be implemented separately from a numerical description of small-scale turbulence with full account of the molecular viscosity, the latter being at present feasible only for $Re \leq 10^3$ (see Ref. 46).

The original approach to the LES problem has been developed by Belotserkovskii and his school.[12–14] The approach is based on a direct numerical solution of the ideal-fluid hydrodynamic equations combined with the approximate description of energy dissipation due to small-scale turbulence. The turbulent flow under study is represented as a relatively ordered slow motion of coherent structures — slightly unstable large eddies transporting portions of a fluid with developed turbulence from place to place.

In particular, this approach may be used to describe the "residual" coherent structures of bifurcational origin (in Refs. 12–14, we call them dynamic structures). However, the structures may also be described analytically by using model equations derived from the hydrodynamic equations by expanding the "slow" amplitudes of wave packets appearing due to bifurcations into series, in powers of the supercriticality parameter (which are analogous to the "epsilon-expansions" in the Landau–Ginzburg theory of the second kind phase transitions). The rest of Monin's article is dedicated to the model equations (see Ref. 40).

## 1.4. General problem formulation

What was said above is in essence the ideology of organizing the computational process for the direct numerical simulation of large-scale wake flows with developed free shear turbulence for very high Reynolds numbers $Re$. Generally speaking, the whole of the analysis splits into the following two *interrelated problems.*

**Problem 1.** *Calculation of unsteady motion of ordered and large-scale turbulent structures.*

Since the motion is large scale and ordered, it can be described by numerical schemes based on the nonstationary hydrodynamic equations (the integral laws of conservation) for the ideal-medium model (Euler's difference equations) and possessing an approximate (efficient) dissipation mechanism. The major objective of the dissipation mechanism is to *smooth* (extinguish) small-scale oscillations and to ensure both the stability of the solution and the required resolution of the flow microstructures. Both experimental and numerical studies have shown that "oriented" compact (asymmetric) differences in the integral laws of conservation used *without* semi-empirical models of turbulence satisfy these conditions because the properties of large-scale motions are mainly determined by *volume convection* (i.e., they have a wavy or dynamic nature) and depend on the solution as a whole. Hence, the calculations should be carried out throughout the flow field on real ("coarse") difference grids and should be followed by the determination of the required averaged characteristics of turbulent flow (say, moments of different orders) by an appropriate statistical reduction of the results.

It should be noted that the possibility of determining the statistical characteristics of an unsteady (oscillatory) large-scale turbulent flow by smoothed equations is far from evident and the *correctness* of the problem formulation should be carefully analyzed (see below).

For the present, the problem solution allows us to simulate the motion of ordered structures, large (energy-transferring) turbulent vortices, the process of turbulence generation and other processes based on large-scale transfer. Also, by using an appropriate representation of the dissipation mechanism, we may study the cascade process of energy transfer, i.e. the descent to small-scale, locally isotropic turbulence. If the dissipation rate corresponding to large structures is determined correctly, then the energy characteristics of turbulence during the descent will evidently be simulated correctly since the process is rather conservative.

This deterministic approach allows us to single out the ordered and large-scale formation characteristics of structural turbulence; however, one has to specify the method for averaging oscillatory fields.

**Problem 2.** *Numerical simulation of the stochastic component of a turbulent shear flow (small-scale turbulence).*

This problem deals with the simulation of the local resolution, i.e. the process of dissipation, the way in which energy propagates along the turbulent core of a jet, etc. We believe such flows should be calculated using statistical methods (or by introducing appropriate turbulent-viscosity coefficients phenomenologically). We may also use algorithms based on oscillatory equations,[47] for which the averaged-flow parameters are determined by solving Problem 1.

It should be noted that at present the above flows should only be calculated within limited subregions by cutting off large-gradient zones out of the general flow pattern. This approach is based on the assumption that if the turbulence scale is small as compared with the averaged-flow scale, then the local structure of turbulence is sufficiently universal for various flows and is determined by the local conditions only (in this case turbulent transfer is determined by *gradient diffusion*).[5] Thus, the flows may be calculated by using clearly defined models and sufficiently fine numerical grids, and hence the required computer capacity may be lowered.

Usually the Navier–Stokes equations are averaged with respect to the Reynolds number *over all scales of turbulence* simultaneously and *over large time intervals* (the regular averaged motion is singled out), and this requires all the structures to be simulated at once. Hence, it is impossible to construct a model of turbulence simulating different classes of motion. Unlike this approach, the method proposed here is based on splitting the general motion into large- and small-scale structures (the nonrandom ordered motion of macrostructures is singled out). The motion of ordered and large-scale turbulent vortices (with dimensions $\lambda \geq h^*$, where $h^*$ is the spacing of a difference grid) is determined by the direct integration of the full hydrodynamic equations, while simulation (smoothing) is only used for the subgrid small-scale fluctuations which cannot be resolved explicitly by numerical integration and possess, as noted above, rather universal properties.

This concept is of great significance for the numerical simulation of turbulence. It is of interest to note that necessity for the splitting was indicated by Dryden as early as in 1948,[48] who when analyzing data from hot-wire experiments conducted in boundary layers concluded that *considerable masses of liquid move more or less as coherent structures.*

The present approach handles subgrid-closure models (see Refs. 24 and 25). However, we must organize the splitting (smoothing) correctly as well as choose properly the models for constructing approximate (smoothed) equations describing the motion of ordered and large-scale turbulent structures. Naturally, this raises questions related to the correctness of the problem formulation, the choice of approximations for small-scale vortices and the estimation of their effect on large-scale motions. Moreover, of course, attention should be paid to the various aspects of the realization of the approach.

## 1.5. Simulation of coherent structures in turbulent flows

The construction of a numerical algorithm to calculate large-scale structures (Euler' s difference equations) should be started by deriving a system of equations approximating the laws of conservation in integral form (the methods of large particles and of fluxes; see Refs. 12–15, 36, and 49–51).

The FLUX method[13,14,49–52] was applied to a numerical simulation of the problems considered. The brief description of that method and of the corresponding difference schemes is given below.[52]

The distinguishing feature of the method of fluxes consists in approximation of the nonstationary equations reflecting the conservation laws for each of the additive characteristics of a medium; these laws are expressed in the integral form for everyone of the elementary volume-cells resulting from discretization of the integration area. Consider the conservation laws for an arbitrary, finite, fixed in space volume $\Omega$:

$$\frac{\partial}{\partial t}\oiiint_{\Omega} F d\Omega + \oiint_{S_\Omega} \mathbf{Q}_\Gamma \mathbf{n} dS = \oiiint_{\Omega} J_F d\Omega. \tag{1.13}$$

Here $F = \{\rho, \xi_i, \varepsilon\}$ are densities of distribution for one of the additive characteristics of a medium — mass $M$, momentum components $X_i$, and total energy $E$ of the volume $\Omega$; $\mathbf{Q}_F = \{\mathbf{Q}_\rho, \mathbf{Q}_{X_i}, \mathbf{Q}_E\}$ are the corresponding vectors of the densities of flows; $S_\Omega$ is a side surface of the volume $\Omega$ having the external normal $\mathbf{n}$; $J_F$ are densities of distribution of the sources (if those are present) within the volume $\Omega$.[u]

For a temporal derivative, the simplest first order approximation was used. The surface integrals are calculated by means of a quadrature formula of rectangle having a central nodal point. Thus, Eq. (1.13) (in the absence of sources) is written down in a form

$$\Omega_{l,m,k}\frac{(F^{n+1}_{l,m,k} - F^{n}_{l,m,k})}{\tau} = -\sum (\mathbf{Q}_F \mathbf{n}_i)_{S_i} S_i. \tag{1.14}$$

[u]In Refs. 14 and 49, L. I. Severinov has proposed the deformation-flow method, in which the property of conservativeness concerns not only total energy, but also internal and kinetic ones.

The summation in Eq. (1.14) is realized over all the facets $S_i$ of the volume $\Omega$, while the lower index $S_i$ corresponds to the value of function at the central point of the side $S_i$. To model the transfer velocity, pressure, and derivatives entering the expressions for components of the tensor of viscous stresses and for the heat-conduction law, the symmetrical difference expressions are used.

To improve the characteristics of the monotony of different schemes within the regions of large gradients, or those of the complicated structure of flow, the hybrid schemes are used which are based on the following items. Let $\xi^{(2)}$, $\xi^{(1)}$ be the values of densities of distribution, calculated with the first- or second-order of approximation, taking into account the direction of transfer velocity, while $\xi^{(3)}$ are the similar values determined with the help of symmetrical difference expressions ($(V_n)_{l+\frac{1}{2}m,k}$ is a velocity component normal to the side $S_{l+\frac{1}{2}}, m, k$):

$$[\xi^{(2)}]_{l+\frac{1}{2},m,k} = \begin{cases} 1.5\xi_{l,m,k} - 0.5\xi_{l-1,m,k} & \text{for } (V_n)_{l+\frac{1}{2},m,k} \geq 0, \\ 1.5\xi_{l+1,m,k} - 0.5\xi_{l+2,m,k} & \text{for } (V_n)_{l+\frac{1}{2},m,k} < 0; \end{cases}$$

$$[\xi^{(1)}]_{l+\frac{1}{2},m,k} = \begin{cases} \xi_{l,m,k} & \text{for } (V_n)_{1+\frac{1}{2},m,k} \geq 0, \\ \xi_{l+1,m,k} & \text{for } (V_n)_{l+\frac{1}{2},m,k} < 0; \end{cases}$$

$$[\xi^{(3)}]_{l+\frac{1}{2},m,k} = \frac{\xi_{l,m,k} + \xi_{l+1,m,k}}{2}.$$

Let us construct the nonuniform, two-parameter scheme in the following way:

$$\xi_{l+\frac{1}{2},m,k} = \alpha_1[\xi^{(1)}]_{l+\frac{1}{2},m,k} + \alpha_2[\xi^{(2)}]_{l+\frac{1}{2},m,k} \\ + (1 - \alpha_1 - \alpha_2)[\xi^{(3)}]_{l+\frac{1}{2},m,k}. \tag{1.15}$$

The choice of values $\alpha_1$ and $\alpha_2$ is carried out locally, on the basis of the properties of flow in vicinity of the volume-cell considered, $\Omega_{l,m,k}$.

As soon as the densities of distribution are written down in the form of Eq. (1.15), it is possible for $\alpha_1 = 0, \alpha_2 = 1$ to obtain the traditional oriented scheme of the method of fluxes, for $\alpha_1 = \alpha_2 = 0$ — the symmetrical scheme, and for $\alpha_1 = 1, \alpha_2 = 0$ — the oriented scheme of the first order of approximation. For nonlinear schemes, when the flows of complicated structure are studied, or those with strong interactions, the consideration of properties of a linear model might lead only to the qualitative recommendations. Therefore, the criteria of choice for the optimal difference schemes are based, primarily, on the physical considerations. One of such criteria consists in the following:

$$\begin{aligned} \alpha_1 = 0, \quad \alpha_2 = 1 \quad &\text{for } [\xi^{(2)}]^n_{\frac{1}{2},m,k} \in [\xi^n_{l,m,k}, \xi^n_{l+1,m,k}], \\ \alpha_2 = 0 \quad &\text{for } [\xi^{(2)}]^n_{\frac{1}{2},m,k} \notin [\xi^n_{l,m,k}, \xi^n_{l+1,m,k}], \end{aligned} \tag{1.16}$$

which means that renunciation of the asymmetrical approximation formulae is realized in the case when the value of the density of distribution, calculated on the basis of the oriented differences of the second order, is found to lie outside the range of values at the nearest points. The situations of that kind might occur, for example, within the regions of large gradients and of tangential discontinuities, i.e. within those having local singularities. If the condition of stability becomes to be onerous with a conversion to completely symmetrical scheme ($\alpha_1 = \alpha_2 = 0$), then one might resort to one-sided differences, setting $\alpha_1 = 1$ (formally, the hybrid scheme considered acquires, locally, the first order of approximation).

It is to be noted that, for conservative schemes, conversion to the locally first order does not lead to a decrease of computational accuracy on the real computational grids.[v] This claim is confirmed by the solution of various nonlinear problems of aerodynamics. Moreover, the local conversion to the first order scheme is preferable in such cases, when from the physical considerations one is able to apprehend *a priori* that either taking into account the "remote pre-history" of a convective flow, or its symmetrical representation, might lead to a distortion of its structure.

The setting of boundary conditions and prescription of the initial data for the viscous, as well for inviscid models of medium, prove both to be traditional by the solution of most of the problems, and are described in detail in Refs. 51 and 52.

One of the ways of investigation of properties of the difference scheme is the application of differential approximation method, developed by Yanenko and Shokin.[53,54] This approach is used to study the properties of linear and quasilinear equations. When methods of differential approximation are applied to the system of equations written in integral form, the term "differential approximation" is not exactly correct. Nevertheless, we shall use

[v]Strictly speaking, the estimates of accuracy of the finite difference schemes are obtained not for the real grids, but for those that are asymptotically fine, and mainly for linear (or quasilinear) model equations.

this term, indicating the general methodology of differential approximation that is applied everywhere (Ref. 50).

Starting equations are the *laws of conservation of mass, momentum and total energy*:

$$\frac{\partial}{\partial t}\oiint_{\Omega} f d\Omega = -\oiint_{S_\Omega} \mathbf{Q}_F \mathbf{n} dS, \quad F = \oiint_{\Omega} f d\Omega,$$

$$F = \{M, P_i, E\}, \quad f = \{\rho, \rho v_i, \varepsilon\}. \tag{1.17}$$

Here $f$ is the density of distribution of an additive characteristic $F$.

In view of the fact that for medium's description a model of compressible gas is used, let us write out density fluxes vectors as (terms depending on the viscosity in equations of motion and energy are for simplicity omitted):

$$\mathbf{Q}_F = f\mathbf{V} + \mathbf{P}_F, \quad P_F = \{0, p\mathbf{n}, p\mathbf{V}\}.$$

Term $\mathbf{P}_F$ takes into account pressure in the equations of motion and energy.

We will construct the hybrid scheme with the following approximation of distribution density on volume's sides $\Omega_{lkm}$:

$$\frac{f^{n+1} - f^n}{\tau} = \frac{1}{\Omega}\sum_{S_\alpha}(\mathbf{Q}_F\mathbf{n})_{S_\alpha} S_\alpha + (I_F)_\Omega,$$

$$(\mathbf{Q}_F\mathbf{n})_{l+\frac{1}{2}} = f_{l+\frac{1}{2}}(\mathbf{V}\mathbf{n})_{l+\frac{1}{2}} = f_{l+\frac{1}{2}}(v_n)_{l+\frac{1}{2}},$$

$$(v_n)_{l+\frac{1}{2}} = \frac{1}{2}(\mathbf{V}_{l+\frac{1}{2}} + \mathbf{V}_l)\mathbf{n}_{l+\frac{1}{2}},$$

$$f_{l+\frac{1}{2}} = \alpha_{f^1} f_l + \alpha_{f^2} f_{l-1} + \alpha_{f^3} f_{l+1}, \quad \text{for } (v_n)_{l+\frac{1}{2}} \geq 0,$$

$$f_{l+\frac{1}{2}} = \alpha_{f^1} f_{l+1} + \alpha_{f^2} f_{l+2} + \alpha_{f^3} f_l, \quad \text{for } (v_n)_{l+\frac{1}{2}} < 0,$$

$$\alpha_{f^1} + \alpha_{f^2} + \alpha_{f^3} = 1. \tag{1.18}$$

Let us investigate the complete system of initial nonlinear equations (without preliminary linearization). In the further text, availability of necessary number of derivatives is supposed.

Before choosing the form of approximation, one should take into account some features of its obtaining when regarding difference schemes, approximating conservation laws written in integral form.

**Note 1.** At the first stage of difference scheme derivation out of integral form, the approximation of volume and surface integrals by means of approximating integral formulae is performed. The terms of differential approximation related by this procedure do not change the model of transfer, but are immediately connected with volume's finite size. If differential approximation of finite difference equations, approximating system of differential equations (which are obtained when volume $\Omega$ tends to 0) are considered, the approximation errors drop out of consideration, because the limiting transfer $\Omega \to 0$ was already performed when deriving the system of differential equations from integral form of conservation laws.

**Note 2.** At the following stage, the approximation of density flux vectors at characteristic points of sides of volume $\Omega$ is performed. At this stage by the derivation of differential approximation of difference scheme, which is approximating the differential equations, errors of approximation of the divergence operator are actually taken into account, while in consideration of the integral form of conservation laws these errors are absent (approximation of density fluxes vectors at the centers of sides is performed, which is followed by summation over all sides with computer accuracy).

Dropping out the cumbersome computations, one presents the differential approximation in hyperbolic form:

$$\frac{\partial}{\partial t}\iint_{\Omega}(f+\delta_f+\delta_t)d\Omega+\iint_{S_\Omega}[(\mathbf{Q}_f+\delta\mathbf{Q}_f)\mathbf{n}-\delta_S]dS$$
$$+O[\max(\tau h_k^2,\gamma_1 h_k^3,h_k^4)]=0. \tag{1.19}$$

In differential approximation obtained the hyperbolic term

$$\delta_t=\frac{1}{2}\frac{\partial f}{\partial t}\tau+O(\tau^2)$$

is related to the approximation of time derivative.

Term

$$\delta_f=\frac{1}{24}\sum_k\frac{\partial^2 f}{\partial x_k^2}h_k^2+O\left(\max_{k,m}\frac{\partial^4 f}{\partial x_k^2\partial x_m^2}h_k^2h_m^2\right)$$

is connected with volume integration.

Term

$$\delta_S=\frac{1}{24}\sum_{k\neq i}\frac{\partial^2(\mathbf{Q}_F\mathbf{n}_i)}{\partial x_k^2}h_k^2+O\left(\max_{k,m}\frac{\partial^4(\mathbf{Q}_F\mathbf{n}_i)}{\partial x_k^2\partial x_m^2}h_k^2h_m^2\right)$$

is connected with surface integrals.

Term

$$(\delta\mathbf{Q}_F)=-\left[\frac{\gamma_1}{2}V_i\frac{\partial f}{\partial x_i}h_i+\frac{1}{8}\left(\gamma_2 V_i\frac{\partial^2 f}{\partial x_i^2}+f\frac{\partial^2 V_i}{\partial x_i^2}\right)h_i^2+\delta_{P_F}+O(h_i^3)\right],$$

where

$$\gamma_1 = 1 - 2\alpha_{f^1} - 4\alpha_{f^2}, \quad \gamma_2 = 1 + 8\alpha_{f^2},$$

$$(\delta_P)_{(\rho)} = 0, \quad (\delta_P)_{(\xi_i)} = \frac{1}{8}\frac{\partial^2 p}{\partial x_i^2} h_i^2 \delta_{ij}, \quad (\delta_P)_{(\varepsilon)} = \frac{1}{8}\left(V_i \frac{\partial^2 p}{\partial x_i^2} + p\frac{\partial^2 V_i}{\partial x_i^2}\right) h_i^2,$$

are immediately related to the approximation of density fluxes vector.

Substitution of the proper values of coefficients $\alpha_{f^P}$ shows up the approximation orders of different schemes and allows estimating scheme viscosity. Thus, for oriented scheme of the first order, scheme viscosity has a form $(1/2)V_i h_i$. For schemes of the second order, the structure of differential approach is more complicated, thus it is difficult to distinguish scheme viscosity in general way.

The resulting differential approach indicates sufficiently complicated form of additional terms changing initial mathematical model, reveals the errors' structure, arising at the different stages of approximation of conservation laws, written for finite volume in integral form, and, broadly speaking, allows to estimate it in numerical solution.[50,51]

These uniform finite-difference schemes are dissipation-stable and divergent-conservative. The schemes allow us to calculate both the smooth solution zones and the zones where a solution is discontinuous. The stability of the calculations is ensured by the *internal dissipation* (the roughness of approximation) alone, making it possible to use the schemes in the presence of curvilinear boundaries and to solve three-dimensional unstable problems.

Thus, *Euler's difference equations* (derived using differential approximations for the simplest splitting schemes in the method of large particles) have the form[31]

$$\begin{aligned}
&\frac{\partial \rho}{\partial t} + \nabla(\rho \mathbf{V}) = 0,\\
&\frac{\partial \rho u}{\partial t} + \nabla(\rho u \mathbf{V}) + \frac{\partial p}{\partial x} = \frac{\partial}{\partial x}\left(\rho \varepsilon_x \frac{\partial u}{\partial x}\right) + \frac{\partial}{\partial y}\left(\rho \varepsilon_y \frac{\partial u}{\partial y}\right),\\
&\frac{\partial \rho v}{\partial t} + \nabla(\rho v \mathbf{V}) + \frac{\partial p}{\partial y} = \frac{\partial}{\partial x}\left(\rho \varepsilon_x \frac{\partial v}{\partial x}\right) + \frac{\partial}{\partial y}\left(\rho \varepsilon_y \frac{\partial v}{\partial y}\right),\\
&\frac{\partial \rho E}{\partial t} + \nabla(\rho E \mathbf{V}) + \frac{\partial p}{\partial y} = \frac{\partial}{\partial x}\left(\rho \varepsilon_x \frac{\partial E}{\partial x}\right) + \frac{\partial}{\partial y}\left(\rho \varepsilon_y \frac{\partial E}{\partial y}\right),
\end{aligned} \tag{1.20}$$

where $\varepsilon_x = (1/2)|u|\Delta x$ and $\varepsilon_y = (1/2)|u|\Delta y$ are the coefficients of the toughness of approximation.

These series expansions are accurate to $O(\Delta t, h^2)$. Note that the structure of the coefficients of the toughness of approximation, $\varepsilon \sim |v|h$, in (1.20) is similar to that of the turbulent-viscosity coefficients for the scale $\lambda, v_{t,\lambda} \sim v_\lambda \lambda$. (This is because both coefficients appear due to the nonlinearity of the processes they describe, and the splitting schemes are adequate to the turbulent-mixing mechanism.) If a turbulent eddy of the scale $\lambda$ moving with velocity $v_\lambda$ is replaced by a large particle moving with velocity $v_h$, then the expressions for $v_\lambda$ and $v_{t,\lambda}$ are identical. If $\varepsilon_x \sim \varepsilon_y \sim v_{t,\lambda}$ then Eqs. (1.20) for $\rho =$ const. (an incompressible fluid) transform into the exact Navier–Stokes equations, where the molecular viscosity $v_{\mathrm{m}}$ is replaced by the effective turbulent viscosity $v_{\mathrm{t}}$ (see (1.1)–(1.3)).

The left-hand sides of (1.20) correspond to the *exact Euler's differential equations* and the right-hand sides to the *dissipation terms*, which describe the *perturbation background*, appear due to the smoothing (1.14) of the subgrid fluctuations and the approximation of the original system of differential equations by finite-difference ones and depend on the internal structure of the representations. The dynamics of this background is the source of the fluctuations. It follows that those Eqs. (1.20) in concrete calculations are dissipative (though they were derived from the ideal-fluid model). The specific form of the dissipation mechanism depends on the nature of the approximation, and generally, its structure may be controlled.

In a certain sense, the dissipation mechanism in (1.20) is analogous to the Reynolds stresses in (1.14) due to the averaging over the scales $\lambda < h$ and represents, in generalized form, the contribution of fine (subgrid) vortices. For schemes of the first-order accuracy the coefficient of the scheme (effective) viscosity $\varepsilon \sim |v|h$ depends on the local flow velocity and the dimensions of the difference grid $h$ (however, it is independent of the molecular viscosity). The roughness of approximation in the schemes with second-order accuracy (namely, in the leading terms of the asymptotic expansion)[50,51] has the form $\varepsilon \sim h \cdot \partial v/\partial x$, which also correlates with our notion of the turbulent viscosity. Hence, in order to get the effective viscosity satisfying the stability condition, we should use coarse calculation grids (from the physical point of view, molecular clusters). Hence, the general principle defining both the nature and the role of the dissipation mechanism may be formulated as *the principle of stable solution for a required degree of resolution.*

Let us compare Euler's difference equations with the Navier–Stokes difference ones. Using a splitting scheme from the method of large particles for the Navier–Stokes model, the momentum equation derived in Ref. 36

for a compressible medium has the form

$$\frac{\partial \rho u}{\partial t} + \frac{\partial (p + \rho u^2)}{\partial x} = \left( \mu(A-2) + \rho |u| \frac{\Delta x}{2} \right) + O(\Delta t, \Delta x^2),$$

where $A = \lambda/\mu, \mu = \rho v, \lambda$ is the second viscosity coefficient and $v$ is the molecular viscosity.

It follows from this that scheme effects also have an influence on the approximation of the viscous terms. In order to ensure the calculation stability for flows with large Reynolds numbers, the effective viscosity must be high and usually exceeds the real viscosity. Thus, we see that for high Reynolds numbers, the numerical solution of boundary-value problems based on the Navier–Stokes equations encounters both technical and theoretical difficulties (*denser grids do not fully overcome the difficulties*).

Thus, if the ideal-gas equations are supplemented with dissipation terms, then for rather arbitrary assumptions about the nature of the dissipation, the generalized solution for the flow macrostructures corresponding to the limiting regimes ($v \to 0$) may be obtained, with a certain accuracy, from the equations incorporating an approximate dissipation mechanism instead of the Navier–Stokes equations.

Let us stress once again that our approach to vortex-motion macrostructures *does not employ semiempirical models of turbulence.* In this case the dissipation mechanism incorporated in the averaging operator itself (which arose due to the use of oriented finite-difference representations) acts as a filter that cuts off (or better to say, smoothes) the subgrid fluctuations. It is this dissipation mechanism *in the averaged form* that accounts for the contribution of small-scale subgrid vortices. Direct calculations on the grid of fully averaged dynamic *macroequations* must ensure a *stable solution*, the *build-up* of the energy integral and the *stabilization* of the unsteady flow parameters. This yields the *effective* dissipation mechanism and satisfies the requirement that the splitting into "large" and "small" scales *is invariant.* Other important characteristics of turbulent flow, such as the energy dissipation rate and the integral scale, may also be determined within the domain of large-scale vortices (for instance, by using Onufriev's rule). Thus, the above large-scale approach allows us to study the major properties of free fully developed turbulence.

In principle, the smoothed equations may be used to estimate flow characteristics on smaller scales too, as well as to study the cascade process of energy transfer down the hierarchy of vortex dimensions, etc. However, this cannot be done without introducing semiempirical values of the turbulent viscosity. Let us discuss a possible scheme for this approach.

The exact simulation of unsteady flows leads to a better understanding of the dynamics of a phenomenon. By using the approximate mechanism of rapid dissipation in Euler's difference equations, we can, in essence, find the averaged characteristics of a flow from the stability conditions. In principle, we can control the operation of the dissipation mechanism in such a way that it will simulate, to a certain extent, the process of cascade transfer for various values of $\lambda$.

If the dynamics of large-scale turbulence of an *unsteady ordered structure* is calculated using Euler's difference schemes with viscosity represented in the form

$$\varepsilon \sim |v|h, \tag{1.21}$$

then, while studying the averaged characteristics of large-scale *energy-transferring turbulent vortices* within the regions of large gradients, we may introduce the turbulent viscosity in the form

$$\hat{v} \sim \overline{\Delta u l}, \tag{1.22}$$

where $\overline{\Delta u}$ is the mean velocity variation at the distances of the order of $l$. For the *local-turbulence zones* determined by local gradients the effective viscosity $\hat{\hat{v}}_{t,\lambda}$ which satisfies the Kolmogorov–Obukhov law[9,10]

$$v_\lambda \sim (\varepsilon\lambda)^{1/3} \quad \text{or} \quad \varepsilon \sim \frac{v_\lambda^3}{\lambda} \sim v_{\mathrm{t},\lambda}\left(\frac{v_\lambda}{\lambda}\right)^2 \tag{1.23}$$

should be used in Euler's equations. Whence

$$\hat{\hat{v}}_{\mathrm{t},\lambda} \sim \frac{\varepsilon\lambda^2}{v_\lambda^2}, \tag{1.24}$$

where $\varepsilon$ is the turbulent energy dissipation rate through the cascade of vortices, which may be obtained by a direct calculation of the large-scale formations (see approximations (1.7)–(1.9)), and $v_\lambda$ is the (oscillatory) velocity variation over the distances $\lambda$. Instead of the absolute motion when a cell moves as a single unit (as in the case of large-scale motions; see (1.21)), within local-turbulence zones ($\lambda \gg \lambda_0$) we should consider the relative motion of liquid particles.

Equation (1.24) together with (1.13)–(1.15) simulates the *attenuation* of turbulence for different scales $\lambda$. The effect of molecular viscosity cannot be ignored within the viscous-motion domain, where the total energy dissipation takes place (for $Re_h \approx 1$, at the distances of the order of $h$), and one has to use the full Navier–Stokes equations with the molecular dissipation mechanism.[w]

Thus, we have to use different turbulent-viscosity coefficients corresponding to different scales of motion. The scale of resolution decreases with decreasing scale of the subgrid vortices. Hence, the smaller the resolution the coarser the subgrid fluctuation may be used since most of the spectrum becomes directly resolvable. The type of the effective-viscosity coefficient is changed if the energy flux is conserved down the cascade of vortices within the regions where the old solution becomes unstable and one has to use smaller spacings (scales) of the numerical grid. Besides, the matching allows us to find the required values of the constants in (1.22) and (1.24).

The above *cascade model* seems to ensure an adequate simulation of turbulent transfer: as the resolution decreases, the importance of stability increases and makes it necessary to pass over to more complicated solutions that account for an increase in dissipation.

It should be mentioned that during the determination of the spatial-temporal field of ordered structures in ordered motion, the role of the dissipation mechanism in (1.20) reduces to *regularization* of a solution. (Generally speaking, the detailed structure of the dissipation mechanism is not significant and its optimal form is determined, as mentioned above, from the stability conditions by calculating on a variety of approximation grids.) However, when studying the unsteady (oscillatory) properties of turbulence using Eqs. (1.22)–(1.24), we may only hope to obtain the mean statistical flow characteristics. Hence, the requirements to the models of subgrid turbulence are quite acceptable in this case too. It is evident that in the framework of the above approach we do not have to solve the Navier–Stokes equations for very large Reynolds numbers.

## 1.6. Correctness of the problem formulation

Although the construction of the flow pattern formed by slow large vortices does not encounter principal difficulties (the conclusion being confirmed by calculations), the possibility of determining averaged characteristics of unsteady (oscillatory) large-scale turbulent motion with the help of

---

[w]The author is thankful to Academician A.M. Obukhov for his attention to this work. The major idea behind the approach was elaborated after a seminar on March 30, 1978.

smoothed equations is far from evident and raises the question of the *correctness of the problem formulation.* Calculations carried out using the smoothed equations incorporating the above dissipation mechanism are performed prior to the formation of a stabilized flow, i.e. prior to the appearance of steady characteristics (generally, unsteady structures). After this, to obtain the averaged characteristics of an oscillatory turbulent flow (moments), one has to subject the results to an adequate statistical procession, which may be carried out either straightforwardly or by using probabilistic approaches.

Let us consider this issue in more detail using the results by Ievlev (see Refs. 55 and 56). For hydrodynamic problems the general system of fluid-mechanical equations may be represented in the form:

$$\begin{gathered}\frac{\partial v_i}{\partial t} + v_k \frac{\partial v_i}{\partial x_k} = F_i, \quad i = 1, 2, 3, \\ \frac{\partial \rho}{\partial t} + v_k \frac{\partial \rho}{\partial x_k} = -\rho \frac{\partial v_1}{\partial x_1}, \\ \frac{\partial \xi(m)}{\partial t} + v_k \frac{\partial \xi(m)}{\partial x_k} = \Pi(m), \quad m = 1, 2, \ldots, N,\end{gathered} \tag{1.25}$$

where $\xi(m)$ denote all the independent parameters determining the state and the motion of the medium with the exception of its velocity $\mathbf{V}$ and density $\rho, v_k$ is the projection of the total velocity $\mathbf{V}$ onto the $x_k$-axis (the repeated subscripts show that summation is carried out for all the coordinate axes) and $F_i$ is the right-hand side of the momentum equation. Thus, for an incompressible fluid,

$$F_i = -\frac{1}{\rho}\frac{\partial p}{\partial x_i} + v\frac{\partial^2 v_i}{\partial x_k \partial x_k}.$$

In a turbulent flow the instantaneous values of $v_i, \rho$, and $\xi(m)$ oscillate and are, in effect, *random quantities.* Let us choose a group of $n$ points and denote by $f_n$ the probability density of various values of random quantities of $v_i, \rho$, and $\xi(m)$ at the points at the same time $t$. Let $A_n$ be the set of quantities of $v_i, \rho$, and $\xi(m)$ at all the selected points. A quantity referring to point $\gamma$ will be denoted by the subscript $(\gamma)$. If the probability density $f_n$ is known, then we may find the average values (mathematical expectations) of any functions of $A_n$, in particular, the average of $v_i, \rho$, and $\xi(m)$ themselves at any of the above points, one- and multi-point moments of high orders, etc.

The equation describing the time variation of $f_n$ has the following form:[55,56]

$$\frac{\partial f_n}{\partial t} = \sum_{\gamma=1}^{n} \left\{ -v_k^{(\gamma)} \frac{\partial f_n}{\partial x^{(\gamma)}} + f_n \left\langle \left( \frac{\partial v_1}{\partial x_1} \right)^{(\gamma)} \right\rangle_{A_n} - \frac{\partial}{\partial v_i^{(\gamma)}} [f_n \langle F_i^{(\gamma)} \rangle_{A_n}] \right.$$
$$\left. + \frac{\partial}{\partial \rho^{(\gamma)}} \left[ f_n \rho^{(\gamma)} \left\langle \left( \frac{\partial v_1}{\partial x_1} \right) \right\rangle_{A_n} \right] - \sum_{m=1}^{N} \frac{\partial}{\partial \xi^{(\gamma)}(m)} [f_n \langle \Pi^{(\gamma)}(m) \rangle_{A_n}] \right\}, \tag{1.26}$$

where the brackets $\langle . \rangle$ denote the conditional mathematical expectation of a quantity for fixed arguments $A_n$ and fixed coordinates for the points and time.

Though (1.26) is an exact equation for $f_n$, it is, however, *unclosed*, because the conditional mathematical expectations on the right-hand side cannot be calculated by using the values of $f_n$ only (to calculate the expectations, we also have to know $f_{n+1}$, and, sometimes, higher-order probability distributions).

Alongside the true values of $v_i, \rho$, and $\xi(m)$ we may consider their **approximate values** $\tilde{v}_i, \tilde{\rho}$, and $\tilde{\xi}(m)$ calculated from the equations with smoothed right-hand sides

$$\frac{\partial \tilde{v}_i^{(\gamma)}}{\partial t} + \tilde{v}_k^{(\gamma)} \frac{\partial \tilde{v}_i^{(\gamma)}}{\partial x_k^{(\gamma)}} = \langle F_i^{(\gamma)} \rangle_{\tilde{A}_n}, \quad i = 1, 2, 3,$$

$$\frac{\partial \tilde{\rho}_i^{(\gamma)}}{\partial t} + \tilde{v}_k^{(\gamma)} \frac{\partial \tilde{\rho}_i^{(\gamma)}}{\partial x_k^{(\gamma)}} = -\tilde{\rho}^{(\gamma)} \left\langle \left( \frac{\partial \tilde{v}_l^{(\gamma)}}{\partial x_l^{(\gamma)}} \right)^{(\gamma)} \right\rangle_{\tilde{A}_n}, \tag{1.27}$$

$$\frac{\partial \tilde{\xi}_i^{(\gamma)}(m)}{\partial t} + \tilde{v}_k^{(\gamma)} \frac{\partial \tilde{\xi}_i^{(\gamma)}(m)}{\partial x_k^{(\gamma)}} = \langle \Pi^{(\gamma)}(m) \rangle_{\tilde{A}_n}, \quad m = 1, 2, \ldots, N$$

where $\tilde{A}_n$ is the multitude of the values of $\tilde{v}_i, \tilde{\rho}$ and $\tilde{\xi}(m)$ at all the points $n$.

Let $\tilde{v}_i, \tilde{\rho}$, and $\tilde{\xi}(m)$ in (1.27) acquire random values either due to the instability of the solutions of the smoothed equations or due to randomness in the initial and boundary conditions. Then, according to Ievlev, the equations for the probability density $\tilde{f}_n$ of the set $\tilde{A}_n(\tilde{v}_i,\ \tilde{\rho},\ \tilde{\xi}(m))$ have the form *coinciding* with that of (1.26). If identical initial and boundary conditions are specified for $f_n$ and $\tilde{f}_n$, then the functions $f_n$ and $\tilde{f}_n$ must also be identical, i.e. all the statistical flow characteristics determined using $A_n$

and $\tilde{A}_n$, respectively, are identical (though $\tilde{v}_i^{(\gamma)}$, $\tilde{\rho}^{(\gamma)}$, and $\tilde{\xi}^{(\gamma)}(m)$) as functions of time are quite different for the same initial conditions for the true distributions $v_i^{(\gamma)}$, $\rho^{(\gamma)}$, and $\xi^{(\lambda)}(m)$.

Thus, *the smoothed equations may, generally, yield correct statistical characteristics of a flow determined by the large-scale turbulence, although the detailed space-time pattern of the oscillatory motion does not correspond to any real process.*

Let us call this fundamental result *the Ievlev principle.* In essence, it shows that the smoothed equations of motion can be used for the direct numerical simulation of unsteady (oscillatory) large-scale turbulent motion. Naturally, we must correctly specify the conditional mathematical expectations on the right-hand sides of the smoothed equations of motion (1.27). Note also that the right-hand sides of the latter equations are *uniquely* determined by the values of $\tilde{v}_i^{(\gamma)}$, $\tilde{\rho}^{(\gamma)}$, and $\tilde{\xi}^{(\gamma)}(m)$ at the nodes of the numerical grid (in contrast to the exact equations (1.25) whose right-hand sides also incorporate random oscillations). Thus, in the case under consideration *the closure is ensured by appropriate averaging (smoothing) of the effects of the subgrid small-scale turbulence.* Besides, the Ievlev principle confirms the validity of the hypothesis that large- and small-scale turbulent motions with large Reynolds numbers are statistically independent.

In putting the approximate values for $\langle F_i \rangle_{A_n}$ meeting some conditions in Eq. (1.26) for the probability density $f_n$, we obtain the exact Friedman–Keller equations for all $m$-point moments, where $m \leq n$. It may be shown that in varying (within certain limits) the parameters of the models used for description of the subgrid turbulence, the large-scale motion properties turn out to be practically the same in all the cases.[55,56]

## 1.7. Calculated results for coherent structures in the wake behind a body

Let us consider some results for the motion of ordered and large-scale turbulent structures.

We shall start by considering supercritical flow past a circular cylinder (see Fig. 1.8(d)). Experiments show that at least two flow modes can exist, viz., *steady (metastable) flow* with symmetric separation zones and *unsteady (absolutely stable) flow* with a developed vortex street.

The numerical simulation of this situation has the special feature that the symmetry of the body, the initial and boundary conditions and the approximating equations make the solutions both symmetric and stable.[x] Evidently, the perturbations due to a numerical grid (such as round-off errors) cannot distort the symmetry which under real conditions is easily distorted by natural influences resulting in the emergence of an asymmetric (auto-oscillatory) flow mode with a developed vortex street (see Fig. 1.8(d) and Refs. 1, 57 and 58). In order to obtain the latter flow mode by calculation, *instantaneous perturbations* (such as a jet in the upper portion of the cylinder or vortices ahead of the body) were introduced at the initial instant, and the subsequent evolution of the flow pattern was analyzed. *In all the cases the same stable "periodic" ordered flow pattern* (the *auto-oscillatory mode*) *was observed*, which agreed well with the one obtained experimentally. The calculations were carried out by Babakov,[50–52] who used the conservative scheme of the FLUX method based on the full equations of unsteady compressible inviscid flow.

Let us discuss some results for the separated-flow mode. In the case of a *circular cylinder*, it is characterized by the presence of a vortex sheet shedding from the body surface at a point oscillating between 120° and 240°. Figure 1.10 shows the flow pattern (the constant-vorticity lines for $M_\infty = 0.7$) for successive time intervals $\Delta t = 2.5$. This type of flow may be described as the motion of equilibrium ordered macrostructures developing in the turbulent wake behind a body for large Reynolds numbers. In particular, this conclusion is supported by visual and quantitative comparisons of calculated flow fields[13,14,50–52] (see Fig. 1.11(b); $M_\infty = 0.64$) and experimental data[1] (see Fig. 1.11(a); $M_\infty = 0.64$, $Re = 1.35 \times 10^6$). Thus, for $M_\infty = 0.54$, the experiment[57] and the calculations[50–52] give, respectively, $St = 0.18$, $C_x^u = 0.9$ ($C_x^S = 0.34$) and $St = 0.178, C_x^u = 0.9$ ($C_x^S = 0.45$), where $St = D/Tv_\infty$ is the Strouhal number, and $C_x^u, C_x^S$, are the drag coefficients for the unsteady and steady cases, respectively.

In order to estimate the reliability of the results, the above problems were solved numerically using the semiempirical $K - E$ model of turbulence, where $K$ is the turbulent energy and E the dissipation rate.[59] The latter model makes the integral characteristics more precise, and the effect of small-scale subgrid structures (which cannot be resolved directly) on the large-scale phenomena can be estimated. The model also yields some energy-related characteristics of small-scale turbulence.

[x]In agreement with the experiment[1] the flow separates behind the shock which closes the local supersonic-flow zone. The existence of the steady flow mode is confirmed by experiments conducted in a low-noise wind tunnel.

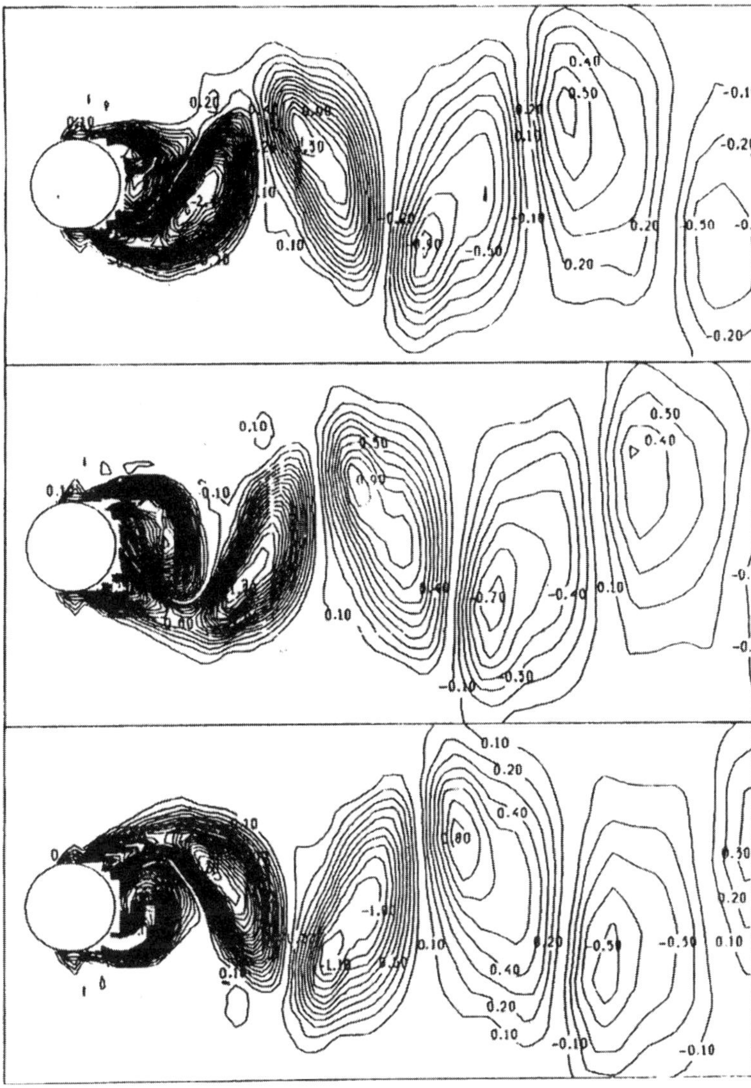

**Fig. 1.10.** Constant-vorticity lines in compressible gas flow past a circular cylinder, $M_\infty = 0.7, \Delta t = 2.5$.

The structure and characteristics of the wake and the visual flow pattern changed *insignificantly*. The same is true for the body drag coefficient $C_x$, which is largely determined by the wave properties of a phenomenon. However, the inclusion of the $K-\varepsilon$ model (using the turbulent viscosity)

(a)

(b)

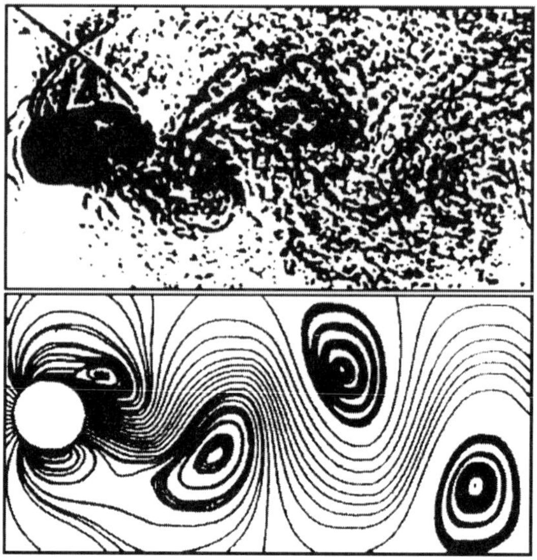

**Fig. 1.11.** Comparison of calculated and experimental results for the turbulent wake downstream of a cylinder: (a) experiment; (b) theory.

results in the Strouhal number varying by 10% as well as in a decrease in the amplitude of the flow-parameter oscillations in the neighborhood of a body due to the formation of a turbulent boundary layer. The flow near the body approaches the steady state and this leads to a certain decrease in the lift coefficient $C_y$.

Insignificant quantitative variations in the flow parameters in the near wake (with the exception of the immediate vicinity of the body) mean that large-scale structures are almost independent of the subgrid approximations (the character of dissipation, the presence of a model of turbulence). By using direct resolution, it is possible to take most of the energy spectrum corresponding to the low-frequency and inertial ranges of the wave number scale. This can be done on realistic grids, thus confirming the hypothesis that large- and small-scale phenomena can be split in fully developed turbulence.

It should be noted that numerical simulation may also be applied to subsonic flows, for which both experimental studies and theoretical analysis are especially difficult. Figure 1.12 shows the calculated flow pattern (the constant-vorticity lines) for $M_\infty = 0.9$ (see Fig. 1.12(b)); for comparison, Fig. 1.12(a) presents experimental data borrowed from Ref. 1 In this case,

(a)

(b)

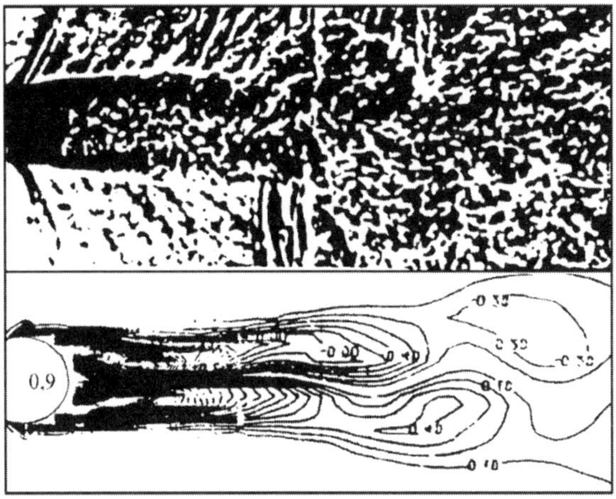


**Fig. 1.12.** Constant-vorticity lines in transonic flow past a circular cylinder, $M_\infty = 0.9$: (a) experimental data[36]; (b) calculated results.

the flow near the body is practically steady, and the vortex structures appear approximately six radii downstream of the cylinder.

Another interesting numerical experiment dealt with a body (an aircraft) moving with a speed corresponding to $M_\infty = 0.5$, into whose fully developed vortex street a screen (modeling a parachute) in the form of an arc $150° < \varphi < 210°$ was inserted seven radii from the cylinder. The restructuring of the flow took approximately 100 time steps. Near the body the flow oscillations died out and the flow became practically steady. However, behind the screen large-scale vortex structures, more intense than before the insertion of the screen (the deployment of a parachute), developed (Fig. 1.13). As a result, the drag coefficient increased by an order of magnitude as compared with the isolated cylinder.

It should be stressed that an oscillatory (random) process is observed in the vicinity of the above ordered coherent structures, which represents the *dynamics of the perturbation background* (related to the averaging of the right-hand sides of (1.20)) and may be isolated as usual. Figure 1.14 shows the averaged characteristics of oscillatory effects — the Reynolds stresses $\langle u'v' \rangle$ (see Fig. 1.14(a)) and the turbulent energy $q = [\langle u'^2 \rangle + \langle v'^2 \rangle]/2$ (see Fig. 1.14(b)) — obtained by Babakov for different Mach numbers $M_\infty$.

Of special interest is numerical solution of three-dimensional unstable problems, which form the basis for *computer-aided-design* systems for complex technological structures. Many CAD-based technological projects

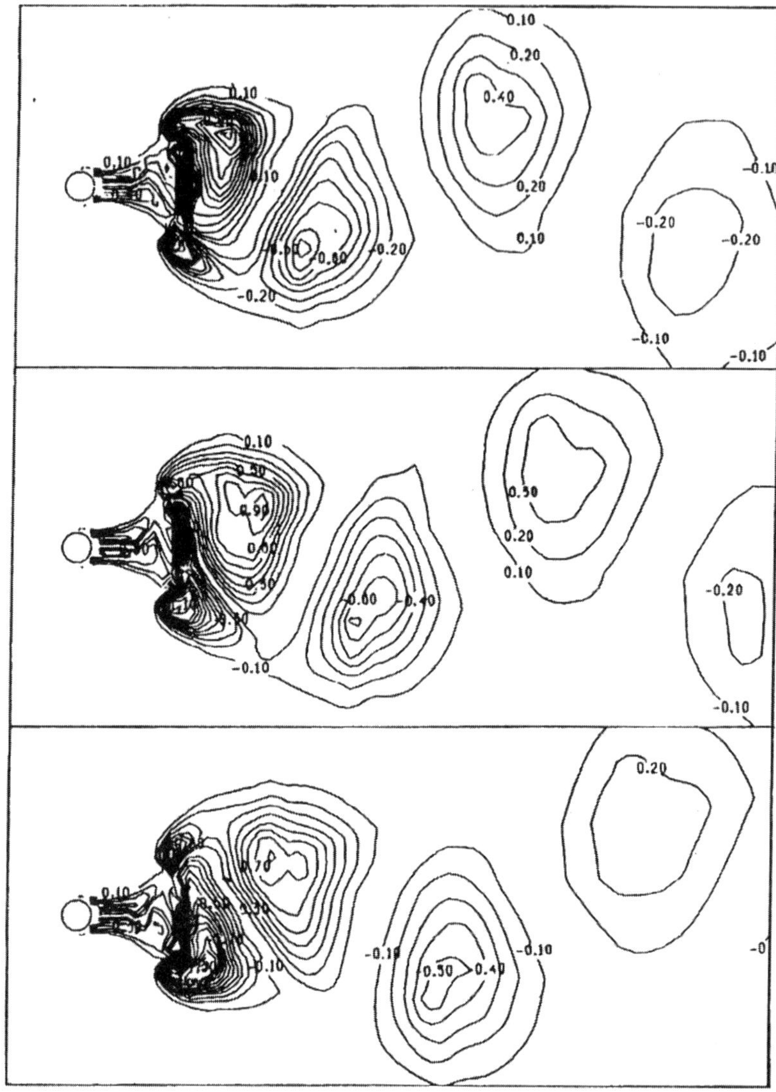


**Fig. 1.13.** Constant-vorticity lines in flow past a three-dimensional body with "parachute", $M_\infty = 0.9$.

have been unrealized due to "inadequate mathematics". These include such practically important problems as the prediction of the velocity field in the wake of a ship or the calculation of the characteristics of an oscillating finite-span wing. Figure 1.15 presents the numerical solution (obtained by Rykov

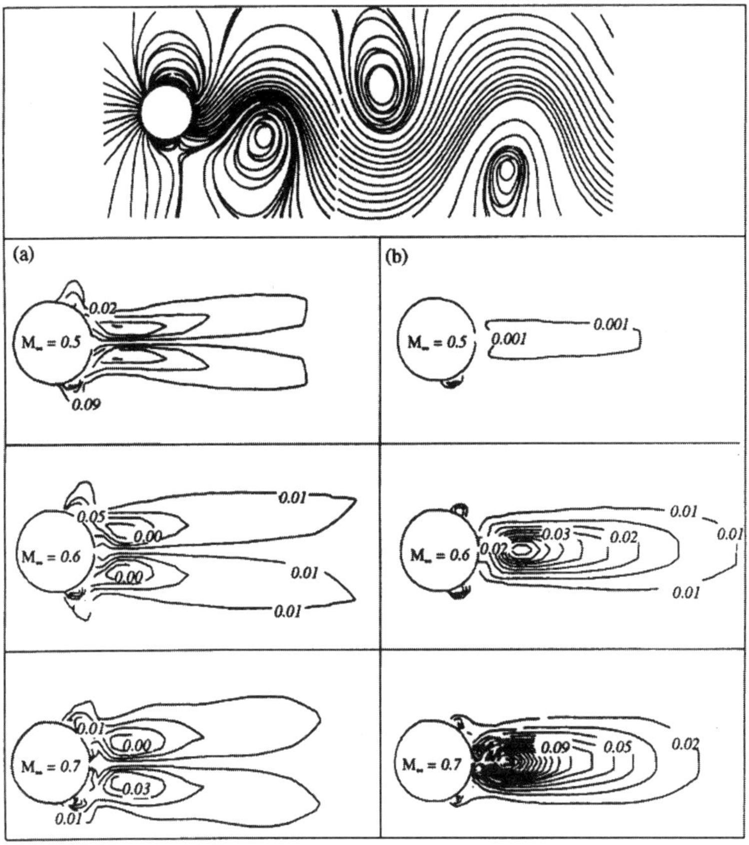


**Fig. 1.14.** Statistical reduction of calculated results on unsteady flow past a circular cylinder: (a) Reynolds stresses $\langle u'v' \rangle$; (b) turbulent energy $q = [\langle u'^2 \rangle + \langle v'^2 \rangle]/2$.

and described in Refs. 13 and 14) for three-dimensional incompressible flow past a rotating plate ($\alpha(t) \gg 0$): Fig. 1.15(a) illustrates the problem formulation and Fig. 1.15(b) shows the velocity fields for a number of cross-sections $y$ = const. The vortices due to rotation of the plate are seen to interact with the vortex braids shedding from the side edges and forming a complicated flow pattern in the turbulent wake downstream of the plate.

Finally, Figs. 1.16–1.18 present some results obtained by Babakov on multiprocessor computer using detailed grids. Figure. 1.16 presents detailed recalculation of large-order structures behind cylinder for different Mach numbers in range 0.5–1. Temperature fields are visualized.

(a)
(b)
(c)

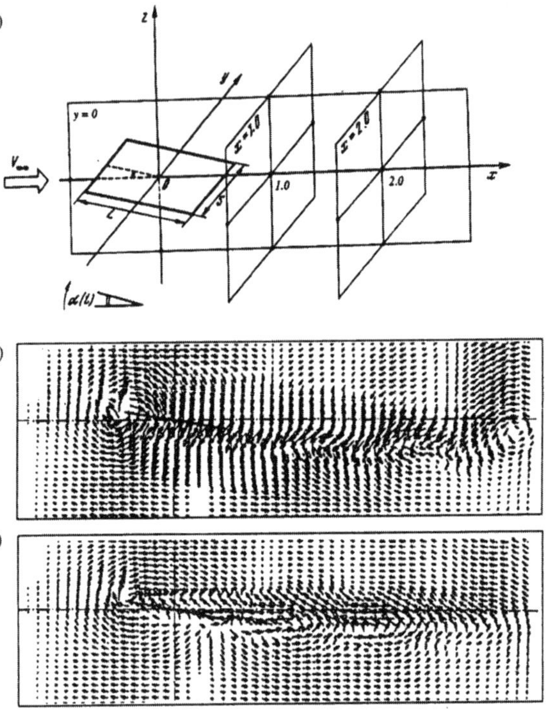

**Fig. 1.15.** Rotation of a flat plate in a three-dimensional flow; velocity vectors in the logarithmic scale in the cross-section $y = \text{const}$: (a) the problem formulation; (b) $y = 0$; (c) $y = 0.25$.

The power of modern parallel supercomputer system allows currently to perform regular three-dimensional calculation of flows past bodies with complex shape in compressible gas. The Euler model is well-proved (and very economic) for such problems at large $Re$ numbers. Figures 1.17 and 1.18 demonstrate numerical simulation of flow around landing modules of different forms in Mars's atmosphere. Density fields are shown. There is a good agreement between numerical and experimental data ($C_x$ =1.35 in experiment and $C_x$ =1.35 in simulation).

## 1.8. On the analysis of spectral characteristics

Considered below are results of simulation of large-scale coherent vortical structures within a near wake of circular cylinder, which are experimentally

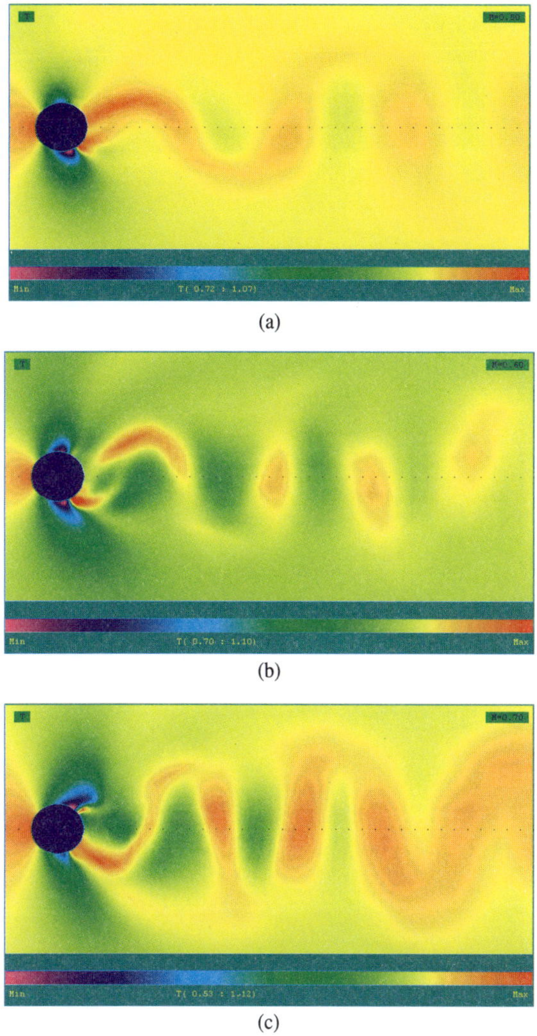


**Fig. 1.16.** Temperature fields in flow past cylinder for different Mach numbers: (a) $M = 0.5$, (b) $M = 0.6$, (c) $M = 0.7$, (d) $M = 0.8$, (e) $M = 0.9$, (f) $M = 1$.

observed in real flows. The numerical investigations are carried out both with a personal computer permitting to use grids having up to 10,000 cells, and with supercomputer, which for two-dimensional problems considered helps to increase number of points within flow-field more than by two orders. For these investigations, specially developed parallel algorithms are used,

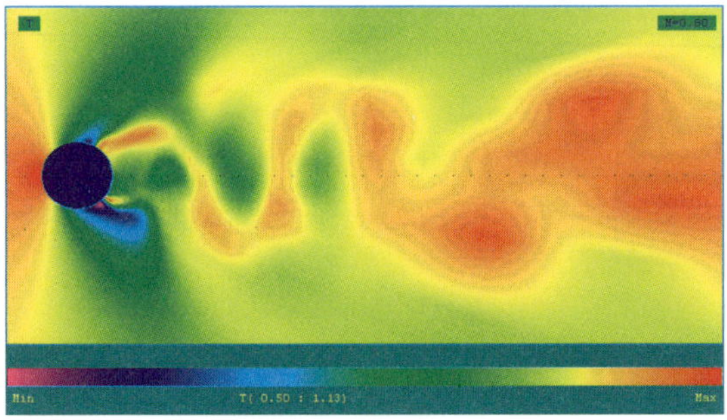


(d)

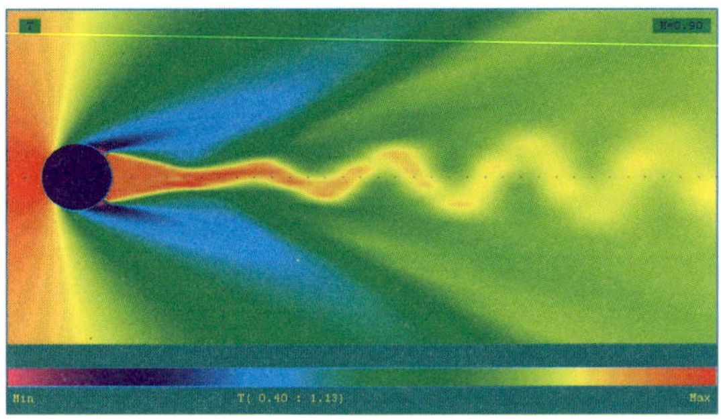


(e)

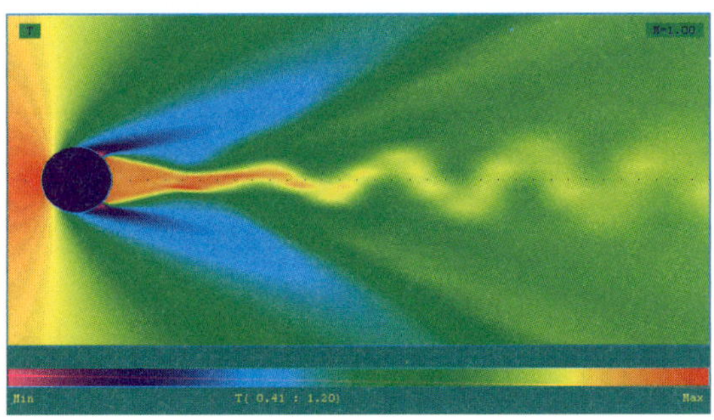


(f)

**Fig. 1.16.** *(Continued)*

which permit to make use of all the resources of supercomputer of a parallel architecture with maximal efficiency.

For a numerical simulation, the nonstationary conservative method of fluxes is used. The complete nonstationary model of inviscid gas is taken as a basis for numerical investigation. The opportunity to obtain, on the basis of this model, the numerical solutions of the type of vortex street,

**Fig. 1.17.** Density field in flow around landing module in Mars's atmosphere ($M = 3$).

was provided earlier[13,51,60] and is associated with the fact that for the case considered, when the oncoming flow is subsonic one, the flow's breakdown is determined not so much by a viscosity, but by a position of the shock closing a local supersonic zone. The investigations, carried out earlier on the basis of complete Navier–Stokes model, have shown that by the increase of Reynolds number (at the large values of it) the position of both the shock wave formed and of the separation point do not, practically, depend on the Reynolds number.

Presented below is analysis of the results of comparison of numerical solutions obtained with grids having the number of cells of the order of 10,000 (personal computer) and of 1,000,000 (multitransputer complex "PARAM" with a parallel architecture).

Shown in Fig. 1.19 are instantaneous streamlines within a near wake of the circular cylinder, obtained at one and the same phase of flow, but with different grids. As it is proved by the comparison, the general pattern of a nonstationary flow and the structure of a flow of that type might be revealed by means of personal computer. As it was found, the main geometrical

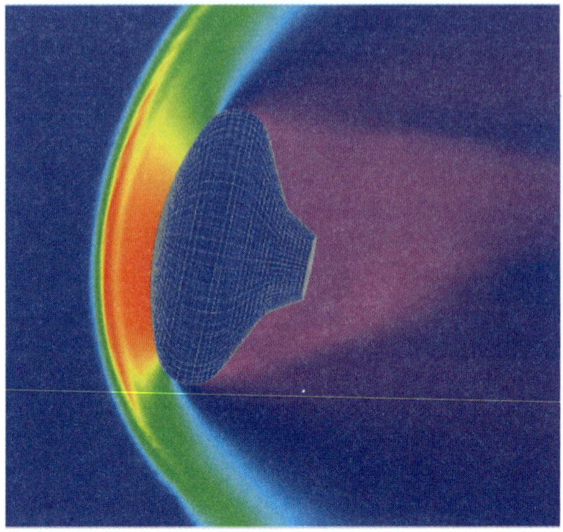


**Fig. 1.18.** Density field in flow around landing module in Mars's atmosphere ($M = 2.5$).

characteristics of the numerically reproduced coherent vortical structures are sufficiently conservative in respect to a grid resolution. The effect of grid's sizes on such integral characteristics as the time-average value of the drag coefficient and on the Strouhal number, is found to be within 5–10% (and this takes place when the number of computational cells is increased by two orders).

However, essentially fine grids permit to obtain more precise local qualitative characteristics of a flow, as well as the corresponding integral characteristics (for example, the lift coefficient). Thus, in spite of the fact that with the rough grids there was not recorded any manifestation of the non-stationarity's effect in the base area, with the fine grids were revealed pulsations of the order of 4%. Presented in Fig. 1.20 is the temporal behavior of the pressure coefficient at the forward critical point, as obtained with a fine grid. Furthermore, presented in Fig. 1.21 is the temporal behavior of the pressure coefficient at the cylinder's surface in the base area, near the non stationary shock closing the supersonic zone. It is evident that the grids, rendered in our disposal by supercomputer, produce a clearly defined

**Fig. 1.19.** Instantaneous streamlines within a near wake of circular cylinder, obtained with different computational grids: $60 \times 90$, $240 \times 480$, $960 \times 1000$.

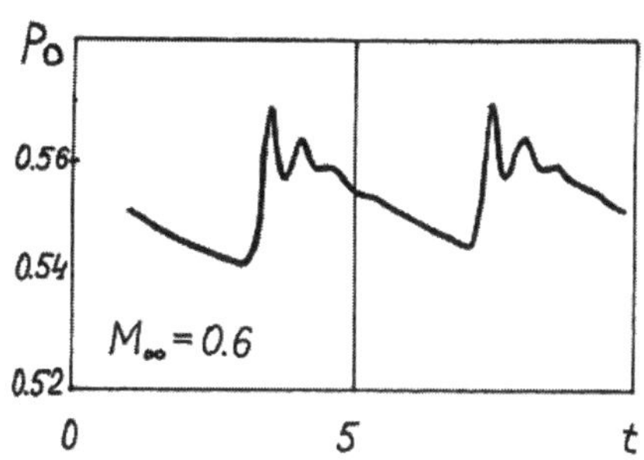


**Fig. 1.20.** Pressure coefficient at the forward critical point (fine grid).

shock front and more correctly reflect its quantitative features, and this exerts some influence on the lift coefficient, which varied up to 25%.

An analysis of the results obtained permits to conclude that the widest picture, which might be seen with rough grids, consists of one or two first harmonics. Moreover, the processing of results obtained with rough grids

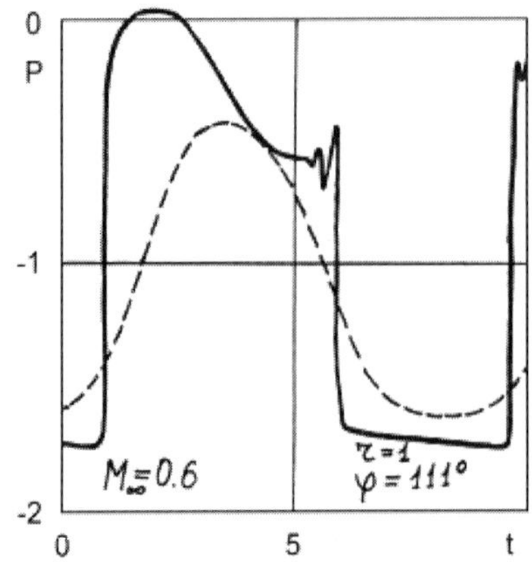


**Fig. 1.21.** Pressure coefficient at the cylinder's surface, within the zone of grid (dotted line — rough grid).

does not permit to carry out the harmonic analysis with sufficient assurance. The use of fine grids permits to reveal a large number of harmonics. Thus, shown in Fig. 1.22 is the result of a harmonic analysis of the temporal behavior of pressure at the cylinder's surface, within its base area, near the shock ($r = 1, \phi = 111°$), near the rear critical point ($r = 1, \phi = 180°$), and beyond the cylinder, at the distance of one radius ($r = 2, \phi = 180°$). Presented in the figure are mean square values of the coefficients of corresponding harmonics.

To make the engineering estimates of the parameters of turbulence by the calculation of large-scale structures for a developed turbulent flow, one might use some semiempirical models. Thus, for the problem under consideration on the nonstationary flow about circular cylinder, but based presently not on the Euler's model but on the $K{-}\varepsilon$ model,[59] the turbulent energy $K$ was calculated and the rate of its dissipation $\varepsilon$. On the basis of these characteristics some other parameters were estimated, such as the integral scale of turbulence $\Lambda$, microscales by Kolmogoroff $\eta$ and by Taylor $\lambda$, and the calculated by Taylor microscale Reynolds number $Re_\lambda$ (further on, all the parameters are dimensionless and referred to the parameters of unperturbed flow):

$$\eta = (\nu^3/\varepsilon)^{1/4}, \quad \lambda^2 = c \cdot k/\varepsilon, \quad c \sim 10\text{–}15,$$
$$Re_\lambda = k^{1/2}\lambda/\nu, \quad \Lambda = \eta \cdot Re_\lambda^{1/2}[2.47 + 0.081(Re_\lambda - Re_\lambda^{1/2})],$$

where $\nu$ is the coefficient of kinematic viscosity.

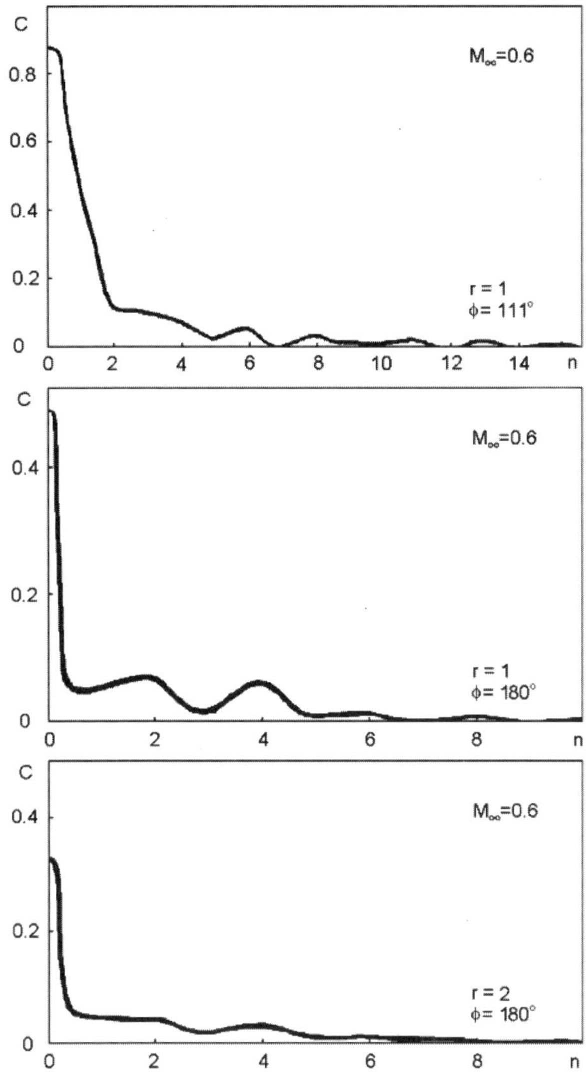


**Fig. 1.22.** Mean square values of the coefficients of corresponding harmonics for the pressure within base area (fine grid).

Presented in Figs. 1.23–1.27 is the variation along the rear cylinder's axis of the corresponding parameters, with the different values of Reynolds number of the oncoming stream. The registered oscillations of functions are associated with a nonstationary vortical structure of the wake behind the cylinder. It might be interesting to note that such a characteristic as

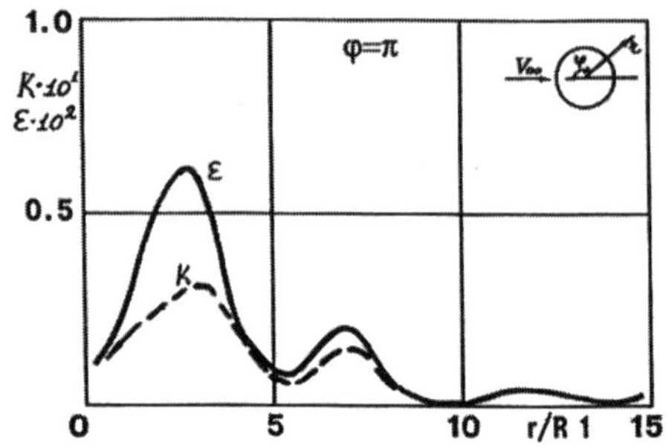


**Fig. 1.23.** Distributions of the turbulent energy ($k$) and of the speed of dissipation ($\varepsilon$) within a nonstationary wake, along the rear cylinder's axis.

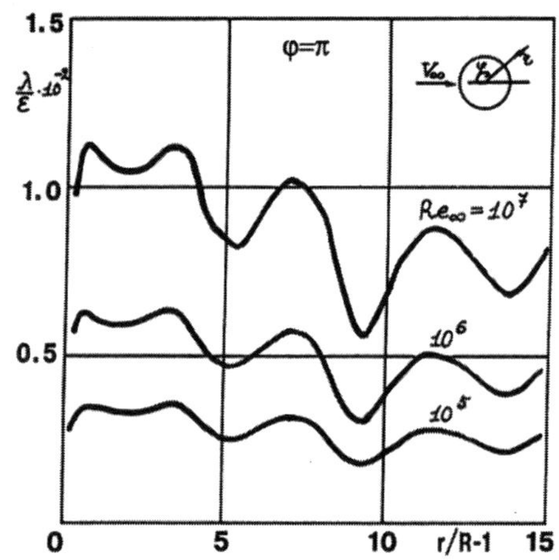


**Fig. 1.24.** Distribution of the Taylorian microscale ($\lambda/\varepsilon$) within a nonstationary wake, along the rear axis of symmetry, for various values of $Re_\infty$.

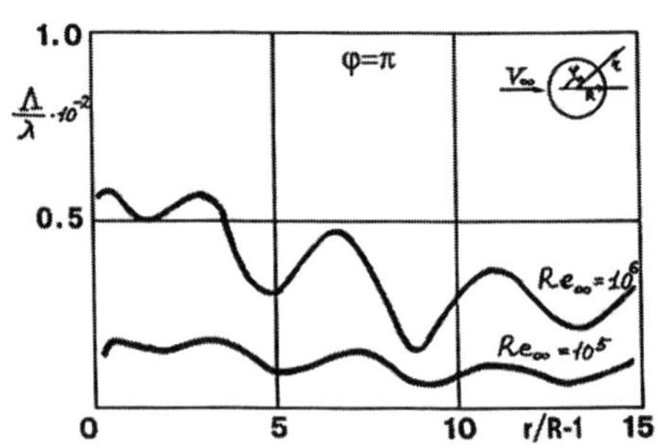


**Fig. 1.25.** Distribution of the integral microscale ($\Lambda/\lambda$) within a nonstationary wake, along the rear axis of symmetry.

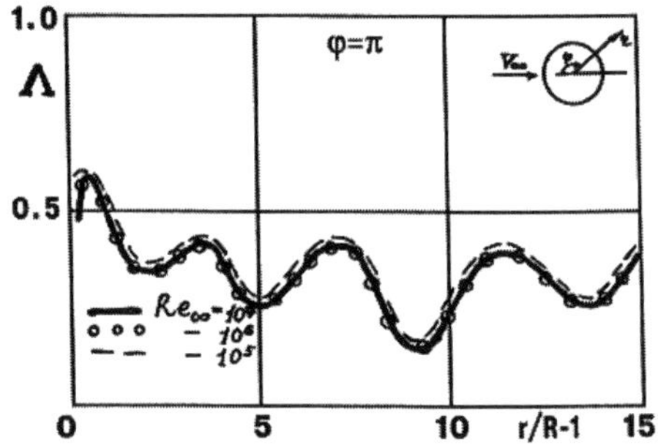


**Fig. 1.26.** Variation of the microscale $\Lambda$ within a nonstationary wake, along the rear axis of symmetry, for different values of $Re_\infty$.

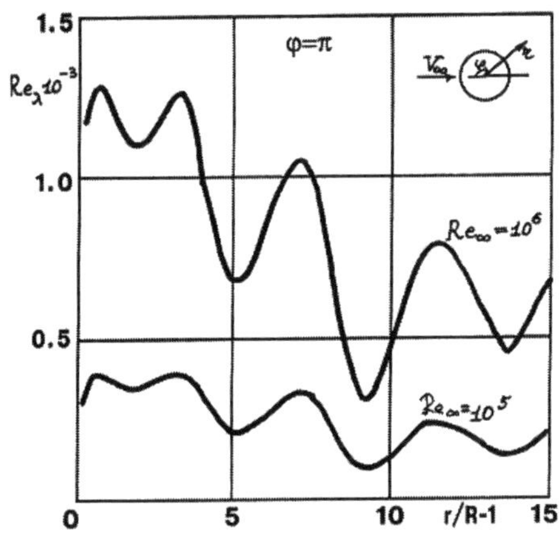


**Fig. 1.27.** Distribution of $Re_\lambda$ within nonstationary wake, along the rear axis of symmetry.

the longitudinal integral scale does not, practically, depend on Reynolds number of the oncoming stream, when this number varies within the range $10^5$–$10^7$, which seems to indicate that the flow within the wake acquired *the developed turbulent regime.*

Based on the estimates indicated above, the use of semiempirical models of spectrum permits to construct the spectral distribution. Thus, for the nonstationary flow behind the cylinder, using the model by Driscall and Kennedy of a one-dimensional spectrum of isotropic flow[29] we have:

$$E_x^0(y) = \int_y^\infty \frac{E^0(y_1)}{y_1}\left(1 - \frac{y^2}{y_1^2}\right) dy_1,$$

$$E^0(y) = \frac{E(k\eta)}{\gamma^2}\eta = A_1\beta\left(\frac{y^2}{z^2+y^2}\right)^2 [(z^2+y^2)^{-5/6} + (z^2+y^2)^{-1/2}]$$
$$\times \exp\left\{-A_3\left[\frac{3}{2}(z^2+y^2)^{2/3} + (z^2+y^2)\right]\right\},$$
$$A_1 = A_3 = 1.65 \text{ (const.)}, \quad \frac{1}{\beta} = 2\int_0^\infty y^2\frac{E^0(y)}{\beta}dy, \quad z \text{ — parameter.} \tag{1.28}$$

Shown in Fig. 1.28 are spectral distributions at the points located at the distances of 0.5 and 10 radii along the rear axis. The distributions are given in respect to coordinate $k\Lambda$ ($k$ is wave number, $\Lambda$ — longitudinal integral scale), while shown in Fig. 1.29 are spectral distributions with respect to

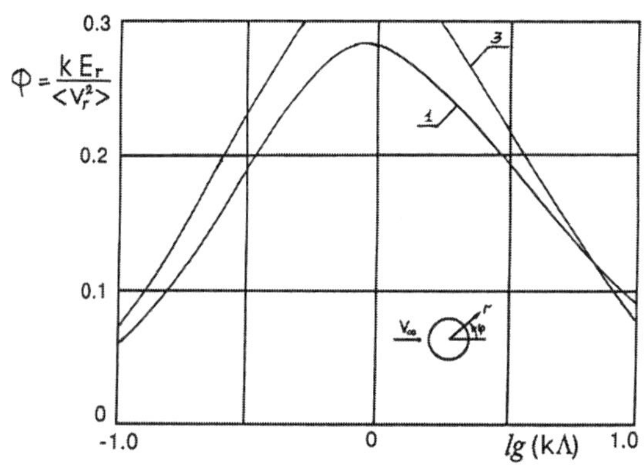


**Fig. 1.28.** Spectral distribution of the function $\Phi$ along the rear axis of symmetry ($M_\infty = 0.6, Re_\infty = 10{,}000$).

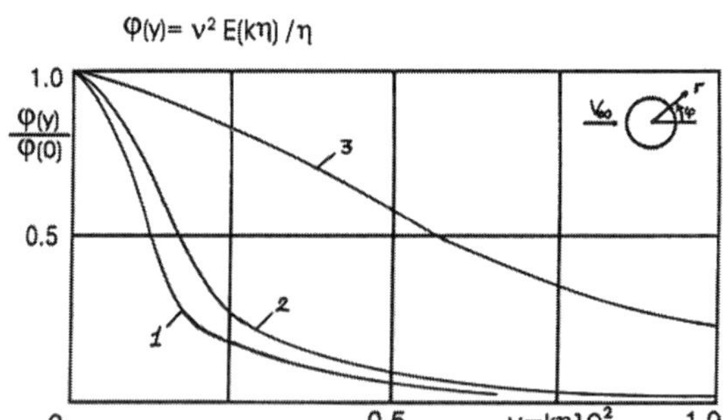


**Fig. 1.29.** Spectral distribution of the function $\phi(V)/\phi(0)$ along the rear axis of symmetry ($M_\infty = 0.6$, $Re_\infty = 10{,}000$).

coordinate $k\eta$, at three points behind the cylinder, located at the rear axis, at the distances of 0.5, 2, and 10 radii from cylinder.

The data of Fig. 1.28 are favorably compared with the experimental results by Onufriev[30,31] (see Sec. 1.3 of the present chapter), as well as with the results by Harlow,[61] who used the spectral method.

## 1.9. Numerical simulation of the random component of turbulence

The *random component of turbulence* in the core of a wake (small-scale turbulence) was simulated numerically with a statistical approach at the kinetic level.

Unlike the above results, which illustrated the simulation of averaged large-scale flow macrostructures, the statistical characteristics (the random component) of turbulence forming the wake core (see Fig. 1.9) may be studied with Monte Carlo methods by constructing imitation models for the corresponding kinetic equations. Here we present some results obtained by Yanitskii,[62–64] who decomposed a turbulent spot considered as a cross-section of a wake.

Similarly to rarefied gas dynamics (RGD), the problem is solved at the level of the distribution function; however, this time the distribution of the instantaneous hydrodynamic velocity $\mathbf{V}$ of a "*fluid*" particle is used. As far as turbulence is concerned, the philosophy of constructing the computational procedure by the splitting method and statistical approaches is the same as for RGD problem. As before, a particle is characterized by its position and velocity; however, in this case it is actually a model of a fluid particle. While constructing the model, the difficulties caused by both the *nonstationarity* of the phenomenon and the absence of a *universal* kinetic equation analogous to the Boltzmann equation in RGD should be overcome. In principle, the simulation may be carried out for different kinds of kinetic equations. One such attempt will be demonstrated below. The main objective is to preserve the basic principles of the method of direct statistical simulation while solving the unstable problems of turbulence[13–14,62,63]; however, they have to be formulated in different terms (see Table1.1).

At this stage of the work the imitation model was based on the relaxational kinetic equation proposed by Landgren and Onufriev[65] for the one-point distribution function $f(t, \mathbf{x}, \mathbf{V})$ of the instantaneous hydrodynamic

**Table 1.1.**

| Object | In RGD | In turbulence |
|---|---|---|
| Particle | Model of atom: Position $\mathbf{r}_i$, Velocity $\mathbf{c}_i$ | Model of a fluid volume: Position $\mathbf{x}_i$, Instantaneous velocity $\mathbf{V}_i$ |
| Distribution functions | One-particle: $f = f(t, \mathbf{r}, \mathbf{c}) \int f\, d\mathbf{c} = \mathbf{u}$ is density | One-point: $f = f(t, \mathbf{x}, \mathbf{V}) \int f\, d\mathbf{V} = 1$ is normalization |
| Moments and velocities: macroparameters | Velocities: $\rho^{-1} \int \mathbf{c} f\, d\mathbf{c} = u$ is gas velocity $(\mathbf{c} - \mathbf{u})$ is thermal rate | Velocities: $\int \mathbf{V} f\, d\mathbf{V} = \mathbf{u}$ is flow average $(\mathbf{V} - \mathbf{u}) = \mathbf{V}'$ is fluctuation |

velocity

$$\frac{\partial f}{\partial t} + \mathbf{V}\frac{\partial f}{\partial x} + \frac{\partial}{\partial \mathbf{V}}\left[\left(-\frac{\mathbf{V}'}{2\tau_1} - \frac{\partial p}{\partial \mathbf{x}}\right) f\right] = \frac{f_m - f}{\tau_2}, \qquad (1.29)$$

where the normal-distribution law has the form

$$f_m = \left(\frac{3}{4\pi E}\right)^{3/2} \exp\left(-\frac{3(v')^2}{4E}\right), \quad \mathbf{V}' = \mathbf{V} - \mathbf{u}$$

$\tau_1 = L_1 E^{-1/2}$, $\tau_2 = L_2 E^{-1/2}$ are the relaxation times, $L_1 E^{(2\gamma_1 - 1)/2} =$ const., $L_2 E^{(2\gamma_2 - 1)/2} =$ const. and $E$ is the turbulent energy density. In a certain sense, kinetic equation (1.29) is analogous to the Boltzmann and Crook equations.

The main empirical constants of the model are the exponents $\gamma_i$ and $Y_i$ in the equations for the variation integral scales depending on the turbulence strength $E$

$$L_1^{-1} \sim E^{(2\gamma_1 - 1)/2} \quad \text{and} \quad L_2^{-1} \sim E^{(2\gamma_2 - 1)/2}$$

and the initial values of the integral scales $L_1^{(0)}$ and $L_2^{(0)}$.

The *diffusion of turbulence* whose energy is initially concentrated within a characteristic radius $r_0$ (a "turbulent spot", see Fig. 1.30) was simulated numerically. Later on the spot characteristic radius $r_{1/2}(t)$ and the density of turbulent energy $E_t(t)$ at the center of the spot vary; the mean pressure $p$ within the turbulent flow zone is assumed to be constant. The initial data (corresponding to $t = 0$) are represented in the form

$$E_0(r) = E_t^{(0)} \exp(-0.69 r^2 / r_0^2), \quad E_t^{(0)} = E_t(0),$$

and distribution function is given in the form

$$f_0(r, \mathbf{V}) = \left(\frac{3}{4\pi E_0(r)}\right)^{3/2} \exp\left(-\frac{3(v')^2}{4E_0(r)}\right).$$

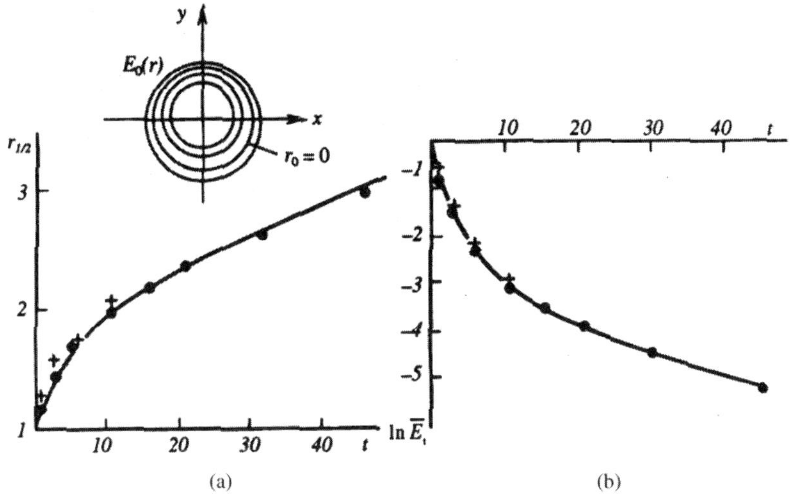


**Fig. 1.30.** Decomposition of uniform turbulent spot: (a) transformation of the characteristic radius; (b) turbulent-energy spot center, $\bar{E}_t = E_t/E_t(0)$.

Figure 1.30 compares calculated results and experimental data[66] for a uniform spot: Fig. 1.30(a) shows the dimensionless characteristic radius of the spot $r_{1/2} = r_{1/2}/r_0$ over time, where $r_{1/2}$ is the radius for which the energy density $E_t(t, r_{1/02}) = E_t(t)/2$, and $E_t(t)$ is the energy density at the center of the spot, and Fig. 1.30(b) shows the turbulent-energy density $\bar{E}_{(t)} = E_t(t)/E_t(0)$ over time at the center of the spot. By comparing the numerical and physical experiments (shown by small crosses and the solid curve, respectively), one can choose the values of empirical constants of the model. The distribution of energy along the spot radius is self-similar (Fig. 1.31), ($\bar{E} = f(\xi), \bar{E} = E(t, r)/E_t(t)$, where $\xi = r/r_{1/2}(t)$ is the dimensionless radius). Using data from a physical experiment,[66] we may represent the self-similar dependence as $f(\xi) = \exp(-0.69\xi^2)$. In numerical calculations, this relationship is usually used to specify the initial data. Also, it is approximated by the expression $f(\xi) \sim J_0(a\xi^2)$, where $a \approx 1.5$ and $J_0$ is the zero-order Bessel function (see Fig. 1.31).

The discrepancy between the numerical simulation and experimental data is typical of most numerical models and is a consequence of a poor understanding of the nature of the phenomenon. For instance, the numerical models fail to explain *why the turbulence-intermittence coefficient differs from unity*, this being the general deficiency of the models causing the discrepancy.[67]

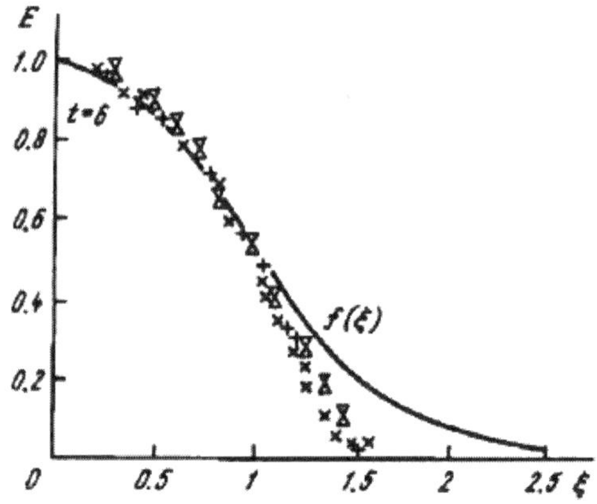


**Fig. 1.31.** Radial distribution of turbulent energy for a uniform spot: Naudasher's experiment[66]; $X, +, x$ are calculated data for $t = 3, 6$ and 11 respectively.

Statistical simulation yields a one-point function of the fluctuation distribution as well as all of its moments. Hence, kinematic models are more informative than approaches using a finite sequence of Reynolds equations. Let us compare the two approaches.

Usually in turbulence problems the mean-velocity fields $u_i$ and the second-order covariances for fluctuation velocities $\langle v_i' v_j' \rangle$ have to be compared. The complete set of third-order covariances $\langle v_i' v_j' v_k' \rangle$ is rarely used. The fourth- and higher-order moments are represented via lower-order moments by employing appropriate hypotheses. For example, the assumption that the velocity fluctuations are isotropic means that all the odd moments, $\langle v_i' v_j' v_k' \rangle$, etc., are actually zero. Another well-known hypothesis is *the Millionshchikov hypothesis, according to which the cumulants of the fourth- and higher-orders are zero.* The fourth-order cumulant for the velocity fluctuations in the radial direction is defined as the combination of moments $\mu_r^4 = \langle v'^4_r \rangle - 3\langle v'^2_r \rangle^2$. The Millionshchikov hypothesis allows us to express all the moments of the fourth- and higher-orders via the second- and third-order moments.

In the framework of the kinetic description the covariances of velocity pulsations are defined using the functions of the distribution of pulsations in the following way:

$$u_j = \langle v_i \rangle = \int v_i f \, d\mathbf{V}, \quad \langle v_i' v_j' \rangle = \int v_i' v_j' f \, d\mathbf{V}, \quad \langle v_i' v_j' v_k' \rangle = \int v_i' v_j' v_k' f \, d\mathbf{V}.$$

The general technique for deriving evolutionary equations for the moments using the kinetic equation is well known: (1.29) is multiplied by the function $\psi$ ($\mathbf{V}$) and integrated over the entire space $\{\mathbf{V}\}$. Thus, (1.29) results in an infinite sequence of moment-transfer equations, the first three of which are

$$
\begin{aligned}
&\frac{\partial}{\partial x_i}u_i = 0,\\
&\frac{\partial u_i}{\partial t} + u_j\frac{\partial u_j}{\partial x_j} + \frac{\partial}{\partial x_j}(\langle v_i' v_j'\rangle + \rho\delta_{ij}) = 0,\\
&\frac{\partial\langle v_i' v_j'\rangle}{\partial t} + u_k\frac{\partial}{\partial x_k}\langle v_i' v_j'\rangle + \langle v_i' v_k'\rangle\frac{\partial u_i}{\partial x_k} + \langle v_j' v_k'\rangle\frac{\partial u_j}{\partial x_k} + \frac{\partial}{\partial x_k}\langle v_i' v_j' v_k'\rangle\\
&\qquad = -\frac{1}{\tau}\langle v_i' v_j'\rangle + \frac{1}{\tau_2}\left(\frac{2E}{3} - \langle v_i' v_j'\rangle\right).
\end{aligned}
\tag{1.30}
$$

The equations for the third- and higher-order moments may be derived in a similar way. The kinetic approach (to derive (1.30)) does not employ complicated hypotheses to justify the truncation of the sequence of equations, and their validity can be verified during the numerical simulation. We believe that this is an important advantage of the approach.

Figure 1.32 shows the results for the turbulent spot problem in the form of the curves for the asymmetry $\alpha_r$ (Fig. 1.32(a)) and excess $\beta_r$ (Fig. 1.32(b)) coefficients of the probability distribution density $f(\bar{v}_2')$ of the fluctuations in the radial velocity as functions of the relative radius $\xi = r/r_{1/2}(t)$. The asymmetry coefficient $\alpha_r = \langle v_r'\rangle/\langle v_r'^2\rangle^{3/2}$ is related to the third moment and characterizes the anisotropy of $f$ within the fluctuation space $v_r'$. The excess coefficient $\beta_r = \langle v_r'^4\rangle/3\langle v_r'^2\rangle^2 - 1$ characterizes the degree of agreement with the Millionshchikov hypothesis. We see that near the boundary of the spot neither quantity is zero. A more detailed analysis[62] of the function $f(v_r')$ has shown that the Millionshchikov hypothesis is valid, to a certain extent, for the center of the spot (for $r \le r_{1/2}(t)$). However, this conclusion is based on calculations, which ignore the phenomenon of intermittency. Perhaps, by taking intermittency into account, the applicability of the above hypothesis may be extended.[67]

Thus, statistical methods yield solutions to problems of anisotropic and nonequilibrium turbulence when the distribution function differs considerably from the normal law. Wider use of these methods is mainly obstructed by the absence of universal kinetic equations of turbulence. However, the work in this direction is presently under way.

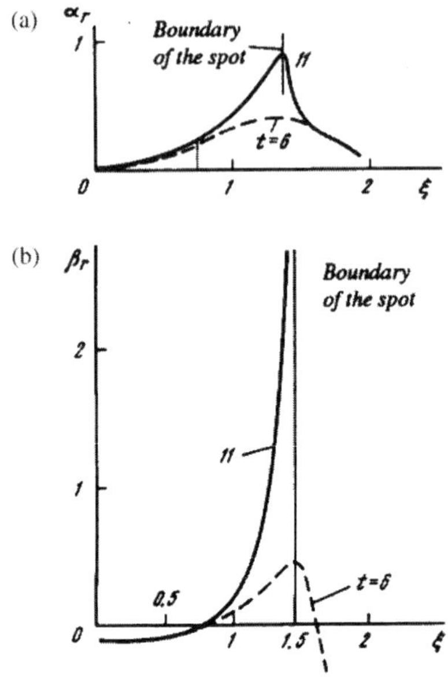


**Fig. 1.32.** Uniform turbulent spot: (a) asymmetry coefficient $\alpha_r$; (b) excess coefficient $\beta_r$.

## 1.10. Laminar–turbulent transition. Simulation of three-dimensional flows in clean rooms

We have seen that the large-scale (global) characteristics of flows with large Reynolds numbers in the wake behind a body (where dynamic processes dominate) may be studied with dissipative models based on conservation laws for inviscid media and on the methods of rational averaging described in the first part. However, in the case of laminar flows, the interaction of perturbations is governed by *molecular-viscosity effects*, and as a consequence the detailed structure of the flow and the nature of its evolution at increasing Reynolds numbers (successive bifurcations, the generation of secondary eddies, and so on) must be studied with Navier–Stokes models.

The nonlinearity of the Navier–Stokes equations and the presence of small parameters as coefficients associated with the higher-order derivatives hinder both analytical study (which can be pursued only with model equations and in the case of special problems) and the numerical integration of the equations. It should be stressed that most current techniques fail to provide reliable results for viscous flows past complicated shapes (such as modern aircraft) and for large Reynolds numbers, especially in the case of turbulent flows. Accurate quantitative results consistent with data from physical equations have been obtained only for two-dimensional laminar flows. Thus, it is still unclear how numerical algorithms should be constructed in order to obtain accurate solutions of the Navier–Stokes equations (especially for three-dimensional flows). Several approaches have been used to develop numerical algorithms for solving the Navier–Stokes equations describing incompressible flows. Most of these studies have dealt with boundary-layer problems. Very accurate numerical schemes employing fixed grid templates (in particular, three-point schemes with fourth-order accuracy in the transverse coordinate) and iterative methods for solving difference equations have attracted most interest.

The classical problems of a viscous incompressible flow past a body of finite dimensions were analyzed using both explicit and implicit schemes with different orders of accuracy. To obtain correct solutions for large-gradient flow regions, it is of great practical importance that the difference scheme be monotonous.[13,68] However, it should be noted that it is difficult to obtain a monotonous difference scheme with a high order of accuracy. It must be borne in mind that explicit schemes can accept only a limited range of Reynolds numbers. On the other hand, implicit schemes with symmetric approximation of convective terms are subject to the stringent requirement that they must be monotonous in each grid step, while schemes with asymmetric approximations of the terms are subject to considerable approximation viscosity, of a magnitude comparable with molecular viscosity.

Most numerical methods have been developed for systems of equations in stream functions $\psi$ and vorticities $\omega$ and have the common deficiency that a boundary condition for vorticity is required on the solid surface — a condition that is unrelated to the physical problem being considered. The additional iterative process associated with this condition limits the rate of convergence of the numerical solution. In addition, the methods for solving the $\omega, \psi$-system cannot be easily extended to solving problems of three-dimensional viscous and compressible flows. As a consequence, there has been much recent attention to the numerical solution of Navier–Stokes equations written in terms of natural variables such as velocity and pressure, e.g.

$$\frac{\partial \mathbf{V}}{\partial t} + (\mathbf{V}\nabla)\mathbf{V} = -\nabla p + v\nabla\mathbf{V}, \quad \nabla\mathbf{V} = 0, \tag{1.31}$$

where $p$ is the pressure, $\mathbf{V}$ is the velocity vector, and $v$ is the kinematic viscosity. Here and below, the density of an incompressible uniform fluid is assumed to be equal to unity, $p = 0$.

Gushchin and Shchennikov have developed numerical algorithms for an extensive class of viscous incompressible flows. These may be used in various problems that are solved by the methods of relaxation and splitting. This work was aimed at creating general and rational techniques for solving multi-dimensional Navier–Stokes equations (1.31) for moderate Reynolds numbers, ensuring an accurate description of the local flow properties. The unified principle for constructing numerical algorithms allows us to analyze a wide range of planar, axisymmetric, and three-dimensional problems of the dynamics of viscous fluids (see Refs. 13, 68–70).

Unsteady transition flows are characterized by nonlinear processes and involve a complicated interaction mechanism. A detailed analysis of a viscous incompressible flow past a finite body was made in the works cited above.

The above approaches also allow us to study a broad range of viscous flows, ranging from fully laminar flows to the laminar–turbulent transition. Just as in the analysis of coherent structures in fully developed turbulent flows based on the Euler equations, the objective in this paper is to solve the full Navier–Stokes equations without having to resort to semiempirical models of turbulence. The generation of a variable numerical grid ensures the necessary balance between the approximations of the convective and viscous terms.

In essence, the technique singles out a laminar boundary layer near the body surface, outside which molecular-viscosity effects may be ignored. For small and moderate Reynolds numbers the energy and dissipation ranges overlap (or come very close together), and we must account for the viscous terms even when considering large-scale structures only.

We shall consider the numerical simulation of generation of the secondary bifurcations in the wake behind a body, flows in laminar–turbulent transition zones, and three-dimensional flows.

Figure 1.33 illustrates the appearance of secondary effects in a viscous incompressible flow past a circular cylinder for different Reynolds numbers $Re$. For $Re = 100$ vortices are shed in turn from the top and the bottom of the body; however, for $Re = 200$, the boundary layer at the bottom separates, and the Kelvin–Helmholtz instability sets in. For $Re = 300$, secondary disturbances start and cause another separation when $Re$ reaches 500 (half a period later the whole sequence is repeated in the other half-plane). Only a numerical experiment can provide such detailed information!

Figure 1.34 presents the results by Gushchin obtained using the above technique. Numerical and experimental results by other authors for the

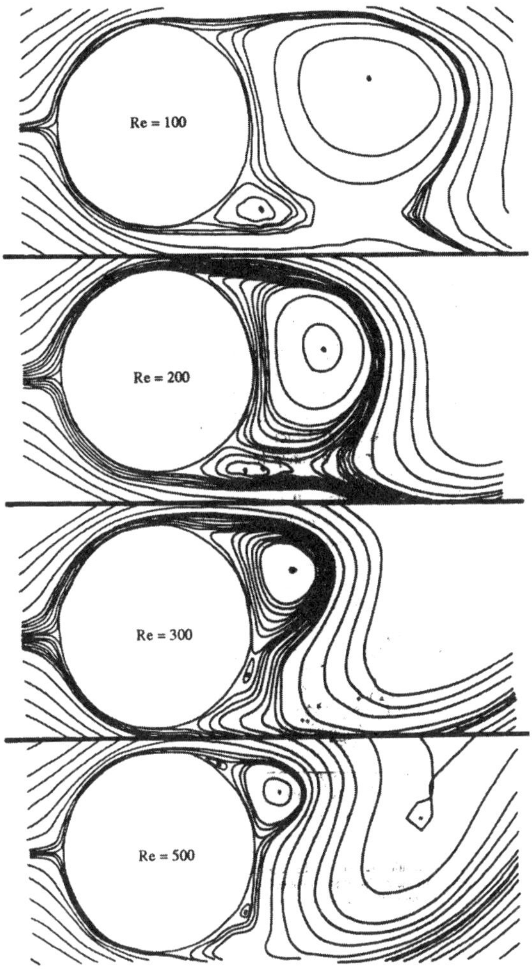


**Fig. 1.33.** Viscous incompressible flow past a circular cylinder (secondary effects in a periodic flow).

*laminar–turbulent transition* and *Re* ranging from $2 \times 10^5$ to $4 \times 10^5$ are shown (Fig. 1.34(a) contains a plot of the "drag crisis", and Fig. 1.34(b) the Strouhal number "spike"). These complicated flows were calculated using the same numerical technique for various Reynolds numbers, and the two-dimensional results agree satisfactorily with experimental data.[y] Both

[y]This result was given special attention by Prof. G. K. Batchelor during seminar at his laboratory (Cambridge, July 1990).

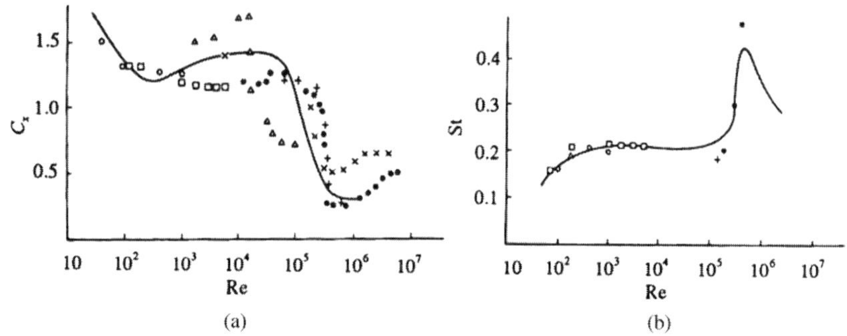


**Fig. 1.34.** Laminar–turbulent transition; (a) the total coefficient versus the Reynolds number; (b) the Strouhal number versus the Reynolds number. Experimental data: × — results: ○ — Jordan, Fromm (1972); Δ — Kanamura *et al.* (1985); □ —Braza *et al.* (1986).

calculation and experiment show there is a sharp increase in the frequency at which vortices shed when approaching the critical Reynolds number; two characteristic frequencies ($St = 0.34$ and $0.42$) were detected.

At present, the simulation of complex three-dimensional *unsteady flows* is very important for the design of "*clean rooms*" (selecting their configuration, positioning equipment, calculating the interaction of various flows, etc.).

In the mid-1970s, many industries, such as microelectronics, pharmaceutics, medicine, and microbiology, encountered a serious problem that hindered their growth, namely the necessity to decrease the concentrations (or in some cases, to eliminate entirely) of microscopic particles, including microorganisms, in gaseous media and to regulate precisely the temperature, humidity, ion content, as well as the direction and the velocity of air flows, the pressure, static electricity, and the vibrational frequency and amplitude. Neither standard production areas nor industrial ventilation techniques could meet the ever-increasing demands on the environment required by these precise technologies.

The most pressing need was in modern microelectronics, where an inadequate microclimate not only negatively affects the quality of manufactured goods, but often makes the development and manufacture of new products (such as superlarge-scale integrated circuits, disk drives, etc.) altogether impossible.

Improved air conditioning, primarily air purification in operation theatres and postoperative wards and the regulation of the content of their inner atmospheres, diminished the occurrence of complications caused by infection and shortened the duration and hence lowered the cost of post-operative treatment. In some cases, such as the treatment of extensive skin burns or premature babies, there is no alternative to clean rooms.

Laboratory work and experimental production in virology, immunology, pharmacy, and the manufacture of microbiological products are only possible in specific and precisely controlled atmospheres. In all these cases, a *clean room* not only protects the manufactured object and increases production efficiency, but also ensures the protection of both the operators and the environment from contamination by dangerous living organisms. The importance of developing and constructing clean rooms is obvious, and the simulation of air flows within them is of great importance in their computer aided design. Forced and convective air flows, heat and mass transfer in the presence of distributed sources, and the control of conditioned air and rational distribution of technological equipment must all be simulated. Alongside laboratory tests, these problems are solved by the *numerical simulation of the gas-dynamics of clean rooms.* As a rule, physical processes in clean rooms are three-dimensional. However, a three-dimensional flow requires several hours of processor time even on the fastest computers to simulate. Therefore, most work in the area involves two-dimensional approximation (Kern, 1989; Bussnaina, 1989; Yamamoto *et al.*, 1988; Deaves *et al.*, 1985, etc.). Nevertheless, some authors have studied three-dimensional flows.[71,72] In this section we present the results by Gushchin and Kon'shin, who studied air flows in a clean room containing a point source of contamination, using the Belotserkovskii splitting method with respect to the physical factors.[70]

Let $u, v$, and $w$ be the projections of the velocity vector $\mathbf{V}$ onto the Cartesian axes $x$, $y$, and $z$. The Navier–Stokes equations will have the form

$$u_{\mathrm{t}} + (u^2)_x + (uv)_y + (uw)_z = -p_x - \nu[(v_x - u_y)_y - (u_z - w_x)_z s],$$

$$v_{\mathrm{t}} + (uv)_x + (v^2)_y + (vw)_z = -p_y - \nu[(w_y - v_z)_z - (v_x - u_y)_x],$$

$$w_{\mathrm{t}} + (uw)_x + (vw)_y + (w^2)_z = -p_z - \nu[(u_z - w_x)_x - (w_y - v_z)_y],$$

$$u_x + v_y + w_z = 0.$$

Let us consider an incompressible flow in a clean room in the form of a rectangular parallelepiped with a square base, whose dimensions are $5 \times 5 \times 4.5$ (Fig. 1.35). In the center of the ceiling there is an orifice measuring $1 \times 1$, through which air enters the room with the velocity $U = 1$. At the bottoms of two opposite walls there are two pairs of exit orifices measuring

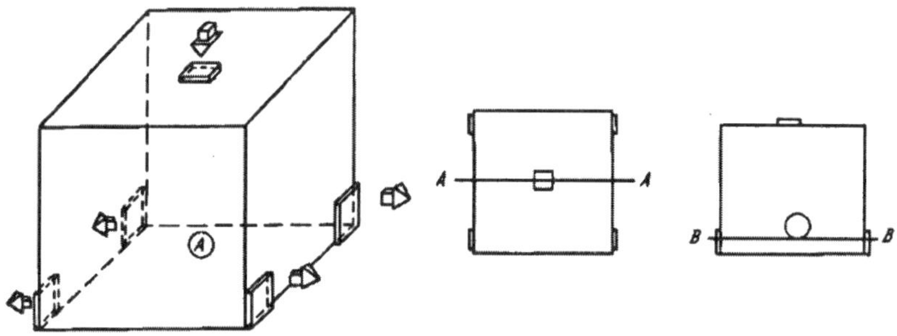


**Fig. 1.35.** Clean-room model.

$1 \times 1$, through which air is pumped out with the velocity $U = 0.25$. The flow is described by the full system of nonstationary hydrodynamic equations for $v = 0$, with the condition that the walls are impermeable. The source of a passive additive $A$ with the concentration $c = 5$ is placed at the axis of vertical symmetry of the room, at the distance $H = 1$ from the floor. The dynamics of the passive additive is described by the transport equation

$$\frac{\partial c}{\partial t} + (\mathbf{V} \cdot \nabla)c = 0.$$

The flow region under study was covered with a uniform grid consisting of $20 \times 20 \times 18$ nodes, the distances between which (the grid spacings) were $h_x = h_y = h_z = 0.25$.

The difference problem was solved by a standard variation of the splitting method with respect to the physical factors,[68] and the finite-difference scheme was based on a hybrid scheme for the convective terms. A similar scheme was used for approximating the convective terms in the equation for the passive additive. The integration time step was chosen automatically subject to the stability condition. By symmetry, the calculations needed to cover only one quarter of the region.

The results of the stable solution are presented in the form of a velocity vector field, and the constant-concentration lines for the cross-sections are marked in Fig. 1.36. The left half of Fig. 1.36 has the velocity vector fields in the vertical cross-section $AA$, and the right half has the constant-concentration lines. The two lower diagrams present the results obtained by the splitting method. The experimental data and calculations obtained on the same grid[71] are also shown. The latter work used the $K-\varepsilon$ model of turbulence. The subsequent figures compare the results of Gushchin and Kon'shin with those of Murakami's. Figure 1.37 compares the three sets of

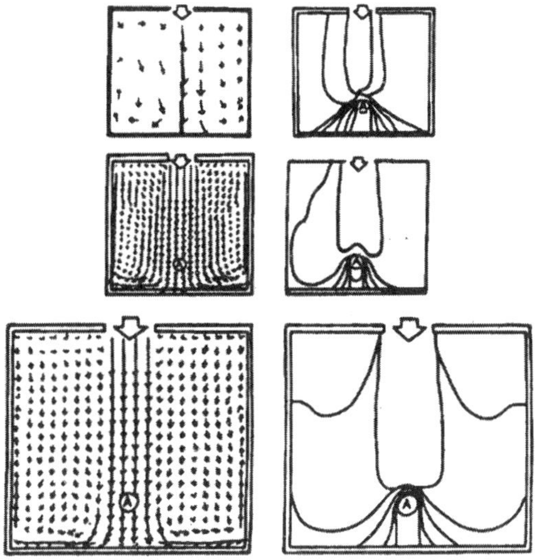

**Fig. 1.36.** Velocity vector fields and the lines of constant additive concentration in the cross-section $AA$ (see Fig. 1.35). The upper and middle portions demonstrate Murakami's[71] experimental and calculated results, while the lower portions correspond to our calculations.

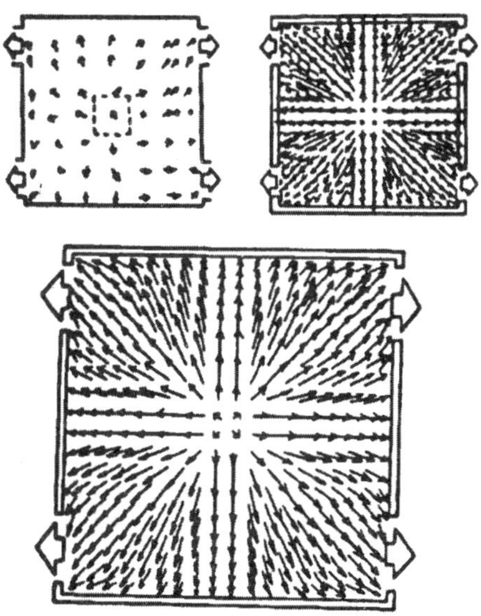

**Fig. 1.37.** Velocity vector fields in the cross-section $BB$ (see Fig. 1.35). The upper and middle portions demonstrate Murakami's[71] experimental and calculated results, while the lower portions correspond to our calculations.

results for the velocity vector field in the horizontal cross-section $BB$ near the floor of the room. The comparison confirms the qualitative agreement between the results obtained by different methods and available experimental data. However, the good agreement may have much to do with the simple geometry of the flow. Finer grids and more sophisticated mathematical models should be used for more complicated geometries (say, in the case of the presence of equipment).

The air flows and dispersion of a passive additive in a clean room were thus simulated. Since the results agree favorably with those obtained by other authors and experimental data, this approach seems to be an effective means for analyzing gas-dynamics, as well as heat and mass transfer in clean rooms.

## 1.11. Transition to chaos (numerical experiments)

### *1.11.1. General aspects*

A present trend in modern nonlinear mechanics is the study of *structural randomness* (chaos born of order), which may develop due to the randomization of deterministic structures existing in the subcritical zone. Randomness is a consequence of the complicated dynamics of a system (it is not induced by noise or fluctuations) and may be treated, in a certain sense, as the process of turbulence generation.[z] A number of scenarios for the transition to turbulence have been proposed (see Refs. 7, 8, and 73), which are usually related to dynamics of the Navier–Stokes equations. Let us describe some typical patterns of the transition, which we shall later compare with numerical studies.

*The pattern proposed by Leray (1934)* is related to the assumption that the solutions of three-dimensional Navier–Stokes equations collapse upon turbulence. The two-dimensional Euler and Navier–Stokes equations possess global solutions in $t$, and it is known that for many turbulent flows the solutions of three-dimensional Navier–Stokes equations do not collapse.[7]

*The pattern proposed by Landau and Hopf (1950)* is based on the idea of a continuous transition to turbulence via an endless cascade of bifurcations; in other words, an ever-increasing number of sequential oscillations, the ratios of whose frequencies are irrational numbers, are excited, a limiting

[z]Here hydrodynamic turbulence is treated as the "unpredictable" chaos motion of a fluid.[7,8] Turbulence should be defined in Ref. 73 as *stochastic evolution of a swirling flow of a (viscous) fluid.*

cycle with the frequency $\omega_1$ is generated for the Reynolds number $Re_1$, the mode with the frequency $\omega_2$ becomes unstable for $Re_2 > Re_1$ and double-periodic *auto-oscillations* arise, then new frequencies appear, and so on. In fact, Landau proposed that turbulence can be simulated by a set, viz., an *attractor* in the form of a torus with a quasi-periodic coil.[10]

*The Feigenbaum model (1978)* of the transition proceeds by doubling and is the most frequently used one. The original limiting cycle becomes unstable because of the interaction between the main-period oscillations and the double-period disturbances, and a new, temporally stable cycle is generated in its vicinity, etc. Recently, Feigenbaum has discovered the universality of transition to randomness in the case of an infinite sequence of bifurcations associated with the doubling of the period of original motion.[8,73]

Various scenarios for the transition to turbulence have been described in detail by Monin[73] (see also Ref. 74).

Recent progress in the qualitative theory of ordinary differential equations (*the theory of dynamic systems*) has yielded a new approach to the onset of randomness. It is based on the assumption that the transition can be qualitatively explained by studying the dynamics of models, which are much *simpler* than the full hydrodynamic equations. The most characteristic are the Lorentz equations and one-dimensional mappings demonstrating hierarchical bifurcations. At the same time, it is hypothesized that turbulence may be treated as a *strange attractor*.[7,8]

Leaving aside the issue of "reality", we shall limit ourselves to noting that "no single turbulent solution is known for the full Navier–Stokes equations; indeed, we are not sure that such solutions exist at all. The form of the stochastic attractor for a turbulent flow is also unknown. The only thing we are sure of (at least we think so) is the structure of the Lorentz attractor. Any conclusion concerning turbulent solutions of the Navier–Stokes equations made by considering this simplified model may seem to be marginal; indeed, some researchers even think that "the truncation of the original equations itself results in the elimination of turbulence" (see Ref. 20).

Thus, in accordance with Ref. 8, the hypothesis that turbulence may be fully described by the Navier–Stokes equations remains mathematically unproved since there is no theorem which would guarantee global existence of solutions to the Navier–Stokes equations in the framework of the initial-value problem. Most probably (though no such examples are presently known), these solutions may prove to be singular, so that the equations will become invalid, and to construct a comprehensive theory would require new physical *principles lying outside the scope of classical hydrodynamics.*

From the physical viewpoint, the following three questions seem to be most significant for the discussion of randomness of dynamic systems[8]:

(i) Which mechanisms are responsible for the onset of chaos in deterministic systems, in other words, what is the mathematical nature of chaos?
(ii) In what ways does chaos set in due to a variation in the parameters of a physical system?
(iii) What are the properties of random motion after it becomes stable?

Next, we shall try to analyze some of the scenarios of the transition to chaos using numerical experiment. In doing so, we shall employ *realistic* and *complete models* (not simple nonlinear systems).

Thus, we shall consider

- Kolmogorov's flow in the presence of a periodic disturbing force;
- wind-induced ocean flows;
- stratified flows;
- the evolution of the Rayleigh–Taylor instability during the laminar–turbulent transition.

A major objective in studying these flows is to analyze secondary (supercritical) flow modes and to realize the irregular and unstable processes of "randomization".

### *1.11.2. "Kolmogoroff's flow" of the viscous fluid at subcritical and supercritical regimes. Transition to chaos*

To analyze the problem exposed in the present section, the complex approach is used, which combines the analytical investigation with the numerical one. The problems to investigate deal with hydrodynamic stability of the flow of a viscous incompressible fluid subjected to the action of the force which is periodical in space. For the weakly supercritical regime of such a flow, the direct numerical simulation is carried out. Moreover, some problems of the transition to chaos are discussed. Representative results of these investigations were obtained by Belotserkovskii *et al.* (see Refs. 75–77).

The study of a stability of laminar flows in respect to small perturbations, which always exists in nature, is of the indubitable interest both for theoretical analysis and for practical applications. This is primarily due to the fact that just the phenomena of stability are responsible for proper explanation of diverse complicated motions of fluid, as well as for the problems of the turbulence initiation.

Thus, according to Landau and Lifshitz,[10] as soon as the Reynolds number (or some other parameter characterizing the flow) passes over the critical value, the stationary flow loses its stability and the new regime is established, which in the general case is a periodical one and auto-oscillatory (secondary flow). This regime, in its turn, proves to be stable only within a certain range of values of the control parameter and loses its stability outside of this range, and so on. It is assumed that as the control parameter grows, the new forms of perturbations might successively appear (bifurcations), and in this process, the conditionally periodical motion will be initiated, which possesses the ever-increasing number of frequencies. Therefore, with a sufficiently large value of $Re$ the flow might be considered as a *turbulent* one.

In Ref. 10, creation of a periodical auto-oscillatory regime is considered as a first step of the transition of a fluid flow from a laminar state to the turbulent one. As it is known, for the planar case there is, generally speaking, no strictly determinated point of transition: laminar flow is transformed into completely turbulent one always along the extent of some *transitional area*, within which superimposed over the main flow are secondary, and, as a rule, oscillatory motions. These motions do not modify the laminar, as a whole, character of main stream, but essentially affect its mean characteristics.

It seems to be of a special importance to get the general pictures of flows within transitional area both for the stationary regimes and for the auto-oscillatory ones, as well as to find the critical values of the flow parameters; of no less importance is the identification of periods of the auto-oscillatory periodical flows and of possibilities to move up to the chaos. For geophysical applications, it would be interesting to study a stability of such a class of flow for which the scales of unstable perturbations are commensurable with the spatial scale of a basic flow. It is likely that such a study might render a clue to the explanation of evolution of the atmospheric cyclones and anticyclones, of the oceanic synoptic vortices, and to elucidate the patterns of interaction of these phenomena with a mean flow, etc.

The mathematical models for the flow of the type indicated above are represented by the boundary-value problems for nonlinear equations in partial derivatives (as it was noted earlier, in a general case these equations are assembled into the multi-dimensional, nonstationary set of Navier–Stokes equations). Considered below is *the problem of a plane flow of the incompressible viscous fluid generated by the force, which is periodical in transverse direction.* This problem was formulated by Kolmogoroff, and in its linear version was studied in many papers.[aa] The authors of the papers cited have proved the fact of the loss of stability of main laminar flow in respect to the spatially periodical perturbations with a large length of waves in direction of the flow. They treated the problem of a nonlinear evolution of perturbations and of an appearance of secondary (stationary or periodical) flows because of the loss of stability of a laminar flow of viscous, incompressible fluid. The experimental studies of a plane periodical flow are described in Ref. 80. However, *the question of existence of the stable, either steady or auto-oscillatory, regimes of secondary flow remains still unanswered; in a large extent, the same might be said on the possibility of a transition to chaos.* Moreover, it was necessary to study the influence of initial flow on the motion under consideration.

Let us consider the problem of a plane flow of the viscous incompressible fluid subjected to the action of external force, periodical in space, directed along the $x$-axis and equal to $\gamma \cdot \sin \eta y$ (parameter $\gamma > 0$). The motion of fluid is described by a set of Navier–Stokes equations. After the introduction of scales of length $\eta^{-1}$, of velocity $\eta^{-2}\nu^{-1}\gamma$ and of time $\eta\nu\gamma^{-1}$, and switching to dimensionless variables, one might, for a problem formulated, write down the set of equations of motion and continuity in the form

$$\begin{aligned}
&\frac{\partial u}{\partial t} + \frac{\partial u^2}{\partial x} + \frac{\partial uv}{\partial y} = -\frac{\partial p}{\partial x} + \frac{1}{Re}\Delta u + \frac{1}{Re}\sin y,\\
&\frac{\partial u}{\partial t} + \frac{\partial uv}{\partial x} + \frac{\partial v^2}{\partial y} = -\frac{\partial p}{\partial y} + \frac{1}{Re}\Delta v,\\
&\qquad\frac{\partial u}{\partial x} + \frac{\partial v}{\partial y} = 0.
\end{aligned} \tag{1.32}$$

Here the symbols $u, v$ corresponds to the components of velocity vector along the axes $x$, $y$; $p$ is pressure, $\nu$ — kinematic coefficient of viscosity; $Re = \gamma/(\nu^2\eta^3)$ — Reynolds number. The flow is studied within area $\Omega$, at the boundaries of which the conditions of periodicity are set — with a period $2\pi/\alpha$, $\alpha > 0$, along the axis $x$, and with a period $2\pi$ along the axis $y$.

[aa]Yudovich,[78] Meshalkin and Sinay[79] (see the bibliography in Ref. 14).

The mean velocity $\mathbf{q}$ of the fluid's motion will be considered as a prescribed quantity:

$$\mathbf{q} = \frac{\alpha}{4\pi^2} \int_\Omega \mathbf{V}\, dx\, dy. \tag{1.33}$$

It is to be noted that the formula (1.32) is equivalent to a prescription of the fluid's mass flow rate though the sides of a rectangle $\Omega$.

The set of Eqs. (1.32), with a mean velocity $\mathbf{q} = 0$ prescribed in accordance to (1.33), has a stationary solution which corresponds to the laminar flow along the $x$-axis with constant pressure:

$$u = \sin y, \quad v = 0, \quad p = \text{const}. \tag{1.34}$$

Considered below are the stationary solutions of the set of Eqs. (1.32), belonging to the type of Eq. (1.34), as well as their stability in respect to the perturbations, which are harmonical in $x$-direction with the wavelength $2\pi/\alpha$ ($\alpha$ is wave number), and in $y$-direction — with the wavelength $2\pi$, within the area

$$\Omega = \{(x, y), |x| \le \pi/\alpha, |y| \le \pi\}. \tag{1.35}$$

Further on, the problem, which was initially formulated for the elementary cell of (1.35), might be conveniently extended along the axes $x$ and $y$, with the periods $2\pi/\alpha$ and $2\pi$, correspondingly. The boundary conditions (conditions of periodicity) have a form

$$f(-\pi/\alpha, y) = f(\pi/\alpha, y), \quad f(x, -\pi) = f(x, \pi), \tag{1.36}$$

where $f \equiv (u, v, p)$. After the introduction of a stream function $\Psi$,

$$u = \partial\Psi/\partial y, \quad v = -\partial\Psi/\partial x,$$

one comes to the conclusion that this function satisfies the equation

$$\left(\frac{\partial}{\partial t} - \frac{\Delta}{Re}\right)\Delta\Psi + J(\Delta\Psi, \Psi) = \frac{1}{Re}\cos y, \tag{1.37}$$

where

$$J(\Delta\Psi, \Psi) = \frac{\partial\Psi}{\partial y}\frac{\partial\Delta\Psi}{\partial x} - \frac{\partial\Psi}{\partial x}\frac{\partial\Delta\Psi}{\partial y}$$

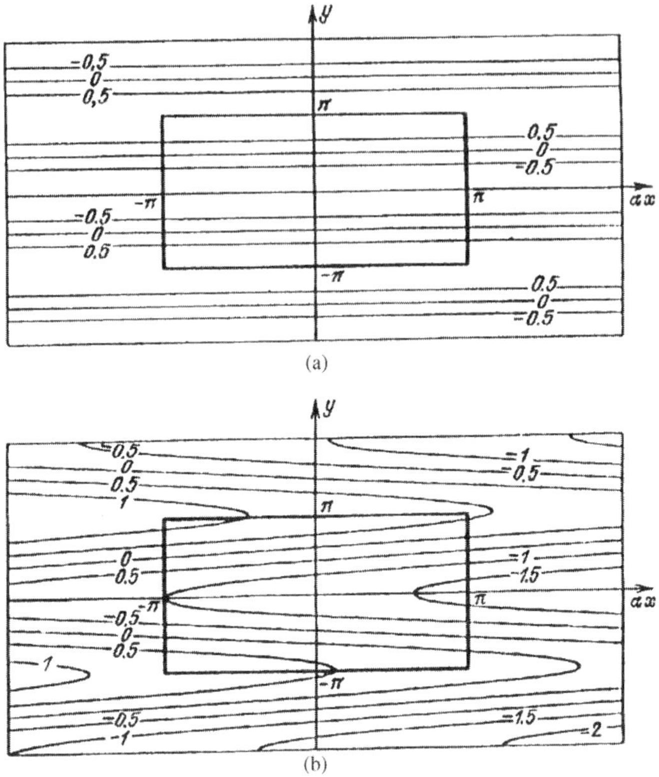

**Fig. 1.38.** Basic flow in a subcritical regime, with $Re = 0.5$: (a) $q(0,0) = 0$, (b) $q(0.05, 0.05) \neq 0$.

is a Jacobian. The stream function of a stationary planar flow, equations (1.34), has a form

$$\Psi = -\cos y. \tag{1.38}$$

The flow pattern corresponding to (1.38) (the lines $\Psi = \text{const.}$) with $\alpha = 0.5$ is shown in Fig. 1.38(a). The similar pattern of a flow was obtained as a result of numerical solution of the problem (1.32), (1.33), and (1.36), with $Re = 0.5, \alpha = 0.5$ and $\mathbf{q} = 0$.

The solution of the form of (1.38) is unique and stable for arbitrary Reynolds numbers, if $\alpha \geq 1$.[78] Thus, the case most interesting for investigation corresponds to a contribution $\alpha < 1$. As it is was shown in papers[78,79] (for the linear version of a problem), in the case of $\alpha < 1$, the stationary solution (1.38), corresponding to a laminar flow, is unstable in respect to the long-wave longitudinal perturbations by a certain value of the parameter $Re$. This value will be called a *critical Reynolds number* $Re_{\rm cr}$. Meanwhile, when $\alpha \to 1\text{–}0$, then $Re_{\rm cr}(\alpha) \to +\infty$, and when $\alpha \to +0$, then $Re_{\rm cr}(\alpha) = \sqrt{2} + 0$.

It will be of interest to find the dependence on $\alpha$ of the critical Reynolds number for small (but finite) values of $\alpha$, and to investigate the properties of solutions both for subcritical regime and for secondary flows, depending on $\mathbf{q}$.[76,77]

To solve the corresponding problem of Orr–Sommerfeld one might resort to asymptotic technique developed by Yudovich. By this way it is possible to find an expression for $Re_{\rm cr} = Re_{\rm cr}(\alpha)$ at small values of $\alpha$[76,77]:

$$Re_{\rm cr} = \sqrt{2} + \frac{3}{2}\sqrt{2}\alpha^2 + \mathrm{O}(\alpha^4). \tag{1.39}$$

The solution is looked for within the rectangular area $\Omega = \{|x| \le 2\pi/\alpha, |y| \le 2\pi\}$, at the boundaries of which the periodical conditions are set in the form of (1.36). As initial data, a certain field of velocity and pressure is prescribed:

$$\begin{aligned} u(x,y,0) = q_1(x,y), \quad v(x,y,0) = q_2(x,y),\\ p(x,y,0) = q_3(x,y), \end{aligned} \tag{1.40}$$

where $q_i(x,y)$ are the given functions satisfying the boundary conditions of Eqs. (1.36), $i = 1,2,3$.

The role of *characteristic parameters* of the flow investigated is played by Reynolds number $Re$, wave number $\alpha$, and by mean velocity of the fluid's motion $\mathbf{q}$ prescribed through (1.33); in the numerical experiment this last quantity is prescribed with the help of an initial field, $q_1$ and $q_2$ from Eqs. (1.40). The area's size in $y$-direction is set to be constant, $H = 1/\eta = 2\pi$, while the size in $x$-direction is equal to $L = 1/(\eta\alpha) = 2\pi/\alpha$.

For Eqs. (1.32), the following scheme of splitting[76,77] is used:

$$\begin{aligned} \frac{\tilde{\mathbf{V}} - \mathbf{V}^n}{\tau} &= -(\mathbf{V}^n\Delta)\mathbf{V}^n + \frac{1}{Re}\Delta\mathbf{V}^n + \mathbf{F},\\ \Delta p = \nabla \cdot \frac{\tilde{\mathbf{V}}}{\tau}, &\quad \mathbf{V}^{n+1} = \tilde{\mathbf{V}} - \tau\nabla p, \end{aligned} \tag{1.41}$$

where $\mathbf{F} = ((1/R)\sin y,\, 0)$ is an external force, periodical in space.

In distinction of the schemes considered earlier, here, at the stage I of splitting, the transfer is realized not only through diffusion and convection, but also through the external force. This question is considered in more detail in Refs. 14, 75–77.

At first, one is going to investigate analytically the behavior of solution of the problems (1.32) and (1.36) at the subcritical regime. Let the mean velocity of fluid's motion be $\mathbf{q} \neq 0$. The flow itself is assumed to be stationary, and the solution of the set of Eqs. (1.32) is presented in the form

$$u(x, y, t) = u(y) + U, \quad v(x, y, t) = V, \quad p(x, y, 0) = \text{const.},$$

where $U$ and $V$ are components of the vector $\mathbf{q}(U, V)$, with $U = \text{const.}$ and $V = \text{const.}$

After the substitution of the above solution into Eqs. (1.32), one obtains the ordinary linear differential equation of the second-order, with constant coefficients. Taking into account the condition of periodicity, Eq. (1.36), and the form of solution of Eq. (1.34) at $\mathbf{q} = 0$, the solution with a prescribed $\mathbf{q}(U, V) \neq 0$ will look as[76,77]:

$$\begin{aligned} u(x, y, t) &= \frac{1}{1 + (Re\, V)^2}(\sin y - Re\, V \cos y) + U, \\ v(x, y, t) &= V, \quad p(x, y, t) = \text{const.}, \end{aligned} \tag{1.42}$$

or, for stream function, as

$$\Psi = \frac{1}{1 + (Re\, V)^2}(\cos y + Re\, V \sin y) + Uy - Ux, \tag{1.43}$$

It is evident that the solution of Eqs. (1.42) is *invariant* with respect to a transformation of shift along the $x$-axis. The general solution, presented above and describing the laminar flow in subcritical regime at arbitrary $\mathbf{q}$, will be called the *basic* one. If $U = V = 0$ ($\mathbf{q} = 0$), then the flow described by Eq. (1.43), coincides with the flow of Eq. (1.38) which is shown in Fig. 1.38(a). The pattern of a basic flow, described by Eq. (1.43) (lines $\Psi = \text{const.}$), with $U = V = 0.05$ ($Re = 0.5, \alpha = 0.5$) is demonstrated in Fig. 1.38(b) ($\mathbf{q} \neq 0$). The numerical solution of a problem on the basis of a complete set of Eqs. (1.32), which was obtained in Refs. 76 and 77 with the same values of parameters $U$, $V$, $\alpha$ and $Re = 0.5$ (which corresponds to a subcritical regime — according to Eq. (1.39) $Re_{\text{cr}} \approx 1.9, \alpha = 0.5$), proves to be stable and stationary.[bb] The flow patterns (obtained analytically and numerically) completely coincide in this case, too.

---

[bb]In order to check the stability of the flows considered, over the solution obtained with $q = 0.05$, at eight central points, were superimposed the perturbations for $u$ and $v$ of the order of 0.05. Such a field was taken as an initial one, and once again the solution was found, which practically coincides with the one obtained earlier, and this fact testifies the *stability* of the flows considered.

When the mean velocity motion $\mathbf{q} = 0$, the investigation of flows at *supercritical* regimes might be conducted analytically, as well.[76,77] One is going to compare the results of analytical and numerical solutions for supercritical conditions, obtained by the loss of stability of the basic flow with $\mathbf{q} = 0$, in the form of Eq. (1.34) or (1.38), which is shown in Fig. 1.38(a). The results of a linear theory of stability permit to find the critical values of the flow's parameters, upon the attainment of which the small perturbations, superimposed over the main flow, begin to grow and, consequently, their amplitudes might reach the finite values. This means that, starting with that moment, the processes which go on within hydrodynamic system, cannot already be studied with the help of linear theory, and aiming to describe the evolution of finite perturbations it would be necessary to resort the complete, nonstationary set of nonlinear equations of hydrodynamics, Eqs. (1.32).

The procedure of investigation of secondary flow is based on the approximation of the hydrodynamic equations by the finite set of ordinary differential equations — Galerkin's method. Chosen as the basic functions were here the most unstable modes, found according to the linear theory of stability of the flow (1.34). Found in Refs. 75–77 was the *general expression for a stream function of the secondary flow*, which took into account the general form of dependence $Re_{\text{cr}}(\alpha)$, Eq. (1.39):

$$\Psi(x, y) = -\frac{Re_{\text{cr}}}{Re}\cos y - \frac{\sqrt{Re_{\text{cr}}(Re - Re_{\text{cr}})}}{Re} \times \left(\frac{1}{\alpha}\sin\alpha x + \frac{2}{Re_{\text{cr}}}\cos\alpha x \sin y\right). \tag{1.44}$$

Here the first term of the expression corresponds to the stream function of a basic flow, and the second term — to that finite perturbation.

The solution's analysis has shown that with $Re < Re_{\text{cr}}$ the laminar regime of Eq. (1.34) is established. By the transition of $Re$ over the critical values, this regime becomes to be unstable, and the secondary stationary flow is originated. The expression (1.44) describes the various regimes of secondary stationary flow, which will be investigated further on, both analytically and numerically.

The secondary flow of Eq. (1.44), obtained with account of interaction with the mean flow of only first harmonics of perturbation, contains several parameters: $\alpha$, $Re_{\text{cr}}(\alpha)$, $Re$. It seems to be worthwhile to study the structure of this flow at $Re > Re_{\text{cr}}$ and at some fixed value of $\alpha < 1$, which, as it might be easily seen, embraces all the possible versions of variation of the control parameters. With such an aim, it might be convenient to change for a new stream function

$$\Psi_1 = -\frac{Re}{b}\Psi = \Psi_0 + \Psi_{\mathrm{B}}, \tag{1.45}$$

where

$$\Psi_0 = \frac{Re_{\mathrm{cr}}}{b} + \cos y, \quad \Psi_B = \frac{1}{\alpha}\sin\alpha x + \frac{2}{Re_{\mathrm{cr}}}\cos\alpha x \sin y,$$
$$b = [Re_{\mathrm{cr}}(Re - Re_{\mathrm{cr}})]^{1/2}.$$

Here $\Psi_0$ is a stream function of a pure shear flow, $\Psi_{\mathrm{B}}$ — stream function of finite perturbations, which actually depends on only one parameter $\alpha$ (and does not depend on $Re$).

An analysis of expression (1.45), carried out in Ref. 75, shows that here exist certain characteristic values of Reynolds number, $Re_1(\alpha)$ and $Re_2(\alpha)$, and to each one of the intervals $Re_{\mathrm{cr}} < Re \leq Re_1, Re_1 < Re \leq Re_2$, $Re > Re_2$ corresponds a certain pattern of flow. As soon as the boundaries indicated are crossed, occurs the qualitative re-organization of a structure of motion.

The values $Re_1$ and $Re_2$ are determined through the formulae

$$Re_1(\alpha) = Re_{\mathrm{cr}}(\alpha)(1+\alpha^2), \quad Re_2(\alpha) = Re_{\mathrm{cr}}(\alpha)\left(1 + \frac{Re_{\mathrm{cr}}^4}{16\alpha^2}\right). \tag{1.46}$$

In the case considered here, with $\alpha = 0.5$ one has $Re_{\mathrm{cr}}(\alpha) \approx 1.9$, $Re_1(\alpha) \approx 2.4, Re_2(\alpha) \approx 8.8$. Below, each one of the range of variation of Reynolds number is investigated analytically and numerically (further on, all the calculations will be carried out with $\alpha = 0.5$).[75–77]

*The first range* ($Re_{\mathrm{cr}} < Re \leq Re_1$): stream function of the finite perturbations $\Psi_{\mathrm{B}}$ is smaller that $\Psi_0$, and therefore the system of vortex chains is originated, the chains being parallel to $x$-axis, while the system is parted by the zones of horizontal jet flows. The flow pattern, obtained analytically from Eq. (1.45), is shown for the case considered in Fig. 1.39(a), with $Re = 2$ (it is to be reminded that for $\alpha = 0.5$, one has $Re_{\mathrm{cr}}(\alpha) \approx 1.9, Re_1(\alpha) \approx 2.4$). It is seen that under the action of shift there appears an incidence of the vortices' axis, while the directions of the angles of turn are interchanged by successive transition from one vortical band to another. As the Reynolds number is increased (with $Re \to Re_1$), the maximal width of the vortical band is also increased, while the zones of jet-type flow are diminished.

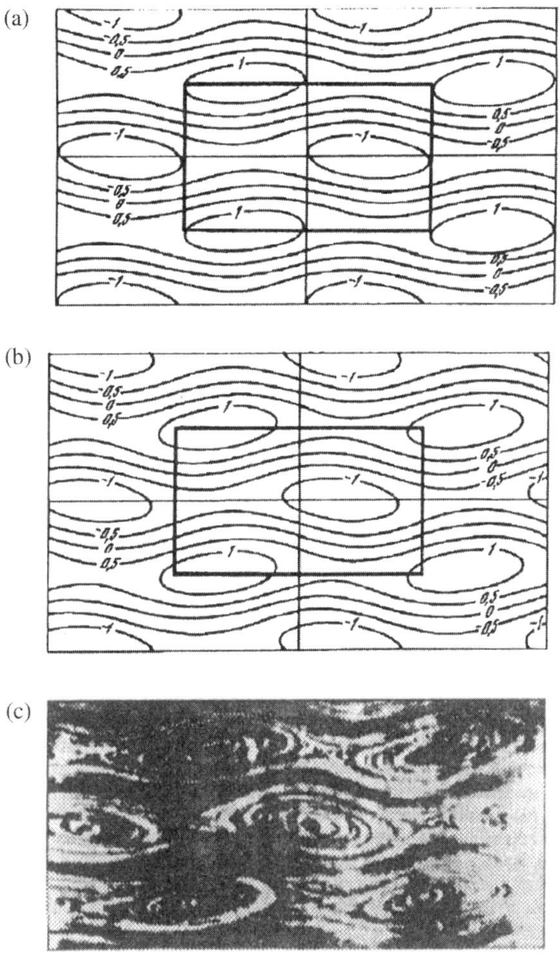


**Fig. 1.39.** Stationary secondary flow with $\alpha = 0.5$, $q(0,0) = 0$; in *the first* range ($Re_{\rm cr} < Re \leq Re_1$): (a) $Re = 2$, analytical solution, (b) $Re = 3.1$, calculation, (c) experiment for $Re > Re_{\rm cr}$.

For comparison, in Fig. 1.39(b) is shown a pattern of stationary flow in supercritical regime, which was obtained as a result of direct numerical solution of the complete original problem, corresponding to Eqs. (1.32); this pattern was created after the loss of stability of a basic flow depicted in Fig. 1.38(a). In spite of the fact that the numerical solution was obtained with $Re = 3.1$, the flow pattern (see Fig. 1.39(b)) corresponds to the first type of the flow. Presented in Fig. 1.39(c) are experimental data with

$Re > Re_{\rm cr}$ (weakly supercritical conditions).[80] The agreement between the results is sufficiently good.

It is to be observed that the numerical solution for $Re \leq 3$ corresponds to the plane laminar flow and coincides with analytical solution obtained for $Re = 0.5$ (see Fig. 1.38(a)). As soon as $Re > 3$, the flow pattern *is changed abruptly*, and the system of alternating vortices is formed. Thus the solution, constructed numerically for $Re = 3.1$, is also *stationary, but corresponding already to the supercritical regime*; this solution is illustrated in Fig. 1.39(b). For a proper comparison with this solution one should take the flow pattern obtained as a result of analytical investigations (Fig. 1.39(a)), and constructed for $Re = 2$ (which, once again, corresponds to the first type of a flow in supercritical regime). The qualitative agreement of results is sufficiently good.

Thus, *with the Reynolds number varying in the range between* 3 *and* 3.1, *the basic flow, which was constructed numerically for a complete set of Navier–Stokes equations, Eqs.* (1.32), *loses its stability, and is established the new (first) stationary regime corresponding to a secondary flow.* This does not agree with a linear theory, according to which the critical Reynolds number expressed through Eq. (1.39), with $\alpha = 0.5$, is equal to 1.9. Such a disagreement in determination of $Re_{\rm cr}$ seems to be explained by the fact, that the expression (1.39) for critical Reynolds number is obtained in linear approximation, for finite (though small) values of $\alpha$, while by the numerical simulation one solves the complete set of nonstationary equations, and the wave number $\alpha$ is equal to 0.5. Moreover, one should observe that, as differentiated from the analytical investigation, which was carried out taking into account only the first harmonics, the numerical simulation is realized with a consideration for both the harmonics with the wavelength above $2h$ (where $h$ is the step of a spatial grid in $x$ and $y$), and for their nonlinear interactions between themselves.

*The second range* ($Re_1 < Re \leq Re_2$): the value of $\Psi_{\rm B}$ is analytical solution of Eq. (1.45), it exceeds $\Psi_0$, and therefore the system of vertically posed vortical bands is formed, the bands being separated by the areas of vertical jet-type flows. Within one and the same vortical band the vortices' axes have one and the same incidence, while by going over to the neighboring band, the direction of angles of incidence is changed for the opposite one; at the same time, the vortices are found to be somewhat extended in transverse direction (Fig. 1.40(a), $\alpha = 0.5, Re = 5$).

For a comparison, presented in Fig. 1.40(b) is the pattern of *stationary flow*, constructed numerically on the basis of Eqs. (1.32) ($\alpha = 0.5, Re = 5$),

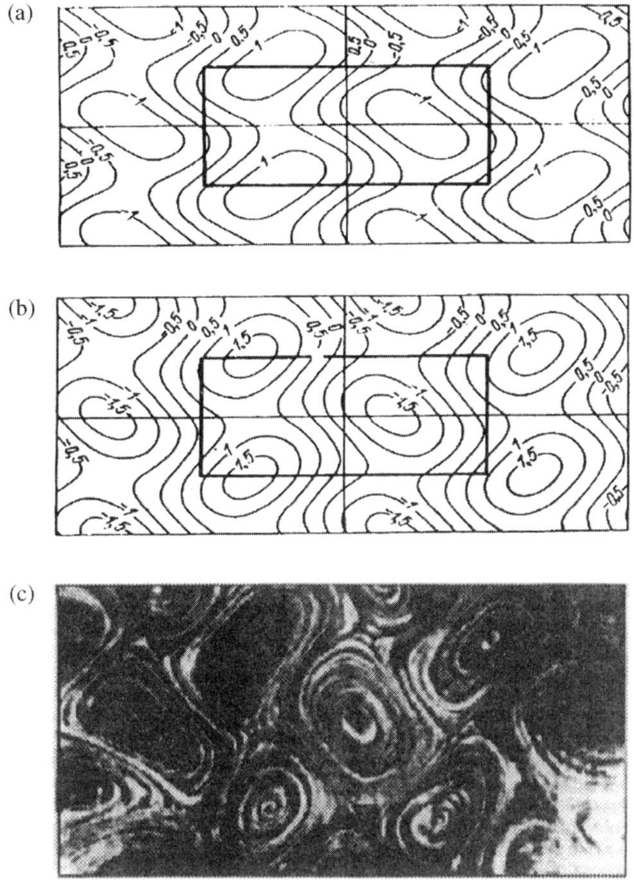


**Fig. 1.40.** Stationary secondary flow with $\alpha = 0.5, q(0,0) = 0$; in *the second* range ($Re_1 < Re \leq Re_2$): (a) $Re = 5$, analytical solution; (b) $Re = 5$, calculation; (c) experiment for $Re > 1.25 Re_{\rm cr}$.

while in Fig. 1.40(c) are demonstrated the experimental data for $Re > 1.25 Re_{\rm cr}$.[80] It is seen that the qualitative agreement between the results is also sufficiently good.

Thus, it is numerically proved that by the variation of $Re$ between 3.1 and 5, the reorganization of a flow structure takes place. This renders a confirmation to the fact, which was proved analytically, and studied experimentally[80] and declares *the existence of qualitatively different regimes of a stationary flow in supercritical range of conditions.* Sufficiently good agreement between the flow patterns, obtained through the numerical

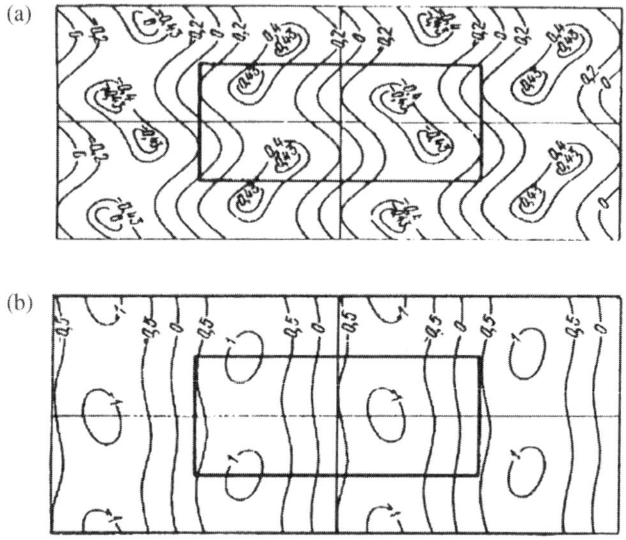

**Fig. 1.41.** Stationary secondary flow with $\alpha = 0.5, q(0,0) = 0$; in *the third* range.

solution of a problem (on the basis of complete set of Eqs. (1.32)), and the similar patterns of supercritical flow, obtained analytically with consideration of interaction of only the main harmonic (for perturbations in $x$-direction) with the mean flow, confirms the correctness of analytical description through Eq. (1.45) of a secondary flow, in the case of $Re$ being close to $Re_{\text{cr}}$ (though formula of Eq. (1.39) was deduced under assumption of small $\alpha$, while here is taken a finite value $\alpha = 0.5$).

*The third range* ($Re > Re_2$) (where taken from Eq. (1.46) is $Re \approx 8.8$ for $\alpha = 0.5$): one more reorganization of the structure of flow takes place, and the flow pattern corresponding to Eq. (1.45) is drastically complicated (Fig. 1.41(a), $Re = 50$). The system of vertical vortical bands is similar here to that of the second range, but the incidence of main vortices increased, and formed within these bands is the new subsystem of vortices of the type of "eight". As it is seen from Fig. 1.41(a), the present regime differs essentially from all the preceding cases by the fact that the flow possesses the fine vortical structure.

The numerical solution of the problem of Eqs. (1.32), (1.36) carried out for $Re = 50$ is also stationary (Fig. 1.41(b)), but is already qualitatively different from the analytical one (see Fig. 1.41(a)), — for example, there is no fine structure. It seems that this might be explained, on the one hand, by the insufficient capacity of resolution of the difference grid, and, on the other hand, by the fact that at large Reynolds numbers some new forms of perturbations might be originated, and by the analytical investigations one cannot already restrict oneself by the consideration of the only harmonic in $x$-direction and of those harmonics in $y$-direction, which have the numbers $n = 0, \pm 1$. Thus, *when the mean velocity of fluid's motion is set to be equal to zero, all the numerical solutions for supercritical regime are stationary.*

In conclusion, the following observation should be made. The analysis of a secondary flow's stability for the case of finite $\alpha$ and of large Reynolds numbers is rather complicated. Because of that, by the analytical study of stability of flows with $\mathbf{q} \neq 0$ (of the type described by Eqs. (1.42) and (1.43)) with respect to small perturbations, the serious difficulties arise. In supercritical conditions the auto-oscillatory regimes might appear, for description of which one should take into account all the types of perturbations. Thus, if the treatment of a secondary flow, springing up because of loss of stability of a laminar flow, is corresponding to expressions of Eq. (1.42), in the paper[78] was made a suggestion of the appearance of nonstationarity in the form of auto-oscillatory regime with a period $T = 2\pi/(\alpha U)$. It seems that in the present case just a numerical approach on the basis of complete nonstationary Navier–Stokes equations (Eqs. (1.32)) permits to conduct the detailed investigation of the class of flows under consideration.

Presently, one is going to consider the flow within a weakly supercritical range, which results upon the loss of stability of the basic flow (Eqs. (1.42) and (1.43)) with $\mathbf{q} \neq 0$, and which is shown in Fig. 1.38(b) (the solution presented here was obtained starting with an initial field having $q_1 = q_1 = 0.05, \alpha = 0.5, Re = 0.5$). Such an analysis might be successfully carried out only numerically.[76,77]

Solution of the problem corresponding to Eqs. (1.32), (1.33), and (1.35) with $Re = 5(\mathbf{q}(0.05; 0.005) \neq 0)$ proves to be stable but nonstationary, and the flow pattern should be considered in its dependence on time (Fig. 1.42). As initial conditions, Eqs. (1.40), one takes the field of velocity and pressure of the basic flow presented in Fig. 1.38(b). As it is seen from Fig. 1.42, the streamlines form a system of vertically placed vortical bands, which are separated from each other by the areas of vertical jet-type flow. Within the limits of one and the same vortical band, the vortices' axes have one and the same incidence, while by going over the neighboring band, the direction of angles of incidence is changed for the opposite one.

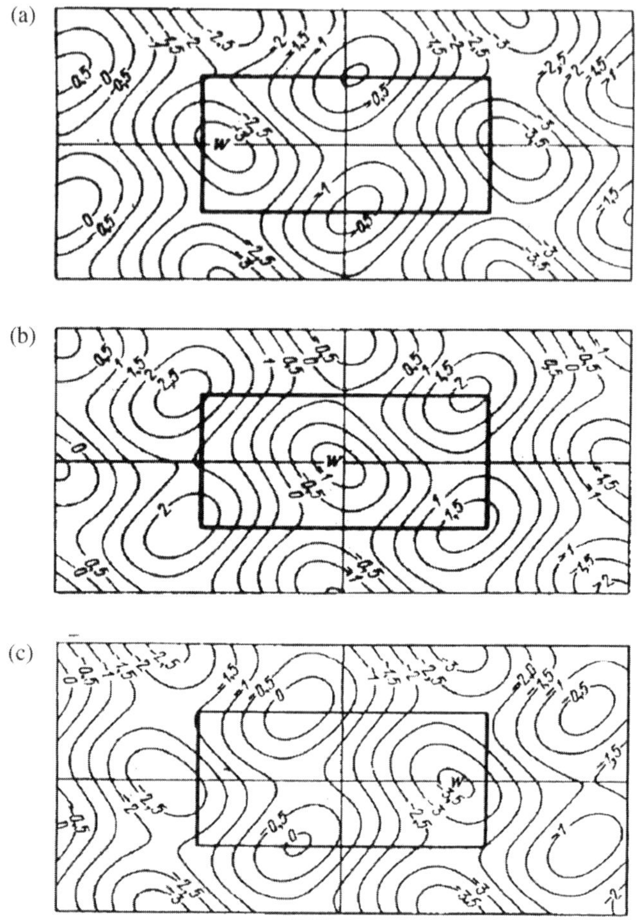

**Fig. 1.42.** Nonstationary secondary flow (calculation) with $Re = 5, \alpha = 0.5$, $q(0.05, 0.05) \neq 0$, period $T = 250$: (a) $t = 75$, (b) $t = 175$, (c) $t = 325$.

As differentiated from the pattern of a stationary flow, which is shown in Fig. 1.40 for the same $Re = 5$ and $\mathbf{q} = 0$, the pattern obtained in the present case ($\mathbf{q} \neq 0$) possesses more complicated structure, and the stream function does not satisfy the periodicity condition in $x$- and $y$-directions. Moreover, this flow proves to be nonstationary: the system of vortices is moving in a positive direction of $x$-axis. In this process, the intensity of a central vortex $W$ (see Fig. 1.42) (as well as of the whole vortex band to which belongs the vortex $W$) is diminished as the vortex goes on, if only this vortex happens to be within the area

$$(2k-1)\pi/\alpha \leq x \leq 2k\pi/\alpha \quad (k = 0, \pm 1, \ldots),$$

and is increased if it turns up to be in the area

$$2k\pi/\alpha \leq x \leq (2k+1)\pi/\alpha \quad (k = 0, \pm 1, \ldots).$$

As it might be seen, the pattern of streamlines is repeated in time in, approximately, 250 dimensionless units, that is, the regime of flow is auto-oscillatory, with a period $T \approx 250$. Now, it is necessary to find out, how will be modified the character of nonstationary flow by the variation of components $q_1$ and $q_2$ of the vector of mean velocity $\mathbf{q}(q_1, q_2) \neq 0$, and of the Reynolds number $Re > Re_{\mathrm{cr}}$.[cc]

In the case, when the components of the vector of mean velocity $\mathbf{q}$, determining the basic flow, are $q_1 = q_2 = 0.005$ (i.e. ten times smaller than in the preceding version), with the same $Re = 5$, one obtains, once again, nonstationary flow pattern presented in Fig. 1.43(a). Here, the temporal interval between the first pattern and the third one is equal to $\Delta t = 50$ (dimensionless units).

Now, one is going to compare the flow patterns obtained in the preceding case (Fig. 1.43(b)) ($Re = 5, \mathbf{q}(0.05; 0.05), \Delta t = 5, T \approx 250$) with the pattern presented in Fig. 1.43(a) ($Re = 50, \mathbf{q}(0.005; 0.005), \Delta t = 50$). The upper and lower patterns in Fig. 1.43(a) and 1.43(b) are closely resembling each other (for example, when judged by the size and intensity of a central vortex $W$ located within the area $0 \leq x \leq \pi/\alpha$). From that, one might come to a conclusion that the period of auto-oscillatory motion the second version will be $T \approx 2500$, i.e. ten times longer than the first one (in the same proportion, as $q_i$ in the second case are smaller than $q_i$ in the first one). Therefore, when $q_1 = q_2 \to 0$ for $Re > Re_{\mathrm{cr}}$, *the period of auto-oscillatory motion tends to infinity and the numerical solution turns to become stationary.* The patterns of such a flow for $Re = 3.1$; 5; 50, with $\mathbf{q}(0,0) = 0$, are presented, correspondingly, in Figs. 1.39(b)–1.41(b).

The next to be considered is the flow pattern in supercritical conditions, after the loss of stability of the basic flow, when $q_1 = 0.05$ and $q_2 = 0$. In the case of $Re = 5$, the flow will be nonstationary; the pattern of streamlines ($\Psi = \text{const.}$) in dependence on time is shown in Fig. 1.44. Comparing the present results with a flow pattern shown in Fig. 1.42, where the period was

[cc]It is to be reminded that the results of numerical simulation give $Re_{\mathrm{cr}} = 3.1$.

**Fig. 1.43.** Nonstationary secondary flow (calculation) with $Re = 5, \alpha = 0.5$, $q(0.005, 0.005) \neq 0; \Delta t = 50; T = 2500$.

$T = 250$, one is able to see that the corresponding patterns are close to each other. In both cases, as the time goes on, one could observe the identical shifts of complete patterns along the $x$-axis, as well as the reduction of intensity of a central vortex $W$.

Similar to what was observed in Fig. 1.42, one could come to a conclusion that in the last case the flow will be, once again, auto-oscillatory one, and its period is equal to $T \approx 250$.

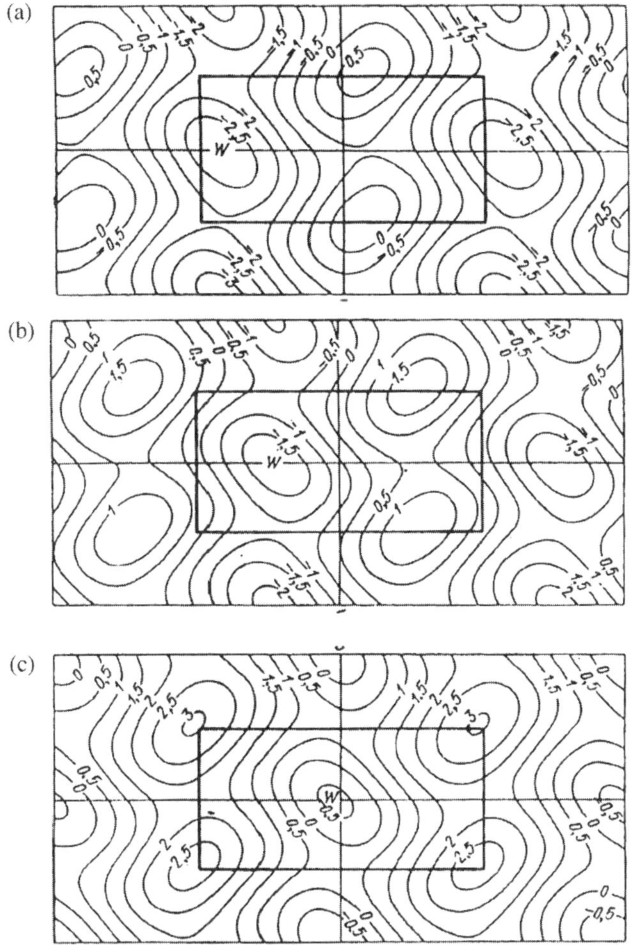

**Fig. 1.44.** Nonstationary secondary flow (calculation) with $Re = 5$, $\alpha = 0.5$, $q(0.05, 0) \neq 0$, $T = 250$: (a) $t = 100$, (b) $t = 150$, (c) $t = 200$.

When the components of the vector of flow's velocity, which determinates the basic flow, are equal to $q_1 = 0$ and $q_2 = 0.05$, then after the loss of stability of the basic flow, the supercritical regime with $Re = 5$ proves to be stable and stationary (Fig. 1.45).

Thus, the period of auto-oscillatory motion does not depend on initial flow $v(x, y, 0) = q_2$, which determinates the basic flow, but depends only on $u(x, y, 0) = q_1$ (see formulae (1.40)). The initial field of velocity ($q_1$,$q_2$)

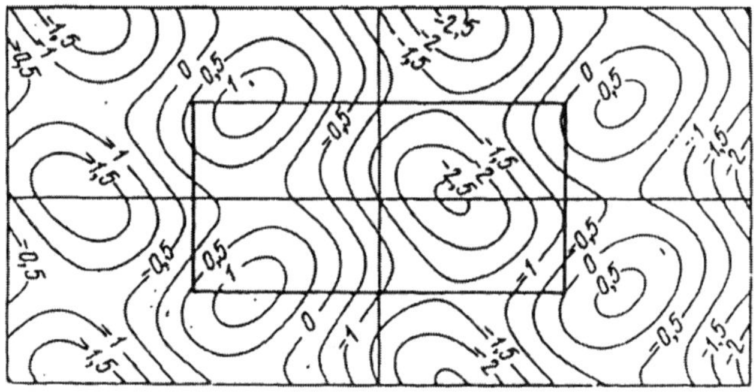

**Fig. 1.45.** Stationary secondary flow (calculation) with $Re = 5$, $\alpha = 0.5$, $q(0, 0.05) \neq 0$.

permits to determine the mean velocity of flow $\mathbf{q}(U, V)$ (Eq. (1.33)) by the calculation of subcritical regimes of flow.

The correctness of results obtained in this section is confirmed by the theorem formulated in paper.[78]

**Theorem 1.** *For those values of Reynolds number, which provide the nontrivial solutions of Eqs.* (1.32), *in the case* $q = 0$, *the same Eqs.* (1.32), *in the case* $q(U, 0) \neq 0$, *have nontrivial solutions, periodical in time,*

$$T = 2\pi/(\alpha U). \tag{1.47}$$

This, with a mean velocity of initial flow prescribed in the form $\mathbf{q}$(0,05;0), the period of auto-oscillatory flow in a numerical calculation is equal to $T = 250$, while the same period, calculated through the formula of Eq. (1.47), is equal to $T = 251$ (our version is considered with $\alpha = 0.5$). It seems that the coincidence of the values of $T$ calculated through Eq. (1.47), with the numerically determined values of the periods of auto-oscillatory motion, gives one the right to come to such a conclusion: *formula of Eq.* (1.47) *is applicable to the weakly supercritical regimes of flow.*

Thus, on the prescription of the mean velocity $\mathbf{q}$ of initial flow, which determines the basic flow, depends the acquisition in supercritical range of the stable (all the regimes mentioned in this chapter, are stable in respect to small perturbations), stationary (when $T \to \infty$), or auto-oscillatory regimes with a period $T = 2\pi/(\alpha U)$.

The numerical results cited here, as well as their comparison with analytical solution and with experimental data, show that by way of a direct numerical solution of the complete set of Navier–Stokes equations (Eqs. (1.32)), one is able to succeed in construction of both stationary and nonstationary stable flows for a supercritical regime of motion. However, the approach described did not permit to get closer to chaos.

It seems that the approaches, described above, provide a possibility to simulate only *initial* (*weakly critical*) stage of the evolution of a process, and within the frame of *the first* bifurcations to estimate the character of formation of the corresponding "cell-like" stationary structures. However, as it follows even from these estimates, the presence of a mean motion ($\mathbf{q} \neq 0$) leads at the supercritical regime to the qualitative change — the flow becomes to be nonstationary (periodical).

Naturally, it would be of a great interest to study the more complete pattern of the phenomenon's development (dynamics of the "large" vortices by the increase of Reynolds number, stochasticity at the small scales, the asymptotic process of transition to chaos, etc.). Just that was done in the recently published papers by Platte, *et al.*[81] and in some others. In the paper[81] with the help of a spectral approach was conducted a direct detailed calculation of the Kolmogorov's flow (KF) (with $\mathbf{q} = 0$). As the bifurcational parameter (connected with $Re$) was increasing, there were observed two main regimes of flow — *small-scale formations* and *large-scale* ones. The sequence of bifurcations was studied for each of these regimes.

The first three bifurcations for small-scale structures were of *a stationary* character (which in fact is in a good agreement with the investigations presented). Of a special interest is the fact that here are discovered three "windows" of the flow's stochastization having the *nonsingular* stochastic attractor. The regime of large-scale formations (large values of the bifurcational parameter) also possesses the "windows" of periodical (laminar) and of stochastical behavior of the solution (with nonsingular attractor). Prior to the formation of large-scale structures the flow was during the considerable time in the state of a "meta-stable chaos". One is going to dwell on this aspect in more detail.

*The description of a model.* The Kolmogorov's flow is defined as a solution of Navier–Stokes equations

$$\frac{d\mathbf{u}}{dt} + \nabla P = \frac{1}{\Omega}\nabla^2\mathbf{u} + \frac{n^2}{\Omega}\mathbf{e}_x \sin ny, \quad \nabla\mathbf{u} = 0 \tag{1.48}$$

with periodical boundary conditions in respect to both independent variables

$$0 \leq x, \quad y \leq 2\pi. \tag{1.49}$$

Equation (1.48) are presented here in dimensionless form; $e_x$ is along the $x$-axis; the bifurcational parameter $\Omega = nRe$ ($n$ is the number of harmonics of the stimulant force), while the critical value of $\Omega$ is

$$\Omega_{\text{cr}} = nRe_{\text{cr}} = n\sqrt{2},$$

where $Re_{\text{cr}}$ is the critical Reynolds number equal to $\sqrt{2}$.

These solutions allow for the following *groups of transformation*:

$$\mathbf{g}^k\mathbf{u} = \left[(-1)^k u\left((-1)^k x, y + k\frac{\pi}{n}\right), v\left((-1)^k x, y + k\frac{\pi}{n}\right)\right],$$
$$k = 0, 1, \ldots, 2n-1; \quad \mathbf{R}\mathbf{u} = [-u(-x,-y), -v(-x,y)], \tag{1.50}$$

$$\mathbf{T}_e\mathbf{u} = [u(x+l,y), v(x+l,y)], \quad 0 \le l \le 2\pi.$$

The properties of symmetry, Eq. (1.49), make it possible to reduce considerably the number of the effective degrees of freedom, which is of especial importance by the use of the spectral method for a solution of the problem, stated in Eqs. (1.48)–(1.49). This method is based on the following type of presentation:

$$u(x,t) = \sum_m u_m(t)B_m(x), \tag{1.51}$$

where

$$\mathbf{B}_m(\mathbf{x}) = [(m_2 - m_1)/|\mathbf{m}|]\exp(i\mathbf{m}\cdot\mathbf{x}), \quad \nabla\cdot\mathbf{B}_m = 0.$$

In the paper[81] it was taken that $n = 4$, so that one obtains $\Omega_{\text{cr}} = 4\sqrt{2}$, and the main flow described by the formula

$$\mathbf{u} = \mathbf{v} = \frac{1}{\Omega}\sin(4y)\mathbf{e}_x. \tag{1.52}$$

The initial velocity field was chosen according to Eq. (1.52) plus small perturbations superimposed on each harmonica (participating in the computational process). Thus, one has

$$\mathbf{u}_0 \equiv (u_0, v_0) = \Omega^{-1}\mathbf{e}_x \sin 4y + 0.510^{-4}$$
$$\times \sum_m [(1+i)\mathbf{B}_m(\mathbf{x}) - (1-i)\mathbf{B}_{-m}(\mathbf{x})]. \tag{1.53}$$

The role of the problem's main bifurcational parameter is played by the ratio $\omega = \Omega/\Omega_{\text{cr.}}$. In the paper cited this number, while gradually increasing, was getting close to 12.5. The calculations were conducted with the number of harmonics $32 \times 32$ (the use of a double resolution $64 \times 64$ did not lead to any essential changes in final results).

*The results of calculations.* Listed below are the critical situations arising by the continuous growth of the parameter $\omega \equiv \Omega/\Omega_{\text{cr}}$.

(i) For $0 < \omega \leq 1$, one obtains the regime of Eq. (1.51).

(ii) For $1 < \omega < 1.97$, the first bifurcation of the flow takes place: there appears the stationary regime of the cell-like structure; as this takes place, the curves of separatrices enveloping the vortices and concurrent with the increase of Reynolds number, change their horizontal position for a vertical one.

(iii) For $1.97 \leq \omega < 2.2$, there appears a new stationary symmetrical regime, which is characterized by the redoubling of the cell's number in horizontal direction.

(iv) For $2.2 \leq \omega \leq 3.6$, one observes a horizontal drift of the flow, connected with the appearance and growth of perturbations near the saddle points of the flow, and thus occurs the gradual violation of the flows symmetry. For $\Omega < 3.53\Omega_{\text{cr}}$, the regime is still stationary (see Fig. 1.46), but starting with $\Omega = 3.53\Omega_{\text{cr}}$ and on, it might be considered as stationary only in *statistical sense.* Within the cell-like structure of a flow the effects of alternativeness come into existence, which is especially clearly demonstrated by the temporal evolution of one of the harmonics $u_{0,1}$. In this realization the sudden splashes are interchanging with those smooth horizontal sections, which are characteristic for the regular cell-like structure. As it is known, rather convenient practical criterion of stochasticity is rendered through the numerical determination of the Liapunov's exponents, which serves as a measure of exponential divergence of the closely located trajectories. Within this range of Reynolds numbers one might already observe a *slight chaotization* of the flow: Liapunov's numbers, small as they are, become positive.

(v) For $3.6 < \omega < 4.47$, the stochasticity is intensified and the flow is changed in its nature: at the background of cell-like structure, the periodical oscillations are present. Here the flow is accompanied by a slight horizontal drift (without mass transfer). At the temporal energy spectrum the main peak is found, corresponding to the main oscillation, and the secondary peaks of smaller amplitude, responsible for its subharmonics, are also seen (Fig. 1.47(a)). The spectrum contains the maxima, which are identified with oscillations of the largest vortices and with those of their "internal structure", correspondingly.

(vi) For $4.47 < \omega < 4.58$, the secondary chaotic regime appears. The flow's spectrum begins to be filled with harmonics of smaller scale (Fig. 1.47(b)). The Poincaret's cross-section records the appearance

(a)

(b)

**Fig. 1.46.** The stationary cell-like flow according to Ref. 99, for $\Omega = 3\Omega_{\rm cr}$: (a) streamlines; (b) isolines of vorticity.

of a strange attractor, which due to the symmetry conditions (Eq. (1.49)) consists of eight components.

(vii) For $4.58 \leq \omega \leq 4.71$, the flow becomes to be "alternately chaotic", with an *ergodic attractor* (but with eight components); the *sudden "jump"* of the attractor onto the other component is observed (Fig. 1.48); the corresponding spectrum becomes to be even more chaotic (Fig. 1.47(c)).

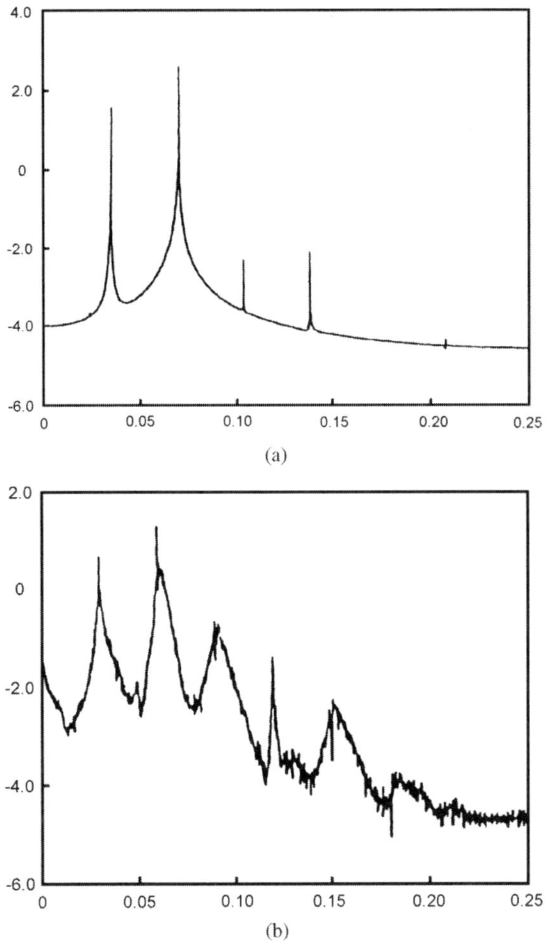


**Fig. 1.47.** The energetical spectrum of velocity $u$: (a) $\Omega = 4\Omega_{\rm cr}$, (b) $\Omega = 4.48\Omega_{\rm cr}$, (c) $\Omega = 4.63\Omega_{\rm cr}$, (d) $\Omega = 10.7\Omega_{\rm cr}$.

(viii) For $\omega \geq 4.72$, the large-scale structure of KF is formed (that is, the small-scale cell-like structure is already completely destroyed), and the regime is created which is classified in literature as "*metastable chaos*". Here are present and remain in coexistence two types of the state: *chaotic* one and *stable laminar* one.

(ix) For $4.72 \leq \omega \leq 5.9$, the former periodicity of a flow is conserved. When $\Omega = 5.9\Omega_{\rm cr}$, the main frequency and the drift of flow are tending to zero, and the appreciable spatial chaotization of KF is initiated.

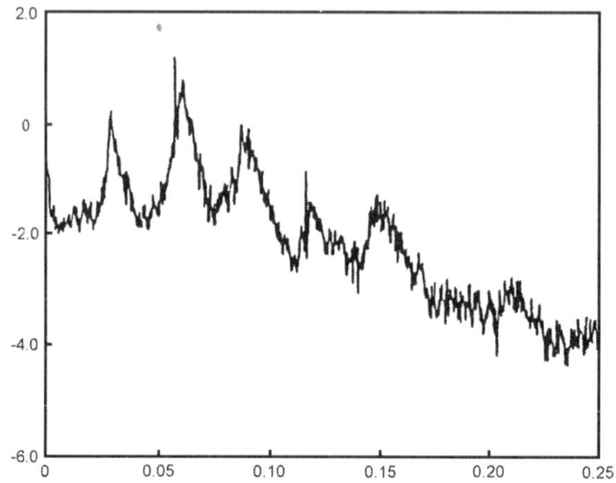


(c)

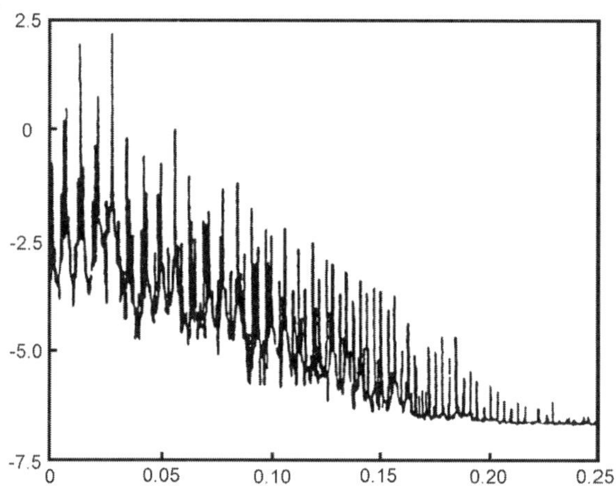


(d)

**Fig. 1.47.** *(Continued)*

(x) For $\omega > 5.9$, one might see the development of the periodic oscillations and of the horizontal drift (Fig. 1.49), and further increase of $Re$ number leads to even stronger chaotization of flow (i.e. to the growth of Liapunov's exponent), to the replacement of a symmetric attractor's structure for a *fractal* one and, finally, to the *continuous* spectrum of energy (Fig. 1.47(d)), the filling of which is realized though the series of the *bifurcations of the redoubling of a period.*

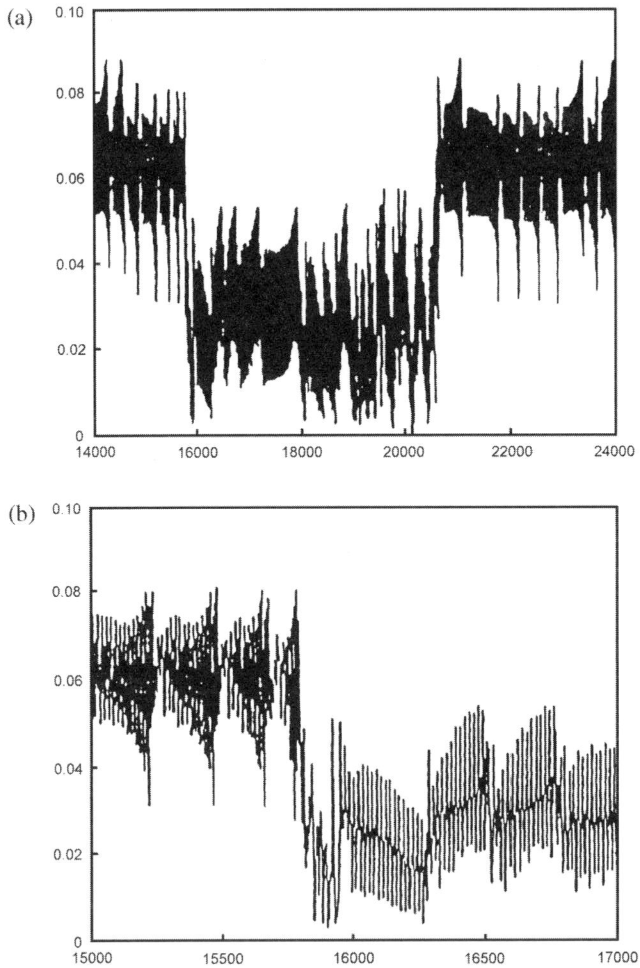


**Fig. 1.48.** The temporal behavior of the stochastic solution for different values of $Re(u_{(0;1)})(t)$, for $\Omega = 4.63\Omega_{\rm cr}$ (the sudden jump from attractor (a) to attractor (b) is observed[99]).

Such was the complicated scenario of the transition from order to "chaos" in the Kolmogoroff's flow.

### *1.11.3. Study of the large-scale turbulence in ocean*

The interest for studies of the oceanic flows is primarily determined by the decisive role of turbulence in the development of hydrodynamic fields in the

(a)

(b)

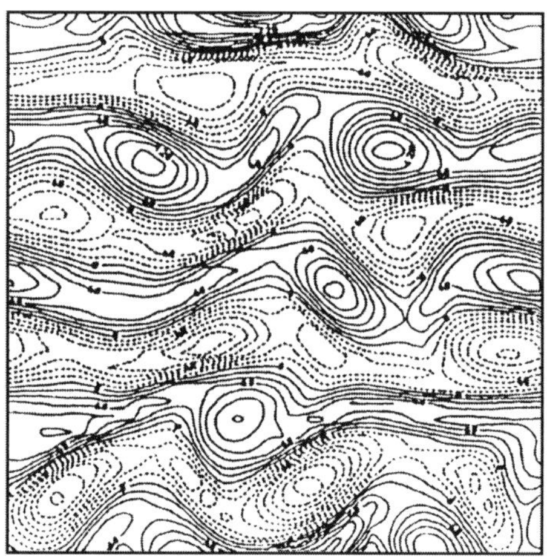

**Fig. 1.49.** The metastable chaotic (turbulent) regime for $\Omega = 10\Omega_{\rm cr}$: (a) streamlines; (b) isolines of vorticity.

ocean. The formerly wide-spread concept of the ocean as a comparatively weakly varying object with a system of stationary flows should be substituted by the newer one, according to which the main energy of oceanic waters in concentrated not in its mean circulation but in an aggregate of vortical formations with different sizes and durations of life (see the bibliography in Ref. 14).

It was found that the studies of the large-scale oceanic turbulence are closely connected with the investigation of the turbulent boundary layer, of which the two domains are most interesting for the researchers. The first of these domains is the wall-adjacent one within which occurs the periodical destroyment of intensely whirled flow serving the source of momentum for the main part of a turbulent flow; the second domain is that of intermittency, which determines the outer boundary of a turbulent boundary layer.

The considerable vorticity, originated at the wall, is developing into clearly pronounced longitudinal vortices, which are afterward carried out of the wall-adjacent sublayer, taking with them their momentum into the external domain. However, the vortices containing energy lose their kinetic energy not only due to the direct action of viscosity; that kind of energy is transmitted also by smaller vortices which, in their turn, convey their energy to even smaller vortices of the next generation, and so on, up to the moment when the sizes of the vortices generated become so extremely small that almost immediately they vanish, being affected by viscosity.

Thus the vorticity variations, which might arise within the flows, should be connected with the viscous properties of the fluid. The cascade process of the vortices' subdivision in three-dimensional turbulent flow determines the energy transmission from the main motion to the small-scale vortical configurations, while in two-dimensional case the flow of energy might assume the opposite value (the effect of negative viscosity).[11] The total energy of large-scale vortices is left in this process approximately constant.

The *intermittency* of flow at the external boundary of the turbulent boundary layer is connected with the large-scale structures in the main domain of flow. By the cascade vortical process, the existence of which in turbulent flows is assumed, such a structure might possess the coherent formations. All these processes are observed by the transition of laminar flow to turbulent one, and the study of that last form may be conducted with somewhat lower Reynolds numbers. Such a regime, which seems to be one of the most complicated in fluid mechanics, happens to be realized everywhere in natural conditions and, in particular, in the large-scale circulation of the ocean and of the atmosphere.

In the paper by Belotserkovskii and Pastushkov,[82] on the contents of which the present section is based, is studied the hydrodynamic stability of the horizontal vortical flows of the viscous, incompressible fluid (taking into account the Earth's rotation and its spherical form), subject to the force of a wind; this study is conducted just for transitional regime and serves as a continuation of paper.[83] Here, using the Arakava's method, the equations of vortex were numerically integrated, and was investigated the question concerning the character of a developed flow by the moderate Reynolds numbers.

The model presented here has no claims for the description of turbulent (chaotic) regime, which is rather complicated both in formulation and in solution, and is very difficult (most probably — impossible) for investigation on the basis of equations of the classical hydrodynamics. The present study was preceded by the numerical investigations of the wide class of problems concerning the loss of stability of the main (laminar) flow and the transition to regular one, stable in respect to the small perturbations in self-sustaining regime. This means that the aim of present study consists in the "approach to chaos" using sufficiently realistic model of flow.

As soon as the numerical experiment in its present formulation was conducted by two different methods, the good agreement between the results of calculations, including the irregular case, permits one to hope for the objectivity of the data obtained. Quite probably, these investigations will help in the understanding of those processes, which go on in the ocean and are connected with the arousal of large-scale turbulence of the atmosphere and of the ocean. Following Ref. 82, the main results are listed below.

When studying the pattern of surface circulation of the World Ocean one could observe that the most striking peculiarity of each particular ocean is a clearly pronounced water circulation within the western domain, as, for example, in the northern part of Atlantic Ocean (moderate latitudes). This circulation, directed mostly clockwise, has comparatively small velocities (from 1 to 10 cm/s) while the intensive and narrow Gulf Stream is pressed to the western shore of America, from Florida to cape Hutteras, from where it turns to the open ocean, has the mean velocity about 100 cm/s (the maximal value might be several times more than that). The width of the Stream's area is, approximately, from 50 to 100 km. The Gulf Stream is not unique. Quite pronounced *western intensification* is inherent also to the Kurosio Stream in Pacific Ocean, Brazilian Stream in South Atlantic and the stream of Agullias near the eastern coast of Africa.

We have every reason to believe that, with exception of a peculiar phenomenon of western intensification, *the pattern of the large-scale oceanic circulation as a whole reflects the pattern of winds*, and thus only the actual tangential stress of wind is, in principle, able to create the total (integrated in vertical direction) horizontal transfer. The important peculiarities of the motions under consideration are the closure of streamlines and the constancy of sign of the external source of vorticity. With the help of simplest estimates Sverdrup[84] has obtained a relation, which is sufficiently well substantiated in the open ocean: $\beta \cdot V = -(1/H)$ rot $\tau$, where $H$ is the ocean's depth and $\tau$ is the vector of the tangential stress of wind.

The Sverdrup's solution has some imperfections inherent to linear methods, but this, actually, does not diminish the value of the analytical studies conducted. Thus, the comparison of those results, which were obtained with the help of various numerical models, with the Sverdrup's data, show that these estimates correctly reflect the mean flow's character in the main area.

Nevertheless, present in the ocean is the wide spectrum of the scales of possible motions, which makes the simultaneous description of all these motions with the help of hydrodynamic equations practically impossible. It might be worthwhile to subdivide these motions into three types: *regular large-scale motions*, which can be described individually, *extremely irregular small-scale motions*, which should be described statistically, and various kinds of *wavy motions* (wavy turbulence), some of which as, for example, planetary waves, are sufficiently regular and allow for an individual description.

The model under investigation pretends to a description of the large-scale turbulence with characteristic linear dimensions considerably surpassing the ocean's depth, so that one can neglect all vertical motions (pulsations). The used here approximation of "the firm roof", as well as disregarding of both stratification of fluid and of its compressibility, leads to the filtration of the surface, internal and acoustic waves. The use of an approximation of $\beta$-plane,[85] needed to take into account the spherical form of the Earth, confines the domain of simulation by the scales of 1000–3000 km and by the moderate latitudes of the ocean. The wind stresses are assumed to be zonal and independent of time.

For the simplest case of the horizontally uniform turbulent flow, having a planar-parallel field of a mean velocity, the coefficient of horizontal turbulent viscosity is introduced. By the numerical realization of the model, for a solution of the doubly averaged Reynolds equations, the scale of averaging (the grid's size) is chosen in such a way that near the western boundary it would not exceed 10–20 km, while near the eastern one — 200–300 km.

Within the frame of a model considered the characteristic scales of length are chosen:

$$\delta_l = (V/\beta)^{1/2}; \quad \delta_m = (A_L/\beta)^{1/3}; \quad \delta_S = r/\beta,$$

where $V = \tau/\beta HL$ (chosen from the Sverdrup's relation); $L$ is the horizontal scale of motion; $A_L$ is the coefficient of horizontal turbulent friction; $r$ is the coefficient of bottom-side turbulent friction, while $L \gg \delta_l, \delta_m, \delta_S$. Depending on the relative value of $\delta_l, \delta_m, \delta_S$, essential part in the equations of motion will be played either by the inertial terms, or by horizontal friction, or by bottom-side friction, and that, eventually, determines the structure of flow. As it will be shown below, the problem's solution is determined by three dimensionless parameters:

$$\varepsilon = \delta_l/L, \quad Re = \delta_l^3/\delta_m L^2, \quad \delta = \delta_S/L.$$

Considered is the evolution of flow in the barotropic ocean having constant depth, under the action of suddenly arising, and furtherly invariable in time, large-scale field of wind. The mathematical formulation of such a problem is sufficiently well known (see, for example, Ref. 86). As it turned out historically, the numerical solutions of the problems of that type were chiefly realized with the help of some or another method of integration of the equations for vorticity. As it was already observed earlier, the imperfections of such an approach are connected with difficulties of the formulation of boundary conditions for a vortex at the solid surface (which are not present in the physical formulation of the problem), as well as with the impossibility of extension to the case of three-dimensional flow. In the present section numerically integrated is the set of Navier–Stokes equations written in *natural* variables, which eliminates the imperfections indicated.

If one chooses the characteristic size of the basin $L$ as the spatial scale, and the quantity $V = \tau_0/\beta HL$ (where $\beta = \partial f/\partial y$ is the latitude variation of the Coriolis' parameter $f = f_0 + \beta y, H$ is characteristic ocean's depth, $\tau_0$ — characteristic value of the tangential stress of wind) is taken as the velocity scale, then the main equations describing the two-dimensional oceanic circulation acquire the following dimensionless form:

$$\partial \mathbf{V}/\partial t + \varepsilon^2(\mathbf{V}\nabla)\mathbf{V} + \mathbf{k}f \times \mathbf{V} = -\nabla P + \gamma^3 \cdot \Delta\mathbf{V} - \delta\mathbf{V} + \tau, \quad \mathrm{div}\mathbf{V} = 0. \tag{1.54}$$

Here the $x$-axis is directed to the east, the $y$-axis to the north, $\mathbf{V}(u, v)$ is vector of the averaged over depth flow's velocity, $P$ is the averaged over depth pressure, $(\tau_x, \tau_y)$ are the components of the wind's tangential stress; $f = (a/L)\mathrm{tg}\,\Theta_0 + Y$ is the Coriolis' parameter in the $\beta$ is the plane approximation, where $a$ is the Earth's radius, $\Theta_0$ is the latitude, $\mathbf{k}$ is the unit vector directed normally to the basin's plane.

The problem's parameters are

$$\varepsilon = \sqrt{v/\beta}/L; \quad \gamma = \sqrt{A_L/\beta}/L; \quad \delta = r/\beta L; \quad V = \tau_0/\beta HL.$$

The parameter $\varepsilon$ characterizes the effect of velocity's convection; the parameter $\gamma$ reflects the presence of a horizontal *turbulent friction* with coefficient $A_L$; the parameter $\delta$ is responsible for the *bottom-side friction* with coefficient $r$. Further on the characteristic value of velocity $\mathbf{V}$ in the open ocean is considered as the controlling parameter of the problem (instead of a controlling parameter $\tau_0/H$). The dimensionless value $x$ varies within the range from 0 to 1; $\tau_x = -1/\pi(\cos \pi y), \tau_y = 0$.

It is to be observed that for the real ocean the parameters $\varepsilon, \gamma$, and $\delta$ are sufficiently small. Since these parameters characterize the process, described by the terms with higher derivatives in Eqs. (1.50), then at the basin's shores should be formed *boundary layers*, outside of which both convection and diffusion are negligibly small. In the case of a stationary problem, by $\gamma = \delta = 0$, the inertial boundary layer of the thickness $\varepsilon$ is formed; by $\varepsilon = \delta = 0$, there is formed the *viscous* boundary layer of the thickness $\gamma$. In the general case, when all the parameters are of one and the same order, should be formed the *inertial-viscous boundary layer.* When the numerical analysis of the equations is conducted, the smallness of the parameters $\varepsilon, \gamma$, and $\delta$ causes certain difficulties due to a necessity of the numerical resolution of the corresponding boundary layers.

It would be convenient to introduce the Reynolds number for boundary layer as a ratio of characteristic values of convection and diffusion: $Re = VL\varepsilon/A_L = V(V/\beta)^{1/2}/A_L$. From here it follows that the parameter $\gamma$, which is used in many papers, might be expressed in terms of $\varepsilon$ and $Re$:

$$\gamma = \varepsilon\sqrt[3]{\sqrt{Re}}.$$

Within the ocean, the values of parameters $L, A$, and $V$ are varied over the wide range. Therefore, the study should be conducted for the parameters $Re, \varepsilon$ and $\delta$ being within certain limits, corresponding to the realistic conditions (in the case, typical for the ocean, one has $\beta = 2\times 10^{13}\,\mathrm{cm}^{-1}\mathrm{s}^{-1}$).

The boundary conditions of the problem are formed in the following way (see Fig. 1.50).

At the *solid boundaries* (shores) the conditions of sticking and nonflow-through are fulfilled:

$$u = 0, \quad v = 0 \quad \text{for } x = 0, \ x = 1, \tag{1.55}$$

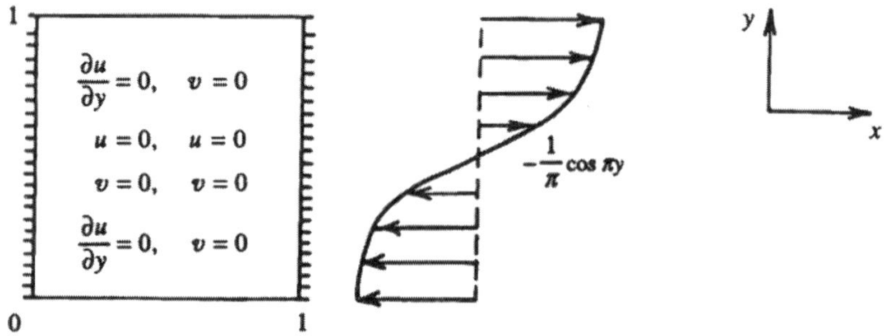


**Fig. 1.50.** Problem's formulation.

while at the liquid boundaries, which are chosen in accordance to the lines of zero rotor for the tangential stress of the wind — the conditions of non-flow-through and of slip:

$$\partial \mathbf{u}/\partial y = 0, \quad v = 0 \quad \text{for } y = 0,\ y = 1. \tag{1.56}$$

At the initial moment of time it is assumed that the fluid is at rest:

$$u = 0, \quad v = 0, \quad p = 0 \quad \text{for } t = 0. \tag{1.57}$$

For the simplicity of the present analysis the bottom-side friction will not be taken into account ($r = 0$). In fact, the introduction of bottom-side friction leads only to the effective increase of the system's viscosity.

In the present formulation simulated are flows, created by the zonal wind in moderate latitudes of the ocean (between the maximum of trade winds and that of western winds), as well as in the open ocean with solid boundaries at the western and eastern shores. The formulation described is not generally accepted. This concerns the boundary conditions at the western and eastern shores (solid boundaries), where the boundary layers might be formed. The substitution of physically satisfactory condition of sticking for that of slip might be justified only for the case when the character of flow outside of the viscous coastal boundary layers would be subject to the insignificant variations within domain. However, in the case of adoption of the slip condition at these boundaries the viscous sublayers are not formed at all, in fact, of course, simplifies the studied class of flows within the whole domain, but seems to be not rendering the correct structure of a flow.

To solve the initial-boundary problem in accordance to Eqs. (1.54)–(1.57), one is using the three-stage method of splitting with respect to the physical factors,[87] which found the successful application to the solution of the problems of gas-dynamics and hydrodynamics.[13,14]

*At the first stage*, it is assumed that the momentum transfer is realized through the convection, diffusion and external forces (those of Coriolis and wind stresses). The intermediate field of $\tilde{\mathbf{V}}$, obtained by that way, does not, generally, satisfy the incompressibility condition (div$\tilde{\mathbf{V}} \neq 0$):

$$\tilde{\mathbf{V}} = \mathbf{V}^n - \Delta t \left\{ \varepsilon^2 (\mathbf{V}^n \nabla)\mathbf{V}^n + \mathbf{k}f \cdot \mathbf{V}^n - \frac{\varepsilon^3}{Re} \Delta \mathbf{V}^n - \tau \right\}_h .$$

*At the second stage*, by way of solution of the Poisson equation, the field of pressure is determined:

$$\{\mathrm{div} \nabla P^{n+1}\}_h = \frac{1}{\Delta t} \{\mathrm{div} \tilde{\mathbf{V}}^n\}_h .$$

*At the third stage* (the transfer is realized only through the pressure gradients), one is able to determine the value of velocity at the next temporal layer:

$$\mathbf{V}^{n+1} = \tilde{\mathbf{V}} - \Delta t \{\nabla P^{n+1}\}_h .$$

Here $\Delta t$ is the step in time, $n$ is the number of the temporal layer, $\{\dots\}_h$ is the difference approximation of the differential operator. The approximation is realized on a "staggered" grid. Convective terms, written in the conservative form, are approximated by the directed, four-point difference. The approximation of all other terms of the first-stage equation and of the second- and third-stage equations, as well as of the boundary conditions, does not create any problems.

The *uniform* divergent-conservative scheme, used here, is in its ideology close to the finite-difference schemes of the FLUX method,[13,49] has the second-order of accuracy with respect to $h = \max\{h_x, h_y\}$, and according to its properties of monotonicity and transportivity the scheme of the first-order of accuracy with upstream differences is approaching. This scheme is dissipatively stable, which property permits to carry out calculations with $\tau \sim h$ even in the absence of a physical viscosity ($Re \to \infty$). Due to the specifical feature of the present problems, connected with a latitude-wise variation of the Coriolis parameter ($\beta$-effect), at the western boundary of the computational domain is formed an intensive boundary layer of the thickness $\gamma = \varepsilon \sqrt[3]{Re}$, for resolution of which one would need a corresponding grid's condensation (for the reason of computer's limited resources). The grid's nonuniformity needed for that is realized through the conversion to a new orthogonal coordinate system $(x', y')$, where

$$x' = (1/s) \ln\{x(l^s - 1) + 1\}, \quad y' = y.$$

The parameter $s$ is responsible for the degree of grid's condensation, and its value is chosen in dependence of the problem's parameters.

The Neumann's problem for the Poisson equation was solved by Fourier's method, using the discrete cos-transformation in $y$ and double-sweep in $x$.

The essential nonlinearity, as well as the presence of several small parameters at the higher derivatives leads to the serious difficulties by the numerical integration of original equations. The presence at the domain's western boundary of the *viscous boundary layer* having a thickness $\gamma = \varepsilon\sqrt[3]{Re}$, and the *inertial boundary layer* of a thickness $\varepsilon$, leads to the necessity of their accurate numerical resolution. However, within the main domain might occur a nonlinear interaction between the Rossby's long, barotropic waves (propagating in western direction), and the shorter ones, reflected from the solid boundary and propagating in eastern direction. The inclusion of this interaction calls for a diminution of the grid's step, and hence, for a considerable expenditure of computer's resources. Thus in the system investigated are present the kinds of motion corresponding to the *vortical* (boundary layer's instability) and to the *wave* (nonlinear interaction of the barotropic Rossby's waves) varieties of turbulence. It seems that there occurs the nonlinear interaction of waves and vortices, which eventually determines the *cascade process* of the energy transfer along the ever-widening range of the wave numbers: the spectrum "diffuses" within the **k**-space, the new wave and vortical structures are formed, and so on.

The model considered here (with $\delta = 0$) contains, as it was noted earlier, two characteristic dimensionless parameters: numbers $\varepsilon$ and $Re$, the actual values of which make important either inertial terms, or those with viscosity. The disbalance between these varieties of terms, as well as the effects of wind stresses and of rotation, leads, for a general case, to the formation of the inertial-viscous boundary layer at the western boundary of the basin; it seems that this layer affects the flow's formation within the domain considered, as a whole. The diminution of $\varepsilon$ (along with the increase of $\beta$) leads to the situation when the inertial terms are prevailing over the viscous ones, and at the western boundary is formed the inertial boundary layer, which, in its turn, leads to the sooner coming of the flow's instability with the same values of Reynolds numbers. With small $Re$ and sufficiently large $\varepsilon$ the viscous terms are prevailing over the inertial ones, and the viscous boundary layer is formed, which stabilizes the flow.

According to the experimental data, the spectrum of values $Re$ and $\varepsilon$ numbers is sufficiently wide. Thus, for example, as the Reynolds number

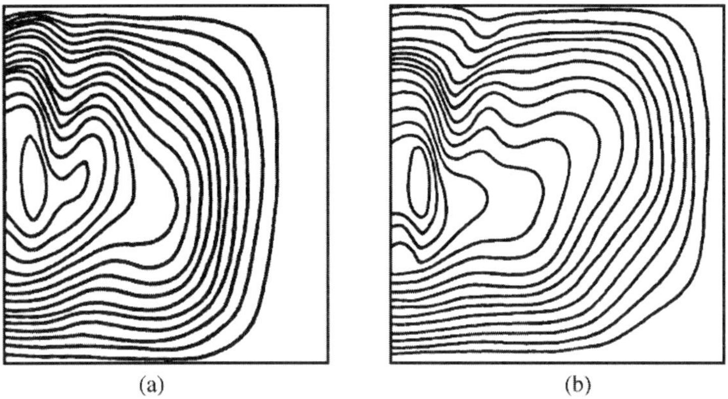

**Fig. 1.51.** Steady flow in barotropic ocean for $\varepsilon = 2 \times 10^{-2}(\psi = \text{const})$. (a) $Re = 0$. (b) $Re = 0.5$.

increases, for rather small value of $\varepsilon = 0.02$, the calculations give, successively, the following regimes:

(i) *steady* ($Re \approx 0.25$–$0.50$, Fig. 1.51);
(ii) *periodic* ($Re \approx 1$–$2$, Fig. 1.52);
(iii) *stable irregular* ($Re \approx 2.5$, Fig. 1.53);
(vi) *unstable irregular* — "the chaos" ($Re \approx 3$, Fig. 1.54).

Moreover, in the last case the sharp increase of system's kinetic energy is observed (Fig. 1.55).

With the increase of Reynolds number (or with the decrease of $\varepsilon$) the flow's asymmetry is intensified, and in the north-eastern corner of a basin the corner vortex is formed, and the "irregularity" becomes stronger. In the stationary case (see Fig. 1.51), the formation of the vortex is just beginning, in periodical regime (see Fig. 1.52), the vortex goes out of the boundary layer in northern direction (upwards) and merges with the corner vortex; here is already initiated the loss of a stability in boundary layer, but the northward motion is strictly periodical (which is evident from the pattern of the kinetic energy's behavior shown in Fig. 1.55). In the *stable irregular* regime, which arises by the intensification of $\beta$-effect (see Fig. 1.53), the vortices are split up and move chaotically within the basin; the periodicity is not observed, and the system's energy oscillates randomly about some fixed value. And, finally, in the *unstable irregular regime* (see Fig. 1.54), arising by the increase of $Re$, the sizes of vortices are essentially decreasing, and they move by random way. The kinetic energy is sharply increased

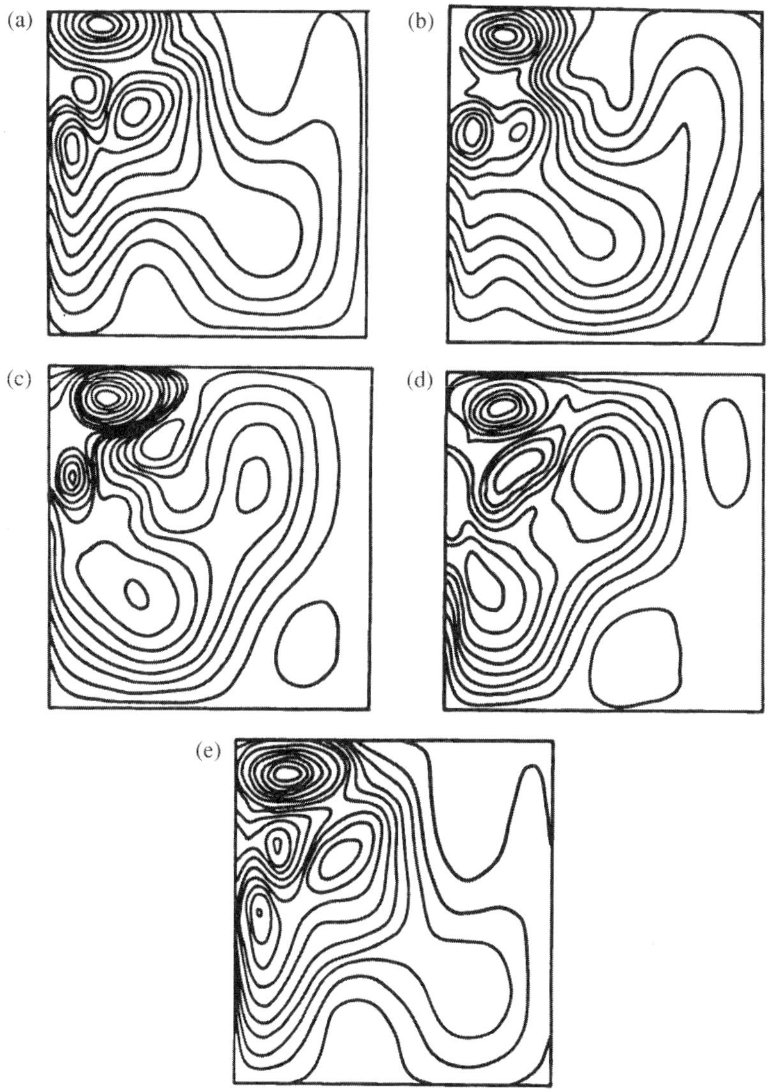


**Fig. 1.52.** Periodic flow for $\varepsilon = 2 \times 10^{-2}$ and $Re = 1$ (period $T \approx 90, \Delta t = 20$, $\psi = \text{const.}$).

and calculations are stopped (see Fig. 1.55). For the values of $Re = 1$ and $Re = 10 (\varepsilon = 0.01)$, the spectral analysis was carried out with respect to the vertical velocity component in the central part of a computational domain, in the lower and upper parts of the boundary layer (Fig. 1.56). As it is

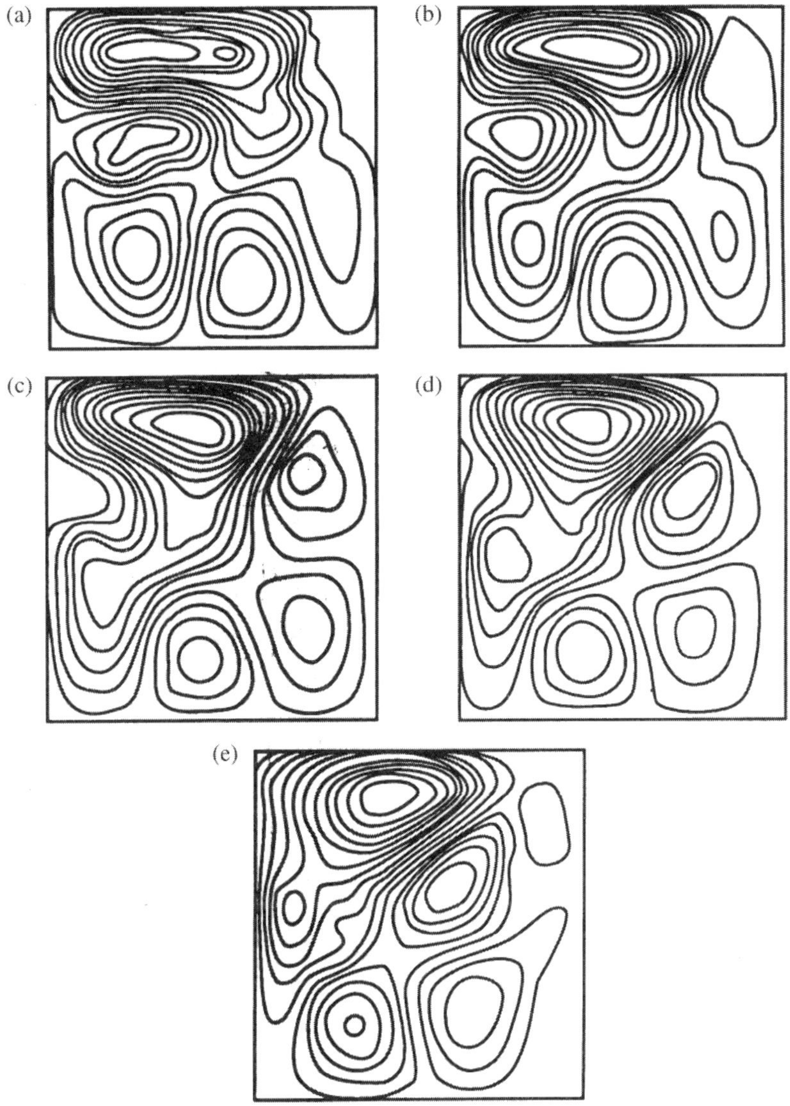

**Fig. 1.53.** Stable irregular flow for $\varepsilon = 2 \times 10^{-2}$ and $Re = 2.5$ ($\Delta t = 20, \psi = \text{const.}$).

seen from Fig. 1.56(a), present within the central domain are, mainly, wave oscillations, corresponding to $T_{1,1} = 55.83$ and $T_{1,3} = 124.84$ for an inviscid linear problem; moreover, characteristic for $Re = 1$ is a high-frequency mode ($T_{1,1}$), whereas for $Re = 10$, both high-frequency and low-frequency

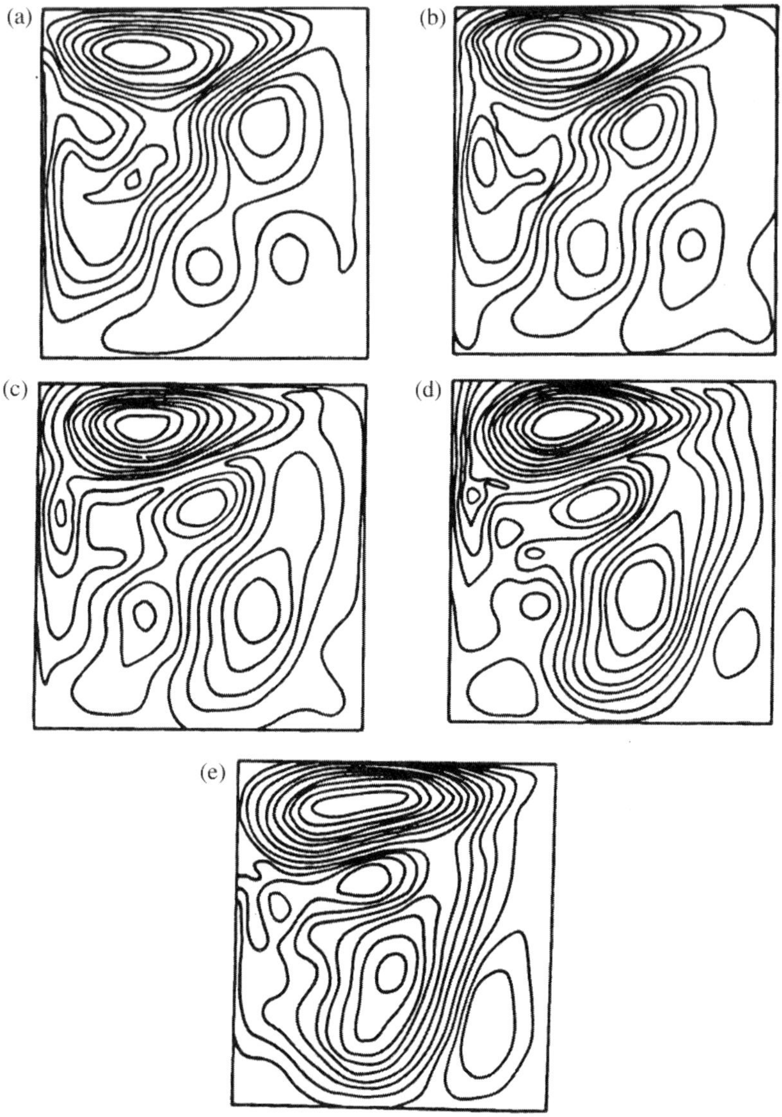

**Fig. 1.54.** Unstable irregular flow for $\varepsilon = 2 \times 10^{-2}$ and $Re = 3$ ($\Delta t = 20$, $\psi = \text{const.}$).

modes are characteristic. Within the lower part of a boundary layer are observed the same oscillations, as are within the main domain (Fig. 1.56(b)). Within the boundary layer's upper part, in addition to the wave modes which are chiefly present in the main domain, are observed the vortical

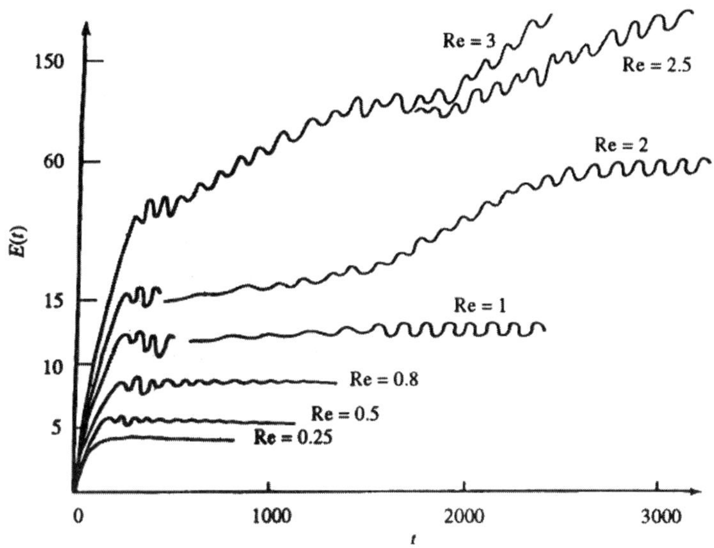


**Fig. 1.55.** Kinetic energy of a system versus time for different values of *Re* and $\varepsilon = 2 \times 10^{-2}$.

modes, too (Fig. 1.56(c)). After filtration of the wave modes one is able to obtain the temporal spectrum of the purely vortical motion. It is seen from Fig. 1.56(d) that for $Re = 1$ within the flow are present the low-frequency oscillations corresponding to the period $T_1 = 130$, while for the same $Re$, in addition to the clearly pronounced $T_1$, are present also high-frequency periods $T_2 \approx 90$ and $T_3 \approx 70$, which means that the spectrum gets smeared toward the high frequencies. From the exposed above data one could conjecture that in the present model, along with the increase of $Re$ number, for the wave mode the lower-frequency harmonics are excited, while for the vortical mode — both low-frequency and higher-frequency ones.

To obtain the more complete notion on the system's behavior, the calculations were carried out for $Re = 1$ and $\varepsilon = 0.03$ and 0.007. As the value of $\varepsilon$ is growing, the flow becomes stationary (Fig. 1.57), just like a flow with small Reynolds numbers. The diminution of $\varepsilon$ (i.e. the intensification of the system's "inertiality") leads to a notable *increase of the intensity of nonstationary motion* both within the boundary layer and outside of it. Along with the reduction of the boundary layer's width, the sizes of vortices are reduced, too, and their motion becomes to be nonstationary and irregular ($Re = 1, \varepsilon = 0.007$, Fig. 1.58). The system's kinetic energy is sharply increased, and the motion acquires a *chaotic* character.

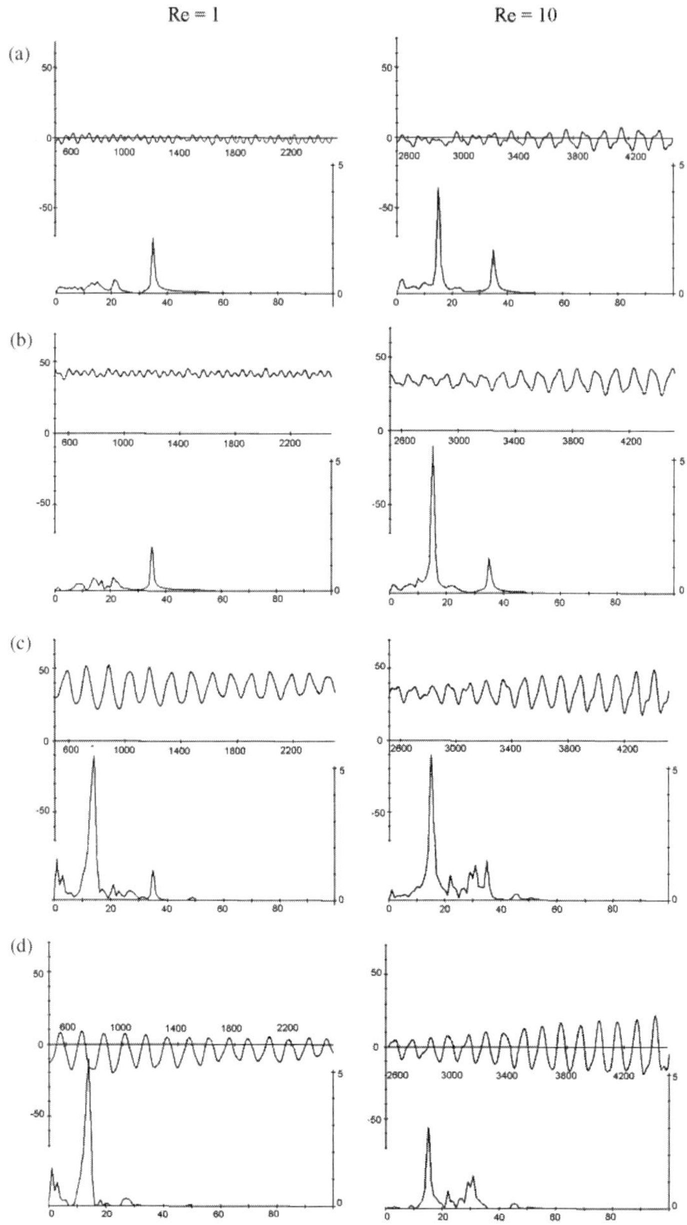


**Fig. 1.56.** Spectral analysis of vertical velocity component for $Re = 1$ (to the left) and $Re = 10$ (to the right) at $\varepsilon = 0.01$: (a) central part of computational domain; (b) lower part of boundary layer; (c) upper part of boundary layer; (d) upper part after filtration of the wave modes.

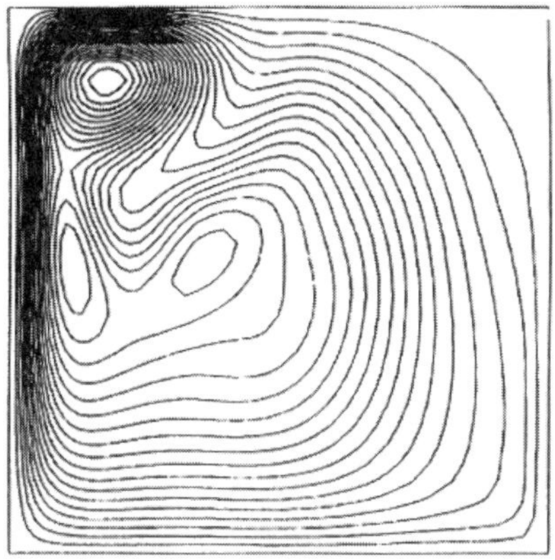

**Fig. 1.57.** Steady regime ($\psi = \text{const.}$) for $Re = 1$, $\varepsilon = 0.03$.

**Fig. 1.58.** Unsteady irregular regime ($\psi = \text{const.}$) for $Re = 1, \varepsilon = 0.007$.

Thus, with the help of a model constructed on the basis of full nonlinear dynamic Navier–Stokes equations, taking into account the turbulent dissipation, Earth's rotation ($\beta$-effect) and in the presence of wind stresses, it is possible to succeed in transition to "chaos". The process of the instability's development qualitatively reminds the scenario proposed by Landau–Hopf. In all the likelihood, it would be of a great interest to carry out the detailed study of a model with $\varepsilon \approx 0.001$ and $Re \approx 10$, which combination is, probably, the most closely related to the real conditions.

It should be noted that the results described were obtained with rather coarse computational grids, with the help of a personal computer (in the real range of time), which fact just accentuates the model's merits, even in spite of its simplicity. Evidently, the going over to the more complicated multilayer model, taking into account the baroclinic effects, the treatment of real coast line, of the bottom's relief, etc., will demand the essential improvement of the operating characteristics of the computer employed.

### *1.11.4. Numerical simulation of the internal waves in a stratified fluid*

The studies of air streams in the mountainous regions of the Earth testify that behind the mountain ridge the nonstationary wave processes are formed (the regularly interchanging cloudy billows), which play the essential role in the problems of *meteorology.* Initially, the problem of the air flow over the Earth's roughness was considered as being exactly similar to that of the incompressible flow of a fluid above the uneven bottom. The approaches based on the method of small perturbations are introduced, along with an assumption of an incompressibility, and with an additional one — that of small sizes of roughness. In the papers by Kochin, published in 1937–1938, the atmosphere was considered already as the two-layer incompressible fluid (possessing different values of density), with the roughness sizes assumed to be finite. These studies permitted to obtain the qualitative notion concerning the flow processes in meteorological scales, but the role of compressibility was still not completely clear, and the problem should be stated in more general form. In the series of works (1938–1950) by Dorodnicyn,[dd] prepared in the "precomputer" time, on the basis of a linear theory was carried out the general investigation on the influence of the Earth surface's relief on the air flows, taking into account the compressibility. It was found that the consideration of compressibility leads to the appearance of new qualitative effects. Thus, in doing so, the streamlines not at all heights do follow the surface's relief (at some levels occurs the "reversal" of air stream); in distinction of the case of incompressible fluid, within the compressible flow behind the mountain ridge spring up several (or even an infinite multitude) wave systems.

The complete problem might be considered in its nonlinear formulation only within the frame of a *numerical experiment*, with the help of modern

[dd]See the bibliography in Ref. 89.

computational technique. On such a basis Belotserkovskii *et al.*[88] have conducted the study of motion of the stratified (with respect to density) fluid in ocean, by the flow over the bottom's roughness, which is closely connected with the widely investigated presently process of the simulation of internal waves. Let us dwell on this study in more details.

The theory of internal waves owes its development mostly to meteorology, due to the directly observed and frequently rather instructive effects, induced by gravitational waves in the atmosphere. The linear theory (for the internal waves of small amplitude) for the problem of the flow of stratified fluid over the obstacle, within the channel of a finite width, was developed by Dorodnicyn. The generalization of a solution by way of tending of the channel's depth to infinity transforms the solution into that for an infinite medium. In this last case the linear solution possesses the similarity of phase patterns and amplitudes with respect to the Froude number $Fr$.

Simulation of the field of internal waves within the flow of a stratified fluid is of a considerable interest. The *internal waves*, which are in an intermediate position with respect to frequency staying between the geostrophic turbulence and the small-scale, three-dimensional one, do exist just due to the stratification and play an important role in the formation of the ocean dynamics as a whole. The internal waves' frequency varies beginning with inertial one $f = 10^{-4}\,\mathrm{s}^{-1}$ ($f$ — parameter of Coriolis) and up to the maximal for seasonal thermowedge (in summer time) Väisälä–Brunt frequency $N = 10^{-2}\,\mathrm{s}^{-1}$. The lengths of internal waves vary within the range from tens of meters up to one or two hundreds of kilometers. The mean amplitudes of internal waves are equal to 10–20 m, but there are examples when the amplitude is larger than 100 m.

The observed overall field of internal waves in the ocean is characterized by the complicated spatial-temporal structure, and therefore the setting of activating factors and mechanisms by the generation of internal waves constitutes a rather complicated problem, which demands for specially planned and very expensive experiments and laborious analysis of the great amount of information.

At the present moment, the main aim of the internal waves' investigation is the data systematization with subsequent substantiation of the spectra of internal waves observed. The eventual goal consists in the *forecast of a complicated spatial-temporal field of internal waves*, but is far from completion,[90] and therefore it would be of certain interest to study the present problem with the help of models having some or another degree of approximation to the real conditions. The role of one of such models is played by the *generation of topographical internal waves* as a result of the stationary flow past the bottom's roughness. The similar problem arose by the description of a flow of air over the mountains, has an ancient history and, as it was already observed, attires an attention up to the present day.

The role of such a process within the ocean is only beginning to be cleared out. Thus, as it follows from Ref. 91, the flow of energy passing from the mean stream to the internal waves might achieve considerable values by the flow past abysmal hills (the parameters of flow are: $N = 10^{-4}–10^{-3}\,\mathrm{s}^{-1}$, mean velocity of the oncoming flow $u_0 = 5–10\,\mathrm{cm/s}$, height of the hills $h = 200–500\,\mathrm{m}$, depth $H = 4–6\,\mathrm{km}$). As it was already mentioned, the life and laboratory investigations play the great role in the understanding of the nature of internal waves. The computational experiment, consisting in a solution of the *complete* set of the nonlinear equations in partial derivatives with the corresponding boundary and initial conditions, not only gives an essential supplement to the results of experimental research, but in some cases (when the proper conditions might not be realized experimentally) becomes the only tool for the investigation of these complicated phenomena.

In the paper[88] the numerical investigation was carried out concerning the complete nonlinear wave pattern arising in the $(x,\ z)$-plane by the transversal flow past a semicircular obstacle of radius $R_0$, with its axis parallel to $y$-axis, the velocity, uniform at infinity and parallel to $x$-axis flow being equal to $u_0$, with the fluid itself being linearly stratified with respect to density (Fig. 1.59).

Under the assumption that the medium is just the incompressible, viscous fluid, stratified with respect to density, the original set of equations

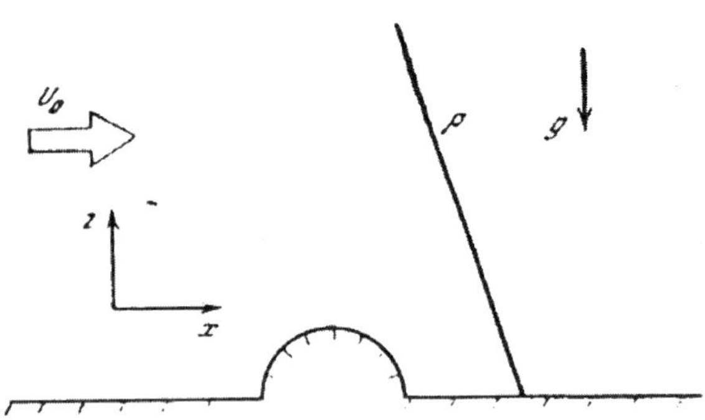


**Fig. 1.59.** Flow with linear stratification with respect to density (problem formulation).

might be written in the following form:

$$\frac{d\mathbf{V}}{dt} = -\frac{1}{\rho}\nabla P' + \frac{2}{Re}\frac{1}{\rho}\Delta\mathbf{V} + \frac{1}{Fr}\frac{\rho'}{\rho}\frac{\mathbf{g}}{|\mathbf{g}|}, \quad d\rho/dt = 0, \quad \nabla\mathbf{V} = 0. \qquad (1.58)$$

Here $\mathbf{V}$ is the vector of velocity, $P'$ is the deviation of pressure from hydrodynamic value; $\rho$, the density; $\rho'$, the density's deviation from equilibrium state; $Re = 2\rho_0 u_0 R_0/\mu$, the Reynolds number ($\rho_0$, the density at the level $z = 0, u_0$, the characteristic velocity of the oncoming flow); $Fr = u_0^2/R_0 g$, the Froude number, expressing the ratio of inertial forces to buoyancy ones; $g$, the acceleration of the gravity.

Boundary conditions: at the solid surface is set the condition of sticking and non-flow-through, at the infinity (the contour located sufficiently far away from body) — the condition of unperturbed flow: $V_x = 1, V_z = 0, \rho = 1 - Fr_f \cdot z$, where $Fr_f \cdot z = N^2 R_0/g$ is the Froude number for frequency, $N = \sqrt{(-g/\rho_0)(d\rho/dz)_0}$, the Väisälä–Brunt frequency. To play the role of initial conditions, the uniform forward flow of the fluid, linearly stratified in density and satisfying the boundary conditions at the solid surface, is prescribed.

For a solution of the problem formulated was used one of the methods of splitting with respect to physical factors,[87] extended to the case of flows of a stratified fluid.[92] By the problem's solution the patterns of flow were restored (i.e. lines $\psi = \text{const.}$, where $\psi$ is stream function), including the lines of equal density deviations from its linear distribution within unperturbed flow ($(\rho'/(d\rho/dz)_0)R_0 = \text{const.}$), as well as phase pictures (i.e. lines $\rho'_x(x, z) = 0$), corresponding to the crests and hollows of internal waves.

The nonuniformity of density's distribution with respect to depth (stratification) leads to the qualitative modification of a problem's solution as compared with the uniform case, which is revealed first of all through the presence of internal waves, generated by the obstacle. The control parameters of the problem are Froude numbers ($Fr, Fr_f$) and Reynolds number.

As it is known, the important influence on the flow's regime is exerted by *the density Froude number* ($Fr_{\mathrm{d}} = (Fr/Fr_f)^{1/2} = u_0/NR_0$). Proceeding from the problem's solution, one could set off two regimes of flow: by $Fr_{\mathrm{d}} \geq 1$ (*weak nonlinearity*) the generation of *stationary* leeward waves occurs, while $\lambda = 2\pi R_0 Fr_{\mathrm{d}}$ ($\lambda$ is the characteristic length of internal wave); by $Fr_{\mathrm{d}} < 1$ (*strong nonlinearity*) the flow acquires the *nonstationary* form, and the additional effects appear: the blocking of a fluid ahead of the obstacle, the collapse of internal waves, the nonmonotonous distribution of the fluid's

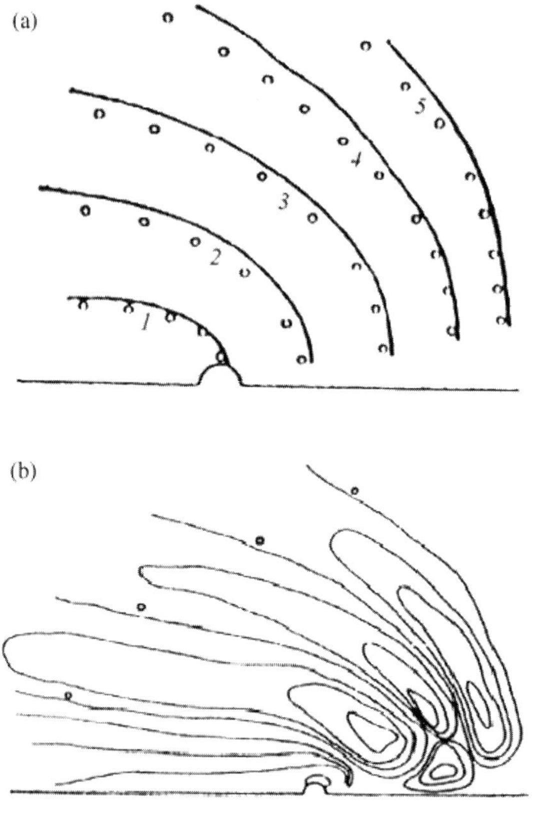

**Fig. 1.60.** Steady stratified flow for $Re = 41.1$, $Fr_{\text{d}} = 1.59$: (a) the phase pictures of the density perturbation's field (solid lines — calculation, points — experimental measurements by Onufriev), (b) isolines of the density's.

velocity within the oncoming stream. Both regimes of the motion will be studied separately.

As concerns the stationary regime, the comparison with experimental measurements was carried out for the following versions:

$$Re = 130, \quad Fr_f = 1.2 \cdot 10^{-4}, \quad Fr_{\text{d}} = 5.38;$$

$$Re = 41.4, \quad Fr_f = 1.4 \cdot 10^{-4}, \quad Fr_{\text{d}} = 1.59.$$

Presented in Fig. 1.60(a) are the phase pictures of the density perturbation's field ($Re = 41.4$), where the solid lines correspond to the numerical calculation, while the points correspond to the experimental measurements.

The phase surfaces 1, 3, 5 correspond to the maxima of the density perturbation ($\rho' > 0$), whereas phase surfaces 2, 4 correspond to the minima ($\rho' < 0$). The distance between the lines of extrema is equal to the half of wavelength. Shown in Fig. 1.60(b) are isolines of the density's deviation $((\rho'/(d\rho/dz)_0)R_0 = \text{const.})$ for the same version. One observes the good agreement with the experimental picture for the density's deviation, which was obtained by the method of holographic interferometry.[92]

As it follows from the linear theory, the non-dimensional length of internal wave, generated by the planar body during its motion within the nonuniform fluid, is defined as $\lambda = 2\pi Fr_{\text{d}}$, $(\lambda = \lambda/R_0)$. For the version with $Fr_{\text{d}} = 5.38$, the values of $\lambda$ obtained through the linear theory both numerically and experimentally, are equal, correspondingly, to 34.8; 32.8; 33.9, and to 9.98; 9.6; 9.9 for the version of $Fr_{\text{d}} = 1.59$. Thus, the results obtained are here in a good agreement with the linear theory, too.

Presented in Fig. 1.61(a) is the graph of dependence of calculated wavelength $\bar{\lambda}$ on the number $Fr_{\text{d}}$. This dependence is a linear one, which fact corresponds both to experimental data and to linear theory. If one introduces a new dimensionless coordinate $\mathbf{r} = \mathbf{r}/Fr_{\text{d}}$ (where $\mathbf{r}$ is a dimensionless distance from the obstacle), then it is possible to construct the *universal phase picture* (the similarity of phase pictures is observed). One of such phase pictures is presented in Fig. 1.61(b) (the markers correspond to the numbers $Fr_{\text{d}}$ 1; 1.59; 3; 5.38).

The agreement between the results of calculations and the experimental data confirms the reliability of a numerical approach, and that permits to hope for a successful investigation of more complicated flows (including the nonstationary ones). In spite of the fact that the problem's solution depends on the ratio of numbers $Fr_{\text{d}}$ and $Fr_f$, it is interesting to clear up the role of each of these parameters individually, by the constant $Fr_{\text{d}}$. The numerical calculations have shown that in the linear case ($Fr_{\text{d}} \gg 1$), the problem's solution practically does not vary, but if $Fr_{\text{d}} \leq 1$, then the solution does, generally speaking, depend on the specific values of $Fr_{\text{d}}$ and $Fr_f$, too.

Shown in Fig. 1.62 are the flow patterns ($Re = 100, Fr_{\text{d}} = 1$) for the three different versions: $Fr_f = 10^{-2}; 10^{-3}; 2 \cdot 10^{-4}$ (a, b, c correspondingly). It proved that by the diminution of $Fr_f$ the amplitude of internal waves is increased, while by the $Fr_f = 2 \cdot 10^{-4}$ under the internal wave's crest is formed the closed vortex (rotor). It is necessary also to point out that the recurrently circulatory motion at the leeward side of the obstacle by the fixed Reynolds number is considerably weaker for the stratified fluid than for the uniform one.

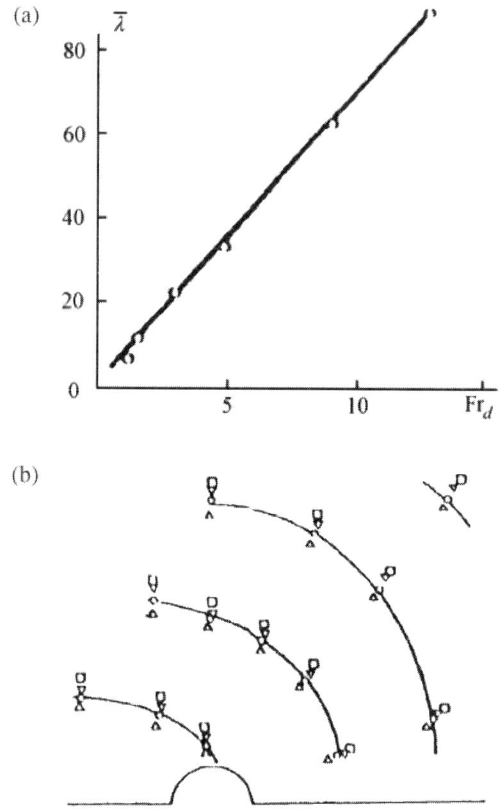


**Fig. 1.61.** Steady stratification flow: (a) linear dependence of calculated wavelength $\bar{\lambda}$ on the density Froude number ($Fr_{\rm d}$); (b) phase picture (solid lines — calculation, points — experimental measurements by Onufriev).

As it was already observed, in the case of $Fr_{\rm d} \ll 1$, the additional effects appear, for example, ahead of the body the area of blocked fluid is formed. The body's form does not exert any essential influence on the phase structure of internal waves, when the longitudinal size of the perturbations' domain does not exceed the half of a wavelength. However, if the fluid ahead of the body stays at rest, or if the vortical formations appear, then the "effective" longitudinal size of the body is growing. This leads to the situation, when by the sufficiently small value of $Fr_{\rm d}$ the blocked fluid comes out as the *secondary source* of excitation of the short internal waves.

Shown in Fig. 1.63 is the flow pattern ($Re = 100, Fr_{\rm d} = 0.5$) for the successive time moments ($\Delta t = 2.5$). The characteristic feature of

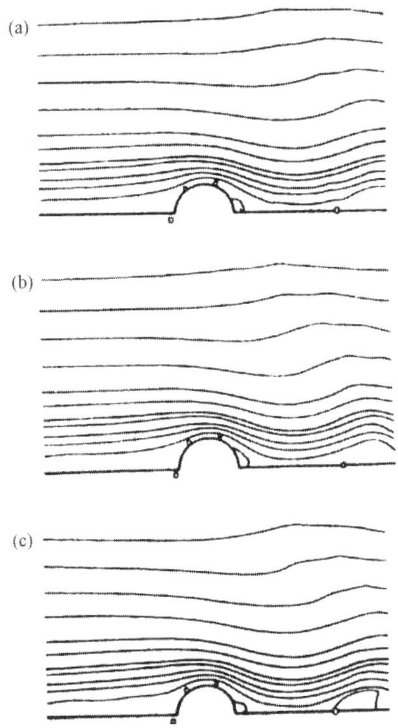


**Fig. 1.62.** Stratified flow past a simicircular obstacle for different $Fr_f (\psi = \text{const.}, Re = 100, Fr_\text{d} = 1)$: (a) $Fr_f = 10^{-2}$; (b) $Fr_f = 10^{-3}$.

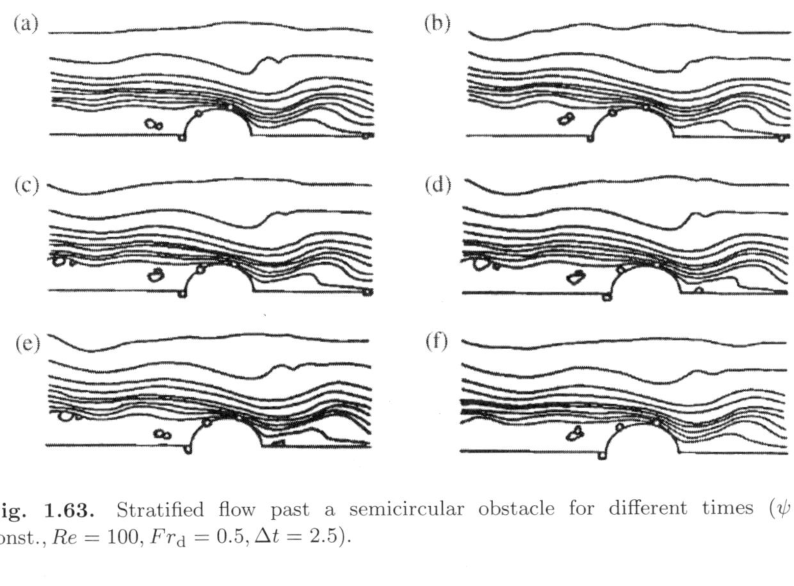


**Fig. 1.63.** Stratified flow past a semicircular obstacle for different times ($\psi$ = const., $Re = 100, Fr_\text{d} = 0.5, \Delta t = 2.5$).

the flow analyzed is the formation of a transitional process from the stationary motion to the nonstationary one. It is seen in the figure that as the time goes on, the steepness of streamlines is increased, and at the certain moment of time, when the streamline comes to the vertical position, the "collapse" occurs. Ahead of the body is formed a stagnant zone, which reveals itself through the presence of a weak vortex, and appear the layers with nonmonotonous distribution of velocity, at the boundaries of which might evolve the Kelvin–Helmhotz instability. It is evident that the smaller are the numbers $Fr$ and $Fr_f$ (by the constant Reynolds number and $Fr_{\rm d}$), the more unstable is such a stratified system. With the exceeding of certain values of $Fr$ and $Fr_f$, the stratification suppresses the instability (the local Richardson number, responsible for the stability of thin layers having large shift of velocity, is growing).

Shown in Fig. 1.64 are the flow patterns for the version with $Fr_{\rm d} = 0.158, Fr_f = 10^{-5}$, by the various Reynolds numbers (10; 100; 250). The value of viscosity exerts an essential influence of the flow pattern in spite of the fact that the parameter $Fr_{\rm d}$ is sufficiently small ($Re = 10$, see Fig. 1.64(a)). As the Reynolds number is increased up to 250 (Fig. 1.64(b)), the qualitative change of the flow pattern takes place. Formed ahead of the body arc several large vortices, which prove to be the secondary sources of the internal waves' excitation, and the motion acquires a clearly pronounced irregular and non stationary character, which just leads to the appearance of "chaos".

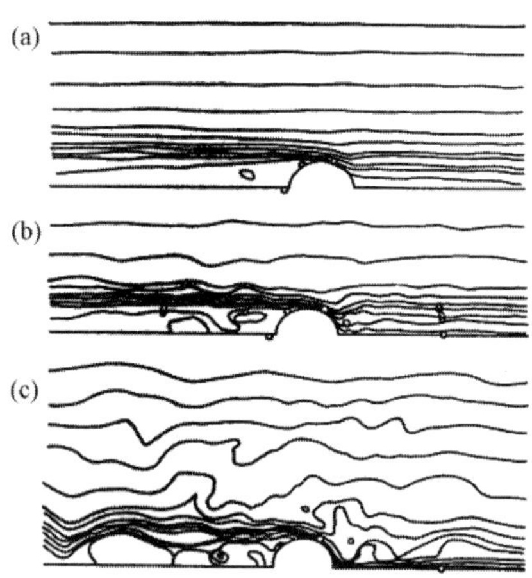

**Fig. 1.64.** Stratified flow past a semicircular obstacle for different $Re(\psi =$ const., $Fr_{\rm d} = 0.158, Fr_f = 10^5)$: (a) $Re = 10$; (b) $Re = 100$; (c) $Re = 250$.

Evidently, here reveals itself an interaction of molecular dissipation, of nonlinear inertial mechanism of a system, and of external gravity force. The conditions created by the large Reynolds numbers and the powerful stratification ($Fr_{\mathrm{d}} \ll 1$), i.e. by the considerable inertial interactions, contribute to the conception of a chaotic motion, which is in a good agreement with experiment.

The results exposed indicate to the perspectivity of using the methods applied for a study of rather complicated, nonlinear and non stationary phenomena. Worth of consideration are the more detailed studies of the processes described, as well as consideration of the motion of air in mountain areas and investigation of multi-dimensional (spatially nonstationary) problems.

### *1.11.5. Rayleigh–Taylor instability: evolution to the turbulent stage*

Let us consider in brief a scenario of the transition to turbulence due to the *Rayleigh–Taylor instability* (*RTI*). We shall use the traditional formulation in which the interface between a heavy and a lighter fluid (with density $\rho_2$ and $\rho_1$, respectively) in a gravity field is perturbed by a variation in the flow velocity field, and after that the dynamics of the interface is analyzed. The situation occurs in shell compression by laser thermonuclear fusion, in generating superstrong magnetic fields, etc.

The RTI evolves in the following stages:

(i) the *linear stage*, when the disturbance amplitude is much smaller than its wavelength, $a < L$;
(ii) the intermediate nonlinear stage, when $a \approx 0.4$L;
(iii) the *regular asymptotic stage*, when fully formed heavy liquid jets ("spikes") fall with a constant velocity;
(iv) acceleration, and lighter fluid "bubbles" rise with a constant velocity;
(v) the *turbulent stage*, which is characterized by intense interaction between disturbances of different wavelengths *and by mixing.*

From the practical point of view, finding the conditions under which the evolution of RTI stabilizes, or at least slows down, and analyzing the mechanism of the RTI evolution at the turbulent-flow stage are important problems.

Extensive theoretical and experimental studies of the evolution of RTI have been carried out. However, they have only considered, as a rule, single-mode initial-data disturbances. On the other hand, theoretical analysis of the developed-turbulence stage (see Ref. 93) has been done. We believe that the *study of multimode interactions and the transition from the nonlinear stage in the RTI evolution to the turbulent one on the basis of full equations* (see Refs. 14, 36, 94 and 95) *is a matter of great importance and interest* (*see Chapter* 3).

Using simple estimates, Inogamov[93] has shown that there exists a mechanism that drives long-wave disturbances and that gives rise to structures of a still larger hydrodynamic scale (the phenomenon has also been observed experimentally). The evolution of instability results in the growth of cavities (bubbles), their coalescence, and the formation of new, larger cavities. However, the resulting flow is also unsteady, and subsequent coalescences give rise to structures on still larger scales. An analysis of the perturbed velocity and pressure fields and of the resulting acceleration field shows that a force parallel to the shear is generated and leads to the skewing and merging of cavities. It would be, of course, of great interest to compare these estimates with calculation results.

A detailed study of the transition to turbulence in this complicated three-dimensional unstable phenomenon is also of great interest. This can be done by considering the full system of ideal compressible nonlinear equations (taking account of the terms related to the gravity acceleration $\mathbf{g}$). The simultaneous multimode interaction of several harmonics over long periods of time (until the rise of developed-turbulence structures) has been studied.[6,14,94,95]

Figures 1.65 show the topology of the interface between two fluids with a density ratio $\rho_2/\rho_1 = 10$ subjected to a two-mode disturbance with $\lambda_1/\lambda_2 = 2$ and 3 (where $\lambda_1$ and $\lambda_2$ are the respective wavelengths). The very *nonlinear stage* of the evolution of RTI, involving the formation of a light-component bubble and the separation of a heavy-component drop (Fig. 1.65(b)), as well as the characteristic bend of the heavy-component jet (Fig. 1.65(a)), which indicates the appearance and evolution of Kelvin–Helmholtz instability (resulting in a still more intense turbulent mixing), are shown in Ref. 94.

Figures 1.66 and 1.67 illustrate the evolution of RTI in the presence of five-mode disturbance for $\rho_2/\rho_1 = 10$ and $\lambda_1 : \lambda_2 : \lambda_3 : \lambda_4 : \lambda_5 = 10 : 5 : 10/3 : 5/2 : 2$.[94] We see that at the beginning of the process the short-wave disturbances develop, moving more quickly than the long-wave ones

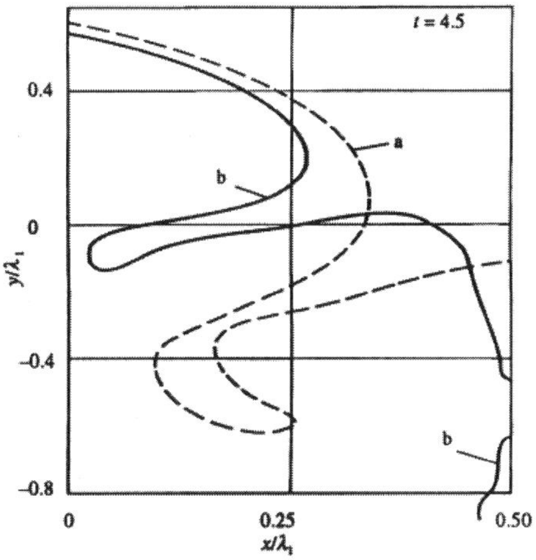


**Fig. 1.65.** Interface for Rayleigh–Taylor instabilty for two-mode disturbance and $\rho_2/\rho_1 = 10$, $t = 4.5$: (a) $\lambda_2/\lambda_1 = 2$; (b) $\lambda_2/\lambda_1 = 3$.

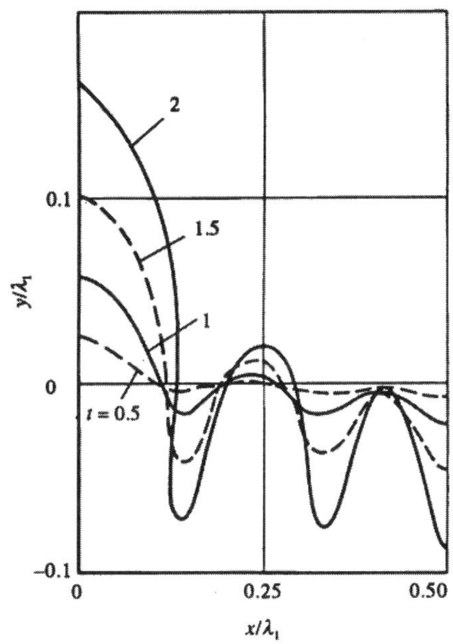


**Fig. 1.66.** Dynamics of the Rayleigh–Taylor instability for the five-mode disturbance (the initial stage).

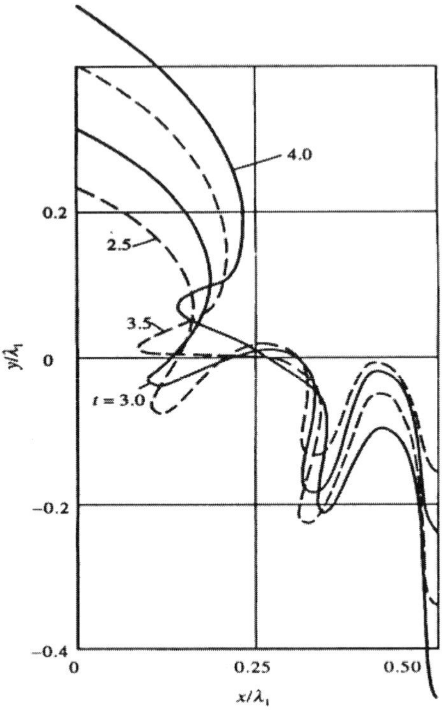


**Fig. 1.67.** Dynamics of the Rayleigh–Taylor instability at the turbulent stage.

(see Fig. 1.66). However, at the nonlinear stage (for $t > 3$, see Fig. 1.67), the long-wave disturbances start suppressing the short-wave ones, the short-wave jets disappear, and the long-wave mode dominates, resulting in the formation of a larger spatial scale (in agreement with the predictions of the theory[93]). Illustrated in Fig. 1.68 is the process of evolution to turbulent stage contact surface's form RTI at the various moments of time (multi-mode initial perturbation). Quite clearly seen are the pictures of the scales enlargement — beginning with 23 bubbles in the first picture ($t \sim 0.5$), and ending with three bubbles in the last one ($t \approx 3.00$). Evidently, this mechanism of the nonlinear interaction of RTI harmonics is typical and is reminiscent of the *Feigenbaum scenario for the transition to turbulence.*

Figure 1.69 shows the interfaces for the *three-dimensional situation.* The calculations were carried out by Belotserkovskii and Davydov by the method of large particles[36] on a coarse grid which, evidently, did not allow the calculations to be continued for large $t$. However, both the fall of rather

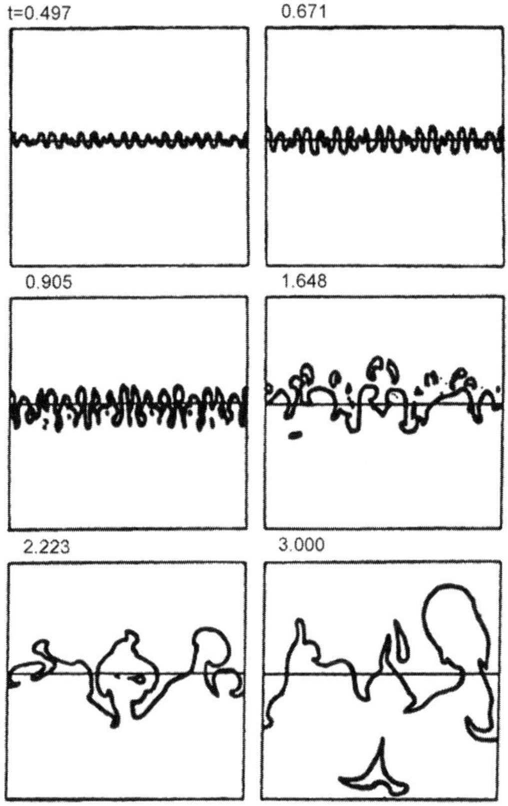


**Fig. 1.68.** The evolution of the contact surface by the RTI with a multi-mode initial perturbation (the scale's enlargement is observed).

narrow jets and the rise of large bubbles can be seen (new results see in Chapter 3).

The Rayleigh–Taylor and the Richtmyer–Meshkov instabilities have been studied extensively on the basis of full dynamic systems by means of advanced numerical techniques and modern computers (Chapters 2 and 3).

Let us draw some qualitative conclusions. We see that the stochastic process may be simulated by using a *strong inertial mechanism* acting within a dissipative dynamic system, composed of an internal mechanism due to nonlinear inertial terms of the equations and external forces causing disturbances. The process develops due to interaction between the inertial mechanism (which facilitates the development of unstable phenomena) and viscous dissipation (which makes the flow steady). A sharp increase in

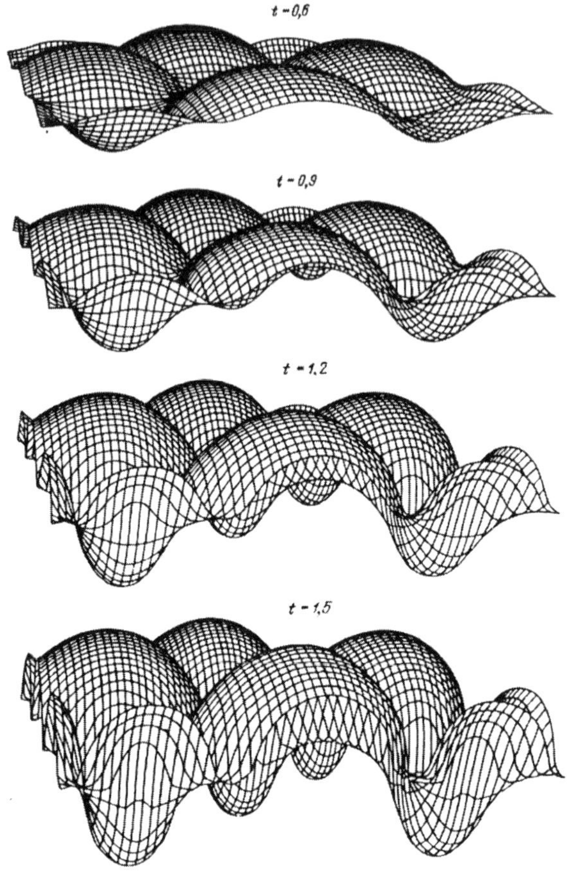


**Fig. 1.69.** Dynamics of the form of partition surface by the three-dimensional RTI, in isometrical projection ($\rho_h/\rho_l = 10$).

unsteadiness and instability ("the transition to chaos") may occur when the inertial mechanism dominates over the dissipative one.

In the above examples the process is manifested *only* in the presence of considerable external perturbations, such as gravity (stratified flows, RTI), wind loading, the Coriolis effects (the Earth's rotation), etc. We believe that the internal inertial mechanism taken alone (as in "pure" Navier–Stokes equations) cannot bring a system to the verge of a stochastic process. In the problems considered above the scenarios of the transition to turbulence were qualitatively similar to those predicted by the Landau (hydrodynamics) and Feigenbaum (RTI) scenarios.

### *1.11.6. Numerical simulation of the convective flow over large-scale source of energy (big fire in the atmosphere)*

One of the results of a big fire consists in the throwing out of a great amount of aerosol particles, ejected in the atmosphere in the form of soot and ashes. Such an ejection of the products of burning into atmosphere might lead to both local and regional climatic consequences, up to the threat of the "*nuclear winter*".[96,97] A review of experimental and numerical simulation of the big fire's consequences has been the subject of consideration of a number of lately published papers.[98] The source of energy's generation by the fire might be simulated by the three-dimensional source, of which the intensity grows in time according to a linear law, up to some maximal value. Speaking of the order of magnitude, the source's intensity corresponded to the case of complete burning of wood materials, as recalculated for the unit area. The investigations described in Ref. 98 permitted to determine the characteristic values of velocity of the flow obtained, of the height of ascent of the cloud of products of burning, of the amount of aerosol ejected into atmosphere and the extent to which this quantity is affected by the atmospheric humidity.

Experiments described in Ref. 99 have shown that the cut-off of the fire's nidus at the later stage (after the hovering of the cloud of fire's products) leads to a reduction of the height of hovering. The results of Ref. 99 are numerically confirmed by the data of Muzafarov and Utyuzhnikov paper.[100]

Moreover, experimentally studied in Ref. 99 is the effect of the fire's area on the height of the convective cloud's spreading. It was found that up to the moment when the nidus' characteristic size is less than the double height of tropopause, the height of cloud's spreading remains to be in a good agreement with the prediction for a point source, while along with the further increase of the fire's area the height of spreading is even somewhat reduced. This result is also confirmed by the data of papers.[100,101]

Following Muzafarov's and Utyuzhnikov's work[100] let us consider the problem's formulation and some results of calculations. The problem is seen as an axisymmetric and, taking into account the wind, essentially three-dimensional. It is interesting to note that although the source has rotational symmetry, the pollution distribution in three-dimensional case loses its symmetry (even without the wind) and the solution for three-dimensional case exists but is not stable.

The pollution distribution above the source is characterized by two types of instabilities. A *Rayleigh–Tailor instability* (while the heavy gas is above the light one as the equilibrium position has been passed) and the *Kelvin–Helmholtz instability*, when the polluted atmosphere is streaming at the so-called altitude of streaming.

The source of the fire for diameters large than 15 km breaks up into *coherent structures* (*the Benard cells*).

In Ref. 100, Muzafarov and Utyuzhnikov did not simulate the source of a fire in details and they considered it as a model one (it may be, for example, the model developed by Grishin[97]).

*The problem's formulation.* The motion of gas within a stratified atmosphere, induced by a fire, will be described based on Reynolds-averaged Navier–Stokes equations. The set of equations is to be presented in cylindrical coordinate system $\{z, r\}$, where $z$ is the coordinate counted out from the Earth's surface. It will be assumed that the problem possesses axial symmetry.

The set of Navier–Stokes equations is written down in the form[100]

$$\begin{aligned}\frac{\partial}{\partial t}\mathbf{q} + \frac{1}{r}\frac{\partial}{\partial x^i}(r\mathbf{e}^i_{\text{inv}}) + \frac{1}{r}\frac{\partial}{\partial x^i}\left(r\frac{\mu}{Re}\mathbf{e}^i_{\text{vis}}\right)\\ = -g\mathbf{f} + \mathbf{q}^* + \frac{1}{r}\mathbf{s}_{\text{inv}} + \frac{1}{r}\frac{\mu}{Re}\mathbf{s}_{\text{vis}}.\end{aligned} \tag{1.59}$$

Here

$$\mathbf{q} = \begin{pmatrix}\rho\\ \rho u\\ \rho v\\ E\end{pmatrix}, \quad \mathbf{f} = \begin{pmatrix}0\\ \rho\\ 0\\ \rho u\end{pmatrix}, \quad \mathbf{q}^* = \begin{pmatrix}0\\ 0\\ 0\\ Q^*\end{pmatrix}, \quad \mathbf{s}_{\text{inv}} = \begin{pmatrix}0\\ 0\\ p\\ 0\end{pmatrix},$$

$$\mathbf{e}^i_{\text{inv}} = \begin{pmatrix}\rho V^i\\ \rho V^i u + p\delta^{i1}\\ \rho V^i v + P\delta^{i2}\\ (E+p)V^i\end{pmatrix}, \quad \mathbf{s}_{\text{vis}} = \begin{pmatrix}0\\ 0\\ \dfrac{2}{3}\dfrac{\partial V^k}{\partial x^k} - \dfrac{4}{3}\dfrac{v}{r}\\ 0\end{pmatrix},$$

$$\mathbf{e}^i_{\text{vis}} = \begin{pmatrix}0\\ \dfrac{2}{3}\dfrac{1}{r}\dfrac{\partial(rV^k)}{\partial x^k}\delta^{i1} - \left(\dfrac{\partial u}{\partial x^i} + \dfrac{\partial V^i}{\partial x^1}\right)\\ \dfrac{2}{3}\dfrac{1}{r}\dfrac{\partial(rV^k)}{\partial x^k}\delta^{i2} - \left(\dfrac{\partial v}{\partial x^i} + \dfrac{\partial V^i}{\partial x^2}\right)\\ \dfrac{2}{3}\dfrac{1}{r}\dfrac{\partial(rV^k)}{\partial x^k}V^i - \left(\dfrac{\partial V^k}{\partial x^i} + \dfrac{\partial V^i}{\partial x^k}\right)V^k - \dfrac{1}{\Pr}\dfrac{\partial h}{\partial x^i}\end{pmatrix},$$

$$i = 1, 2; \quad k = 1, 2,$$

where $\mathbf{q}$ is the vector of independent variables; $\mathbf{e}^i_{\text{inv}}$ are the vectors of inviscid flows; $\mathbf{e}^i_{\text{vis}}$ — vectors of viscous flows; $\mathbf{f}$ is the vector of the external force due to the gravity field; $\mathbf{q}^*$ is the vector of the three-dimensional source of the energy's generation; $\mathbf{s}_{\text{inv}}$, $\mathbf{s}_{\text{vis}}$ are the source-type "non-conservative" terms connected with "non-Cartesian" character of coordinate system $\{z,\ r\}$. Both here and further on, the summation on repeated indices is assumed.

The following notations are used in Eqs. (1.59): $t$, time; $\{x^1, x^2\} = \{z, r\}$; $\rho$, density; $\{V^1, V^2\} = \{u, v\}$, velocity components in $\{z, r\}$-directions; $E = \rho(\varepsilon + \frac{V_k V^k}{2})$, total energy; $\varepsilon$, internal energy; $p$, pressure; $h = \varepsilon + p/\rho$, enthalpy; $g$, gravity acceleration; $\mu$, viscosity; $Q^*$, heat released into the unit volume from external sources.

All the quantities in Eqs. (1.59) are dimensionless. The connection between dimensionless quantities $f$ and dimensional one $f_{\text{phys}}$ is prescribed in the form $f_{\text{phys}} = \bar{f} \cdot f$, where $\bar{f}$ is a non-dimensionalizing factor.

The variables in Eqs. (1.59) are non-dimensionalized by the means of the following quantities: $\bar{x}^i = L_0$, where $L_0$ some linear scale, characteristic for the problem; $\bar{V}^i = V_0$, where $V_0$ is some characteristic velocity; $\bar{t} = L/V_0$, $\bar{\rho} = \rho_0$, where $\rho_0$ is some characteristic density; $\bar{E} = \rho_0 V_0^2$, $\bar{p} = \rho_0 V_0^2$, $\bar{\mu} = \mu_0$, which is some characteristic value of viscosity; $\bar{g} = L g_0 / V_0^2$, $g_0$ being the value of gravity acceleration at the Earth's surface; $\bar{Q}^* = L/\rho_0 V_0^2$.

Introduced into Eqs. (1.59) are dimensionless numbers $Re$ and $Pr$:

$$Re = \frac{\rho_0 V_0 L}{\mu_0}, \quad Pr = \frac{\mu C_p}{\lambda_T},$$

Here $C_p$ is specific heat at constant pressure; $\lambda_T$ is heat conductivity. It is assumed that the equation of state is presented in a form

$$p = (\gamma - 1)\rho\varepsilon, \quad \varepsilon = C_v T,$$

where $\gamma$ is the adiabatic index (for air $\gamma = 1.4$); $C_v$, is the specific heat at constant volume.

The values of coefficients of turbulent transfer were calculated according to an algebraic model,[102] which for the axisymmetric case has a form

$$\mu = \mu_l + \mu_t,$$

$$\mu_t = \frac{\rho l^2}{2} \left\{ \left( \frac{\partial v}{\partial z} + \frac{\partial u}{\partial r} \right)^2 + 2 \left( \left( \frac{\partial v}{\partial r} \right)^2 + \left( \frac{\partial u}{\partial z} \right)^2 + \left( \frac{v}{r} \right)^2 \right) \right\}^{1/2},$$

$$l = K\left\{(u^2+v^2)^{1/2}B^{-1/2} + B^{1/2}\left[\left(\frac{\partial^2 v}{\partial r^2}+\frac{1}{r}\frac{\partial v}{\partial r}-\frac{v}{r^2}\right)^2+\left(\frac{\partial^2 v}{\partial z^2}\right)^2\right.\right.$$
$$\left.\left.+\left(\frac{\partial^2 u}{\partial r^2}+\frac{1}{r}\frac{\partial u}{\partial r}\right)^2+\left(\frac{\partial^2 u}{\partial z^2}\right)^2\right]^{-1/2}\right\},$$

$$B = \left(\frac{\partial v}{\partial r}\right)^2+\left(\frac{\partial v}{\partial z}\right)^2+\left(\frac{v}{r}\right)^2+\left(\frac{\partial u}{\partial r}\right)^2+\left(\frac{\partial u}{\partial z}\right)^2,$$

where $\mu_{\text{l}}$ and $\mu_{\text{t}}$ are the coefficients of laminar and turbulent viscosity, correspondingly; $l$ is the mixing length, $K$ is the empirical constant. By the numerical calculations it was assumed that $K = 0.125$, $Pr = 1$.

The source of energy generation was prescribed in volumetric form, with radius $R$ and height $h$, which was here assumed to be equal to 100 m. The intensity of a space source $Q^*$ varied in time linearly, up to the value of $Q^*_{\max}$, which was reached after 30 min. During the next 30 min, the source's intensity was assumed constant. After 60 min, the source was cut off.

The boundary conditions were prescribed in the following way. At the Earth's surface, three boundary conditions are prescribed (those of non-leakage and of sticking, and for internal energy):

$$u = v = \frac{\partial \varepsilon}{\partial z} = 0.$$

Prescribed at the axis of symmetry are the conditions of symmetry for thermodynamical functions and for $z$-component of velocity, as well as the condition of vanishing of $r$-component of velocity:

$$\frac{\partial}{\partial r}\{\rho, u, \varepsilon\} = v = 0.$$

At the artificial boundaries arising by the substitution of the infinite domains by the finite ones, those characteristical boundary conditions are prescribed which follow from consideration of the set of control equations as being locally one-dimensional near the boundary (one considers the spatial direction normal to the boundary). Then for those characteristics, which are coming into the domain from the boundary, the boundary condition for the corresponding Riemann's invariant is formulated, $\alpha_i|_b = \alpha_i|_{b,\text{unpert}}$, while for those going out — the condition of compatibility is taken. Here index $b$ corresponds to the value of Riemann's invariant calculated by the parameters of unperturbed atmosphere at the boundary point of interest.

Results of the numerical simulation.[100] The investigation was carried out concerning the effect of the fire's area on the dynamics of flow and on the height of the admixture's ascent. The admixture was considered as "passive" one, without taking into account its mass and neglecting the diffusive transfer. As concerns the cloud of the fire's products, its radius and the height of its ascent were determined through the maximal values of $r$ and $s$, which corresponded to the passive admixture.

The following values of radius $R$ for seat of the fire were considered: 5, 8, 11, 22, 33 km. The maximal intensity of the energy generation, as calculated for the unit area of surface, corresponded to $q_m = 5 \times 10^4\,\mathrm{W/m^2}$.

The atmosphere's parameters corresponded to the international standard. In the case of $R = 5\,\mathrm{km}$ (when the total intensity $q_f$ of the energy release is equal to $3.9 \times 10^6\,\mathrm{MW}$) the initial stage of the fire's development is accompanied by the air's influx into the nidus area adjacent to a wall. As this takes place, two vortices are generated, diversely oriented and located one above the other, the lower vortex being slightly shifted to the right. The propagation of admixture is characterized by the formation of two "tongues", each particular of them being formed by "its own" vortex. In Figs. 1.70(a) and 1.70(b) the distribution of admixture is shown, which corresponded to the time moments 20 and 60 min. It is clearly seen that

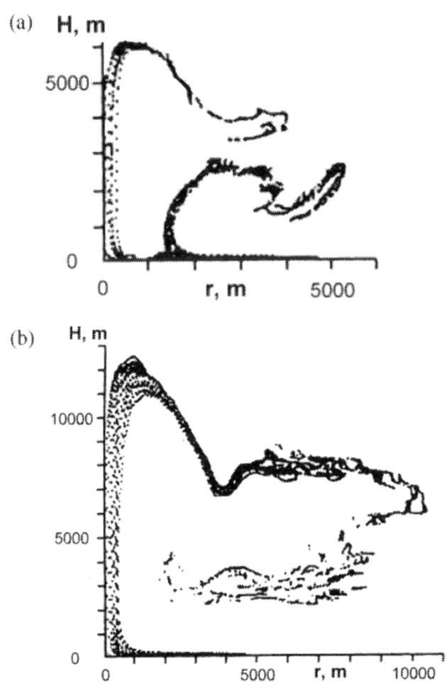


**Fig. 1.70.** Distribution of the pollutions for $R = 5\,\mathrm{km}$ (radius of the source): (a) $t = 20$ min, (b) $t = 60$ min.

the upper "tongue" outstrips the lower one due to the fact that it is permanently and intensively feeded by the upstream air flow in vicinity of the axis of symmetry. Within, approximately 30 min, the cloud of admixture goes up to the maximal height of 12–13 km. As the process goes on, the upper edge of a cloud carries out the oscillations about the equilibrium position, with the period of 5–7 min corresponding to the natural frequency of atmospheric oscillations which is determined by the Brunt–Vaissala relation

$$N^2 = \frac{q}{T_e}\left(\frac{dT_e}{dz} + \gamma_{\mathrm{a}}\right),$$

where $T_e$ is the temperature of air within the unperturbed atmosphere and $\gamma_{\mathrm{a}}$ is the adiabatic gradient of temperature. The temperature at the boundary of a source acquires the characteristic values of 380–420 K.

The temporal dependencies of the radius of the cloud of admixture (the coordinate of the point having the maximal value of $r$) and of maximal height of the admixture's ascent are presented in Fig. 1.71. The formation of three plateaus in the graph of $r_{\max}(t)$ is caused by the reorganization of flow in connection with protrusion of, initially, the lower "tongue", and later the upper one. In spite of the fact that the height of admixture's ascent is equal to 13 km, the main level of propagation of the products of burning is about 6–7 km.

The version, corresponding to $R = 8$ km ($q_f = 10^7$ MW), has the following peculiarities. At the initial period of time the lower vortex, which was already developed, is located not immediately under the upper one, but is proved to be shifted to the right with its center being found in the area of 5 km. Therefore, the propagation of admixture in radial direction is determined just by the lower vortex, which captures the products of burning and throws them out, into the atmosphere. The maximal height of admixture's ascent corresponds to 12–14 km. After 40 min, the edge's radius attains the value of 12 km, which exceeds that of the preceding version by 6 km.

The dynamics of formation of the gas-dynamical flow with $R = 11$ km ($q_f = 1.9 \times 10^7$ MW) is essentially different from that of two preceding versions. At the initial period of time one can observe three vortices formed within the flow, corresponding, approximately, to the values of $r = 1, 4, 9$ km.

It should be noted that the displacement of the right computational boundary from the mark of 30 km toward that of 25 or 20 km did not bring about any modification of the structure of flow within the rest of a domain. This testifies to the fact that the characteristic boundary conditions applied

(a)

(b)

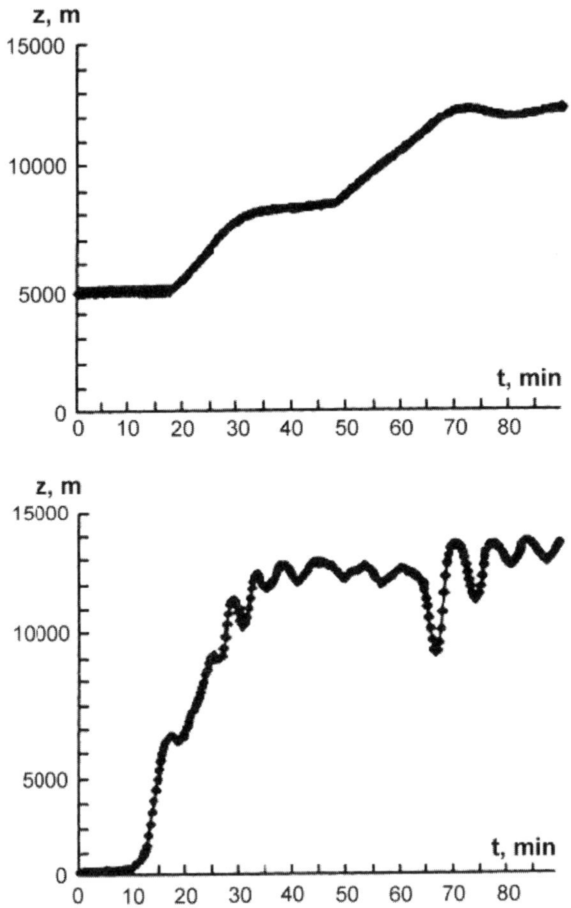


**Fig. 1.71.** The temporal dependencies for $R = 5\,\text{km}$: (a) $r_{\max}$ (the maximum radius of the cloud of admixture), (b) $z_{\max}$ (maximal height of the admixture's ascent).

here permit to let out of the domain some weak perturbations without their reflection from the boundary.

Upon the cut-off of the source of energy generation (60 min), the amplitude of oscillations of the upper edge is appreciably reduced. At that moment, the mean height of the upper edge corresponds to 14 km. The functional dependencies $r_{\max}(t)$ and $z_{\max}(t)$ are presented in Fig. 1.72. While the height of the upper edge with the growth of $R$ is increased just insignificantly, the radius of the cloud of products of burning is growing in such an extent that it exceeds the radius of seat of the fire in, approximately, two

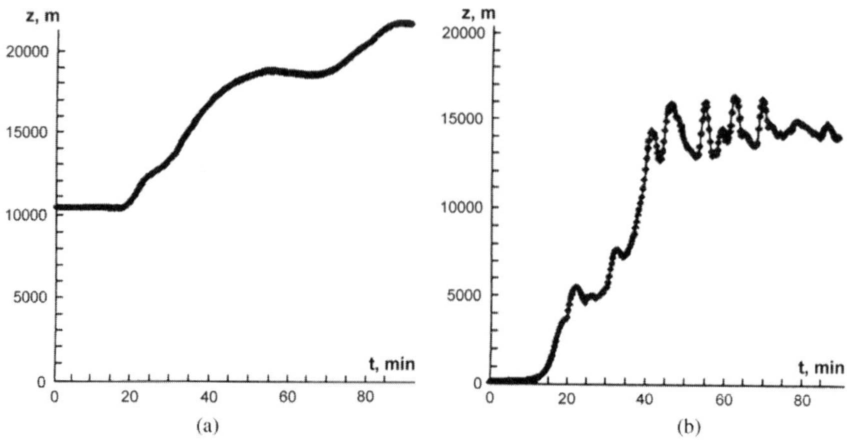


**Fig. 1.72.** The temporal dependencies for $R = 11$ km: (a) $r_{\max}$, (b) $z_{\max}$ (for notation, see Fig. 1.71).

times. The radius of seat of the fire $R = 11$ km corresponds to the maximal height of the admixture's ascent. Along with the further increase of $R$ the maximal height of ascent is somewhat diminishing. The form of the cloud of admixture, which is made up at the time moments of 60 and 90 min, correspondingly, is presented in Fig. 1.73(a) and 1.73(b). As it is seen from these drawings, in the case of $R = 11$ km, the main part of the admixture is propagated at the height of 3 km. After the source's cut-off one is able to observe the noticeable lowering of the cloud in vicinity of the axis of symmetry, which in fact, is in agreement with experimental data.[99]

The $q_f$-dependence of $z_{\max}$ is up to $R = 11$ km in good agreement with a well known formula for the point source[103]

$$z_{\max} = Aq_f^{1/4}, \tag{1.60}$$

where $z_{\max}$ is measured in kilometers and $q_f$ in megawatts.

Beginning with $R = 11$ km, the dependence in the form of Eq. (1.60), corresponding to the point source, is appreciably violated. Thus, for the cases of $R = 22$ km and $R = 33$ km, the maximal height of the admixture's ascent is equal, correspondingly, to 16 and 15 km. Upon the cut-off of the source of energy release the height of ascent of the convective column is reduced by 3 km. The form of the cloud of admixture for $R = 22$ km and $t = 80$ min is shown in Fig. 1.74.

Thus, the maximal height of ascent of the products of burning with area of seat of the fire being varied, measures 16 km, which essentially affects

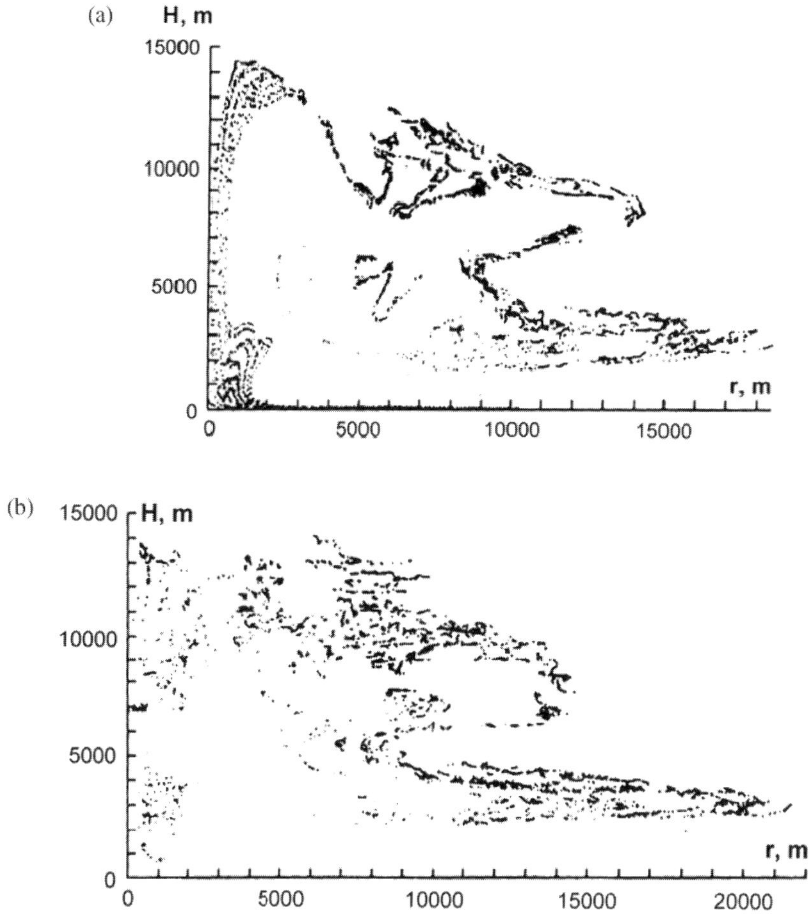


**Fig. 1.73.** Distribution of the pollution for $R = 11\,\text{km}$: (a) 60 min, (b) 90 min.

the quantity of aerosol, ejected into the atmosphere. The formation of the admixture's cloud at the height of a tropopause is explained by the influence of the dissipative processes associated with the flow's turbulization. Presented in Fig. 1.75 are the relationships $z_{\max}(t)$ for the data corresponding to Figs. 1.70 and 1.71, but with prescription of the coefficient of turbulent viscosity $\nu_t$, so that the Reynolds number $Re$ acquired the constant values of $5 \times 10^3$ and $10^4$, correspondingly. It is seen that in the case of a weak influence of the dissipative processes, or of their total absence, the throwing-out of admixture into the atmosphere might be effectuated with $Re$ as small as 5 km.

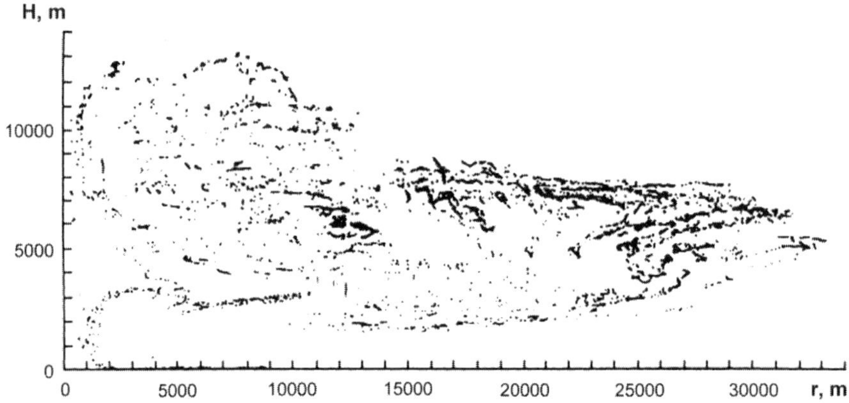


**Fig. 1.74.** Distribution of the pollution for $R = 22\,\text{km}$ and $t = 80$ min.

## 1.12. Axiomatic model of fully developed turbulence

Taking into account the philosophy of the above numerical procedure and the results of calculating coherent structures in turbulent flows, we have created, together with Shifrin,[104] an *axiomatic model of fully developed turbulence* in an incompressible fluid (which may be considered the theoretical foundation of the above approach). Here we present only some basic ideas of the theory. It is suggested that the motion of a fluid for $Re \to Re_{\text{cr}}$ is characterized by the presence of regions within which the velocity field is continuous but undifferentiable.[ee]

This is based on the statement that, from the physical point of view, fully developed turbulence is determined by the mixing of liquid particles, which destroys the order within any system of points in a finite time. In other words, we cannot introduce Lagrangian coordinates in a region of fully developed turbulence in four-dimensional physical "space-time".

The formation of subregions (turbulent spots) in a laminar-flow region (with a differentiable velocity field) is explained by the Gauss least constraint principle, applied to a situation in which the laminar flow loses its

[ee]In this connection, we cite Kochin, *et al.*[105]: "There is some doubt whether the turbulent-flow velocity can be represented at all in terms of continuous functions of the spatial coordinates and time... It may be that we should approximate the paths of turbulent motion by means of a continuos function that has no time derivatives at any of its points, such as the Weierstrass function."

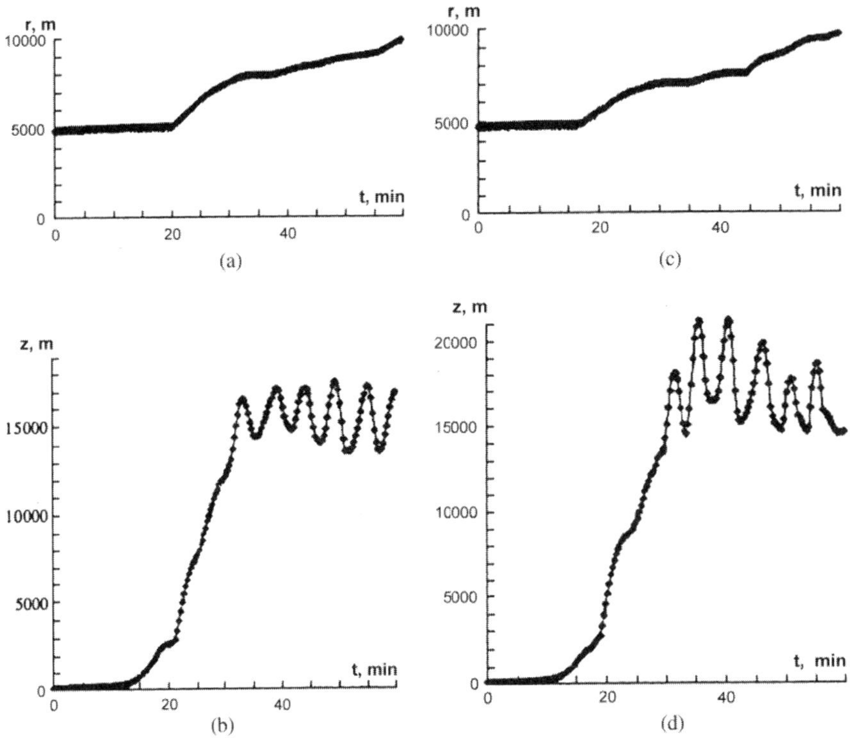


**Fig. 1.75.** The temporal dependences for $R = 5\,\text{km}$ and $\nu_T = \text{const.}$: for $Re = 5 \cdot 10^3$ ((a) $r_{\max}$, (b) $z_{\max}$) and for $Re = 10^4$ ((c) $r_{\max}$, (d) $z_{\max}$).

asymptotic stability, and the transition to a more extensive class of flows becomes energetically preferable.

The flows comprised in the class are described by using an extended formulation based on the following principles:

(i) since the differential Navier–Stokes equations are invalid for describing undifferentiable fields, the integral laws of mass, momentum, and total energy conservation are used in their most general form; they may be adequately represented by differential equations (macroequations) for quantities averaged over a mobile finite subregion or cell; then a fluctuation tensor is introduced in a natural way without employing closure hypotheses;

(ii) the Navier–Stokes and Fourier laws for the viscous stress tensor and heat flux tensor, respectively (reducible to the classical laws on differentiable fields), are extended;

(iii) the macroequations taken together with the laws of equilibrium thermodynamics, the extended Navier–Stokes and Fourier laws, and the averaging operator over a cell form a closed system for the description of both averaged and instantaneous (true) quantities, for which the initial-value problem is formulated with conditions imposed at the body surfaces;
(iv) the ordered structure of a turbulent flow is defined as an attractor of an asymptotically stable solution for averaged quantities; the scale of an averaging cell plays the role of the stable parameter;
(v) ill-posing (nonuniqueness and instability of solutions) of the problem of reconstructing the true fields from the averaged fields obtained for a finite cell scale (subject to the condition that the averaged quantities be stable) is interpreted as randomness of their physical states.

This concept is considered to be both the philosophical basis and verification of the validity of direct numerical simulation of ordered developed-turbulence structures by algorithms approximating (or imitating) the integral conservation equations.

Of greatest importance is fully developed turbulence ($Re \to \infty$), which may be simulated (as the leading term of the boundary-layer-type asymptotic expansion) in the framework of equations describing the laws of conservation (in the Euler form) subject to the impermeability condition.

## 1.13. Conclusion

In conclusion, we formulate the main elements of the approach developed above.

(i) Direct numerical models are constructed, which allow one to analyze an extensive class of nonlinear problems of modern aerodynamics, namely, the phenomena of *fully developed free shear turbulence.*
(ii) Using experimental data that prove the existence of ordered, random structural formations in developed shear turbulence, the processes are subdivided into *random vortex motions* and *ordered motions of large-scale vortices.*
(iii) The large-scale and the ordered nature of the latter motion admit descriptions by numerical schemes based on the nonstationary *hydrodynamic equations* (instead of statistical approaches).

(iv) The individual character of both the unsteady ordered motion and large-scale turbulent macroscopic structures makes it possible to study them within the low frequency and inertial intervals for large Reynolds numbers by *direct* numerical simulation (without employing semi-empirical turbulence models) based on *full smoothed ideal dynamics equations*, namely, the nonstationary Euler equations in the form of integral laws of conservation with an approximate dissipation mechanism (generated by averaging flow parameters over a cell and ensuring the stability of the calculations). *Finite-difference oriented schemes* with high accuracy can identify large-scale formations within a specified range of error by employing an approximate representation for subgrid fluctuations.

(v) The **correctness** of the above problem formulation is demonstrated: there exists, in principle, the possibility of obtaining correct estimates of statistical flow characteristics that depend on large-scale turbulence by using smoothed equations of motion, in which the contribution of small-scale eddies is accounted for approximately (without the requirement that the true fields of the oscillating quantities be correctly calculated).

(vi) Local effects (the *random component*) of turbulence (the Reynolds stress distribution, the turbulent energy density, the rate of the latter's dissipation, etc.) within large-gradient zones are studied using the *statistical approach* and *kinetic models of turbulence.*

(vii) The *laminar-flow modes* and the *laminar–turbulent transition* are studied using *the full Navier–Stokes equations.*

(viii) Numerical experiments are used to analyze "scenarios" of the transition to chaos, i.e., it is assumed that the randomization of dynamic systems (the Navier–Stokes model) can occur only in the presence of external perturbations (the gravitational field, effects of the earth's rotation, wind loads, and the like) that create the necessary inertial mechanism; the interaction of multimode disturbances is used to study the mechanism of the transition to the turbulent stage during the evolution of the Rayleigh–Taylor instability.

It should be noted that the study of turbulence — the problem "with inaccurately specified information" and the "extended" problem formulation introduced above (the original system with an averaging operator) — is analogous to the *parametric extension* of the Tikhonov regularization method.[106] The model should not be considered a mathematical model of a phenomenon, but rather a mathematical model for studying the phenomenon (here, turbulence). The notion of an ordered structure (stably realized as an attractor only for certain values of the parameter $h^*$ (the numerical grid step) seems to be close to Tikhonov's notion of the $\varepsilon$-stability of a solution. In other words, the ordered structure of a turbulent flow is the regularized description of the flow[ff].[4]

The study of ordered structures in turbulent flows is part of the widely discussed topic of *self-organization*, i.e., of the process by which order emerges in complex nonlinear systems and media (see Ref. 107–110). Without discussing the topic in detail, we would like to note that the emergence of order is evidently related to the collective (cooperative) behavior of subsystems comprising a system. Since the models, entities, and mechanisms of the theory of self-organization in complex nonlinear systems and media have not been thoroughly studied, the use of the heuristic approaches of numerical experiment seems to be justified.

However, it must be said that *nonequilibrium, nonlinearity*, and *the dissipativity of a medium* are fundamentally important for the formation of such structures. Self-organization is a consequence of the evolution of spatially nonuniform instabilities, which are stabilized due to the balance between energy dissipation and its addition via the sources of nonequilibrium.[109,110]

In the numerical approach to the study of turbulence, it is important to simulate correctly the process of generation and evolution of structures both in time and over the scales. Coherent structures may only exist far from equilibrium due to a large addition of energy and matter. The evolution of the process is treated as a sequence of transitions in the hierarchy of structures of increasing complexity, with the stable solution being selected against the background of ever-increasing dissipation. In essence, this is the general Prigogine principle of self-organization in nonequilibrium systems.[107]

The philosophy of a multilevel approach to the study of turbulence is based on the development of rational numerical methods that describe a phenomenon considered. At the start of this review, we posed the following question: What is the model that should be used in devising a scheme for the analysis of separated or turbulent flows corresponding to various models of motion? Should the model be that of an ideal fluid, or should it

---

[ff]A turbulent-flow ordered structure is defined as an attractor of an **asymptotically stable solution for average quantities**; in this case, the stability parameter depends on the cell scale $h$.

consist of the Navier–Stokes equations, or a kinetic model? We answer the question as follows: "One should start with a structural representation of the turbulent flows and use the models that give the best agreement with the mechanisms of interaction that occur in the system". Thus, large-scale transport at very large *Re* numbers is analyzed with the averaged dynamic model for an ideal medium, analyses of laminar–turbulent flows should consider the viscous-interaction mechanism (the Navier–Stokes equations), and analyses of stochastic processes should use kinetic-level models. This method of rational simulation (BRAIN-WARE) agrees well with the mechanisms of interaction in structural turbulence and decreases requirements for computer time.

## Acknowledgement

The authors thank A. S. Monin for reading the manuscript and for numerous productive discussions of the issues.

## References

1. M. Van Dyke, *Album of Fluid Motion* (Parabolic Press, Palo Alto (California) 1982.
2. G. K. Batchelor, *Theory of Homogeneous Turbulence* (Cambridge University Press, 1955).
3. S. A. Orszag, *Handbook of Turbulence. Fundamentals and Applications*, W. Frost and T. H. Moulden (eds.), Vol. 1 (Plenum Press, New York, London, 1977), pp. 311–347.
4. J. O. Hinze, *Turbulence* (McGraw-Hill, New York, 1975).
5. A. A. Townsend, *The Structure of Turbulent Shear Flow* (Emmanuel College, Cambridge, 1956).
6. B. J. Cantwell, Organized motions in turbulent flows, *Ann. Rev. Fluid Mech.* **13**, 457–515 (1981).
7. *Strange Attractors* (Mir, Moscow, 1981) (in Russian translation).
8. H. L. Swinney and J. P. Gollub (eds.), *Hydrodynamic Instabilities and the Transition to Turbulence*, Topics in Applied Physics, Vol. 45 (Springer-Verlag, Berlin, Heidelberg, New York, 1981).

---

See also: O. M. Belotserkovskii, Computational experiment: direct numerical simulation of complex gas-dynamics flows on the basis of Euler, Navier–Stokes, and Boltzmann models, Karman's Lecture, von Karman Institute for Fluid Dynamics, March 15–19, 1976; in *Numerical Methods in Fluid Dynamics*, H. J. Wirz and J. J. Smolderen (eds.) (Hemisphere, Washington-London, 1978), pp. 339–387.

9. A. N. Kolmogorov, Equations of turbulent motion of incompressible fluid, *Izv. Akad. Nauk SSSR. Ser. Fiz.* **6**, 56–58 (1942) (in Russian).
10. L. D. Landau and E. M. Lifshitz, *Fluid Mechanics* (Pergamon, London, 1956).
11. A. S. Monin and A. M. Yaglom, *Statistical Fluid Mechanics: Mechanics of Turbulence* (MIT Press, Cambridge, Massachusetts, 1975).
12. O. M. Belotserkovskii, Direct numerical simulation of free developed turbulence, *USSR Comput. Math. Math. Phys.* **25**(12) (1985).
13. O. M. Belotserkovskii, *Numerical Modeling on the Mechanics of Continuous Media*, 2nd edn. (Fizmatlit, Moscow, 1994) (in Russian).
14. O. M. Belotserkovskii, *Numerical Experiment in the Turbulence: From order to Chaos* (Nauka, Moscow, 1997) (in Russian).
15. D. P. Chapmen, Numerical aerodynamics and the prospects of its development, *AIAA J* **17**(12), 1293–1313 (1979).
16. H. Daiguji, Y. Miyake and N. Kasagi, Mathematical modeling of turbulent flows, *Fluid Dynamics Research* **20** (1997).
17. S. M. Belotserkovskii and M. I. Nisht, *Separated and Attached Ideal Flow Past Thin Wings* (Nauka, Moscow, 1978) (in Russian).
18. S. M. Belotserkovskii and A. S. Ginevskii, *Turbulent Jets and Wakes Simulation Using the Method of Discrete Vortices* (Fizmatlit, Moscow, 1995) (in Russian).
19. J. Shetz, *Turbulent Flows: Processes of Injection and Mixing* (Mir, Moscow, 1984) (Russian translation).
20. J. Marsden, Attempts to relate the Navier–Stokes equations to turbulence, P. Bernard and T. Ratiu (eds.) in Lecture Notes in Mathematics, Vol. 615 (Springer-Verlag New York, Heidelberg, Berlin, 1977).
21. A. A. Dorodnitsyn, Review of methods for solving Navier–Stokes equations, Lecture Notes in Physics, Vol. 18, 23–47 (1973).
22. V. V. Struminskii, On the theoretical foundations of turbulent flows, *Dokl. Akad. Nauk SSSR* **280**(3), 570–574 (1984) (in Russian).
23. V. V. Struminskii, Theoretical foundations of turbulence and the simplest example of a turbulent flow, *Dokl. Akad. Nauk SSSR.* **280**(4), 820–826 (1984) (in Russian).
24. J. W. Deardorff, The use of subgrid transport equations in a three-dimensional model of atmospheric turbulence, *J. Fluid Eng.* **95**, 429–438 (1973).
25. J. H. Ferziger, Large-eddy numerical simulations of turbulent flows, *AIAA J.* **15**(9), 1261–1267 (1977).
26. J. Smagorinsky, S. Manabe and I. J. Holloway, Numerical results from a nine-level general circulation model of the atmosphere, *Month. Weather Rev.* **93**, 727–768 (1965).
27. J. R. Herring, S. A. Orszag, R. H. Kraichnan and D. G. Fox, Decay of two-dimensional homogeneous turbulence, *J. Fluid Mech.* **66**, 417–444 (1974).
28. V. R. Kuznetsov and V. A. Sabel'nikov, *Turbulence and Combustion* (Nauka, Moscow, 1986) (in Russian).

29. R. J. Driscall and K. A. Kennedy, A model for the turbulent energy spectrum, *Phys. Fluids* **26**(5), 1228–1233 (1983).
30. P. G. Zaets, A. T. Onufriev, N. A. Safonov and R. A. Safarov, Experimental study of the turbulent one-dimensional spectrum function in rotating pipe flow: importing of the isotropic uniform turbulent model, *Proc. 5th EPS Liquid State Conf.* Moscow October 16–21 (1989), pp. 33–36.
31. P. G. Zaets, A. T. Onufriev, N. A. Safonov and R. A. Safarov, Experimental study of the energy spectrum behavior in turbulent flow through a rotating pipe *Zh. Prikl. Mekh. I. Tekhn. Fiz.* (1991) (in Russian).
32. J. Qian, Variational approach to the closure problem of turbulent theory, *Phys. Fluids* **26**(8), 2098–2104 (1983).
33. U. Shuman, G. Gretzbah and L. Kliser, *Methods for Calculating of Turbulent Flows* (Mir, Moscow, 1984) (in Russian).
34. A. Leonard, Energy cascade in large-eddy simulations of turbulent fluid flows, *Adv. Geophys. A* **18**, 237–248 (1974).
35. A. I. Tolstykh, High accuracy non-centered compact difference schemes for fluid dynamics applications, *Series Adv. Math. Appl. Sci.* **21** (1994).
36. O. M. Belotserkovskii and Yu. M. Davydov, *Method of Large Particles in Gas Dynamics: Numerical Experiment* (Nauka, Moscow, 1982) (in Russian).
37. Structures and mechanisms of turbulence. 1, 2, Lecture Notes in Physics, Vol. 75/76 (1978).
38. M. A. Goldshtik (ed.), *Structural Turbulence* (Nauka, Novosibirsk, 1982) (in Russian).
39. M. A. Goldshtik and V. N. Shtern, *Hydrodynamic Stability and Turbulence* (Nauka, Moscow, 1977) (in Russian).
40. A. S. Monin, Coherent structures in turbulent flows, *Russian J. Comput. Mech.* **1**(1), 5–13 (1993).
41. G. L. Brown and A. Roshko, On density effect and large structure in turbulent mixing layers, *J. Fluid. Mech.* **64**, 775–816 (1974).
42. G. R. Offen and S. J. Kline, A proposed model of the bursting process in turbulent boundary layers, *J. Fluid. Mech.* **70**, 209–228 (1975).
43. A. K. M. F. Hussain, Coherent structures and turbulence, *J. Fluid Mech.* **173**, 303–356 (1986).
44. J. L. Lumley (ed.), Whither turbulence? *Turbulence at the Crossroads*, (Springer, New York, 1990).
45. G. I. Barenblatt, S. I. Voropaev and I. A. Filippov, The model of Fedorov's coherent structures in upper layer of ocean, *Dokl. Akad. Nauk SSSR.* **307**(3), 720–724 (1989) (in Russian).
46. M. V. Melander and F. Hussain, NASA Report (CTR-S 88) (1988).
47. V. V. Struminskii, *Turbulent Flows* (Nauka, Moscow, 1977) (in Russian).
48. H. I. Dryden, Recent advances in the mechanics of boundary layer flows, *Adv. Appl. Mech.* **1**, 1–40 (1948).
49. O. M. Belotserkovskii and L. I. Severinov, Conservative method of "fluxes" and calculation of viscous heat-conducting flow past a finite body, *USSR Comput. Math. Math. Phys.* **13**(2), 385–397 (1973).

50. A. V. Babakov, Numerical simulation of some problems of aerohydrodynamics, Preprint of Computer Center AN SSSR Moscow (1986) (in Russian).
51. A. V. Babakov, On the possibility of numerical simulation of unstable vortex structeres in the near wake *USSR Comput. Math. Math. Phys.* **28**(2), 173–180 (1988).
52. A. V. Babakov, Modeling of large-ordered coherent structures in the near wake, in *Study of Turbulence* (Nauka, Moscow, 1994), pp. 223–245 (in Russian).
53. N. N. Yanenko and Yu. I. Shokin, About first differential approximation of difference schemes for hyperbolic system of equations, *Sib. Mat. Zh.* **10**(5), 1173–1187 (1969) (in Russian).
54. Yu. I. Shokin, *Method of differential Approximation* (Nauka, Novosibirsk, 1979) (in Russian).
55. V. M. Ievlev, *Turbulent Motion of High-Temperature Continua* (Nauka, Moscow, 1975) (in Russian)
56. V. M. Ievlev, *Numerical Modeling of Turbulent Flows* (Nauka, Moscow, 1990) (in Russiaian).
57. O. M. Kuznetsov and S. G. Popov, Vortices in a plane gas dynamic wake behind a cylinder, *Fluid Dynam.* (2), 112–113 (1967).
58. D. V. Bazhenov and L. A. Bazhenova, The influence of acoustic disturbances on wind noise, *Proc. 2nd All-Union Symp. Physics of Acoustic-Hydrodynamic Phenomena in Optoacoustics* (Nauka, Moscow, 1982), pp. 105–109 (in Russian).
59. S. B. Pope, The calculation of turbulent recirculating flows in general orthogonal coordinates, *J. Comput. Phys.* **26**(3), 197–217 (1978).
60. A. V. Babakov, O. M. Belotserkovskii and A. P. Zuzin, About two regimes of the compressible flow around cylinder, *Dokl. Akad. Nauk SSSR.* **279**(2), 315–318 (1984).
61. D. C. Bernard, F. H. Harlow, R. M. Rauenzahn and C. Zemach, Spectral transport model for turbulence, Report LA-UR-92-1666, Los Alamos National Laboratory (1992).
62. V. E. Yanitskii, A statistical method for solving some problems of the kinetic theory of gases and turbulence, Doctoral thesis, Moscow (1984) (in Russian).
63. O. M. Belotserkovskii and V. E. Yanitskii, The statistical method of particles-in-cells for solving problems of rarefied gas dynamics 1, 2, *USSR Comput. Math. Math. Phys.* **15**(5), 184–198 (1975).
64. O. M. Belotserkovskii, A. I. Erofeyev and V. E. Yanitskii, Direct statistical simulation of aerohydrodynamic problems, *Uspekhi Mekh.* **5**(3–4), 11–40 (1982) (in Russian).
65. A. T. Onufriev, On the equations of the semi-empirical theory of turbulence, *Zh. Prikl. Mekh. I. Tekhn. Fiz.* (2), 66–71 (1970) (in Russian).
66. E. Naudasher, Flow in the wake of a self-propelling body and related sources of turbulence, *J. Fluid Mech.* **22**(4), 625–656 (1965).
67. O. V. Troshkin and V. E. Yanitskii, Intermittency in a problem of free turbulence, Preprint of Computer Center AN SSSR, Moscow (1988) (in Russian).

68. O. M. Belotserkovskii, V. A. Gushchin and V. V. Shchennikov, A method of splitting for solving problems of viscous incompressible flows, *USSR Comput. Math. Math. Phys.* **15**(5), 190–199 (1975).
69. V. A. Gushchin, Viscous flows past three-dimensional bodies, *USSR Comput. Math. Math. Phys.* **16**(2), 251–256 (1976).
70. O. M. Belotserkovskii, V. A. Gushchin and V. N. Kon'shin, A method of splitting for analyzing stratified free-surface flows, *USSR Comput. Math. Math. Phys.* **27**(4), 181–191 (1987).
71. S. Murakami, S. Kato and Y. Suyama, Numerical and experimental study on turbulent diffusion field in conventional flow type clean room, *ASHRAE Trans.* **94**(2), 469–493 (1988).
72. T. H. Kuehn, Computer simulation of airflow and particles transport in clean rooms, *J. Environ. Sci.* **9–10**, 21–27 (1988).
73. A. S. Monin, Hydrodynamic instability, *Uspekhi Fiz. Nauk.* **150**(1), 61–105 (1986) (in Russian).
74. W. Frost and T. H Moulden, *Handbook of Turbulence. Fundamentals and Applications.* (Plenum Press, New York, London, 1977).
75. S. O. Belotserkovskii, A. P. Mirabel and M. A. Chusov, On the construction of supercritical modes for a plane periodic flow, *Izv. Akad. Nauk SSSR. Ser. Fiz. Atmosf. I. Okeana* **14**(1), 11–20 (1978) (in Russian).
76. S. O. Belotserkovskii, Simulation of viscous incompressible flows based on Navier–Stokes equations, Doctoral thesis, Moscow (1979) (in Russian).
77. S. O. Belotserkovskii and V. A. Gushchin, New numerical models in the mechanics of continua, *Uspekhi Mekh.* **8**(1), 97–150 (1985) (in Russian).
78. V. I. Yudovich, An example of the origination of a secondary steady or periodic flow due to the loss of stability in a laminar viscous incompressible flow, *Prikl. Mat. I. Mekh.* **29**(3), 453–467 (1965) (in Russian).
79. L. D. Meshalkin and Ya. G. Sinay, Study of stability of steady soluition for one equation system of plane motion of incompressible viscous fluid, *Prikl. Mat. I. Mekh.* **25**(6), 1140–1143 (1961) (in Russian).
80. N. F. Bondarenko, M. Z. Gak and F. V. Dolzhanskii, Laboratory and theoretical models of plane periodic flow, *Izv. Akad. Nauk SSSR. Ser. Fiz. Atmosf. I. Okeana* **15**(10), 1017–1026 (1979) (in Russian).
81. W. Platt, L. Sirovich and N. Fitzmaurice, An investigation of chaotic Kolmogorov flows, *Phys. Fluids.* **3**(4), 681–696 (1991).
82. S. O. Belotserkovskii and A. R. Pastushkov, Modeling of large-ordered turbulence in ocean, Preprint of Computer Center AN SSSR, Moscow (1992) (in Russian).
83. V. M. Kamenkovich, S. O. Belotserkovskii and M. S. Panteleeva, To question about numerical modeling of barotropic flows generated by large-ordered field of wind, *Izv. "POLIMODE"* (IO AN SSSR, Moscow, 1985) 15, 3–32 (in Russian).
84. H. U. Sverdrup, Wind-driven currents in baroclinic ocean: with application to the equatorial currents of the Eastern Pacific, *Proc. Acad. Sci.* **33**, 318–326 (1947).

85. V. M. Kamenkovich, Foundations of ocean dynamics (Gidrometeoizdat, Leningrad 1973) (in Russian).
86. W. R. Holland, Oceanic general circulation models, in *The Sea*, Vol. 6 (John Wiley and Sons, 1997), pp. 3–45.
88. O. M. Belotserkovskii, S. O. Belotserkovskii and A. R. Pastushkov, Numerical modeling of internal waves in flow of stratified fluid around semi-circle, *Probl. Prikl. Mat. I. Inform.* (Nauka, Moscow, 1987), pp. 11–21.
89. *Problems of Applied Mathematics and Informatics* (Nauka, Moscow, 1987) (in Russian).
90. D. J. Olbers, Models of the oceanic internal waves field, *Rev. Geophys. Space Phys.* **21**(7), 1567–1606 (1983).
91. T. H. Bell, Topographically generated internal waves in the open ocean, *J. Geophys. Res.* **80**(3), 320–327 (1975).
92. O. M. Belotserkovskii, S. O. Belotserkovskii and V. A. Gushchin, Numerical modeling of unsteady (periodic) flow of viscous fluid in the wake past cylinder, *USSR Comput. Math. Math. Phys.* **24**(8), 181–191 (1984).
93. N. A. Inogamov, The turbulent stage of the Taylor instability, *Pis'ma v Zh. Tekhn. Fiz.* **4**(12), 743–747 (1978) (in Russian).
94. O. M. Belotserkovskii, Yu. M. Davydov and A. Yu. Dem'yanov, Intermode interaction in the case of the Rayleigh–Taylor instability, *Dokl. Akad. Nauk SSSR.* **288**(5), 1071–1074 (1986) (in Russian).
95. Yu. M. Davydov and A. Yu. Dem'yanov, Application of the method of large particles to numerical simulation of stabilization and unification of the Rayleigh–Taylor instability harmonics, Preprint of Computer Center AN SSSR, Moscow (1987).
96. M. I. Budiko, G. S. Golitsyn and Yu. F. Izrael, Global climate catastrophes. Nature of climate catastrophes. Influence of a nuclear conflict on a climate, Gidrometeoizdat, Moscow (1985) (in Russian).
97. A. M. Grishin, Mathematical modeling of forest fire and new methods of its suppression (Nauka, Novosibirsk, 1992) (in Russian).
98. G. S. Golitsyn, Yu. A. Gostinzev and A. F. Solodovnic, Turbulent buoyancy jet into the stratified atmosphere, *Zh. Prikl. Mekh. I. Techn. Fiz.* **6**, 61 (1989).
99. V. A. Balabanov, M. A. Mechkov, A. T. Onufriev, N. A. Safarov, R. A. Safarov and V. E. Skorovarov, The dynamical pattern of the turbulent convective high-altitude transport generated by large area sources of buoyancy in the stratified atmosphere, *Proc. 1st Nat. Russian Conf.* (1994).
100. I. F. Muzafarov and S. V. Utyuzhnikov, Numerical simulation of convective columns over large conflagration into a atmosphere, *Teplofizika Vysokikh Temperatur.* **33**(4), 594–601 (1995).
101. A. V. Konyuchov, M. V. Meshcheryakov, I. F. Muzafarov, A. T. Onufriev, N. A. Safarov, R. A. Safarov and S. V. Utyuzhnikov, Laboratory and numerical simulation of convective flow over large-scale source of energy, in *Co-ordination Meeting in INTAS/ERFORTAC Project "Pan-European Network on Flow, Turbulence and Combustion"* (1996).

102. J. E. Penner, L. C. Hasselan and L. L. Edwards, Buoyant plume calculations, *AIAA Paper* **459**, 1 (1985).
103. B. R. Morton, G. I. Taylor and J. S. Turner, Turbulent gravitational convection from maintained and instantaneous sources, *Proc. Roy. Soc. A.* **234**, 1–23 (1956).
104. E. G. Shifrin, About turbulence, in *Study of Turbulence* (Nauka, Moscow, 1994) pp. 86–102 (in Russian).
105. N. E. Kochin, L. A. Kibel and N. V. Roze, *Theoretical Hydrodynamics*, Part 2 (Fizmatgiz, Moscow, 1963) (in Russian).
106. A. N. Tikhonov and V. Ya. Arsenin, *Methods for Solving Ill-Posed Problems* (Nauka, Moscow, 1986) (in Russian) (English translation: Halsted-Whiley).
107. G. Nikolis and P. Prigogine, *Self-Organization in Nonequilibrium Systems* (Mir, Moscow, 1979) (in Russian translation).
108. H. Haken, *Synergetics* (Springer Verlag, 1980).
109. M. I. Rabinovich and D. I. Trubetskov, *Introduction to the Theory of Oscillations and Waves* (Nauka, Moscow, 1984) (in Russian).
110. P. Berge, Y. Pomeau and C. Vidal, *L'ordre dans le chaos*, Inst. des Hautes Etudes Scientifiques (1984); *Order in Chaos* (Mir, Moscow, 1991) (in Russian translation).

# Chapter 2

# Modeling of Richtmyer–Meshkov Instability

This chapter summarizes a numerical study of the Richtmyer–Meshkov instability evolving when a shock wave passes from a light medium into a dense one (fast–slow combination). It is shown that apart from the physical parameters of the media and the shock-wave intensity, wave refraction at the perturbed interface is important. In particular, a secondary shock-wave mechanism that accounts for the penetration of light-phase bubbles into the dense phase and of dense jets into the light phase and explains the leveling of the fronts of the refracted and reflected shock wave is discussed. The results of numerical simulation agree with the experimental measurements.

## Introduction

Most problems in modern physics that hopefully will bring practical results, such as that of harnessing nuclear fusion, break down into a variety of smaller problems that must be solved for the success of the project as a whole.

We have already shown[1] that the stability of the fuel compression process is crucial in generating break-even fusion conditions, that is, conditions under which the fuel will be compressed to a density tens of thousands of times, the initial value and the temperature will be raised to several keV. Obviously, such transitions will generate large gradients of pressure, velocity, and other parameters. The experimental difficulties involved in that are well known, but numerical simulation is also difficult.

The most constructive approach is to subdivide the complex overall physical process into specific parts amenable to an analysis that will extract the basic mechanisms behind the observable phenomena, by neglecting the physical processes that are not significant in a given range of parameters.

Years of experimental and theoretical studies have proved that such hydrodynamic instabilities as the Rayleigh–Taylor, Richtmyer–Meshkov, and Kelvin–Helmholtz types have important effects on stable target compression. These instabilities occur not only in dense high-temperature plasmas, but are also widely observed in continuous media under normal conditions. If the basic properties of these instabilities were retained under such normal conditions, then it would be much simpler to study them, even though that still doesn't lead to a problem's solution.

In this chapter, we shall analyze the problems generated by the evolution of the Richtmyer–Meshkov instability induced by the interaction of a traveling shock wave and a perturbed interface between two media of different densities.

When a plane shock wave is incident on such a plane interface, the discontinuity degenerates. Depending on the values of the thermodynamic parameters and velocities in the media on two sides of the interface, the result will be either of the two cases, namely, shock waves will propagate through both media, or a shock wave will propagate in the less-dense medium, while an expansion wave will travel through the denser medium. If the incident shock-wave front and the plane interface extend sufficiently far in the direction perpendicular to the shock, then the process can be treated as one-dimensional. Such a process is well understood and can be described analytically. If, however, the perturbed interface is not plane but undulating, then the process becomes much more complicated. Richtmyer[2] was among the first to show that, and his work was later experimentally confirmed by Meshkov.[3] The evolution of a Richtmyer–Meshkov instability (hereafter referred to as the RMI) is illustrated in Fig. 2.1. This figure, for which we thank Prof. Zaitsev,[4] shows schlieren photographs made by Zaitsev's group in experiments with a shock wave traveling from argon into xenon. This photograph depicts the first of the two one-dimensional cases we have cited above, i.e. that with two shock waves (a refracted and a reflected one). Henceforth, we will deal with this case only. In the second case, i.e. with a shock wave and an expansion wave, there occurs a change in phase of the interface, whereby convexities are transformed into concavities, and vice versa. The interface itself becomes unstable as well. We shall not deal with this case here because it is much more complicated to be described within the space allotted here.

The Mach number of a shock wave incident from argon on undulating interfaces with wavelengths of 7.2, 3.6, 2.4, 1.2, and 0.8 cm was 3.5 in all the Zaitsev's experiments. The initial amplitude of the sinusoidal interface perturbation was 1 cm. The photographs in the columns of Fig. 2.1 correspond to the various periods of the RMI evolution for the wavelengths indicated at the top. The schlieren photographs in the rows refer to the

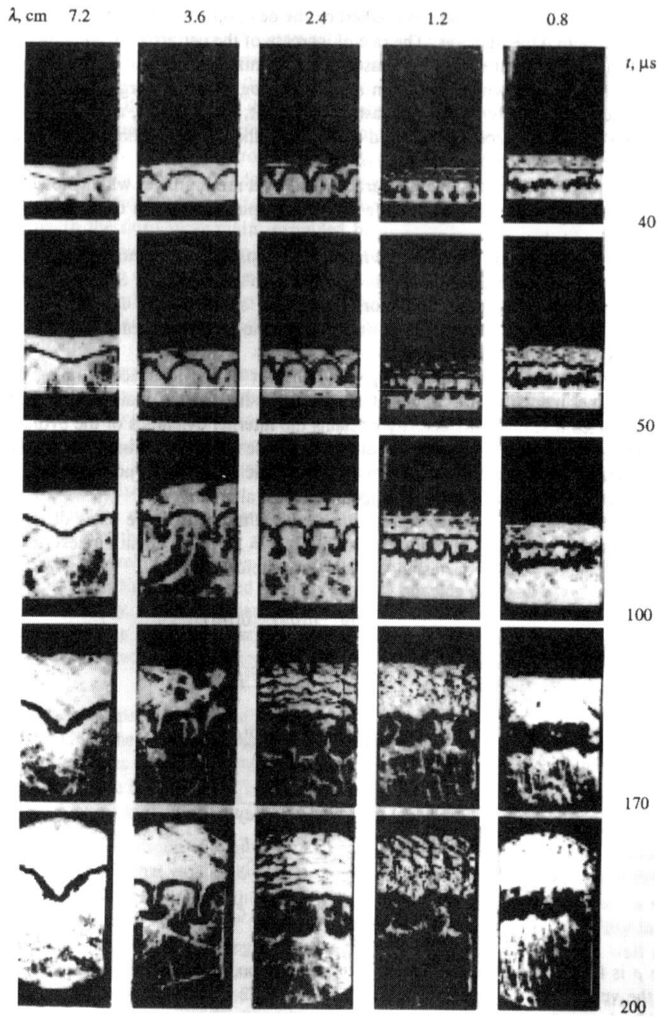


**Fig. 2.1.** Schlieren photographs of the Richtmyer–Meshkov instability evolution.

same time from the beginning of the wave-interface interaction (this time, in microseconds, is indicated at the right) for various wavelengths of the initial interface perturbation. The pressure on the interface between the media before the shock-wave incidence was 0.5 atm. The evolution of the instability can be nominally subdivided into four stages: the linear, nonlinear, transitional, and turbulent stages. The first, or linear, stage is characterized by a linear increase in the perturbation amplitude with time, as was predicted by Richtmyer[2] within the framework of the linear approximation $a \ll \lambda$ (where $a$ is the amplitude, and $\lambda$ is the wavelength of the interface perturbation):

$$\frac{da}{dt} = ka^*\mathrm{At} \cdot V.$$

Here, $a^*$ is the amplitude of the interface perturbation after the passage of a shock wave, $k = 2\pi/\lambda$ is wave number, $\mathrm{At} = (\rho_1 - \rho_0)/(\rho_1 + \rho_0)$ is Atwood number, $\rho_1$ and $\rho_0$ are densities of the gases on the left and the right of the interface, and $V$ is the velocity of the interface between the gases after the interaction. The rates of penetration of the light gas into the heavy one (in the form of bubbles) and of the heavy gas into the light one (in the form of jets) prove to be identical, as demonstrated by the schlieren photographs in the first column corresponding to a wavelength of 7.2 cm.

The second, or nonlinear, stage is distinguished by smaller bubbles and larger jets, accompanied by a slower rate of penetration of one medium into the other. The process ends in the formation of mushroom structures at the tips of the jets, as illustrated by the schlieren photographs for three early instants of time (second and third columns of Fig. 2.1).

The third, or transitional, stage is marked by the development of vortices at the tips of the jets penetrating into the light gas. The rate of increase of the perturbation amplitude is smaller than that in the nonlinear stage. The gases begin to mix, and the interface becomes diffuse. At some stage, the growing mushroom structures progressively merge. This is the end of the transitional stage, after which the last, or turbulent, stage begins, as is clearly evident in the two bottom photographs in the third column and the top three photographs in the fourth column, corresponding to a wavelength of 1.2 cm.

The fourth, or turbulent, stage is characterized by a mixing layer which slowly increases in thickness, as illustrated by the schlieren photographs in the fifth column corresponding to a wavelength of 0.8 cm.

In general, we can make the following conclusions from the photographs of Fig. 2.1. First, the duration of the separate stages decreases with wavelength. Second, the Richtmyer linear approximation agrees satisfactorily with the experimental data for longer wavelengths and for the initial instants of time. Third, the rate of increase in the perturbation amplitudes slows down from one stage to another.

The physical mechanism underlying this behavior of the interface is rather difficult to ascertain experimentally in a way that allows the values of the main physical parameters to be estimated and that satisfactorily explains the internal dynamics of the evolution of instability. A theoretical analysis is also difficult because the processes in question are described by nonlinear partial differential equations, most of which are thus far not amenable to analytic solution. Thus, numerical simulation appears to be the best alternative.[5]

The hydrodynamic processes were numerically simulated on the basis of the model of an ideal multicomponent compressible fluid with $\gamma = 5/3$, which can be described by a set of Euler's equations in a Cartesian coordinate system $(x, y)$, i.e.

$$\frac{\partial \rho}{\partial t} + \frac{\partial(\rho u)}{\partial x} + \frac{\partial(\rho v)}{\partial y} = 0, \quad \frac{\partial \rho\xi}{\partial t} + \frac{\partial(\rho\xi u)}{\partial x} + \frac{\partial(\rho\xi v)}{\partial y} = 0,$$

$$\frac{\partial \rho u}{\partial t} + \frac{\partial(\rho u^2 + P)}{\partial x} + \frac{\partial(\rho uv)}{\partial y} = 0, \quad \frac{\partial \rho v}{\partial t} + \frac{\partial(\rho uv)}{\partial x} + \frac{\partial(\rho v^2 + P)}{\partial y} = 0,$$

$$\frac{\partial \rho E}{\partial t} + \frac{\partial u(\rho E + P)}{\partial x} + \frac{\partial v(\rho E + P)}{\partial y} = 0, \quad P = \rho\xi\frac{R}{\mu_1}T + \rho(1-\xi)\frac{R}{\mu_2}T.$$

Here $\rho$ is density, $\xi$ is fractional concentration, $u$ is velocity in the $x$-direction, $v$ is the velocity in the $y$-direction, $E = e + (u^2 + v^2)/2$ is total specific energy, $e$ is specific internal energy, $P$ is pressure, $T$ is temperature, $e = (C_{v1}\xi_1 + C_{v2}\xi_2)T$, $\mu_1$ and $\mu_2$ are molar weights of the components, $C_{v1}$ and $C_{v2}$ are specific heats at constant volume, and $R$ is the universal gas constant.

The geometry of the domain was rectangular. A shock wave was assumed to propagate in the $x$-direction, and sinusoidal interface perturbations were plotted along the $y$-axis, so that half the wavelength of the initial interface perturbation was arranged in this direction. This enabled symmetry to be used as the boundary condition for fixed values of $y$. The following initial conditions were maintained at the $x$-boundaries of the domain: the parameters behind the incident shock wave were set on the right boundary, while those of the quiescent denser gas were set on the left boundary. At time zero, the integration domain was subdivided into three parts. In the first subdomain, confined between the left boundary and the undulating perturbed interface, the parameters of the quiescent xenon were given at a pressure of 0.5 atm and at normal temperature. In the second subdomain, confined between the interface and the incident shock-wave front, the density, the mass concentration, and the specific energy of the quiescent helium were set at 0.5 atm and at room temperature. In the third subdomain, extended behind the shock-wave front, the physical parameters were calculated by taking into account the Mach number of the shock wave and the parameters of the quiescent helium ahead of the shock wave. The Mach number for the calculations presented in this chapter was 2.5. A stationary finite-difference grid covered the integration domain with 520 cells along the $x$-coordinate and 36 cells along the $y$-coordinate.

## 2.1. Numerical method

For correct definition of the continuous media behavior it is necessary to satisfy three main conservation laws: mass, momentum, and energy. The mathematical formulation of these laws results in the quasilinear partial differential equations system of the hyperbolic type. In 3D Cartesian space this equations system can be written in the vector-matrix form:

$$\frac{\partial \mathbf{W}}{\partial t} + A\frac{\partial \mathbf{W}}{\partial x} + B\frac{\partial \mathbf{W}}{\partial y} + C\frac{\partial \mathbf{W}}{\partial z} = 0, \tag{2.1}$$

$$\mathbf{W}^T = (\rho, \rho u, \rho v, \rho w, \rho E),$$

$$A = \begin{pmatrix} 0 & 1 & 0 & 0 & 0 \\ -u^2 + P'_\rho & 2u + P'_{\rho u} & P'_{\rho v} & P'_{\rho w} & P'_{\rho E} \\ -uv & v & u & 0 & 0 \\ -uw & w & 0 & u & 0 \\ u\left(P'_\rho - E - P/\rho\right) & E + P/\rho + uP'_{\rho u} & uP'_{\rho v} & uP'_{\rho w} & u\left(1 + P'_{\rho E}\right) \end{pmatrix},$$

$$B = \begin{pmatrix} 0 & 0 & 1 & 0 & 0 \\ -uv & v & u & 0 & 0 \\ -v^2 + P'_\rho & P'_{\rho u} & 2v + P'_{\rho v} & P'_{\rho w} & P'_{\rho E} \\ -vw & 0 & w & v & 0 \\ v\left(P'_\rho - E - P/\rho\right) & vP'_{\rho u} & E + P/\rho + vP'_{\rho v} & vP'_{\rho w} & v\left(1 + P'_{\rho E}\right) \end{pmatrix},$$

$$C = \begin{pmatrix} 0 & 0 & 0 & 1 & 0 \\ -uw & w & 0 & u & 0 \\ -vw & 0 & w & v & 0 \\ -w^2 + P'_\rho & P'_{\rho u} & P'_{\rho v} & 2w + P'_{\rho w} & P'_{\rho E} \\ w\left(P'_\rho - E - P/\rho\right) & wP'_{\rho u} & wP'_{\rho v} & E + P/\rho + wP'_{\rho w} & w\left(1 + P'_{\rho E}\right) \end{pmatrix},$$

where $x$, $y$, $z$, $t$ — independent variables; $\rho$ — density; $u$, $v$, $w$ — velocity components; $E$ — total specific energy; $P = P(\mathbf{W})$ — pressure. Let us introduce diagonal matrices, composed of linear-independent eigenvalues of matrices $A$, $B$, $C$ and their absolute values:

$$\Lambda_A = \mathrm{diag}\{u+c, u, u, u, u-c\}, \ |\Lambda_A| = \mathrm{diag}\{|u+c|, |u|, |u|, |u|, |u-c|\},$$
$$\Lambda_B = \mathrm{diag}\{v+c, v, v, v, v-c\}, \ |\Lambda_B| = \mathrm{diag}\{|v+c|, |v|, |v|, |v|, |v-c|\},$$
$$\Lambda_C = \mathrm{diag}\{w+c, w, w, w, w-c\}, \ |\Lambda_C| = \mathrm{diag}\{|w+c|, |w|, |w|, |w|, |w-c|\},$$

where $c$ is the sound velocity, and also matrices $\Omega_A$, $\Omega_B$, $\Omega_C$, consisting of independent eigenvectors of matrices $A^T, B^T, C^T$. Let us multiply from the left equations (2.1) by matrix $\Omega_A^T$ and use the following correlation:

$$\Omega_A^T A = (A^T\Omega_A)^T = (\Omega_A\Lambda_A)^T = \Lambda_A\Omega_A^T. \tag{2.2}$$

The transformed equation takes the form

$$\Omega_A^T\frac{\partial \mathbf{W}}{\partial t} + \Lambda_A\Omega_A^T\frac{\partial \mathbf{W}}{\partial x} + \Omega_A^T B\frac{\partial \mathbf{W}}{\partial y} + \Omega_A^T C\frac{\partial \mathbf{W}}{\partial z} = 0. \tag{2.3}$$

Let us split matrix $\Lambda_A$ into two: nonnegative $\Lambda_A^+ = (\Lambda_A + |\Lambda_A|)/2$ and nonpositive $\Lambda_A^- = (\Lambda_A - |\Lambda_A|)/2$. Then let us introduce the two-point approximation of the first derivatives with respect to $x$, taking into account the inclination of characteristics, and multiply the left part of equation (2.3) by matrix $(\Omega_A^T)^{-1}$, which always exists, because the matrix $\Omega_A^T$ consists of linear-independent eigenvectors of matrix $A^T$, and therefore, is nonsingular, having in the result

$$\begin{aligned}\frac{\partial \mathbf{W}}{\partial t} &+ (\Omega_A^T)^{-1}\Lambda_A^-\Omega_A^T\frac{\mathbf{W}_{l+1,m,k}^n - \mathbf{W}_{l,m,k}^n}{h_1} \\ &+ (\Omega_A^T)^{-1}\Lambda_A^+\Omega_A^T\frac{\mathbf{W}_{l,m,k}^n - \mathbf{W}_{l-1,m,k}^n}{h_1} + B\frac{\partial \mathbf{W}}{\partial y} + C\frac{\partial \mathbf{W}}{\partial z} = 0,\end{aligned} \tag{2.4}$$

where $h_1$ is the difference grid step along $x$. From the comparison of equations (2.1) and (2.4), it is seen that the equations elements, which do not depend on the derivation with respect to $x$, have no distinctions. Therefore, the transformations, carried out for approximating the first derivative with respect to $x$, can be used for approximating the first derivatives with respect to $y$ and $z$, if we define eigenvalues and eigenvectors of matrices $B^T$ and $C^T$. Introducing the difference approximation of the derivative with respect to time between the nearest time layers, we will come to the final formulae fit for description of transition from the time layer $n$ to the time layer $n+1$[1]:

$$\begin{aligned} \mathrm{W}^{n+1}_{l,m,k} &= \mathrm{W}^{n}_{l,m,k} \\ &\quad + ((\Omega^T_A)^{-1}\Lambda^-_A\Omega^T_A(\mathrm{W}^n_{l+1,m,k} - \mathrm{W}^n_{l,m,k}) + (\Omega^T_A)^{-1} \\ &\qquad \Lambda^+_A\Omega^T_A(\mathrm{W}^n_{l,m,k} - \mathrm{W}^n_{l-1,m,k}))\,\frac{\tau}{h_1} \\ &\quad + \left(\left(\Omega^T_B\right)^{-1}\Lambda^-_B\Omega^T_B\left(\mathrm{W}^n_{l,m+1,k} - \mathrm{W}^n_{l,m,k}\right) + \left(\Omega^T_B\right)^{-1}\right. \\ &\qquad \left.\Lambda^+_B\Omega^T_B\left(\mathrm{W}^n_{l,m,k} - \mathrm{W}^n_{l,m-1,k}\right)\right)\,\frac{\tau}{h_2} \\ &\quad + \left(\left(\Omega^T_C\right)^{-1}\Lambda^-_C\Omega^T_C\left(\mathrm{W}^n_{l,m,k+1} - \mathrm{W}^n_{l,m,k}\right) + \left(\Omega^T_C\right)^{-1}\right. \\ &\qquad \left.\Lambda^+_C\Omega^T_C\left(\mathrm{W}^n_{l,m,k} - \mathrm{W}^n_{l,m,k-1}\right)\right)\,\frac{\tau}{h_3}, \end{aligned} \tag{2.5}$$

where $h_2$, $h_3$, and $\tau$ are the difference grid steps along variables $y$, $z$, and time $t$, respectively.

## 2.2. Model calculations

The model calculation set[6] has been performed to examine the developed schemes' properties, in which the possibility of comparison with already known analytic solutions was provided. The first task formulation is connected with the realization of "Vega" project — the space vehicle flight to the Galley comet, in the course of which the relative coming close velocity between vehicle and comet was equal to 80 km/s. So, there are two infinite planar layers, made of the same material, for instance, of aluminum, the first of them has 10 $\mu$m thickness and the second — 50 $\mu$m. At the first moment of time their temperature is the same and equal to 250 K. The layers are surrounded by the same material stationary vapor with a density of $10^{-5}$ g/cm$^3$ and at the same temperature. The thicker layer collides with the thinner one at the relative velocity of 80 km/s. The state equation with the isentropic exponent 5/3 was used. At the beginning such formulated task has three discontinuity planes in the integration domain: there is the density discontinuity on the boundary of vapor — stationary layer; there is the velocity discontinuity on the boundary of two layers; the density and velocity are subjected to discontinuity in the contact place of moving layer — vapor. The discontinuity values achieve six–seven orders. On the

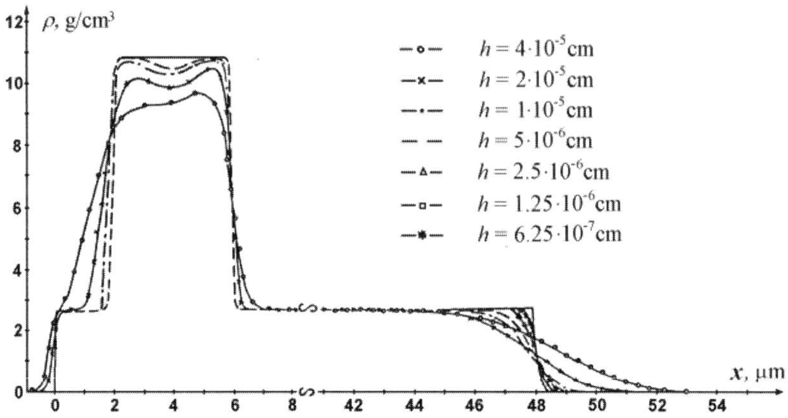


**Fig. 2.2.** The density at 150 ps after the beginning of the discontinuity decays at the boundaries of both the two layers and the moving layer — vapor calculated on the sequentially decreased difference grids.

two-layer boundary the discontinuity decay will be accompanied by the formation of two shock waves propagating in the opposite directions and the centered rarefaction wave will arise on the boundary of the moving layer — vapor. The results of numerical calculations on the sequentially decreased difference grids for these two cases are presented in Figs. 2.2 and 2.3. The different curves correspond to various calculations with the scheme (2.5) at the varying grid steps along space coordinates. In all cases uniform space grid was used, but it is doubled from one variant to another. The corresponding curves are noted by dotted and dot-and-dash lines. The solid line marks the analytic solution. Solving the problem of the discontinuity decay with the formation of a shock wave and a rarefaction wave leads to the results, presented in Fig. 2.4. At the initial moment at the left of the discontinuity gas-dynamic parameters had the following values: density — 2.7 g/cm$^3$; velocity — 80 km/s; pressure — 9000 atm, and at the right — density — $10^{-5}$ g/cm$^3$; velocity — zero; pressure — 0.033333 atm. The gas was assumed to be an ideal one with the isentropic exponent $\gamma = 1.4$.

Presented in Fig. 2.4, density and velocity distributions correspond to the time moment 15 ns after the beginning of the discontinuity decay. In the whole, in spite of the problem of the initial formulation complexity, that is expressed in the harsh density and velocity gradients present in the integration domain, when value jumps at a single mesh reach six or seven orders, there is the satisfactory correlation between exact solutions and results, obtained by means of the scheme (2.5). At the sequential grid

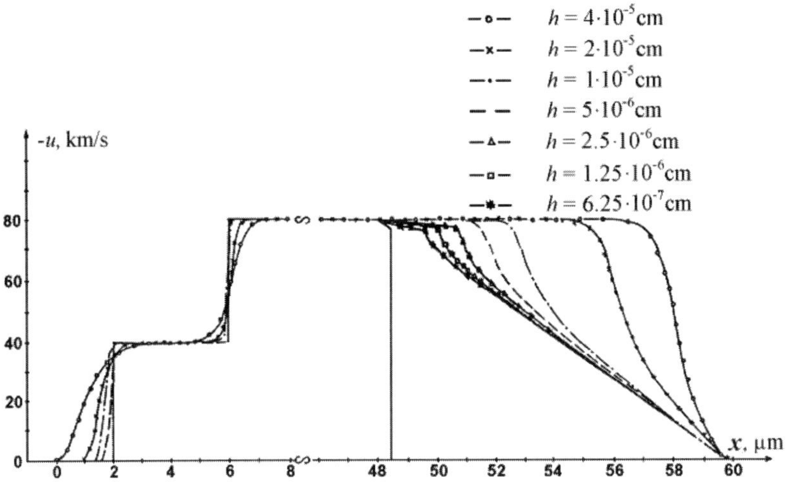


**Fig. 2.3.** The velocity at 150 ps after the beginning of the discontinuity decays at the boundaries of both the two layers and the moving layer — vapor calculated on the sequentially decreased difference.

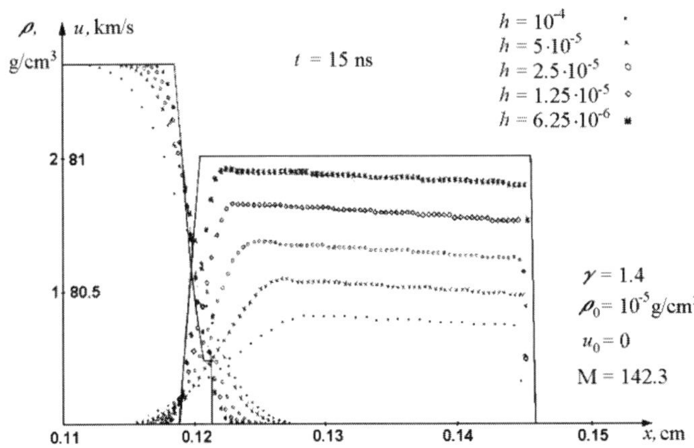


**Fig. 2.4.** The density and velocity at 15 ns after the beginning of discontinuity decays with shock wave formation and rarefaction wave calculated on the sequentially decreased difference grids.

reduction in size it was noted that the integral curves uniformly converge for all parameters.

Another set of model calculations was connected with the problem of inertial confinement fusion (ICF).[7,8]

Recently considerable attention has been paid to the problem of the interaction of the radiation from powerful lasers with matter. In the simulation of the physical process it is necessary to allow for such diverse phenomena as hydrodynamic motion, electron heat conduction, the exchange of energy between the electron and ion components of the plasma, the absorption of laser radiation, spontaneous magnetic fields, etc. The result is a complex mathematical model consisting of a system of nonlinear partial differential equations.

Another difficulty is that the results of the numerical calculations performed on "coarse" meshes are strongly dependent on the features of the specific physical model.[9,10] The problem arises of an additional estimation of the results of simulation in order to exhibit the most important parameters. This can be done by a physical experiment, but because of the extremal conditions of the course of the physical processes in it, as a rule integral values of the secondary phenomena are measured: the neutron flux, the velocity of motion of the region of maximum radiation, etc.[11] There is therefore an increase in the role of exact self-similar solutions obtained on simplified models, but describing some particular aspects of the process.

Analytic solutions can be successfully used to check out the programs. Moreover, they permit a rapid and economical investigation of the qualitative features of the processes involved and find out the most efficient models.

In this section, a comparative study is made of one hydrodynamic compression process. In Sec. 2.2.1, we briefly describe the solution of the Cauchy problem for a system of Eulerian equations with specially chosen data. In Sec. 2.2.2, a comparison is made with the numerical solution obtained and considered in Ref. 7. In Sec. 2.2.3, a comparison is made of the results of calculations by different models, one of which is generally accepted, and the other is based on the solution presented in Sec. 2.2.1.

### *2.2.1. The Couchy problem for one-dimensional isotopic flow of an ideal gas*

We consider a spherically-symmetric, one-dimensional, isotropic flow of an ideal gas. The general theory of these flows was developed in Refs. 12 and 13. In Ref. 14, an exact solution of the hydrodynamic equations was given for the case where the velocity satisfies the relation $u = r\Phi(t)$. However, to obtain the test version it was necessary to make specific the statement of the problem and choose a complete definite function on which the solution depends. For this purpose the Cauchy problem was formulated with specific initial profiles of velocity and density. The method of finding the unknown relations was not connected with the general solution mentioned above; nevertheless, the actual solution can be obtained from it with additional assumptions.

The method used is also suitable for the analysis of more complex problems, where, in addition to the hydrodynamic phenomena account is also taken of the nonlinear thermal conductivity, or of the exchange of energy between the electron and ion components in the plasma, or of both together. This is easily verified by making similar calculations. There are a series of papers where related problems are investigated.[15–23] A distinguishing feature of this task is that the solutions obtained in explicit form are directly connected with the initial density and velocity profiles.

We formulate the Cauchy problem:

$$\frac{\partial\rho}{\partial t}+\rho\frac{\partial u}{\partial r}+u\frac{\partial\rho}{\partial r}+2\frac{u\rho}{r}=0,\quad \frac{\partial u}{\partial t}+u\frac{\partial u}{\partial r}=-\frac{\nabla P}{\rho},$$
$$P=c_0\rho^{\gamma},\quad \rho|_{t=t_0}=Ar^3,\quad u|_{t=t_0}=Br,$$
$$\rho(r_0,t_0)=\rho_0,\quad u(r_0,t_0)=u_0. \tag{2.6}$$

To be specific, let $\gamma=5/3$. We will seek a solution in the form $\rho=r^3f(t)$ and $u=rg(t)$. This form of relations for density and velocity is specified because after substituting them in the initial system (2.6) it is transformed into two ordinary differential equations for the functions $f(t)$ and $g(t)$, and at the same time the initial conditions

$$\frac{df(t)}{dt}+6f(t)g(t)=0\quad\text{and}\quad\frac{dg(t)}{dt}+g^2(t)+5\,c_0f^{2/3}(t)=0 \tag{2.7}$$

are satisfied. Expressing $g(t)$ in terms of $f(t)$ and substituting into the second equation of (2.7), we obtain

$$\frac{d^2\ln f(t)}{dt^2}-\frac{1}{6}\left[\frac{d\ln f(t)}{dt}\right]^2-30\,c_0f^{2/3}(t)=0. \tag{2.8}$$

Solving (2.8), we find

$$\rho=c_2^3r^3\left[\left(c_1-\frac{c_2t}{6}\right)^2-180c_0\right]^{-3},$$
$$u=\frac{c_2r}{6}\left(\frac{c_2t}{6}-c_1\right)\left[\left(c_1-\frac{c_2t}{6}\right)^2-180c_0\right]^{-1},$$

$$r = \frac{c_3}{c_2^{1/2}}\left[\left(c_1 - \frac{c_2 t}{6}\right)^2 - 180c_0\right]^{-\frac{1}{2}}. \tag{2.9}$$

The solution possesses the interesting property that when $(c_1 - c_2t/6)^2 -180c_0 \to 0$, the paths of all the particles intersect at the same instant at the same point of space. We will determine the characteristics of the flow considered. For this we have to integrate the two equations $dr_+ = u + C$ and $dr_- = u - C$, where $C$ is the local speed of sound. As a result, we obtain

$$r_+ = c_4\left[\left(c_1 - \frac{c_2 t}{6}\right)^2 - 180c_0\right]^{1/2} \cdot \left|\frac{(180c_0)^{1/2} + (c_1 - c_2t/6)}{(180c_0)^{1/2} - (c_1 - c_2t/6)}\right|^{12^{-\frac{1}{2}}},$$

$$r_- = c_5\left[\left(c_1 - \frac{c_2 t}{6}\right)^2 - 180c_0\right]^{1/2} \cdot \left|\frac{(180c_0)^{1/2} + (c_1 - c_2t/6)}{(180c_0)^{1/2} - (c_1 - c_2t/6)}\right|^{-(12^{-\frac{1}{2}})}.$$

Since $1/2 \geq 12^{-1/2}$, all the characteristics intersect simultaneously at one point, the density $\rho$ and velocity $u$ tending to infinity. The explanation is that all the particles gather in the central region independently of the distance at which they are situated at the initial instant. Consequently, if a physical process described by the system of equations (2.6) can be organized, then compression to extremely high densities and temperatures becomes possible.

### *2.2.2. Boundary conditions*

It is complicated to maintain the initial conditions at very great distances with a sufficient degree of accuracy. This implies that for a practical realization, instead of the Cauchy problem considered, we must formulate another one which will take into account the finiteness of the domain of integration. This is possible if the velocity and density decrease rapidly with distance, but leads to more unwieldy calculations. Another approach is used to take into account the effect of the boundary conditions on the solution as a whole with unchanged initial data.

It is known from the general gas-dynamic theory that the surface of the perturbation front corresponds to a characteristic hypersurface. If the characteristics emerge from points where a noticeable deviation from the theoretical values is observed, and the characteristics of the region situated close to the center do not intersect before some characteristic time $t'$, then

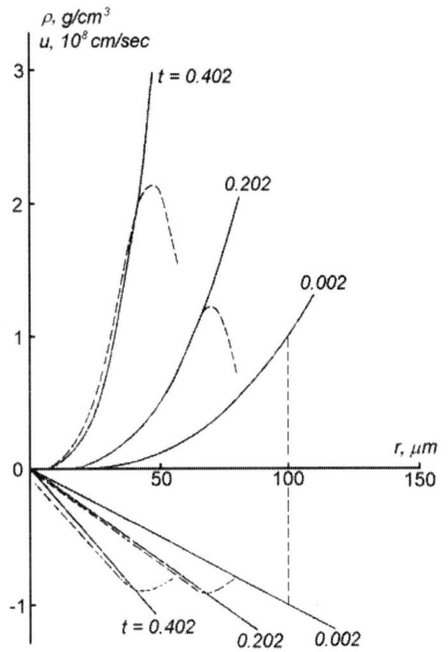


**Fig. 2.5.** — analytic solution, - - - numerical solution.

relations (2.9) found will be valid for part of the time. In each specific case the characteristic time $t'$ is determined from the equation $(c_1 - c_2t'/6)^2 = 180c_0$ and is connected with the compression time.

Consequently, it may be expected that taking the finiteness of the region of integration into account will not essentially change the nature of the flow. A series of numerical calculations was carried out to check this assumption. The computational algorithm described in Ref. 1 was fundamental for the program. It was appropriately modified for the problem to be solved. The direct hydrodynamic part of the equations was approximated by the grid-characteristic method,[5] to which "conservative" properties were imparted. The results of the comparison are shown in Fig. 2.5. In the region of positive values of the ordinate axis, graphs of the density at various instants are shown, and in the region of negative values — graphs of the velocity. In complete agreement with formulae (2.9), the steepness of the curves increases. The observed difference can be explained by the effect of the boundary conditions.

### 2.2.3. Comparison of results by different models

Of definite interest is the comparison of the results of calculations performed for various physical models. One of them is based on the considerations discussed above, when at the initial instant some profiles of physical quantities are specified, and then their variation, taking into account the hydrodynamic motion, the electronic thermal conductivity, and the exchange of energy between the electronic and ionic components of the plasma, is observed.

Another model is a complete spherical shell onto which radiation with large flux density falls uniformly. In both cases the same physical processes are taken into account and the physical constants are the same. The shell is homogeneous with density of matter $1\,\mathrm{g/cm^3}$. The energy of laser pulse in the second case was the same as the total energy of the plasma for the first model, and the total masses were equal. Figure 2.6 shows the results of a numerical calculation by the first model at the instant of attainment of the maximal compression. The analog of Lowson criterion is $\rho R \approx 0.05$.

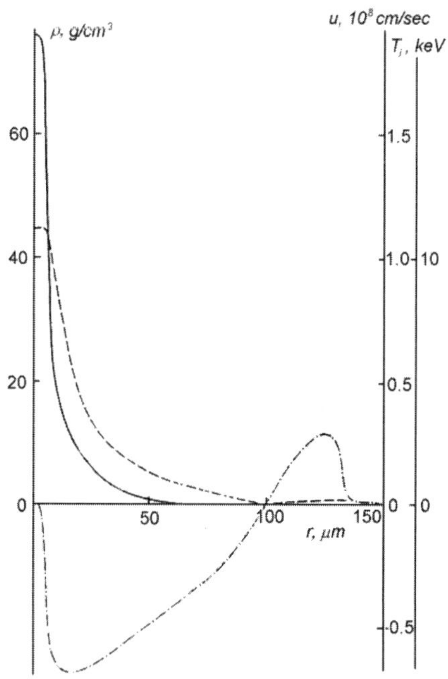


**Fig. 2.6.** —— density, - - - - ion temperature, - · · · - velocity.

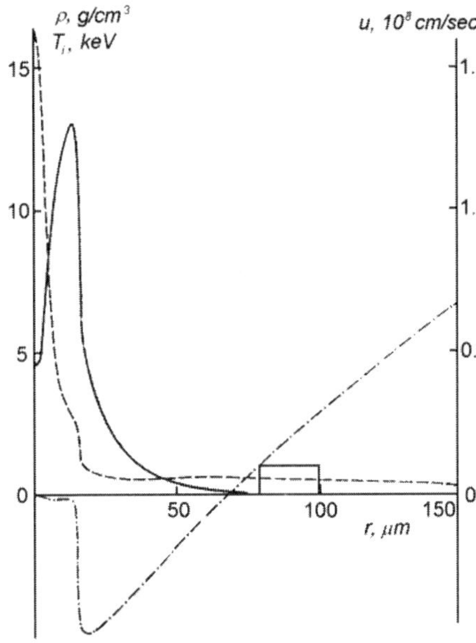


**Fig. 2.7.** —— density, - - - - ion temperature, - · - · - velocity.

Figure 2.7 shows the extremal values for the second case. It is seen that $\rho^1_{\mathrm{max}}/\rho^2_{\mathrm{max}} \approx 5$. Consequently, an approach based on the solution of problem (2.9) may be considered.

The examined problems were one-dimensional; therefore, the next test variant is connected with the studying of the plane shock wave refraction on the plane inclined interface and has analytical solution also.[24,25]

Various regimes of refraction of the planar shock wave at the planar oblique contact boundary during the "fast–slow" transition are studied in the present section. The analytical and numerical results, obtained for the regular refraction's regime, coincide within the range of admissible error, up to the "critical" value of the angle of inclination of the original plane of contact discontinuity. The numerical simulation of the irregular regime for "supercritical" angles was carried out.

The process of the shock wave's travel through the boundary of partition between two media depends on numerous factors of both the physical origin — Mach number of the incident shock wave, Atwood number, thermodynamic properties of the gases, etc., and of the geometrical one — the perturbation's wavelength, its amplitude, form, etc. Presently there exists a number of actual problems of scientific and technical character, which need to be solved, and in which the behavior of contact boundaries after their interaction with shock waves, and of shock waves between themselves, is of major importance for the development of the physical process as a whole. The processes of such a type include the controlled inertial nuclear fusion, the chaotic turbulent flows, and the detonation within gaseous mixtures.

The theoretical and experimental research in this direction, which is conducted in numerous scientific laboratories in different countries of the world,[4,26–29] permitted to accumulate the necessary practical material and to systematize it. Thus, Ref. 24 proposed a classification providing the two distinct cases of shock wave's refraction at the contact boundary of partition between two gases: that of a regular refraction and of irregular one, within the frame of which further concretization might be done. The first type of refraction — the regular one is most completely investigated, when all the waves are planar and intersect in a single point at the contact boundary. Such a quality of investigation is connected with the fact that for this type of refraction one is able to carry out an exact mathematical analysis, which reduces the problem to a solution of the algebraic equation of the 12th order. The theoretical results obtained allow their direct experimental check. Alternatively, an attempt to consider the irregular refraction of the nonplanar original contact boundary leads to the essential difficulties in the way to the analytical solution of the problem. There arises a necessity in a more general formulation of the problem by means of its description in the form of a set of partial differential equations, which would reflect three main laws of conservation of mass, momentum, and energy. As a result of that, the means of the problem's solution are changed too — the approximate or numerical methods of solution of the problems formulated appear as the principal instrument of analysis. The weakness of such an approach consists in the fact of existence of error in the solution obtained, of which the value is determined approximately. Therefore, in the first place, the analytical solutions, obtained by the numerical methods with necessary accuracy, not only will confirm the correctness of a result, but also will indicate the well-foundedness of the preliminary assumptions concerning the solution's character, which means that the answer is quasi-analytical. In the second place, getting such solutions gives an evidence to the reliability of the computational algorithms worked out, and to the prospect of their application to more complicated problems of which there is no analytical solution at the moment. In the third place, the data of numerical calculations might be immediately compared with the experimental observations, and in the case of their agreement might seriously elucidate the physical mechanism of the phenomenon investigated.

As is exposed below, by means of a model problem of a shock wave's passage from He to Xe crossing the planar contact boundary, the analytical and numerical analyses permit to clear up the question concerning the critical angle between the direction of shock wave's propagation and the plane of a contact surface, when the regular regime of refraction is changed for the irregular one. The value of this angle is rather important in the case of an original contact boundary being curvilinear, when it is necessary to find out at which portions of a surface the regular refraction is realized, and at which portions — the irregular one; upon doing that, sufficiently small portions of a curvilinear surface might be treated as planar ones.

Let us consider a specific version of the physical formulation, which might be found in experimental investigations.[5] In this version the shock wave with a Mach number $M = 2.5$ passes from He to Xe by the initial pressure at the contact surface being equal to 0.5 atm, the density of an unperturbed He $\rho_{\mathrm{He}} = 8.21 \times 10^{-5}$ g/cm$^3$, the initial density Xe $\rho_{\mathrm{Xe}} = 2.71 \times 10^{-5}$ g/cm$^3$, isentropic exponent of both gases $\gamma = 5/3$. Before the shock's arrival the planar contact boundary stays at rest, and its plane makes up an acute angle $\psi$ with the direction of shock's propagation.

### 2.3. The analytical approach

Let us consider first the range of inclination angles $\psi$ in the vicinity of 90° and assume that upon refraction of the shock wave $i$ at the contact boundary $m$–$m$ all the discontinuity surfaces, i.e. the reflected shock $r$ and the refracted one $t$, and the shifted tangential discontinuity, remain to be planar (see Fig. 2.8).

This assumption is substantiated by consideration of the one-dimensional decay of a discontinuity, when the planar shock wave interacts with a parallel planar contact surface. For small deviations of $\psi$ from 90°-value one might expect that the physical picture, as a whole, will change just insignificantly, and only at the contact boundary will appear a refraction point, where all the discontinuity surfaces intersect.

An additional assumption consists in the uniformity of distribution of the physical quantities between the discontinuity surfaces. As soon as the initial state of both gases at the contact boundary and the Mach number of the incident shock are known, one might write down the set of equations

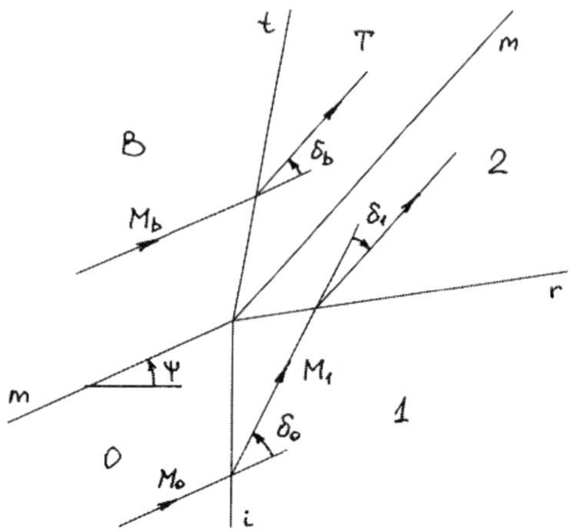


**Fig. 2.8.** The regular configuration for the case of "fast–slow" transition; $m$–$m$ — contact boundary, $i$ — incident shock, $t$ — refracted shock, $r$ — reflected shock.

interrelating the discontinuity surfaces. The main relations needed to obtain the analytical solution for the case of regular refraction in perfect gases are presented in Ref. 26. These include three equations of shock polars:

$$\mathrm{tg}^2\delta_0 = \left[\frac{P_1/P_0 - 1}{1+\gamma_0 M_0^2 - P_1/P_0}\right]^2 \frac{\frac{2\gamma_0}{\gamma_0+1}M_0^2 - \frac{\gamma_0-1}{\gamma_0+1} - P_1/P_0}{\frac{\gamma_0-1}{\gamma_0+1} + P_1/P_0}, \tag{2.10}$$

$$\mathrm{tg}^2\delta_1 = \left[\frac{P_2/P_1 - 1}{1+\gamma_0 M_1^2 - P_2/P_1}\right]^2 \frac{\frac{2\gamma_0}{\gamma_0+1}M_1^2 - \frac{\gamma_0-1}{\gamma_0+1} - P_2/P_1}{\frac{\gamma_0-1}{\gamma_0+1} + P_2/P_1}, \tag{2.11}$$

$$\mathrm{tg}^2\delta_b = \left[\frac{P_t/P_b - 1}{1+\gamma_b M_b^2 - P_t/P_b}\right]^2 \frac{\frac{2\gamma_b}{\gamma_b+1}M_b^2 - \frac{\gamma_b-1}{\gamma_b+1} - P_t/P_b}{\frac{\gamma_b-1}{\gamma_b+1} + P_t/P_b}, \tag{2.12}$$

the relation between the flow's Mach numbers before the incident shock and after it:

$$\frac{1+\frac{\gamma_0-1}{2}M_0^2}{1+\frac{\gamma_0-1}{2}M_1^2} = \frac{\frac{\gamma_0+1}{\gamma_0-1} + P_1/P_0}{\frac{\gamma_0+1}{\gamma_0-1} + P_0 P_1}, \tag{2.13}$$

condition of the velocities' equality along the original contact boundary:

$$M_b = M_0\sqrt{\frac{\gamma_0}{\gamma_b}\frac{\mu_b}{\mu_0}}, \tag{2.14}$$

relation between the pressures at the tangential discontinuity:

$$\frac{P_t}{P_b} = \frac{P_1}{P_0}\cdot\frac{P_2}{P_1}, \tag{2.15}$$

and relation between the angles of flow's deviation:

$$\delta_0 + \delta_1 = \delta_b. \tag{2.16}$$

Used here are the following notations (see Fig. 2.8): $\delta_0$, $\delta_1$, and $\delta_b$ — the angles of flow turn in its passage across the incident, reflected, and refracted shocks, correspondingly (in the coordinate system connected with the refraction point), $M_0$, $M_1$, and $M_b$ — Mach numbers of the same flows before the turn, $P_0$, $P_1$, $P_2$, $P_b$, and $P_t$, — pressures in the appropriate domains, $\gamma_0$ and $\gamma_b$ — isentropic exponents of the light gas and heavy one (for general case those are different), $\mu_0$ and $\mu_b$ — molecular masses of the gases. Let us show that this set of equations might be reduced to a single algebraic equation of the 8th order.

One introduces the notations $a_0 = \operatorname{tg}\delta_0$, $q = P_1/P_0 - 1$ (known parameters), and $y = P_t/P_b - 1$ (the unknown variable). Equations (2.11) and (2.12) are rewritten in the form

$$\operatorname{tg}^2\delta_1 = \frac{A_1}{B_1}, \quad \operatorname{tg}^2\delta_b = \frac{A_b}{B_b}, \tag{2.17}$$

where $A_1$, $B_1$, $A_b$, and $B_b$ are the cubic polynomials of $y$:

$$\begin{aligned}
A_1 &= -y^3 + y^2\left[3q - \frac{2\gamma_0}{\gamma_0+1}(q+1)(M_1^2-1)\right] \\
&\quad -y\left[3q^2 + \frac{4\gamma_0}{\gamma_0+1}q(q+1)(M_1^2-1)\right] + \frac{2\gamma_0}{\gamma_0+1}q(q+1)(M_1^2-1) \\
&\quad +q^2\left(q + \frac{2\gamma_0}{\gamma_0+1}(q+1)(M_1^2-1)\right) \equiv -y^3 + C_1, \qquad (2.18) \\
B_1 &= y^3 - y^2\left[3q - \frac{2\gamma_0}{\gamma_0+1}(q+1) + 2\gamma_0(q+1)M_1^2\right] \\
&\quad +y\left[2(\gamma_0 M_1^2(q+1)+q)\left(q - \frac{2\gamma_0}{\gamma_0+1}(q+1)\right)\right. \\
&\quad \left. + (\gamma_0 M_1^2(q+1)+q)^2\right] \\
&\quad -(\gamma_0 M_1^2(q+1)+q)^2\left(q - \frac{2\gamma_0}{\gamma_0+1}(q+1)\right) \equiv y^3 + D_1, \\
A_b &= -y^3 + \frac{2\gamma_b}{\gamma_b+1}(M_b^2-1)y^2 \equiv -y^3 + Ay^2, \\
B_b &= y^3 - y^2\left[2\gamma_b M_b^2 - \frac{2\gamma_b}{\gamma_b+1}\right] + y\left[\gamma_b^2 M_b^4 - 2\gamma_b M_b^2\frac{2\gamma_b}{\gamma_b+1}\right] \\
&\quad +\frac{2\gamma_b}{\gamma_b+1}\gamma_b^2 M_b^4 \equiv y^3 + D_b.
\end{aligned}$$

After the expression of $M_1$ from Eq. (2.13) and the insertion of a result into Eq. (2.18) one might readily obtain

$$A_1|_{y=0} = a_0^2 B_1|_{y=0}.$$

From here it follows that $A_1 - a_0^2 B_1$ has a zero free term:

$$A_1 - a_0^2 B_1 = yE_1, \tag{2.19}$$

where $E_1$ is a quadratic trinomial.

By means of Eq. (2.17), Eq. (2.16) is transformed into

$$\begin{aligned} Q_0(y) \equiv & \left[A_b(B_1 - a_0^2 A_1) + B_b(A_1 - a_0^2 B_1)\right]^2 \\ & -4A_1 A_b B_1 B_b (a_0^2 + 1)^2 = 0. \end{aligned} \tag{2.20}$$

The degree of polynomial $Q_0(y)$ at the left-hand side of Eq. (2.20) is not higher than 12. After substituting here the expressions of Eqs. (2.18) and (2.19), one can see that

$$Q_0(y) = y^2 Q_1(y).$$

($Q_1(y)$ is polynomial of degree not higher than 10), and due to the fact that zero roots are not interesting for us, the order of equation is reduced to the value of 10, at least:

$$\begin{aligned} Q_1(y) \equiv & \left[y(A - y)(B_1 - a_0^2 A_1) + B_b E_1\right]^2 \\ & -4(A - y) A_1 B_1 B_b (a_0^2 + 1)^2 = 0. \end{aligned} \tag{2.21}$$

Actually, the disclosing of brackets and singling-out of the highest-order components reveal the zero factors at $y^{10}$ and $y^9$, and thus Eq. (2.21) is of 8th order.

With the earlier made assumptions concerning the character of interaction between the shock wave and the contact boundary, only positive roots of Eq. (2.21) have physical sense. Considered below is the behavior of these roots depending on the original angle of the contact surface's incidence $\psi$, by earlier formulated initial conditions. The corresponding plot of $y$ as dependent on $\psi$ is shown in Fig. 2.9. It is seen that at the angles smaller than some critical value $\psi_{\text{cr}}$ (in the present case it is 48.18°), there are no positive roots, while at larger angles there might be from two to four such roots. Without additional assumptions one is not able to determine which of these solutions is realized in practice. In Refs. 24 and 26, the hypothesis was put forward elaborating the Férmat principle for linear wave system — of the possible solutions is realized that one, which corresponds

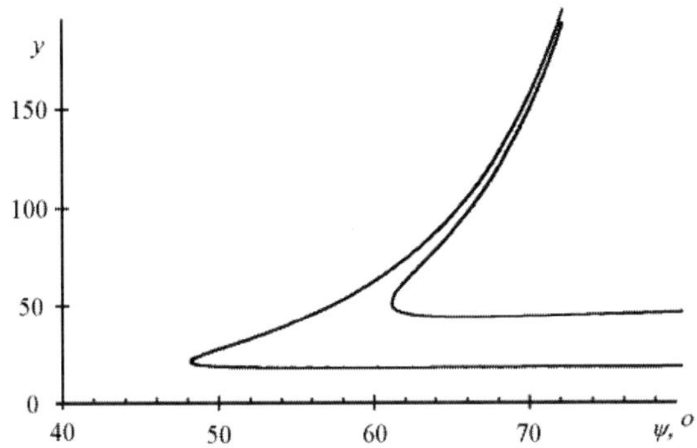


**Fig. 2.9.** Behavior of the roots of Eq. (2.21) as dependent on the angle of inclination of the contact surface $\psi$.

**Table 2.1.**

| $\psi$ | $\delta_0$ | $\delta_b$ | $P_t/P_b$ | $\phi$ | $\theta$ | $\chi$ |
|---|---|---|---|---|---|---|
| 50 | 22.75 | 15.53 | 19.775 | 18.48 | 57.23 | 60.32 |
| 55 | 20.48 | 14.16 | 19.058 | 33.01 | 61.31 | 64.03 |
| 60 | 17.94 | 12.43 | 19.090 | 43.69 | 65.51 | 67.87 |
| 65 | 15.21 | 10.53 | 19.277 | 52.82 | 69.68 | 71.68 |
| 70 | 12.33 | 8.52 | 19.493 | 61.05 | 73.81 | 75.43 |
| 75 | 9.34 | 6.44 | 19.688 | 68.70 | 77.90 | 79.13 |
| 80 | 6.27 | 4.31 | 19.838 | 75.99 | 81.95 | 82.78 |
| 85 | 3.15 | 2.16 | 19.932 | 83.05 | 85.98 | 86.40 |

to the minimal time of the perturbation's propagation between two arbitrary points. In our case, corresponding to this criterion is the curve for smallest positive value of a root (see Fig. 2.9). The results of calculations for several values of angle $\psi$, exceeding the critical one, are presented in Table 2.1, which contains results of the analytical calculations for a number of values of angle $\psi$. The angles are indicated in degrees; $\phi$, $\theta$, and $\chi$ are, correspondingly, the angles of inclination of reflected shock wave, of the shifted contact boundary, and of the refracted shock wave.

To substantiate the theoretical deductions on a critical value of $\psi$, on the flow generated and on the character of interaction, it would be necessary to conduct an independent investigation. The role of that might be played either by the full-size experiment, or by the problem's solution in more general formulation, not subject to the additional assumptions — for example, on the basis of Euler's equations, using the numerical methods.

## 2.4. Computational experiment

The problem was solved in the two-dimensional Cartesian space and consisted of the mathematical formulation of three main conservation laws — those of mass, momentum, and energy. At the initial moment of time the integration domain was divided into three parts by the planar oblique contact surface and the front of an incident shock wave. Within the first of these domains, located at the left of the surface of partition of two media, were prescribed the values of gas-dynamical parameters of unperturbed Xe, within the second domain, located between the contact boundary and the shock's front — the values for unperturbed He. The point of intersection of the contact boundary and of the shock's front was at that moment at the upper boundary of the integration domain. At the right of the shock's front the values of gas-dynamical parameters corresponded to the analytical values for the prescribed Mach number of the incident shock wave. The analytical solutions on the basis of the above theoretical approach were used as the boundary conditions at the upper boundary and at the right one. For the possibility to evaluate the effect of a computational error there was conducted a series of calculations with sequentially redoubled computational grids. The results of comparison of the analytical and numerical solutions, obtained for the angle of inclination of a planar contact boundary, equal to 50° and close to the critical value, are presented in Fig. 2.10 in the form of pressure distribution. The analytical solution is depicted by a solid line, while dashed line corresponds to the numerical solution on the spatial grid with $h_x = h_y = 7.3 \times 10^{-4}$, and dotted line — to the similar grid with $h_x = h_y = 1.8 \times 10^{-4}$. It is evident that the location of both refracted and incident shock waves is practically coincident for all the cases, a certain smearing of the front, observed for the reflected shock, is connected with

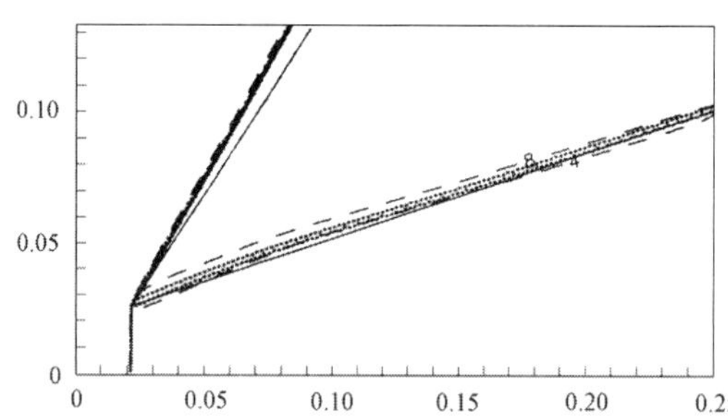


**Fig. 2.10.** Numerical solution for the case of regular refraction.

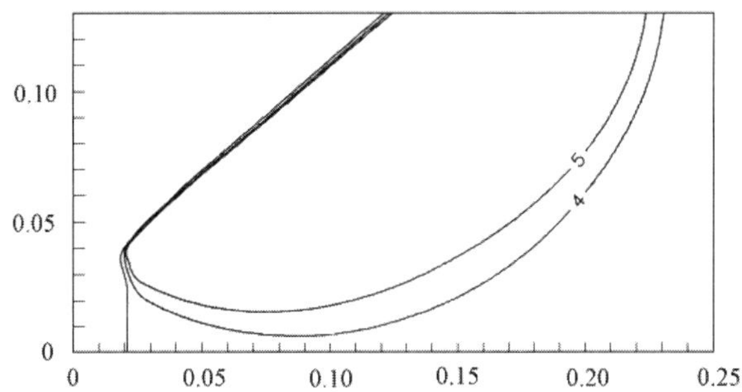


**Fig. 2.11.** Numerical solution for the case of irregular refraction.

the diffusion properties of the difference method and is diminished by the refinement of a difference grid. At the prescribed value of angle, $\psi = 50°$, the regular regime of shock wave's refraction is realized.

The calculations, aimed to check the convergence of the difference solutions, were conducted by means of the electronic computer PARAM 8000 of multi-transputer type, at the grid consisted of $1371 \times 1634$ cells. As an example of irregular refraction is presented the pressure distribution corresponding to the original angle of inclination of the contact surface $\psi = 30°$, obtained by the numerical calculation for the grid with $h_x = h_y = 1.8 \times 10^{-4}$ (see Fig. 2.11). For this version it was not possible to obtain the analytical solution. One is able to observe the curvilinear form of the reflected shock, as well as the Mach's reflection. The values of pressure indicated at the isobars are given in atmospheres.

The results presented in this section and in the number of some other papers,[5,30] are in accordance with the theoretical and experimental notions on the specific features of the process of shock's refraction at the planar contact boundary, and demonstrate the effectivity of a numerical simulation for the study of such complicated hydrodynamical phenomena as the Richtmayer–Meshkov's instability in a three-dimensional spatial case.[30]

Of special interest are the studies of the complicated nonlinear processes and of mechanisms of interaction for the transformation stages and for the irregular regimes (which are of great practical importance). The results presented correspond to the first stage of the development of investigations in the direction indicated, which presently goes on to the further elaboration of specific parallel algorithms for computations with the help of multi-processor computers.

On the basis of the developed programs the problem of cumulating of the shock waves from energy-emitting point sources in 3D axially symmetric space was studied.[31] In the strong shock waves interaction the cumulative effects enable to realize high plasma temperatures and pressures. These effects are observed in laboratory experiments on the momentary energy generation by a laser source or electrical breakdown. Assume that the 3D axially symmetric space is filled with a stationary homogeneous ideal gas (density 1.29 mg/cm$^3$, specific internal energy 0.19 kJ/g, isentropic exponent $\gamma = 1.2$), and a momentary 3D-distributed energy emission takes place in the gas. Two variants were analyzed: in the first one the energy equal to 180 J was released along a thin plane ring; in the second one, the energy was divided into equal parts between six points uniformly arranged on the circumference of the same radius (0.25 cm), as that of the plane ring. The specific internal energy increased up to 23.2 MJ/g; so the arising shock waves were strong. The isobars, corresponding to these two variants, obtained at the numerical simulation, are presented in Figs. 2.12 and 2.13 for the time moment of 90 ns. Pressure values (in kbar) are noted near the corresponding curves. In these figures, plane $XY$ coincides with that of the initial energy generation; two energy-generating centers (out of six) are shown with dots, where the ring is denoted by a dotted line. It is seen that at this shock waves interaction stage the 3D variant differs from the axial one by the presence of increased pressure, density, and temperature local areas, that is connected with the collision of shock waves at first from the

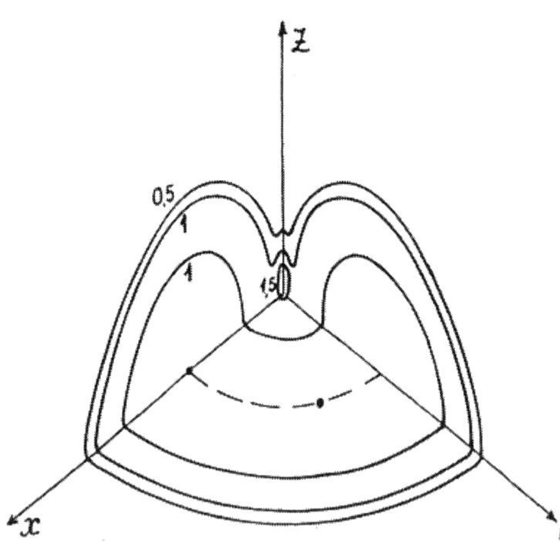


**Fig. 2.12.** Isobars (in kbar) at 90 ns corresponding to the ring version.

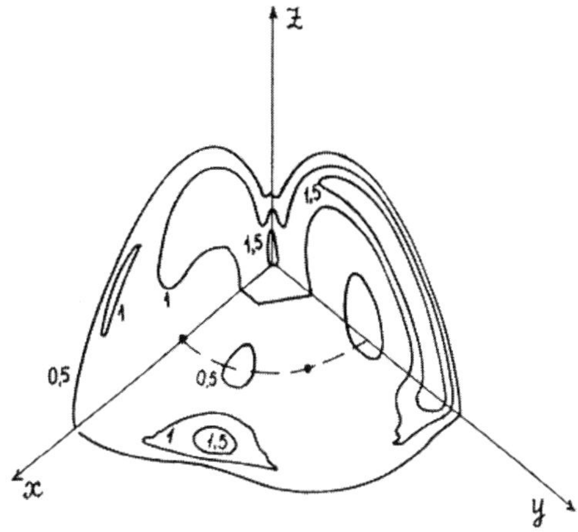


**Fig. 2.13.** Isobars (in kbar) at 90 ns corresponding to the version of six points.

nearest and then from the more distant energy-generation centers. In both cases, cumulative effects are observed in the center; there are jets, developing along the symmetry axis, as well.

The program for mathematical modeling of a Richtmyer–Meshkov instability was based on a divergent version of the grid-characteristic method, which is somewhat different from that discussed in a previous publication.[1] A feature of this version is its conservative nature for supersonic flows when the mass velocity in a cell exceeds the local speed of sound, with no intermediate interpolation at cell boundaries being used. The technique does not require the explicit isolation of the interfaces, and the finite-difference scheme admits a through computation over the integration domain as a whole. This feature is also of particular interest for other problems, where physical mechanisms of gas-dynamic processes are not obvious beforehand and are subjects of study in themselves. A uniform Euler grid fixed in space meets the same requirement. It also enables us to analyze, with approximately equal accuracy, all the peculiarities arising in the flow at any point in the computation region. This fact may serve as a basis for more exact future computations, in which finite-difference grids adapted to the specific solution may be used. By means of such grids, it should be possible to

(i) reveal other aspects of physical processes;
(ii) bring to light the details of the gas-dynamic pattern;
(iii) use the schemes of second- or higher-order accuracy;
(iv) single out more intricately shaped interfaces;

(v) solve problems in multiply-connected domains or with curvilinear boundaries.

All of these considerations were taken into account in formulating a mathematical model of a Richtmyer–Meshkov instability that would give satisfactory agreement with the experimental data.

## 2.5. Physical mechanisms of the RMI evolution

Of fundamental importance for the problem at hand is the interaction between a shock wave and a perturbed interface. The evolution of a Richtmyer–Meshkov instability in time depends in many respects on the degradation of the resulting discontinuity, as well as on the generation of waves, their shape and direction.

The simplest case in which a shock wave is refracted by an interface occurs when a plane shock wave is incident on a plane interface between two originally quiescent media inclined at an angle $\alpha$ to the direction of the shock-wave front. At the limit as $\alpha \to 0$, we obtain the one-dimensional case mentioned above. When two shock waves are generated due to interface degradation, if the angle $\alpha$ is close to zero, then refracted and reflected shock waves, as well as the contact surface, are rotated through small angles relative to each other, while still remaining planar. As a consequence of this situation, a point forms at the interface at which the incident, the refracted, and the reflected waves converge. The point moves with time over the original interface in the direction of the incident shock wave. Similar problems have already been studied in the literature; of special mention is a theoretical paper published by Henderson[24] and numerical experiments conducted by Hawley and Zabusky.[27] As the angle $\alpha$ increases, the shock waves are gradually attenuated and then transformed into compression waves, or, under certain conditions, into expansion waves.[24] It follows that the evolution of a Richtmyer–Meshkov instability and the attendant phenomena depend on the initial experimental setup.

An undulating perturbation of the interface over small regions (for short arc lengths) can be treated as a broken surface composed of a number of planes, each inclined at a certain angle to the direction of the incident shock wave. This means that the results obtained for a constant angle $\alpha$ hold for a sinusoidal interface perturbation, but only for a small region of the interface, in which its curvature can be neglected. Consequently, during the initial stage of the refraction of a shock wave by an undulating surface, one can expect a gradual attenuation of the generated shock waves followed by their degeneracy. This fact should be typical for small shock wavelengths.

Now let us consider in more detail the case of a sinusoidal interface perturbation when the maximum angle between the tangential plane and the direction of the incident shock wave is not large enough for the irregular interaction to be observed. Then the vector of the mass flow velocity behind the front of the incident shock wave will, while traveling through the surface of the reflected shock wave, be deflected through some angle from its original direction toward the noninteracting region of the discontinuity. This is due to the fact that the normal component of the momentum flux vector relative to the shock-wave front undergoes a discontinuity, while the tangential component is conserved. This gives rise to the velocity vector component normal to the velocity behind the incident shock-wave front. A similar deflection, but in the opposite direction, occurs in the adjacent symmetrical part equal to half the wavelength of the initial interface perturbation. The condition of the velocity field as a result of refraction by half the wavelength of a sinusoidal interface perturbation is presented in Fig. 2.14. When the flow pattern extends toward the negative values of $y$, a mirror-symmetry pattern with velocities deflected toward the $x$-axis is obtained. As can easily be seen from the isobar profiles (Fig. 2.15) corresponding to $5\,\mu$s after the onset of refraction, a local increase in pressure occurs in the interaction region between the two symmetrically deflected flows. The velocity vector field in Fig. 2.14 also corresponds to this instant time. The deceleration process continues for $2\,\mu$s and is accompanied by an increase in the maximum pressure, which attains values of about 17 atm, as well as by a shift of this region along the symmetry axis from the interface toward

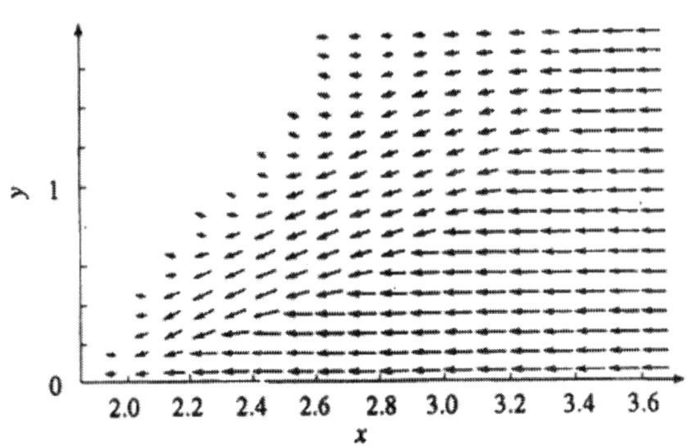


**Fig. 2.14.** Velocity vector field at $t = 5\,\mu$s.

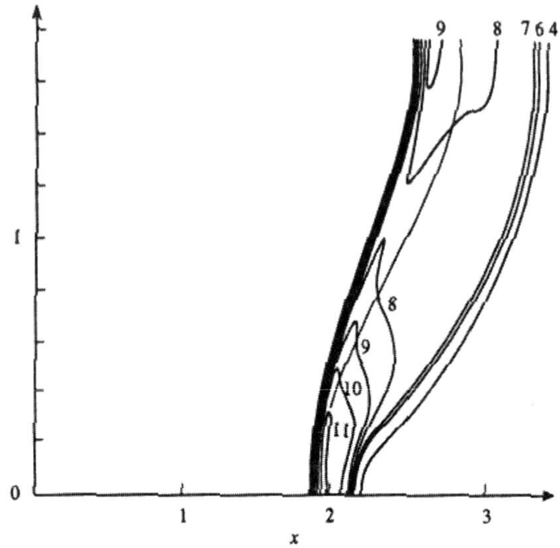


**Fig. 2.15.** Isobar profiles (in atm) at $t = 5\,\mu$s.

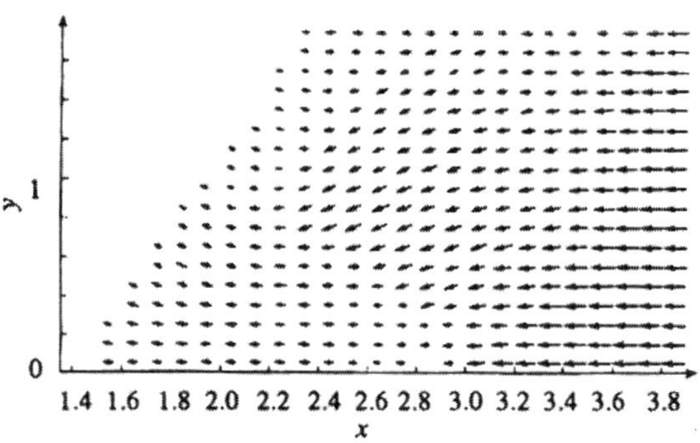


**Fig. 2.16.** Velocity vector field at $t = 10\,\mu$s.

the reflected shock-wave front. This process ends in the generation of a secondary shock wave propagating in the space between the refracted and the reflected shock waves at right angles to them. Figures 2.16 and 2.17 illustrate the velocity-vector field and the isobar profiles at time $t = 10\,\mu$s after the beginning of interaction. It can be seen that the velocities behind the secondary shock wave acquire their initial direction, but are smaller than those behind the incident shock wave. Another higher-pressure region is also formed in xenon in the initial interaction region, since the refraction process

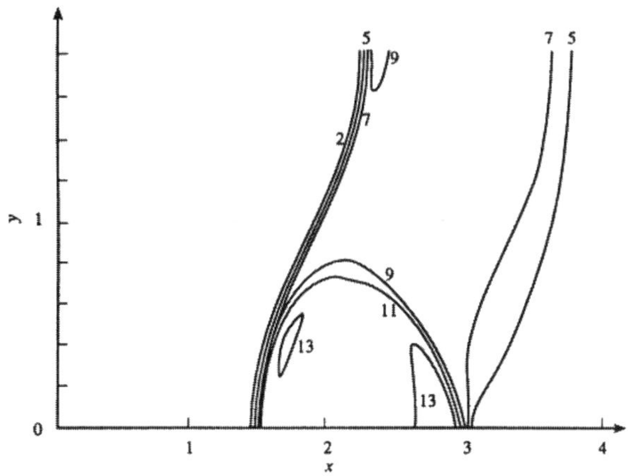


**Fig. 2.17.** Pressure distribution at $t = 10\,\mu$s.

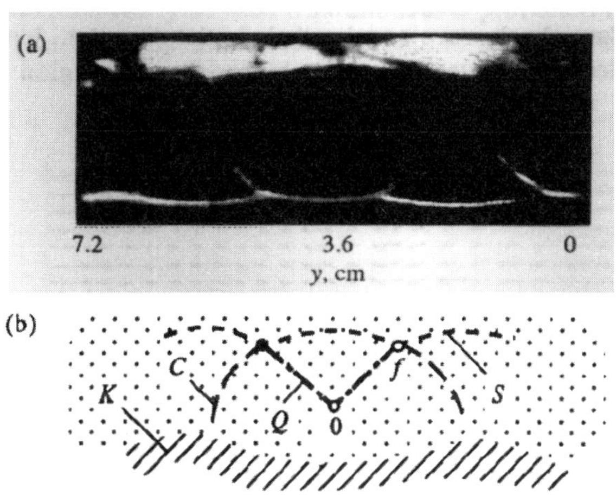


**Fig. 2.18.** Schlieren photograph of the secondary shock waves: (a) Schlieren pattern of $S_0$ passage from Xe into He ($P_0$ = 0.5 atm, $M$ = 3.5); (b) Wave diagram: $S$ — transmitted shock wave, $C$ — secondary shock wave, $Q$ — tangential discontinuity, $f$ — point of fracture, and $K$ — interface.

here is close to one-dimensional, and the pressure behind the refracted shock-wave front is maximum and may be about 10 atm.

The generation of secondary shock waves is not only manifested in numerical calculations, but is also confirmed in experiments (Fig. 2.18). The generation and propagation of secondary shock wave result in the

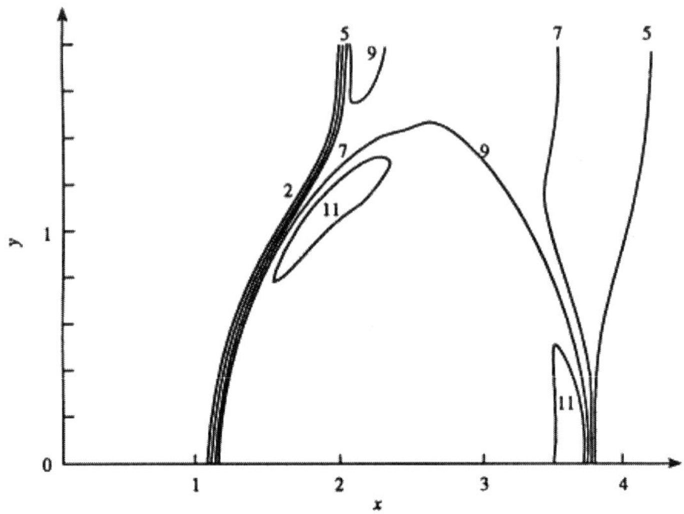


**Fig. 2.19.** Isobar profiles (in atm) at $t = 15\,\mu$s.

penetration of helium into xenon in the form of a bubble and in leveling the reflected shock-wave front when it merges with the secondary shock-wave front. This is confirmed by the isobar profiles (Fig. 2.19) for $t = 15\,\mu$s. After the transmission of half the wavelength of the initial interface perturbation, the secondary shock waves start interacting with one another, and the linear stage of the Richtmyer–Meshkov instability evolution ends. The initial phase of the nonlinear stage is illustrated in Fig. 2.20, which represents the isobar profiles for $t = 20\,\mu$s. It can be seen that the pressure in xenon increases due to the interaction between the secondary shock waves.

The process lasts for several microseconds and terminates when the pressure in xenon exceeds 20 atm, resulting in the generation of another series of shock waves, which level out the refracted shock-wave front. Pressure reduction in the higher-pressure region of the xenon causes a strong inflow of xenon into the helium as a jet. The fluid begins to streamline about the expanding higher-pressure region, as illustrated by the velocity-vector field in Fig. 2.21, as well as by the isobar profiles for $t = 25\,\mu$s in Fig. 2.22. By this time, the reflected shock-wave front has leveled out and becomes plane. At a later stage, the next series of secondary shock waves flattens the refracted shock-wave front. Propagating through the expanding xenon jet, these waves generate vortices at its tip. This marks the termination of the nonlinear stage and the beginning of the transitional stage.

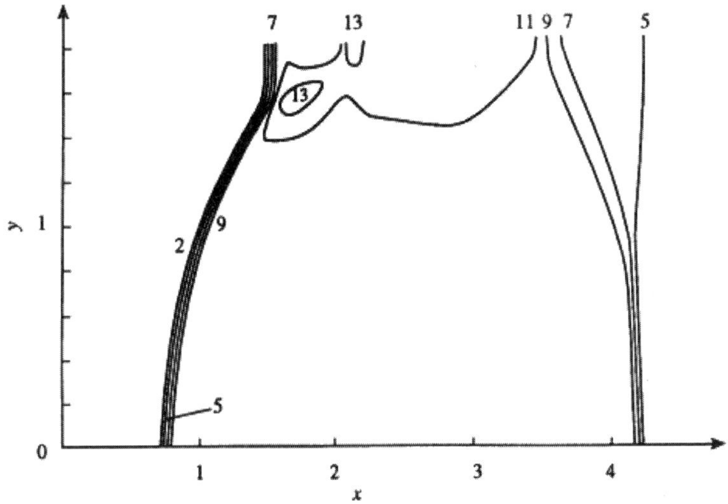


**Fig. 2.20.** Pressure distribution at $t = 20\,\mu s$.

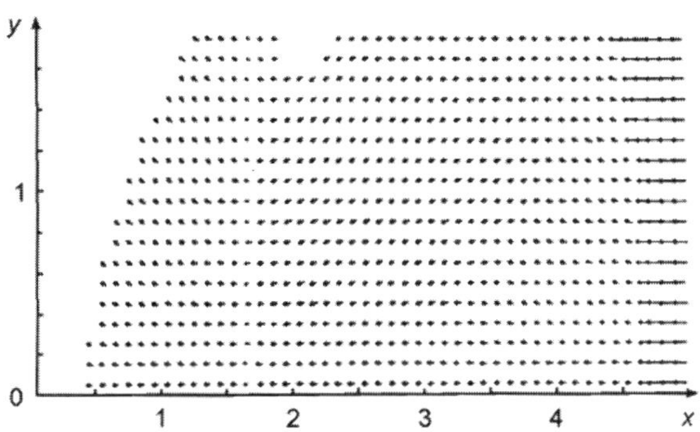


**Fig. 2.21.** Velocity vector field at $t = 25\,\mu s$.

Figure 2.23 illustrates three isodensity surfaces at $t = 10$, 40, and $70\,\mu s$ after the beginning of interaction. The surfaces illustrate the growth of helium bubbles, the formation and development of xenon jets, secondary shock waves in xenon, and the generation of vortices.

In collaboration with Prof. Zaitsev and his colleagues, a detailed correlation was made between the numerical and experimental data, as shown in Fig. 2.24. Here, the following notation is used: $S$ is the front of the refracted shock wave in xenon, $S_0$ is the incident shock wave in helium, $R$ is the front

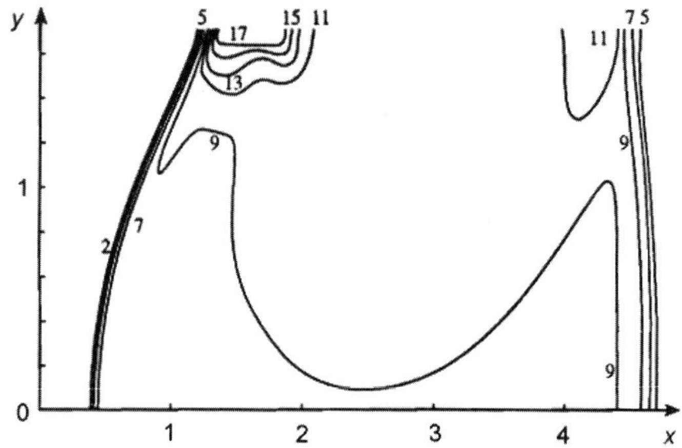


**Fig. 2.22.** Isobar profiles (in atm) at $t = 25\,\mu$s.

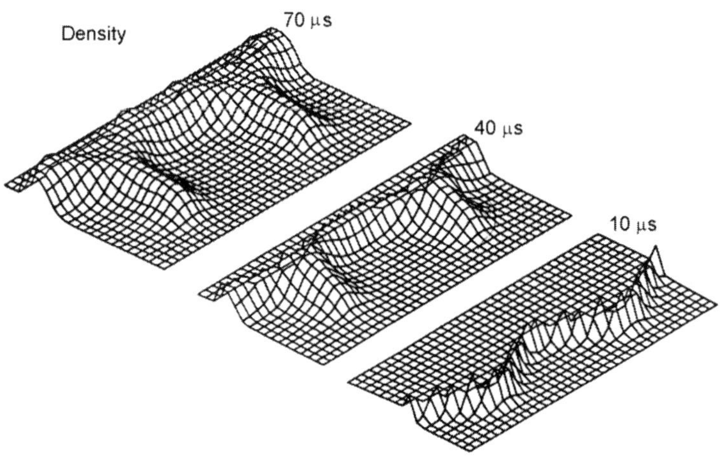


**Fig. 2.23.** Isodensity surfaces during Richtmyer–Meshkov instability evolution.

of the reflected shock wave in helium, $K$ is the interface after refraction of a shock wave, $K_0$ is the unperturbed interface, and $Q$ are tangential discontinuities. The superscript "$ex$" denotes the curve corresponding to the experimental data.

Figure 2.24 shows the isobar profiles for the region covering half the wavelength of the initial interface perturbation at $t = 1,\ 7,\ 30$, and $50\,\mu$s

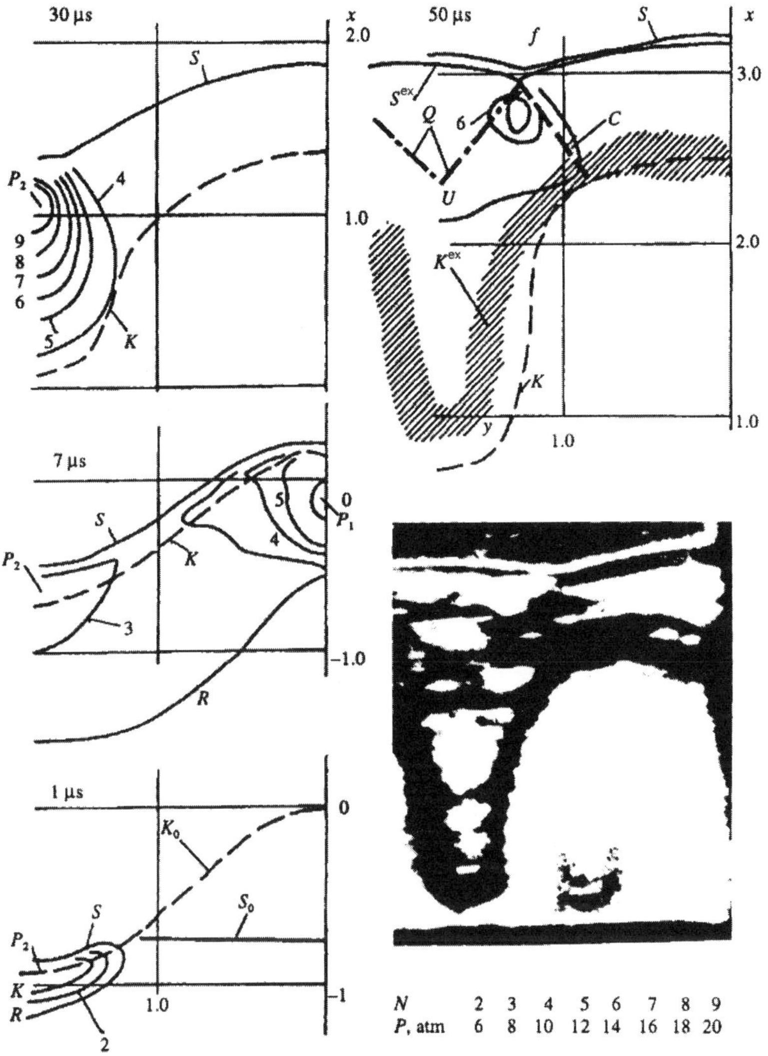


**Fig. 2.24.** Comparison between the numerical and the experimental data for the Richtmyer–Meshkov instability evolution.

after the onset of the refraction. The curves illustrate both the linear and nonlinear stages of the Richtmyer–Meshkov instability evolution. The labels on the isobars indicate their sequential numbers. The corresponding pressures (in atmospheres) are given to the right at the bottom of the figure.

A comparison is also made between the numerical and the experimental data for $t = 50\,\mu s$. The experimental data were obtained by processing the Schlieren photographs, one of which is shown in Fig. 2.24. The calculated and the experimental curves are seen to be in good agreement both in shape and in arrangement. This may be regarded as an indirect indication that the proposed mechanism of evolution of the Richtmyer–Meshkov instability is valid.

## 2.6. A sequential transition to turbulence in RMI instability

The continuum mechanics problem of development of the unsteady chaotic flows from ordered laminar flows arises in many problems of science and technology ranging from fluid motion in channels to controlled fusion. Despite its widespread occurrence in nature and exposure to laboratory studies, a common understanding of the causes and mechanisms of this phenomenon is currently lacking. This is due to both the diversity of physical processes involving a significant role played by turbulence and the insufficient elaborateness of mathematical tools for the quantitative treatment of the appropriate equations describing the phenomena observed.

It is known from the experimental study of Richtmyer–Meshkov instability development[4] that the turbulence stage begins earlier when the ratio of the perturbation amplitude to the wavelength is more than unity. So let us consider a theoretical analysis of the physical mechanisms of transition to turbulence in RMI, based on the results of numerical simulations and their comparison with the experimental data obtained by Zaitsev's group.[5]

In what follows, the passage of a plane shock wave (SW) at Mach 2.5 from helium to xenon at the pressure of 0.5 bar across a sinusoidally perturbed interface, with the perturbation wavelength 0.8 cm and amplitude 1 cm, serves as an example of the mechanism of transition to turbulence to be studied. The problem is solved in a two-dimensional Cartesian formulation for a model of an ideal non-heat-conducting inviscid gas described by the system of Euler equations.

The interaction of the plane SW with the perturbed interface is referred to as the refraction process of the SW at the interface. In the case under study, at any instant during the refraction, one can mark a point along the interfacial line where the incident SW intersects the interface. We call this the refraction point. Apart from the incident SW and the tangential discontinuity, it issues a refracted SW propagating in xenon and reflected

SW traveling in helium counter to the incoming flow behind the incident SW. Note the importance of the refraction process for the subsequent course of events. In this case, the process has an irregular character; the reflected SW has a highly curved shape, and, near the refraction point, part of its front merges with the incident SW, enhancing it.

In the direction perpendicular to the motion of the primary SW, the most curved part of the reflected SW advances beyond the refraction point, which leads to the interaction between parts of the SWs reflected from the two adjoining half-wavelengths well before the end of the refraction process. This implies that the linear stage of evolution of the instability does not occur in the case under consideration.

At the very beginning of refraction, the first high-pressure region appears near the crest value of the perturbation of xenon in helium, due to the normal incidence of SW onto the interface. Since the curved reflected shock front makes an obtuse angle with the incoming flow almost everywhere, the flow deflects toward the minimum value of the perturbation of xenon in helium after passing the front. For similar reasons, the flow behind the tangential discontinuity in xenon turns toward the perturbation maximum. The explanation is that xenon ahead of the refracted SW is quiescent in the laboratory frame of reference, and the tangential components of a velocity vector are conserved across the shock front. A collision of flows deflected symmetrically toward the perturbation maximum gives rise to the second local high-pressure region in xenon.

In the vicinity of the tangential interface, a low-pressure region appears, due to the deflection of one part of the flow toward the perturbation maximum and the other part toward its minimum. By the end of the refraction process, the primary incident SW enhanced by the reflected SW, after interacting with the interface near the perturbation minimum, generates another reflected SW, which starts moving toward the perturbation maximum and counter to the flow behind the reflected SW. At the same time, the curved part of the SW reflected from the adjoining half-wavelength of initial perturbation reaches the perturbation maximum of xenon and, after its interaction with local regions of excessive and reduced pressure, initiates the formation of a mushroom-shaped structure at the leading edge of the xenon jet. Thus, by the end of the refraction process, formation of the mushroom-shaped structure of xenon in helium has already started, which is characteristic of the transition stage (see Fig. 2.25). Like the linear stage, the nonlinear stage does not occur in this case.

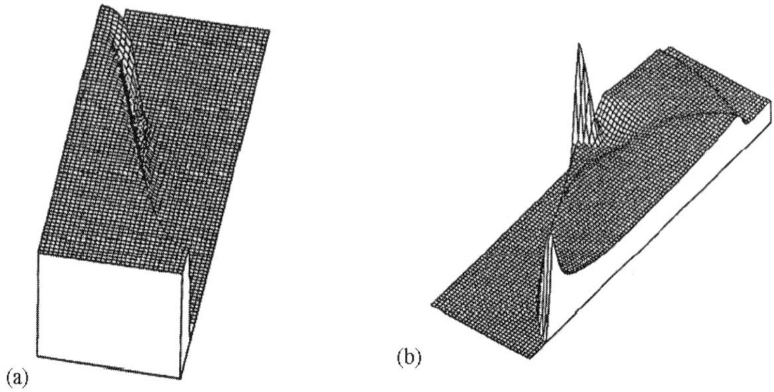


**Fig. 2.25.** Three-dimensional plots of mass concentration of (a) Xe and (b) pressure at time $T = 4\,\mu$s after the beginning of refraction of a shock wave at the interface.

Due to the unboundedness of the initial interface in the direction perpendicular to the motion of the primary SW and its periodicity, the number of secondary SWs emerging after the end of refraction is unlimited, as well. They propagate within the space between the diverging fronts of the refracted and reflected SWs. Their periodic interaction with one another inside the mushroom-shaped structures of xenon in helium promotes the formation of regions of excessive and reduced pressure and leads to the expansion, merging, and mixing of these structures. Note that the interaction of the secondary SWs with the refracted SW, involving displacement of the tangential discontinuity, leads to the formation of a train of alternating local regions of positive and negative vorticity, making up a vortex street located between the refracted shock front and the turbulized interface (Fig. 2.26).

This appears to be the physical mechanism of transition to turbulence in the variant of Richtmyer–Meshkov instability that has been considered.

## 2.7. Three-dimensional numerical simulation of the RMI

The passage from the one-dimensional case, characterized by the classical discontinuity degradation, to the two-dimensional case, has led to the discovery of a fundamentally new phenomenon, namely, the Richtmyer–Meshkov instability. Will an analysis of three-dimensional flows contribute significantly to our understanding of the physical mechanism of instability? To answer this question, the model of the Richtmyer–Meshkov instability for

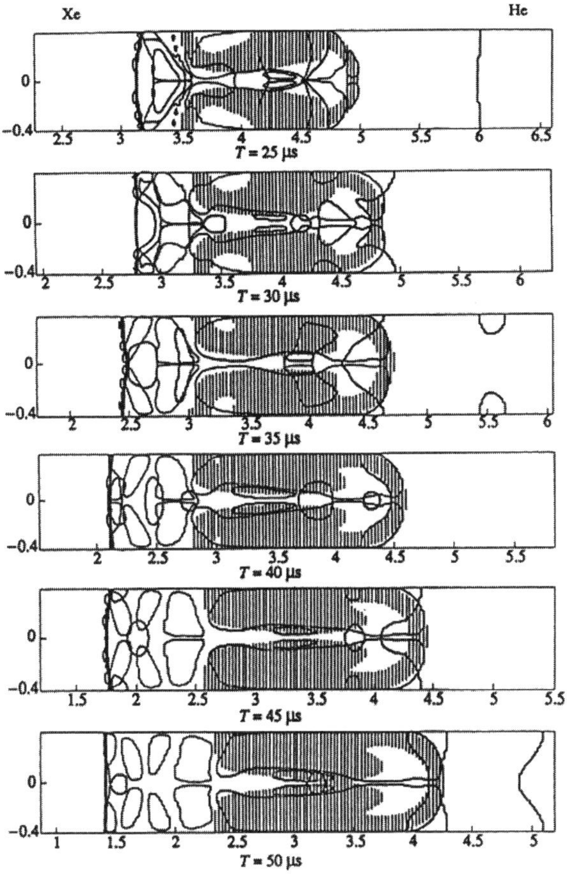


**Fig. 2.26.** The integration domain of one wavelength of the initial perturbation (8 mm) at consecutive instants. Hatching highlights the mixing zone (mass concentration of He from 0.1 to 0.9). Thick lines are isobars; thin lines are the isolines of vorticity.

the three-dimensional unsteady flows is required. A suitable initial approach would be needed to model a cellular sinusoidal structure in which the peaks of the harmonic perturbations of the interface in two mutually perpendicular directions ($x$ and $y$) coincide, while a plane shock wave affects the cellular structure in the third ($z$)-direction.

During the mathematical modeling, it would be sufficient to consider only a quarter of a cell, since the symmetry planes normal to the $(x, y)$-plane pass through maxima and minima of interface perturbation, both in the $x$- and the $y$-directions. The calculations for one of such three-dimensional

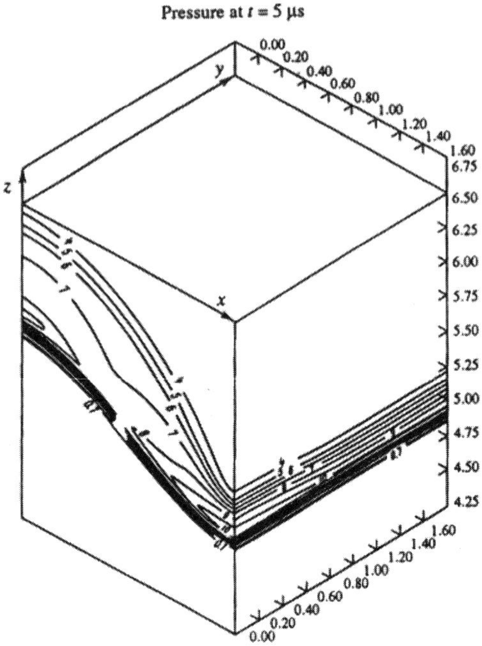


**Fig. 2.27.** Three-dimensional pressure distribution. The secant $(x, y)$-plane corresponding to $z = 6.5$ cm.

versions with the amplitude perturbation of 1 cm and wavelength of 3.6 cm at $t = 5\,\mu$s are presented in Figs. 2.27 and 2.28. By this time, the refraction of the shock wave by the interface between the two media has already been completed, and the reflected and refracted shock waves have been formed. In Figs. 2.27 and 2.28, the cross-sections of the three-dimensional space by the Cartesian coordinate planes are illustrated. The $z$-axis passes through a maximum of the interface perturbation and is normal to an incident plane shock wave propagating in the negative $z$-direction. The $z$-scale (in centimeters) is plotted parallel to the $z$-axis on the right-hand side of the figures. Half the wavelength of the initial interface perturbation fits into both the $x$- and the $y$-directions. For the individual perturbations taken separately, the $z$-axis appears to be a symmetry axis. The $x$-coordinates are indicated on the straight line parallel to the $x$-axis at the top of the figures. Similarly, the $y$-scale is shown at the bottom. In Fig. 2.27, the $(x,\ y)$-plane is parallel to the front of the incident plane shock wave. This cross-section corresponds to the coordinate $z = 6.5$ cm and is in the region occupied by

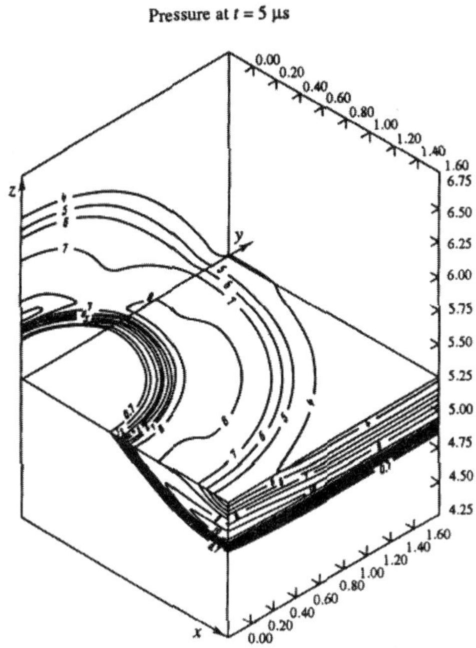


**Fig. 2.28.** Three-dimensional pressure distribution. The secant $(x, y)$-plane corresponding to $z = 5.25$ cm.

helium perturbed by the initial shock wave. The $(x,\ z)$- and the $(y,\ z)$-planes shown in Figs. 2.27 and 2.28 are symmetry planes and contain the isobar profiles (in atmospheres). As shown in Fig. 2.28, the $(x,\ y)$-plane corresponding to $z = 5.25$ cm partially intersects the region occupied by xenon, which lies closer to the $z$-axis. The reflected shock wave in helium propagates through the other part of the plane. As also shown in Fig. 2.28, the curves plotted in the $(x,\ z)$-plane illustrate the pressure distribution on the other side of the rectangular parallelepiped, which is invisible in Fig. 2.27. The behavior of the 8 atm isobar in the $(x,\ y)$-plane (see Fig. 2.28) indicates that, in this case, the flow pattern becomes essentially three-dimensional.

When studying complicated physical phenomena that take place either at the galactic scale, for example, in astrophysical investigations of the substance-mixing processes in supernovas, or at the microscale, inherent in nuclear physics, for example, in realizing the inertial thermonuclear fusion, there arises the necessity of the analysis of physical mechanisms and their adequate description for various hydrodynamical instabilities. These are the Rayleigh–Taylor, Kelvin–Helmholtz, and Richtmyer–Meshkov instabilities. As numerous experimental and theoretical investigations[32–46] show, the change of the dimensional characteristics of phenomena, namely, the transition from two-dimensional to three-dimensional flows, is accompanied by the appearance of new physical effects. In lower-dimension problems, these effects are either absent or manifest themselves in a degree inaccessible for observation. Among the hydrodynamical instabilities, the three-dimensional flows formed when developing the Richtmyer–Meshkov instability are the least studied because, for their numerical simulation, it is necessary to have not only a large-volume difference grid, but also higher quality algorithms for considering violent discontinuities, in particular, shock waves and their interactions. It is this fact that substantially complicates experimental diagnostic investigations. In the numerical simulation of the Rayleigh–Taylor and Kelvin–Helmholtz instabilities,[36,47] it was established that, for identical initial amplitudes of perturbations and wavelengths, the growth rate for the perturbations is higher in the three-dimensional case as compared to the two-dimensional one, while the process of formation of mushroom-shaped structures proceeds more slowly.

Similar results for the Richtmyer–Meshkov instability were obtained in other experiment.[48] In this connection, there arise two types of questions. They are associated, first, with studying physical mechanisms that lead to the observable phenomena, and, second, with establishing the relationship between the growth rates for the perturbation amplitude and a number of geometrical and physical quantities. These are the amplitude and the duration of the perturbation, its shape, the Mach and Atwood numbers, the thermodynamical properties of substances, etc., which require systematic investigations (both theoretical and experimental).

Now, we present the results of comparative computational studies of the development of the Richtmyer–Meshkov instability for the two-dimensional Cartesian case[49] and for the corresponding axially symmetrical cylindrical and three-dimensional variants.[50]

As an example, we choose the passage of a shock wave from helium into xenon at an initial pressure of 0.5 atm on the contact boundary and a Mach number of 2.5. The shape of the interface between the two media before the beginning of the interaction with the shock wave for the two-dimensional plane and three-dimensional formulations of the problem is presented in Fig. 2.29.

In the two-dimensional plane case, an initial contact boundary was perturbed by a sinusoidal wave with an amplitude of 1 cm and a wavelength

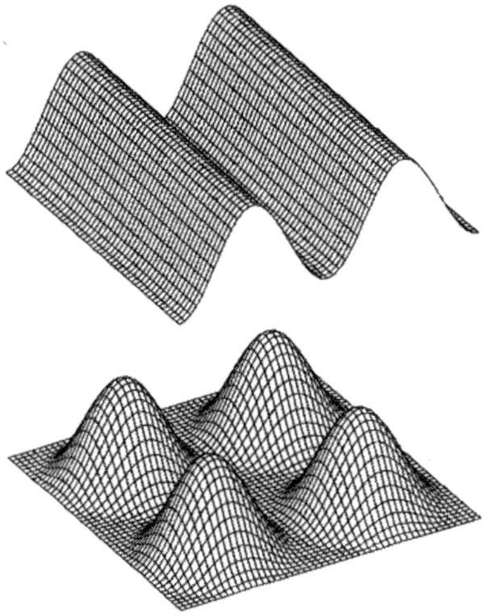

**Fig. 2.29.** Fragment of the contact surface at an initial instant of time for the two-dimensional plane and three-dimensional formulations of the problem.

of 0.8 cm. In the axially symmetrical case, the calculated region represents a round-section cylinder in which the interface between two gases was inside and had a longitudinal axially symmetrical sinusoidal shape with an amplitude of 1 cm and a diameter $0.8\sqrt{2}$ cm. On the lateral cylinder boundary, we set the impenetrability conditions. In the three-dimensional case, the entire plane contact boundary was divided into identical $0.8\sqrt{2}$ cm side squares into which the circles were inscribed. The part of the surface that turned out to be inside the inscribed circle was perturbed in a similar way to that in the axially symmetrical case with a 1 cm amplitude, whereas the remaining part retained a plane shape. The shock wave began to interact with the contact boundary from the side of the maximum of the perturbation amplitude. The passage of the shock wave through the interface between two gases is termed further as the shock-wave refraction on the contact boundary.

As is well known,[24] the shock wave refraction can have both a regular character and an irregular one. In the plane case for the earlier-set initial conditions, an irregular refraction accompanied by the generation of the refracted and reflected shock waves takes place on the larger part of the interface. This manifests itself in the curvilinear shape of the reflected shock wave and in the presence of the Mach wave in the vicinity of the refraction point (the point on the contact boundary where the refracted, reflected, and incident shock waves intersect). The character of the refraction defines the shape and position of the reflected and the refracted shock waves, the features of the flow near the contact boundary, and, in the following time, the intensity of generated secondary shock waves, formation of local regions of elevated and lowered pressures, the positive and negative vorticities, etc.[49] A substantial distinction of the axially symmetrical and three-dimensional cases from the plane one is the presence of additional flows beyond the fronts of the refracted and the reflected shock waves, which corresponds to the larger spatial dimension of the problem. The interaction between flows beyond the refracted shock wave on the symmetry axis leads to the appearance of the elevated pressure regions. As a result, maximum pressures in these regions turn out to be several times higher than those in the two-dimensional plane case. As time elapses, this results in more intense penetration of a heavy gas into the lighter one beyond the front of the arising secondary shock wave. As the front of this secondary shock wave goes into the boundary with the lighter gas, the process of formation of the mushroom-shaped structure is accelerated, although this fact does not lead to an accelerated growth of the mushroom heads as is observed in the plane case. The transformation of the initial sinusoidally perturbed contact boundary into the mushroom-shaped one occurs as a result of the existence of the basic local vortex structure arising in the process of the refraction.[49] The trajectories of the points of the interface between two media, which are realized in the process of developing the Richtmyer–Meshkov instability in the three-dimensional case and clarify the dynamics of deformation of the surface, are presented in Fig. 2.30. The oscillatory character of the behavior of a number of trajectories is associated with a multiple passage through this region of secondary shock waves. The distinction between the three-dimensional and axially symmetrical variants consists in the accelerated motion of the plane part of the perturbation base in the direction of propagation of the incident shock wave and in the time lag of neighboring axially symmetrical perturbations in the contact region. The dynamics of modification of the contact boundary for all three cases is shown in Fig. 2.31. In the calculations, we used the $1131 \times 75$ and $800 \times 75$ grids for the corresponding two-dimensional variants and the $800 \times 75 \times 75$ grid for the three-dimensional variant.

As the perturbation amplitude used below, we took the distance between the contact-boundary points that penetrated most deeply into each of the gases under consideration. The change in the perturbation amplitude with time for the three variants studied is presented in Fig. 2.32. For these

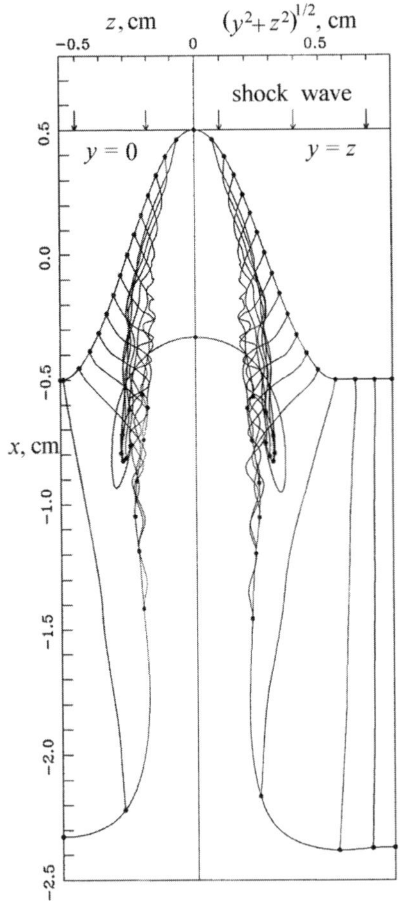


**Fig. 2.30.** Calculated trajectories for a number of points of the interface from the moment $t = 0$ to $t = 20\,\mu$s for the three-dimensional formulation of the problem. The trajectories that lie in the planes $y = 0$ and $y = z$ are shown. The interface at the corresponding moments and the position of the shock wave at the initial moment are also presented.

variants of the initial formulation of the problem, the anticipatory growth of the perturbation amplitude is observed in the axially symmetrical and three-dimensional cases as compared to the plane two-dimensional variant from the completion of the refraction of the shock wave, until the completion of the calculations. In the first 15–20 $\mu$s, we observe a more accelerated amplitude growth in the axially symmetrical cylindrical case as compared to the three-dimensional one. At later instants of time, the situation

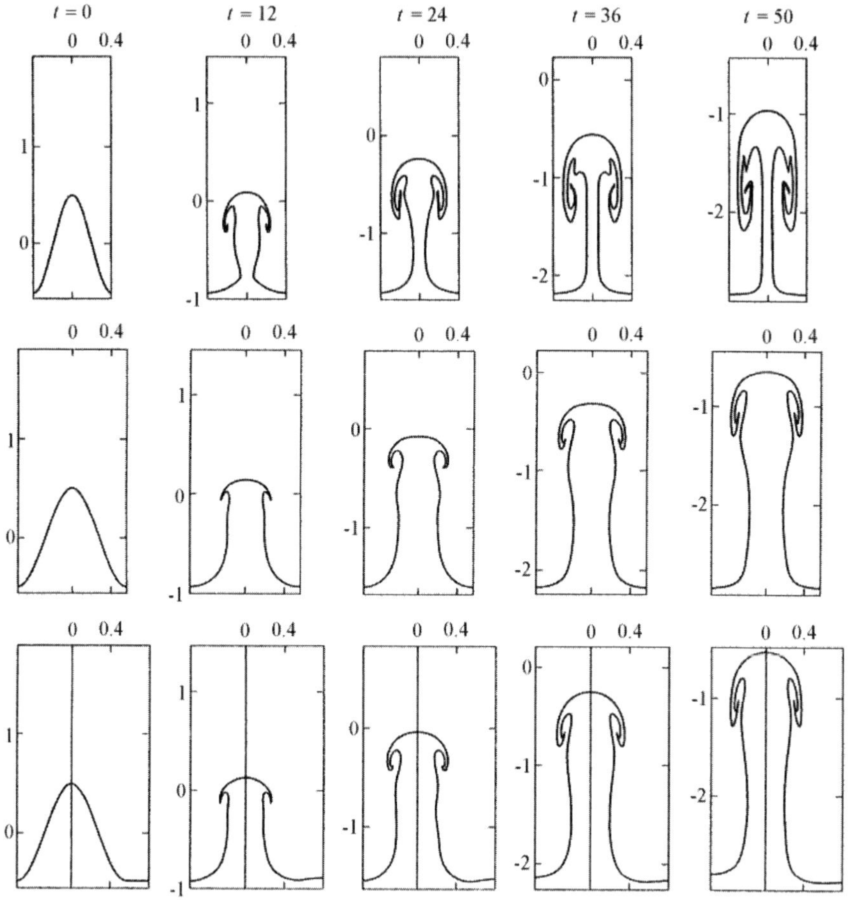

**Fig. 2.31.** Calculated interface at various moments (in $\mu$s) for three formulations of the problem under investigation: two-dimensional plane, two-dimensional axially symmetrical, and three-dimensional flow motion. In the last case, the projections onto the planes $y = 0$ and $y = z$ are shown, as in Fig. 2.30. The scale of the axes is given in cm.

changes to the opposite one, and the amplitude of the three-dimensional perturbation exceeds that for the cylindrical case.

## 2.8. Conclusion

The results presented in this chapter prove that numerical simulations can be used not only to explain physical phenomena and facts observed in

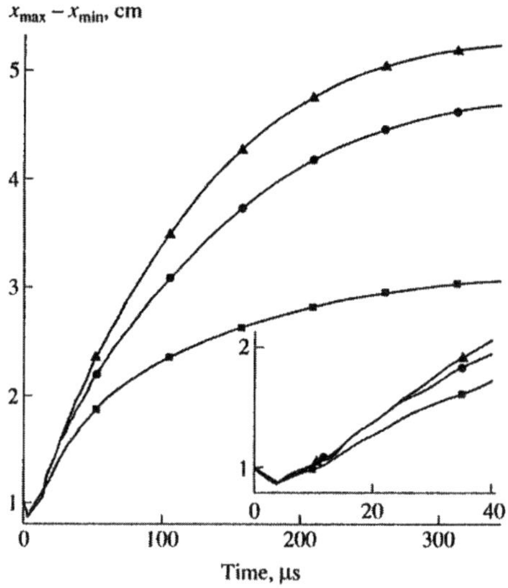


**Fig. 2.32.** Change of the perturbation amplitude with time for the three variants under investigation: (■) two-dimensional plane, (•) two-dimensional axially symmetrical, and (▲) three-dimensional variants.

experiments, but also to predict new effects, and generalize the well-known data. The successful theoretical and numerical study of physical process requires:

(i) an exact and comprehensive theoretical model;
(ii) a correct mathematical formulation of the model;
(iii) numerical methods capable of obtaining solutions of the specified accuracy in a reasonable time on computers currently in use.

## Appendix

The results of the three-dimensional numerical simulation can be presented in the form of a set of slides that display quantitative information about each of the physical parameters of interest at any time for an arbitrary point in the domain of integration. The preliminary information about the development of physical processes during the examination of the slide-film promotes more accurate experimental studies (Figs. A.1(a)–A.1(d)).

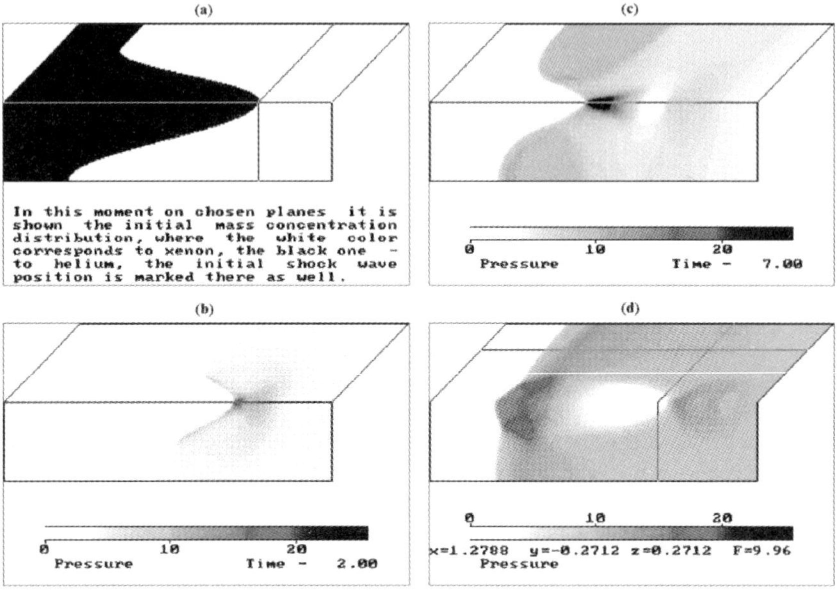


**Fig. A.1.** Results of the three-dimensional numerical simulation.

## References

1. O. M. Belotserkovskii, V. V. Demchenko, V. I. Kosarev and A. S. Kholodov, Numerical simulation of some problems on laser compression of shells, *USSR Comput. Math. Math. Phys.* **18**(4), 117–137 (1978).
2. R. D. Richtmyer, Taylor instability in shock acceleration of compressible fluids, *Commun. Pure Appl. Math.* **13**, 297–319 (1960).
3. E. E. Meshkov, Instability of an interface between two gases accelerated by a shock wave, *Mekh. Zhidk. i Gasa — Izv. Acad. Nauk SSSR* **5**, 151–157 (1969) (in Russian).
4. V. B. Rozanov, J. G. Lebo, S. G. Zaitsev *et al.*, Experimental studies of gravitational instability and turbulent mixing of stratified flows in an acceleration field in connection with problems of inertial-confinement fusion, Preprint No. 56 FIAN, Moscow (1990).
5. A. N. Aleshin, V. V. Demchenko, S. G. Zaitsev and E. V. Lasareva, The interaction of shock waves with undulating tangential discontinuity (in Russian), *Mekhanika Zhidkosti i Gaza — Izv. RAN* **5**, 168–174 (1992).
6. V. V. Demchenko and A. M. Oparin, Multidimensional numerical simulation of strong shock waves interaction and Richtmyer–Meshkov instability development *Proc. 19th ISSW*, Marseille, France (1993).

7. V. V. Demchenko, On some exact solution of Euler equations system accounting of nonlinear heat conduction, *Trans. MIPT* 114–117 (1977).
8. V. V. Demchenko, The comparative studying of some hydrodynamic compression processes, *USSR Comput. Math. Math. Phys.* **19**(2), 286–292 (1979).
9. S. I. Anisimov *et al.*, Numerical simulation of the processes of laser compression and heating of simple shell targets, *Fiz. Plasmy* **3**(4), 723–732 (1977).
10. P. P. Volocevich *et al.*, The process of super-high compression of matter and the initiation of thermonuclear reactions by a powerful pulse of laser radiation, *Fiz. Plasmy* **2**(6), 883–897 (1976).
11. E. G. Gamalii *et al.*, On the possibility of measuring the characteristics of laser plasma by neutrons of the DT-reaction, *Lett. Zh. ETF.* **21**(2), 156–160 (1975).
12. L. I. Sedov, *Methods of Similarity and Dimensions in Mechanics* (*Metody podobiya i razmernosti v mekhanike*) (Nauka, Moscow, 1972).
13. K. P. Stanyukovich, *The Unsteady Motions of Continuous Media* (*Neustanovivshiesya dvizheniya sploshnoi sredy*) (Gostekhizdat, Moscow, 1955).
14. L. I. Sedov, On the integration of the equations of one-dimensional motion of a gas, *Dokl. Akad. Nauk SSSR* **90**(5), 735 (1955).
15. A. A. Samarskii and I. M. Sobol, Examples of the numerical calculation of temperature waves, *USSR Comput. Math. Math. Phys.* **3**(4), 945–970 (1963).
16. P. P. Volocevich *et al.*, Travelling waves in a medium with non-linear heat conduction, *USSR Comput. Math. Math. Phys.* **5**(2), 199–217 (1965).
17. S. I. Anisimov and N. A. Inogamov, The development of instability and loss of symmetry in the isentropic compression of a spherical drop, *Lett. Zh. ETF.* **20**(3), 174–176 (1974).
18. P. P. Volosevich and E. I. Levanov, *Self-Similar Solutions of the Gas-Dynamic Equations Allowing for Non-Linear Heat Conduction* (*Avtomodel'nye resheniya uravnenii gazovoi dinamiki s uchetom nelineinoi teploprovodnosti*) (Institut prikl. matem. TGU, Tbilisi (1977).
19. L. V. Ovsyannikov, A new solution of the equations of hydrodynamics, *Dokl. Akad. Nauk SSSR* **111**(1), 47–49 (1956).
20. R. E. Kidder, Theory of homogeneous isentropic compression and its application to laser fusion, *Nucl. Fusion* **14**, 53–60 (1974).
21. V. A. Belocon *et al.*, An example of ultrahigh compression, Preprint No. 39, *IPM Akad. Naul SSSR* (1978).
22. N. V. Zmitrenko and S. P. Kurdyumov, N- and S-modes of self-similar compression of a finite mass of plasma in particular modes with peaking, *Prikl. matem. i tekhn. fiz.* **1**, 3–23 (1977).
23. Ya. M. Kazhdan, On the question of the adiabatic compression of a gas under the action of a spherical piston, *Prikl. matem. i tekhn. fiz.* **1**, 23–30 (1977).
24. L. F. Henderson, On the refraction of shock waves, *J. Fluid Mech.* **198**, 365–386 (1989).
25. V. V. Demchenko, Analytical and numerical study of sequential contact surface turbulization in Richtmyer–Meshkov instability, *Proc. 5th Int.*

*Workshop on the Physics of Compressible Turbulent Mixing*, New York, USA (1995), pp. 331–337.
26. L. F. Henderson, The refraction of a plane shock wave at a gas interface, *J. Fluid Mech.* **26**, 607–637 (1966).
27. J. F. Hawley and N. J. Zabusky, Vortex paradigm for shock accelerated density stratified interfaces, *Phy. Rev. Lett.* **63**, 1241–1244 (1989).
28. D. L. Youngs, Modeling turbulent mixing by Rayleigh–Taylor instability, *Physica D* **37**, 270–287 (1989).
29. J. Grove and R. Menikoff, The anomalous reflection of a shock wave at a material interface, *J. Fluid Mech.* **219**, 313–336 (1990).
30. V. V. Demchenko, Numerical simulation of Richtmyer–Meshkov instability, *Russ. J. Comput. Mech.* **1**(2), 51–66 (1993).
31. V. V. Demchenko and I. V. Nemchinov, Three-dimensional gas motions by break-down at several points on circle, *Fiz. Gorenija i Vzryva.* **6**, 131–134 (1990).
32. D. L. Youngs, Three-dimensional numerical simulation of turbulent mixing by Rayleigh–Taylor instability, *Phys. Fluids A* **3**(5), 1312–1320 (1991).
33. R. P. J. Town and A. R. Bell, Three-dimensional simulations of the implosion of inertial confinement fusion targets, *Phys. Rev. Lett.* **67**(14), 1863–1866 (1991).
34. J. Glimm, X. L. Li, R. Menikoff *et al.*, A numerical study of bubble interaction in Rayleigh–Taylor instability for compressible fluids, *Phys. Fluids A* **2**(11), 2046–2054 (1990).
35. B. A. Remington, S. W. Haan, S. G. Glendinning *et al.*, Large growth, planar Rayleigh–Taylor experiments on Nova, *Phys. Fluids B* **4**(4), 967–978 (1992).
36. D. Ofer, D. Shvarts, Z. Zinamon and S. A. Orszag, *Phys. Fluids B* **4**, 3549 (1992).
37. K. I. Read, *Physica D* **12**, 45–58 (1984).
38. E. E. Meshkov, One approach to the experimental studies of hydrodynamic instability evolution. Creation of gas-gas interface using dynamic technique, Preprint No. 44–96, VNIEF, Sarov, Russia (1996).
39. V. A. Andronov, S. M. Bakhrakh, E. E. Meshkov *et al.*, Turbulent mixing at shock-accelerated interface, *ZhTE* **71**(8), 806–811 (1976).
40. V. V. Bashurov, Yu. A. Bondarenko *et al.*, Experimental and numerical evolution studies for 2-D perturbations of the interface accelerated by shock waves, *Proc. 5th IWPCTM*, New York, USA (1995) pp. 285–293.
41. Yu. A. Kucherenko, V. E. Neuvazhaev and A. P. Pylaev, Behavior of a region of gravity-induced turbulent mixing under conditions leading to separation, *Phys. Doklady* **39**(2), 114–117 (1994).
42. K. O. Mikaelian, Numerical simulation of Ricthmyer–Meshkov instability in finite-thickness fluid layers, *Phys. Fluids* **8**(5), 1269–1292 (1996).

43. L. Hallo, R. L. Morse, J. M. Charisse, N. M. Hoffman and N. Toqie, Modeling of linear perturbation growth in gas dynamics: from incompressible to compressible flows, *Proc. 6th IWPCTM*, Marseille, France (1997), pp. 191–196.
44. R. H. Cohen, W. P. Dannevik *et al.*, Three-dimensional high-resolution simulations of Richtmyer–Meshkov mixing and shock-turbulence interaction, *Proc. 6th IWPCTM*, Marseille, France (1997), pp. 128–133.
45. V. V. Nikishin, V. F. Tishkin, N. V. Zmitrenko, I. G. Lebo, V. B. Rozanov and A. F. Favorsky, Numerical simulation of nonlinear and transitional stages of Richtmyer–Meshkov and Rayleigh–Taylor instabilities, Preprint No. 30 FIAN, Moscow (1997).
46. Q. Zhang and S. Sobin, An analytical nonlinear theory of Ricthmyer–Meshkov instability, *Phys. Lett. A* **212**, 149 (1996).
47. T. Yabe, H. Hoshino and T. Tsuchiya, Two- and three-dimensional behavior of Rayleigh–Taylor and Kelvin–Helmholtz instabilities, *Phys. Rev. A* **44**(4), 2756–2758 (1991).
48. S. Zaytsev, A. Aleshin, E. Lazareva *et al.*, *Proc. 4th IWPCTM*, Cambridge, England (1993), pp. 291–296.
49. O. M. Belotserkovskii, V. V. Demchenko and A. M. Oparin, *Dokl. Akad. Nauk.* **334**(5), 581–583; [*Phys. Dokl.* **39**(2), 118–120 (1994)].
50. O. M. Belotserkovskii, V. V. Demchenko and A. M. Oparin, Unsteady three-dimensional numerical simulation of the Richtmyer–Meshkov instability, *Phys. Dokl.* **42**(5) 273–276 (1997).

# Chapter 3

# Rayleigh–Taylor Instability: Analysis and Numerical Simulation

The chapter consists of two parts. The first one has mainly theoretical character. The results of Anisimov, Chekhlov, Dem'yanov, and Inogamov,[1] describing the instability of periodic solutions, the initiation and development of structures, are used here. Discussed in the second part are the results of some versions of numerical simulation, performed during a long time period by many authors by help of various hydrocodes.

## 3.1. The theory of Rayleigh–Taylor instability: modulatory perturbations and mushroom-flow dynamics

Two aspects of the theory of Rayleigh–Taylor instability (RTI) are considered. The first one deals with the instability of periodic solutions, which causes an increase in the amplitude of subharmonic perturbations. The second one describes the initiation and development of mushroom-shaped flows.

It is known that after a spatially periodic stimulation of a boundary that is Rayleigh–Taylor unstable, the asymptotic solution looks like a stationary chain of periodically alternating bubbles and jets. This part of the chapter gives an analysis of the evolution of longer-wave perturbations in this chain, which are the modulations of the periodic structure.

The analysis is based on theoretical and numerical methods. Theoretically, it is rigorously proved that when the modulation has a period equal to the doubled period of the chain, then:

(i) the natural modes are expanded into a spatial series in half-integer harmonics;
(ii) only two different modes are possible, corresponding respectively to the merging and competition of the bubbles.

The numerical calculations confirm the theoretical prediction that the stationary chain is unstable with respect to the class of such modulation perturbations, and also determine the growth rates of this instability.

After that the problem of mushroom-shaped flows is considered. Two analytical models have been developed. The first one describes the initiation of mushroom structures and the second one — their development. The analytical and numerical results are compared and appear to be in good agreement.

The computer simulations were made by a new and promising numerical method developed for simulating the dynamics of nonhomogeneous media. This is a new version of the method of pseudo-compressibility, which is extended to the nonhomogeneous case.

### 3.1.1. Introduction

The importance of the RTI problem is widely recognized. This instability is encountered in such widely different situations as inertial confinement fusion (ICF),[2–4], astrophysics,[5] magnetohydrodynamics,[6,7] cavitation, boiling, and detonation.[8] In addition to the applied problems, the RTI is involved in many basic nonlinear phenomena of turbulent flow and randomness in distributed Hamiltonian systems.[8–18]

Earlier investigations of the RTI concentrated on the following subjects:

(i) the linear analysis of the stability of hydrostatic equilibrium, taking into account various physical factors that complicate these equilibrium, such as small-scale effects that determine the minimum perturbing wavelength and are nondissipative (the linear analysis surface tension) or dissipative in character (viscosity diffusion, heat conduction and, for magnetic fields, finite conductance), as well as large-scale effects such as vertical homogeneity of the density profile and compressibility;

(ii) the analysis of periodic stationary solution[19–25];

(iii) the averaged description of turbulent mixing by means of equations of turbulent diffusion. The latter equations were set up by closing, by means of gradient ratios, the chain of momentum equations for correlators, and were derived by means of the model that is based on the concept of the mixing length.

The earlier physical and computer experiments examined either periodic or averaged statistical dynamics. The averaged dynamics did not contain any information about coherent structures generated in the mixing layer in the direction transverse to the vector of free fall acceleration. The periodicity was analyzed on the basis of experimental data obtained by observing the rise of bubbles in tubes and on the basis of numerical simulations. In experiments on turbulent mixing, the structures transverse to the acceleration were commonly neglected, being regarded as fluctuations. The aim was to average the physical parameters over the longest possible transverse distance that would give a "good" fit, and thus to extract from these experiments as much information as possible about the stratification of the average density profile and its rate of expansion.

At present, the focus of attention is gradually shifting toward analysis of the interaction of harmonics, sometimes called "multimode dynamics". This shift of focus is motivated by a desire to understand better the complicated physical mechanisms involved in such phenomena and to apply this understanding in a new approach to the dynamics of regular and random phenomena.

This section analyzes the functional neighborhood of the periodic stationary solution, also briefly called a stationary solution or "stationary", and considers the linear interaction between the periodic stationary solution and the longer-wave perturbations. This topic is directly related to the interaction of harmonics.

This stationary solution is an important factor in the analysis of the RTI, where further progress is predicated on progress in the complicated problem of stability of the stationary solution. The analytical approach to this problem faces great difficulties due to the singularity in the resultant integral equations caused by the existence of jets that tend to infinity and are asymptotically free falling. The numerical approach is also complicated.

The difficulties can be overcome by combining the theoretical and numerical approaches. The instability of the stationary solution to modulatory perturbations of periodic structure will be proved in this part of the chapter. Note that this is the first time that a linear analysis of the problem has been undertaken.

Modulatory perturbations are of a great importance in the dynamics of the RTI since they lead to an increase in the average scale of coherent structures and to an acceleration of the expansion of the mixing region.[10]

### *3.1.2. Periodicity and symmetry of modulatory perturbations*

Consider a hydrostatic equilibrium between two liquids, one of them filling the upper half-space and the other — the lower one. At equilibrium the liquids are at rest and the interface between them is horizontal. If the lighter liquid is at the bottom, the equilibrium is unstable. This is called the RTI.

Assume that we are dealing with homogeneous incompressible ideal liquids, and that the ratio of densities of the bottom (light) to the top (heavy) liquid is $\mu = \rho_1/\rho_2 = 0$. This is the classical setting for the analysis of the evolution of the RTI.

Let there occurs, at time $t = 0$, a perturbation of the interface and/or of the velocity potential, describable by a single harmonic with period $\lambda$. The perturbation builds up exponentially during the linear stage with an increment $y = \sqrt{gk}$, where $k$ is the wave number. Then, the exponential increase slows down, and the transient process starts, as a result of which a steady state is established. An analytical description of this transition can be found in Ref. 19.

The stationary solution[19–24] is actually a periodic chain of bubbles and jets (Fig. 3.1(a)). Down to a depth of 40% of $\lambda$, where $\lambda$ is measured from point $S$, the boundary of the top surface of a bubble is fairly well approximated by a circular arc of radius $R$, which is equal to about $2/k$. In the laboratory reference frame in which the liquid is at rest at $y = +\infty$, the bubbles rise at a velocity of $u_+ = (0.23 \pm 0.01)\sqrt{g\lambda}$. But in the reference frame moving with the bubbles, there occurs a stagnation point at which the velocity is zero. This stagnation point is labeled $S$ in Fig. 3.1(b). Symbolically, the solution may be presented in the form of a chain of points $S$ separated by the distance $\lambda$, Fig. 3.1(c).

Now let us consider the periodicity and symmetry conditions. We begin with perturbations of the stationary solution. Let the perturbed solution have a period of $m\lambda(m > 1$ is an integer) and satisfy the symmetry conditions on the vertical straight lines $S_1$ and $S_2$ separated by the period $m\lambda$ as shown in Fig. 3.2(a). The perturbed potential $\varphi$ and boundary displacements $\eta$ are subject to the evenness conditions on the symmetry lines. These perturbations are easiest to analyze if we use a two-dimensional

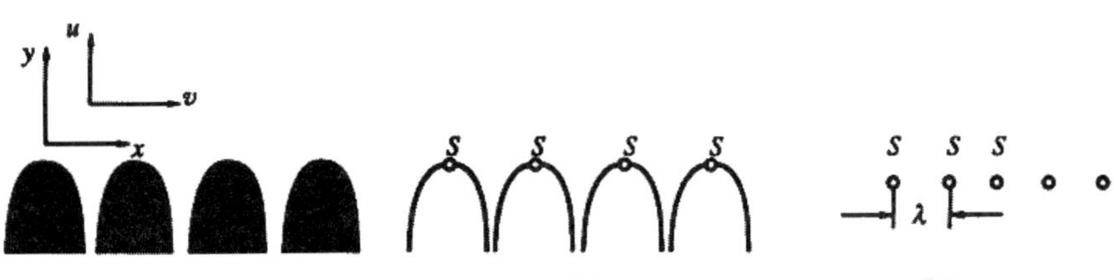


**Fig. 3.1.** (a) Structure of stationary solution; (b) position of stagnation points $S$; (c) periodic chain of points $S$.

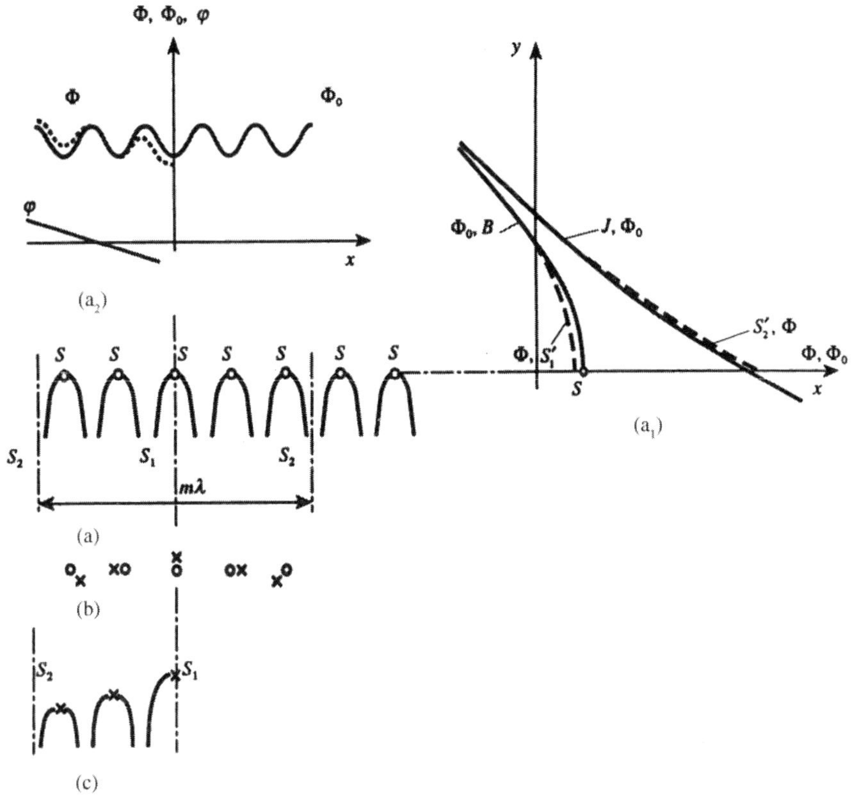


**Fig. 3.2.** (a) Perturbation periodicity and symmetry; ($a_1$) and ($a_2$) are potentials in a vertical section $y$ and a horizontal section $x$, respectively; (b) unperturbed (circles) and perturbed (crosses) chains of top points of bubbles; (c) perturbed boundary at $m = 5$. Crosses correspond to top points.

numerical simulation in which the computation region is bounded by a rectangular frame. In this case, straight lines $S_1$ and $S_2$ serve as the lateral boundaries of the region. This makes it possible to investigate perturbations with wavelength $m\lambda$ that is half the length of the computational region, thereby economizing on the memory and the processor time. We take as the boundary conditions the symmetry conditions on the lines $S_1$ and $S_2$. The initial conditions are similarly chosen.

The effect of the boundary conditions at the upper and the lower boundaries of the computation region was studied in detail and the results were compared with the cases of free and rigid boundary conditions. We followed the general principle that the upper and lower boundaries should be set as far as possible from the interaction region in order to minimize their effect on flow dynamics. The distances to the boundaries were chosen with reference to available processor time and memory. The calculations were performed in vertically elongated rectangles. In these rectangles the ratio of the vertical dimension to the horizontal one ranged from 3:1 to 6:1, while the ratio of the initial heights of the top and bottom liquids ranged from 2:3 to 3:2.

The main physical properties of the solutions analyzed are given in Fig. 3.2(a), 3.2(a$_1$), and 3.2(a$_2$). The axis $y = 0$ in Fig. 3.2(a$_1$) is shown as if it was the extension of the straight line with the stagnation points $S$ in Fig. 3.2(a). The solid curve $B$ in Fig. 3.2(a$_1$) is the unperturbed potential $\Phi_0$ plotted against height $y$: $\Phi(x = n\lambda, y)$, $n = 0, \pm 1, \pm 2, \ldots$. The verticals $x = n\lambda$ pass through the points $S$. The curve $B$ terminates at point $S$. Since $S$ is a stagnation point, $\partial\Phi_0/\partial y$ on this curve vanishes at this point. The solid curve $J$ corresponds to $\Phi_0(x = \lambda/2 + n\lambda, y)$. The straight lines $x = \lambda/2 + n\lambda$ pass through the axes of symmetry of the jets. As $y \to +\infty$, the potential $\Phi_0$ degenerates into an asymptote corresponding to free fall. The unperturbed potential $\Phi_0$ is the stationary solution. The potentials are given in the reference frame moving with point $S$ (see Fig. 3.2(a)). As $y \to +\infty$, the potentials $\Phi$ and $\Phi_0$ tend asymptotically to a descending straight line. This corresponds to homogeneous flow incoming at a velocity of $\nabla\Phi_0$.

The full potential $\Phi$ in the linear approximation is the sum of the stationary potential $\Phi_0$ and the perturbed potential $\varphi$. Variation of potential $\Phi$ as a function of $y$ on the straight lines $S_1$ and $S_2$ is shown in Fig. 3.2(a$_1$) by the dotted curves $S_1'$ and $S_2'$, respectively. Figures 3.2(a$_1$), 3.2(a$_2$), 3.2(b), and 3.2(c) show the case when the perturbation is such that the velocity of the incoming flow of the bubbles on the straight line $S_1$ is less than the unperturbed rate at which the steady-state bubbles rise. Since $\mathbf{u} = \nabla\Phi$, this means that the bubble travels upward against an incoming flow (see Figs. 3.2(b) and 3.2(c)), and therefore the curve $S_1'$ in Fig. 3.2(a$_1$) passes below the curve $B$. A similar reasoning shows why the curve $S_2'$ passes above the curve $J$ and why the bubbles, located with period $m\lambda$ and near the straight lines $S_2$, lag behind the unperturbed bubbles (see Figs. 3.2(b) and 3.2(c)).

Figure 3.2($a_2$) illustrates how the potentials on level $y = y_1$, $\Phi(x, y = y_1, t)$, $\Phi_0(x, y = y_1)$, and $\varphi(x, y = y_1, t)$ depend on $x$ at time $t$. The level $y_1$ seems to be located in the $(x, y)$-space in Fig. 3.2(a). The minima of the potential $\Phi_0(x)$ occur above the tops of the bubbles. The height of the peaks of potential

$$\frac{\left[\Phi_0\left(x = \frac{\lambda}{2} + n\lambda, y_1\right) - \Phi_0\left(x = n\lambda, y_1\right)\right]}{\sqrt{g\lambda^3}},$$

tends to zero as $y_1/\lambda \to +\infty$. The potential $\varphi$ behaves as shown in Fig. 3.2($a_2$). Its minima are located at $x = (m\lambda)$ and its maxima at $x = m\lambda/2 + n(m\lambda)$, $n = 0, \pm 1, \ldots$.

In accordance with the above, the potentials $\Phi$, $\Phi_0$, and $\varphi$ are symmetric relative to the respective points $x$ (see Fig. 3.2($a_2$)). The perturbed boundary is a modulation of the initial periodic solution (see Figs. 3.2(b) and 3.2(c)). The central bubble is accelerated vertically, the middle bubble is shifted horizontally (see Figs. 3.2(b) and 3.2(c)), the lateral bubbles are decelerated, and the jets passing through the vertical straight lines $S_2$ are intensified. As a result, the top of the lateral bubble located near the straight line $S_2$ in Fig. 3.2(c) is shifted downward and to one side.

The perturbation dynamics of the stationary solution will be considered on the basis of these perturbation properties.

### *3.1.3. Cutting off the singularities associated with jets*

Section 3.1.2 dealt with the modulatory perturbation of the stationary solution in which jets of infinite length are postulated.

Modulation instability is numerically simulated as follows: first, the Navier–Stokes equations are integrated with respect to variables $x, y$, and $t$ over a rectangular grid at a low viscosity $\nu \ll \sqrt{g\lambda^3}$ and, second, the dynamics of the boundary $y = N(x, t)$ and the potential $\Phi(x, y, t)$ are defined.

If we adhere strictly to the description given in Sec. 3.1.2, at $t = 0$, we must subject the hydrostatic equilibrium to a perturbation in the form of a single harmonic and wait until $t = +\infty$ before $N$ and $\Phi$ degenerate into $N_0$ and $\Phi_0$. During this period of time, infinite jets form. Then, we take this moment as a new initial time $t_0$ and impose the initial perturbation $\eta(x, t_0)$,

$\varphi(x, y, t_0)$ on the stationary solution, representing it in the form of a chain

$$\eta(x, t_0), \varphi(x, y, t_0) = Re \sum_{-\infty}^{+\infty} © \ln(z - z_n),$$

where $\Sigma © = 0$ with weak sources $|æ_n| \ll \sqrt{g\lambda^3}$ and small displacements $|z_n - y_* - n\lambda| \ll \lambda$, $n = 0, \pm 1, \ldots$. Then let us take, for instance, $y_* \approx \lambda/2$, that is, we position the source near the center of the circle drawn about the stagnation point $S$ and inscribed as $N_0(x)$.

In fact, this procedure cannot be implemented because it is impossible to get a stationary solution with infinite jets by the selected integration method due to limitations on both the computational time and domain. The key point of this section is that a rigorous stationary solution is not necessary; this is because the steady state is heterogeneous in space (Fig. 3.3(a)). By time $t_1$ the boundary section above the line $a_1b_1$, which is marked with dots in Fig. 3.3(a), and the potential $\Phi$ above $a_1b_1$ are close to the stationary solution. As the time proceeds, the line $a_1b_1$ shifts downward at approximately the velocity of the flow. In essence, due to the ellipticity of the equation, the effect of the nonstationary region on the potential in an incompressible liquid decreases exponentially outside the region, the scale

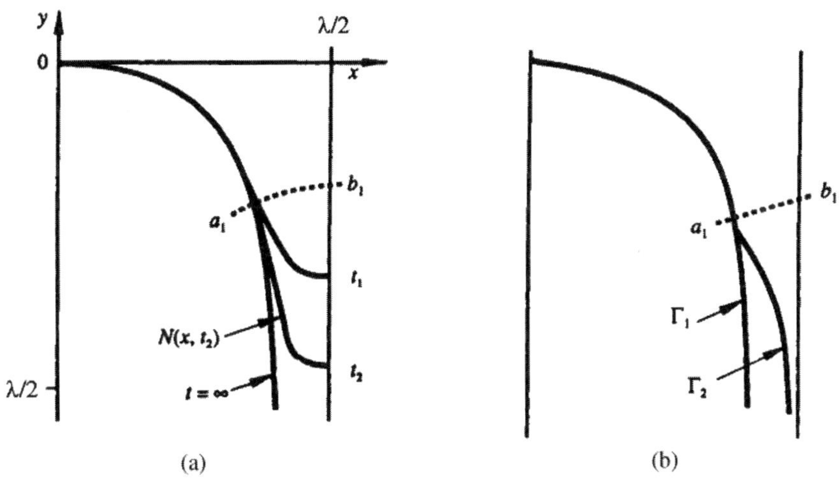


**Fig. 3.3.** (a) Succession $t_1$, $t_2$, $t = \infty$ of boundary positions $y = N(x, t)$. At time $t_1$ the nonstationary region is below the line $a_1b_1$. Above this line the solution is close to the stationary solution: $N, \Phi \to N_0, \Phi_0$. (b) Above the potentials $\Phi_1$ and $\Phi_2$ associated with boundaries $\Gamma_1$ and $\Gamma_2$, respectively, are very close.

of the decrease corresponding to the size of the region

$$\Phi(x,y,t) = \Phi_0(x,y) + \delta\Phi(x,y,t),$$
$$\frac{\delta\Phi}{\sqrt{\delta\lambda^3}} = \frac{\chi_1}{\exp\left(\frac{\chi_2(y-N(x=\lambda/2,t))}{D}\right)},$$

where

$$\frac{D}{\lambda} \approx \frac{0.23}{\sqrt{2}}\sqrt{\frac{\lambda}{N_0\left(x=\lambda/2,t\right)}}$$

is the width of the steady-state jet at a depth of $N_0$ (an unperturbed boundary), and both functions $\chi_1(x,y,t) \cong 1$ and $\chi_2(x,y,t) \cong 1$ are in the region occupied by the liquid.

Let the potential $\Phi_0$ be induced by barriers $\Gamma_1$ and $\Gamma_2$, with the barriers coinciding above the line $a_1b_1$. Accordingly, the potential $\Phi_0$ will be very close above the line $a_1b_1$ (Fig. 3.3(b)) despite the fact that the barriers $\Gamma_1$ and $\Gamma_2$ differ significantly below the line.

In this way, problems involving an infinite length of jet are solved because, first, the nonstationary region is localized and, second, the jets are narrow. As a result, we obtain the following important conclusion. The dynamics of the bubbles can be considered irrespective of the behavior of the jets. Therefore, parameters such as the rate of rise and the curvatures of the steady state and perturbed bubbles, and the distribution of the potential in their vicinity are weakly related to the dynamics of distant sections of the jets. In practice, while studying the distance from a section of jet, we come to the following conclusion: the effect of the section of jets separated by a distance greater than $\lambda/2$ from the top of the bubble is insignificant.

Taking this into account, we consider the hydrostatic equilibrium between two liquids. Let us subject this equilibrium to a biharmonic perturbation of the following type:

$$N(x,0) \equiv 0,$$

$$\Phi(x,y,\,0) = [\theta(y)-\theta(-y)]\left[-\frac{A_1}{k}\cos kx\exp(-k|y|) - \frac{mA_1}{k}\cos\frac{kx}{m}\exp\left(-\frac{k|y|}{m}\right)\right].$$

Let $A_1 = (0.4 \div 0.7) \times 0.23\sqrt{g\lambda}$. In this case, during the period of time $t_* = (1 \div 2)\gamma^{-1}$, $\gamma = \sqrt{gk}$ the jets develop amplitudes $\geq \lambda/2$, and at $A_2 = 0$, the solution becomes stationary. We select $A_2(0) \ll A_1(0)$. Usually, $A_2$ amounts to 10–20% of $A_1$. During the period of time $t_*$, the amplitude $A_2(t)$ varies. Let us consider the amplitudes $A_2(t)$ in the reference frame moving with the stagnation point of the steady-state bubbles. The changes are such that, at $t = t_*$, the amplitudes $A_2(t)$ are very small compared with $0.23\sqrt{g\lambda}$. Therefore, the modulation amplitude associated with this harmonic may be still considered linear. As a result, we obtain the flow shown in Fig. 3.2(c) but with finite jets.

To sum up, the above considerations mean that we can expect that such characteristics of modulatory perturbations as the potentials and instability increments will be obtained with the same accuracy as the values of $u_+$ and $R$ for steady-state bubbles calculated from the results of numerical simulation of the flow with finite jets.

### 3.1.4. Classification of perturbations

The class of perturbations in question is described in Sec. 3.1.2. Sections 3.1.3 and 3.1.4 are devoted to the two main principles needed to solve the problem. The first is that the distant sections of jets that have been transformed into a state close to free fall have a diminishing effect on the dynamics of both the steady state and perturbed groups of bubbles. There is, thus, no need to wait for ages until the jets grow very long during a numerical simulation (see Sec. 3.1.3). The other principle concerns the classification of the perturbations. It so happens that we can rigorously prove that only two linearly independent modes are possible. These modes are closely related. For $m = 2$, these modes describe the merging and competition of bubbles, respectively.

We begin the analysis of these problems by expanding the perturbations considered in Sec. 3.1.2 as a Fourier series. Let the $x$,$y$-frame be chosen as shown in Fig. 3.1(a). In this case, the unperturbed boundary $N_0(x)$ and the potential $\Phi_0(x, y)$ are expanded as a cosine Fourier series

$$N_0 = \sum_{n=0}^{\infty} B_n C_n, \quad \Phi_0 = \sum_{n=0}^{\infty} A_n e_n C_n, \tag{3.1}$$

where

$$C_n = \cos(knx), \quad e_n = \exp(-kny).$$

Now let us consider a modulatory perturbation. Since an increase in $m$ requires a larger computational grid, it is evident that the computational time increases with $m$. Here, we shall restrict ourselves to the first nontrivial case when $m = 2$.

From the periodicity and symmetry it follows that for an arbitrary integer $m$ there are only two ways to obtain the lateral boundary conditions. In one case, the straight line $S_2$ pass along the axes of symmetry of the unperturbed jets (see Fig. 3.2(a)). In the other case, the line passes along the axes of symmetry of the bubbles. For $m = 2$, the straight line $S_2$ passes through $kx = \pi + 4\pi n$, $n = 0, \pm 1, \ldots$, in the first case, and through $kx = 4\pi n$ in the second one.

We separate the time and space variables as usual. At this time, the perturbations in the boundary and the potential are $\exp(\gamma t)\eta(x)$ and $\exp(\gamma t)\varphi(x, y)$, respectively. The straight line $S_2$ is equivalent to rigid walls on which any velocity component normal to the wall vanishes. From the periodicity and symmetry it follows that the Fourier series in the first- and second-cases are

$$\varphi(x, y) = \alpha_1 e_{1/2} S_{1/2} + \alpha_2 e_1 C_1 + \alpha_3 e_{3/2} S_{3/2} + \alpha_4 e_2 S_2 + \cdots, \tag{3.2}$$

$$\varphi(x, y) = \beta_0 + \beta_1 e_{1/2} C_{1/2} + \beta_2 e_1 C_2 + \beta_3 e_{3/2} C_{3/2} + \beta_4 e_2 C_2 + \cdots, \tag{3.3}$$

$$e_{m/2} = \exp\left(-\frac{kmy}{2}\right),$$

$$S_{m/2} = \sin\frac{kmx}{2}, \quad C_{m/2} = \frac{kmx}{2}.$$

Figure 3.4 shows the unperturbed boundary $N_0(x)$ and the directions of perturbations of the velocity on the symmetrical vertical streamlines. Figure 3.4(a) corresponds to Eq. (3.2). Jets $a$, $c$, etc. are reduced, and jets $b$, $d$, etc. are amplified. The tops of the bubbles are shown as a chain of crosses. It is evident that they approach each other in pairs. Anticipating the trend of evolutions, we should say that the development of this mode leads asymptotically to the decay of jets $a$, $c$, etc. and to the attraction of the pair of bubbles $a_1$, $a_2$, etc. and the pair of bubbles $c_1, c_2$, etc. Figure 3.4(b) corresponds to Eq. (3.3). In this case, bubbles $a$, $c$, etc. contract and bubbles $b$, $d$, etc. expand. This mode is called the bubble competition mode. The chain of crosses marks the position of the tops of the bubbles as the mode develops. It is evident that bubbles $b$, $d$, etc. shift forward, while bubbles $a$, $c$, etc. lag behind.

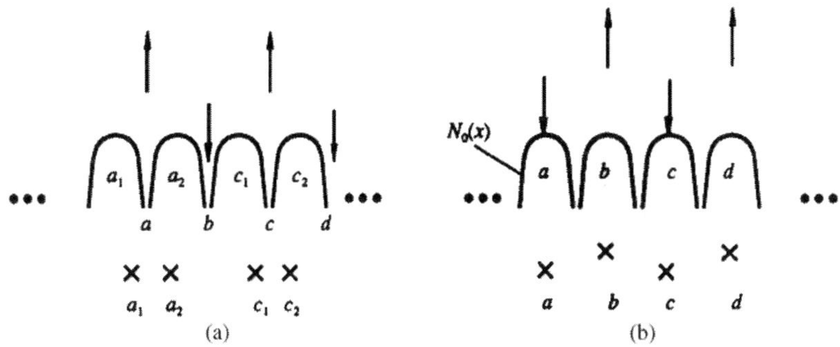


**Fig. 3.4.** (a) Perturbations causing every other jet to decay; (b) perturbations causing every other bubble to decay. The solid line shows the unperturbed boundary, the arrows show the directions of the velocity perturbations.

The kinematic and dynamic boundary conditions that are nonstationary, stationary, and linearized about the stationary solution have the following form:

$$N_1 = \Phi_y - N_x \Phi_x,$$
$$\Phi_1 = -\frac{(\nabla\Phi)^2}{2} - |g|\, N,$$

$$\Phi_{0y} - N_{0x}\Phi_{0x} = 0, \quad (\nabla\Phi_0)^2 + 2\,|g|\,N_0 = 0,$$
$$\gamma\eta = \varphi_y + \Phi_{0yy}\eta - N_{0x}\varphi_x - \Phi_{0x}\eta_x - N_{0x}\Phi_{0xy}\eta, \tag{3.4}$$
$$\gamma\varphi = -\Phi_{0x}\varphi_x - \Phi_{0y}\varphi_y - \Phi_{0x}\Phi_{0xy}\eta - \Phi_{0y}\Phi_{0yy}\eta - |g|\,\eta, \tag{3.5}$$

where all the potentials are taken at the boundary, for instance,

$$\Phi(x,y,t)|_{y=N(x,t)}, \quad \Phi_0(x,y)|_{y=N_0(x)}, \quad \varphi(x,y)|_{y=N_0(x)}.$$

The potentials $\Phi$, $\Phi_0$, and $\varphi$ are harmonic

$$\Delta\Phi = \Delta\Phi_0 = \Delta\varphi = 0.$$

Let us drop the function $\eta(x)$ from the boundary conditions (3.4) and (3.5), and express $\eta$ by means of $\varphi$ and (3.5). We obtain

$$\eta = G|_{N_0}(\gamma\varphi + \Phi_{0y}\varphi_y)|_{N_0},$$
$$-\frac{1}{G} = |g| + \Phi_{0x}\Phi_{0xy} + \Phi_{0y}\Phi_{0yy}.$$

We substitute this expression into (3.4). After cumbersome manipulations, we obtain

$$\begin{aligned}\gamma(\Phi_x G_x + \Phi_x G_y N_x - AG)\varphi + [\Phi_x\Phi_y(G_x + G_y N) + \gamma\Phi_x N_x G \\ + \Phi_x\Phi_{xy}G + \Phi_x\Phi_{yy}N_x G - AG\Phi_y - 1]\,\varphi_y + \Phi_x\Phi_y N_x G\varphi_{yy} \\ + (\Phi_x^2 N_x G + \Phi_x\Phi_y G)\,\varphi_{xy} + \Phi_x^2 G\varphi_{xx} = 0, \qquad (3.6)\end{aligned}$$

where $A = -\gamma + \Phi_{yy} - N_x\Phi_{xy}$, and the functions $\Phi$, $G$, and $A$ are taken at the unperturbed boundary. In order to simplify the cumbersome expressions in (3.6), the functions $N$and $\Phi$ were written without the zero subscript.

In this manner, the perturbed potential $\varphi$ should satisfy the Laplace formula $\Delta\varphi = 0$ and meet the boundary condition (3.6). Dividing (3.6) by $\Phi_x^2 G$, substituting the expansions of (3.1) into (3.6), expanding $\exp(-nN_0)$ in $\cos\,(knx)$, and converting the cosine and sine products into the cosine and sine sums, we finally obtain

$$\begin{aligned}[\Phi_{xx} + S^1\varphi_{xy} + (\gamma S^2 + S^3)\,\varphi_x + C^4\varphi_{yy} + (\gamma C^5 + C^6)\,\varphi_y \\ +\gamma(\gamma C^7 + C^8)\,\varphi]_{y=N_0(x)} = 0, \qquad (3.7)\end{aligned}$$

where the functions $S^j(x)$ and $C^j(x)$ are

$$S^j = \sum_{n=1}^{\infty} S_n^j S_n, \quad C^j = \sum_{n=0}^{\infty} C_n^j C_n, \quad S_n = \sin(knx), \quad C_n = \cos(knx).$$

The coefficients $S_n^j$ and $C_n^j$ are related in a complicated way to the coefficients $A_n$ and $B_n$ of the expansions in (3.1), viz.,

$$S_n^j = S_n^j(A, B), \quad C_n^j = C_n^j(A, B).$$

Substituting (3.2) and (3.3) into (3.7), and converting the sine and cosine products into their sums, we can rigorously prove that the even and odd coefficients in (3.2) and (3.3) are unrelated. This means that all the integer harmonics drop out of (3.2) and (3.3). Therefore, the bubble merging and competition modes are expanded into half-integer sine and cosine, respectively. Equations (3.2) and (3.3) can therefore be represented as

$$\varphi(x, y) = \sum_{n=1}^{\infty} e_{\bar{n}} \begin{pmatrix} a_n & S_{\bar{n}} \\ b_n & C_{\bar{n}} \end{pmatrix}, \qquad (3.8)$$

where the upper values in the column refer to the bubble merging mode, and the lower to the bubble competition mode, $\bar{n} = n - 1/2$, $a_n$ and $b_n$ are the expansion amplitudes of these modes.

From this it follows that in the linear stage the upshift of the accelerated bubbles (see Fig. 3.4(b)) from an unperturbed position is equal to the downshift of the decelerated bubbles.

The presentation so far has focused on the basic theoretical principles for solving modulatory instability problems. Before analyzing the results the numerical methods used should be mentioned. One is the method of large particles developed by Belotserkovskii and Davydov,[26] and the other is a new version of the method of pseudo-compressibility,[27–30] which is here applied for the first time to this class of problems.

### 3.1.5. Results

The description above covers perturbations and basic principles for solving tile problem. The aim of our program was to simulate and investigate the dynamics of modulatory perturbations, specifically the linear evolution of the modulation amplitude. As a result, we have obtained a solution of the appropriate spectral problem, its natural modes, and the eigenvalues $\gamma$.

It should be noted that the instability increments could be determined by the methods of direct two-dimensional numerical simulation. For instance, we have obtained the classical instability increment of hydrostatic equilibrium by this method quite accurately.

The evolution of modulatory perturbations is presented in Figs. 3.5 and 3.6. The analysis of the amplitude increase shows that it is approximately exponential in time. This proves the instability of the periodic solutions in

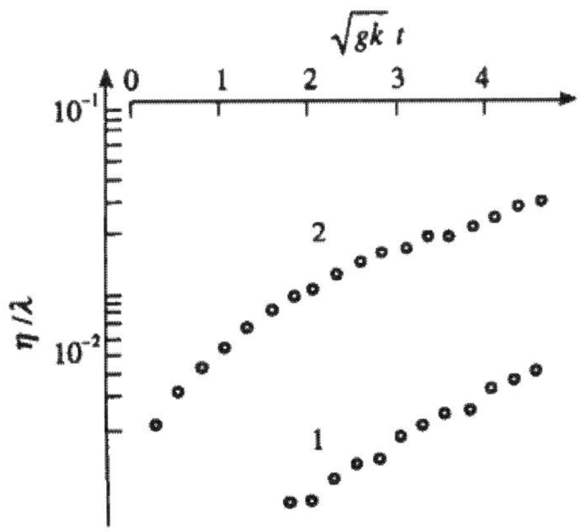


**Fig. 3.5.** Evolution of modulatory perturbation by the method of large particles.

this class of perturbations. Figure 3.5 illustrates the surface displacement from the steady-state position at $x = -0.2\lambda$, $\eta = \eta(-0.2\lambda, t)$, with the circles 1 marking the development of the bubble-merging mode. It should be kept in mind that $\eta(x,t) = \exp(\gamma t)\,\eta(x)$. Our use of the same letter to label both the perturbation amplitude before and after the separation of the space and time variables should not lead to misunderstanding since they have different numbers of arguments. The position of the origin $x = 0$ is shown in Fig. 3.1(a). In the case of the bubble merging mode, the calculation is made in the region $-\lambda/2 \le x \le \lambda/2$; there are two jets at $x = \pm\lambda/2$. In the case of the bubble competition mode, $0 \le x \le \lambda$; there is a single jet at $x = \lambda/2$. Let us say a few words about the selection of point $x^*$ at which the perturbation amplitude $\eta = \eta(-0.2\lambda, t) = \exp(\gamma t)\eta(x^*)$ has been determined. In the case of the bubble merging mode, on the one hand, this point should be outside the region $x = 0$ since the perturbations are expanded in a sine series (see Sec. 3.1.4), and therefore $\eta(0) = 0$ and $\eta(-x) = \eta(x)$. On the other hand, it should be separated from the jets. Due to these considerations, we choose the point $x^* = -0.2\lambda$.

In the case of the bubble competition mode, the dependence $\eta(0,t)$ is shown with the circles 2 in Fig. 3.5. Figure 3.6 shows two curves, both refer to the bubble competition mode. One is the curve $\eta(0,t)$ and the other is the curve $\eta(\lambda,t)$. These curves have different signs (see Fig. 3.4(b) and its caption). That is why the absolute value is taken. The initial sections of

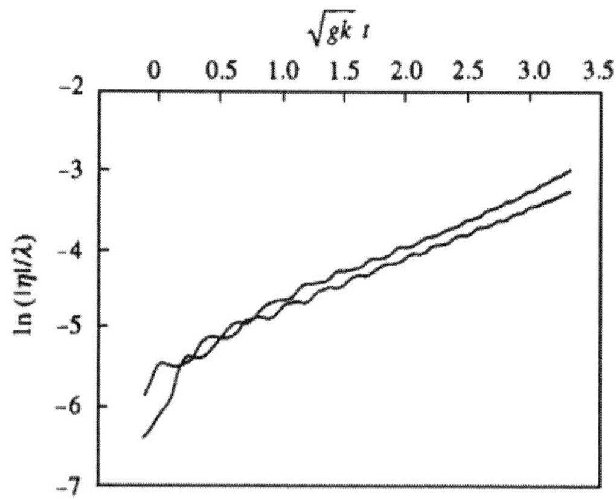


**Fig. 3.6.** Evolution of modulatory perturbations by the method of pseudo-compressibility.

these curves up to the time instance

$$t = \frac{1.5}{\sqrt{gk}},$$

should be neglected. This is because during this period of time, the unperturbed solution in our calculations is in the mode of steady-state rise of the bubbles. It should be emphasized that the end of the transient process and the start of the steady-state mode well agree with the break in the curves $\eta(0,t)$ and $\eta(\lambda,t)$ (see Figs. 3.5 and 3.6).

Sections follow the breaks with roughly linear time dependence, whose duration is $(2 \div 2.5)/\sqrt{gk}$. During the linear section, a significant increase in the amplitudes of the boundary and velocity perturbations takes place. This increase amounts approximately to 3.5. The slope of the curves during the linear section can be used to determine the instability increment $\gamma$. The average values of $\gamma$ are

$$\frac{\gamma'}{\sqrt{gk}} = 0.53 \pm 0.07, \quad \frac{\gamma''}{\sqrt{gk}} = 0.45 \pm 0.05, \quad \frac{\gamma'''}{\sqrt{gk}} = 0.61 \pm 0.05. \tag{3.9}$$

Here, $\gamma'$ and $\gamma''$ refer to the method of large particles in the bubble merging and competition mode, respectively, and $\gamma'''$ to the method of pseudo-compressibility in the bubble competition mode, $k = 2\pi/\lambda$, and $\lambda$ is the period of the unperturbed solution. The results of different methods should coincide. Thus, we should have $\gamma'' = \gamma'''$, which is only approximately satisfied in (3.9). Evidently, this inaccuracy is associated with the errors in the numerical simulation.

We should note here that it is difficult to obtain these results. To increase the accuracy of $\gamma$, it is necessary: (a) to improve the accuracy of the calculations and (b) to increase the duration of the linear sections. In the above calculation, when Eq. (3.9) was derived, this duration was about $1.2(\gamma)^{-1}$. To satisfy condition (a), finer grids have to be used. We used the grid with $N_x = 40$ and $N_y = 60$. To satisfy condition (b), it is necessary to reduce the amplitude of the initial perturbation. We then encounter the following difficulties: first, it is necessary to carry out the calculations at very small perturbation amplitudes and, second, the computational time leads to vertically longer computational regions. In summary, we may say that it is difficult to increase the accuracy.

It follows from the linearity of the perturbations and from the exponential time dependence that the solutions do exhibit dynamics close to those of the natural modes:

(i) the amplitudes of the perturbations are small;
(ii) the values of ln $\eta$ as a function of $t$ are approximately linear;
(iii) from the symmetry for the bubble merging mode (see Sec. 3.1.4), it follows that $\eta(0,t) = 0$; in fact, the amplitude of this mode is much less (in comparison with curves 1 and 2 in Fig. 3.5) than the same initial amplitudes;
(iv) from the symmetry for the bubble competition mode, it is easy to find that during the linear section, we should have $\eta(0,t) = -\eta(\lambda,t)$; in fact, this relation is satisfied with good accuracy (see Fig. 3.6); at the end of the computation, we have

$$\frac{2(|\eta(0,t)| - |\eta(\lambda t)|)}{|\eta(0,t)| + |\eta(\lambda,t)|} \approx 20\%.$$

It should be noted that these calculations have been made only for $m =$ 2. The problem of perturbations for $m = 2$ has been left uninvestigated. However, the results obtained for $m = 2$ leave no doubt as to the instability of the solutions for $m > 2$.

The values of $y$ are quite reasonable. From general observations it follows that $\gamma/\sqrt{gk}$ should be equal to $\cong l$, and the increments of the bubble merging and competition modes should be real, that is, the exponential increase in their amplitudes should occur without any variation in time. Both these assumptions are true.

### *3.1.6. Classification of stability problems*

Theoretical analysis is usually required for the solution of physical problems. As a rule, the derivation of a general solution of a problem is the major difficulty. Usually one can arrive at solutions only by simplifying the problem at a particular point after which the region in which the behavior of the parameters can be analyzed is expanded by analyzing their behavior in neighborhoods of the point. This extension may give rise to stability problems.

Now we classify stability problems. As a rule, they involve a method for studying the effect of heterogeneity on the initial unperturbed solution ("point"), which is a one-dimensional solution, except for a small subclass of problems on the uni-dimensional stability of one-dimensional solutions.

Separating the spatial variables, the direction of the perturbations, and the time variable, we arrive at a one-dimensional spectral problem. As a rule, this problem is of the form $L\Psi_\lambda(x) = 0$ plus the boundary conditions, where $L = L(\partial_x, \ldots, U(x) \ldots, \lambda)$ is a local operator. Its form, the form of "potentials" $U(x)$, follows from the unperturbed one-dimensional problem. These problems form the traditional class of stability problems. This is the simplest class of problems.

At the next stage, a more detailed study leads to spatially heterogeneous solutions. And the next stage in complexity is the problem of studying the stability of these multi-dimensional unperturbed solutions. These problems form a much richer, more interesting, and important class of problems by means of which we can define the dynamics of the systems under consideration. This case is analyzed in this chapter. It should be noted that there is a great gap between the complexity of problems involving the stability of one- and multi-dimensional solutions. These remarks are important for understanding the relationship of the present investigation to others dealing with stability problems.

In our approach, the initial solution is spatially heterogeneous in two dimensions. Since its unperturbed and perturbed functions are solutions of the Laplace equation, its stability may be reduced in principle to $L\psi_\lambda(x) = 0$ plus the boundary conditions, but now with a nonlocal, that is, integro-differential, operator.

There are few such problems. Perhaps the only example of such a problem is that of the Benjamin–Fejer modulation instability in nonlinear gravitational waves, which was studied by V. E. Zakharov. This problem was solved for small values of the nonlinearity parameters by expanding in terms of this parameter. This method cannot be applied to our case since the unperturbed solution is strongly nonlinear.

The only systematic method that can be used in our case is the Wentzel–Kramers–Brillouin (WKB) approximation. However, this cannot be applied in one of the most important cases, namely, when the perturbation scale is of the same order as the scale of the unperturbed solutions. This is why we have had to use direct methods.

We have managed to reduce the problem to modulation instability (see Sec. 3.1.5). Moreover, we have managed to obtain the spatial structure of the eigenfunctions (see Sec. 3.1.4 and 3.1.5) and evaluate the growth rate of the most important unstable mode that doubles the space period (see Sec. 3.1.5).

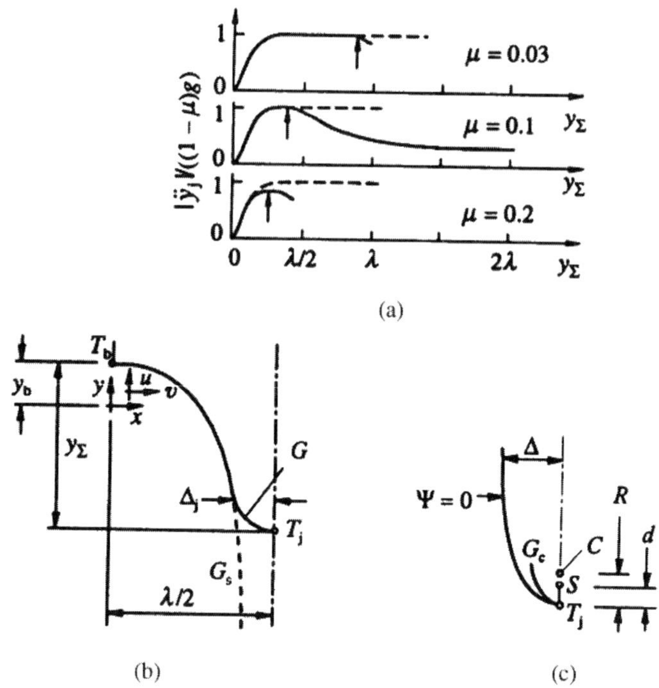


**Fig. 3.7.** (a) Characteristic stages in the evolution and the attendant changes in $y$; the amplitude corresponding to the mushroom initiation is shown with an arrow; (b) shape of the nonstationary boundary $G$ differing from the stationary boundary $G_S$ in the vicinity of point $T_j$; (c) the stagnation streamline $\Psi = 0$ in the case of solving Eq. (3.11). The figures show symmetrical halves of full period.

### *3.1.7. Initiation of a mushroom structure*

Let us consider the dynamics of a small-amplitude sinusoidal perturbation with period $\lambda$. During the linear stage, the instability evolves exponentially with increment $\gamma = \sqrt{At\, gk}$ where $At = (1-\mu)/(1+\mu)$ is the Atwood number, $\mu = \rho_l/\rho_h$ is the ratio of the densities, and $\rho_l$ and $\rho_h$ are the densities of the light and the heavy liquid, respectively.

At $\mu = 0$, during the nonlinear stage, the bubbles begin expanding and the jets constricting. As the bubble amplitude $y_b$ (Fig. 3.7(b)) reaches a value of $(0.07 \div 0.1)\lambda$, steady-state conditions of the ascent of the bubble become to be established.

Here, the bubble ceases to expand, and its shape, the curvature of the apex in particular, is fixed. The jet amplitude $y_j$ reaches $(0.2 \div 0.3)\lambda$ and the acceleration $\ddot{y}_j$ is close to $g$. The jet width amounts to a quarter of $A$. These conditions are analyzed in Refs. 19–25.

The acceleration is plotted against the sum of the amplitude $y_{\sum} = y_b + |y_j|$ in Fig. 3.7(a). The amplitudes are determined in Fig. 3.7(b). The case when $\mu = 0$ is shown by the dashed lines in Fig. 3.7(a). The acceleration unit $(1 - \mu)g$ at $\mu = 0$ is equal to $g$. The explanations for this case are to be found below.

The width of the jet near the stagnation point $T_j$ will be defined as the width of the steady-state jet. A steady-state jet $G_S$ is shown by the dashed line in Fig. 3.7(b). Let us define the half-width of the steady-state jet by $\Delta_j$. The amplitudes are measured from the initial position of the interface. The laboratory reference frame is shown in Fig. 3.7(b).

Let us consider the effect of the density of the lower medium, $\rho_l$, on the dynamics of the heavy liquid. When the velocities are not high, the density $\rho_l$ decreases the acceleration from $g$ to $(1 - \mu)g$. After this reduction, the perturbation evolution in time coincides with that for $\mu = 0$.

The jet velocity $|\mu_j| = |\dot{y}_j|$ increases with time, and the radius $R_j$ of the head curvature decreases. At the same time, the ratio of the hydrodynamic force exerted by the incident flow of the light liquid on the jet of heavy liquid to the weight force increases. Ultimately, at a jet amplitude $y_{\rm cr} = |y_j(t_{\rm cr})|$, this ratio approaches unity.

The moment in time $t_{\rm cr}$, is very important. In the vicinity of the critical amplitude $|y_j| \cong y_{\rm cr}$, the acceleration $\ddot{y}_j$ starts decreasing (the onset of the decrease is shown by an arrow in Fig. 3.7(a)), and tile function $R_j(t)$ or $R_j(y_{\sum})$ is reaching a minimum. These changes are determined by the initiation of the mushroom structure.

It should be noted that the dependence on the amplitude $y_{\sum}$ is more general, because the characteristic time shift is a function of the delay $\cong \gamma_1 \ln(\lambda/a(t = 0))$, which is determined by the perturbation evolution at the initial stage.

Figure 3.7(a) illustrates the dependence $\ddot{y}_j(y_{\sum})$. The dashed lines correspond to the light liquid with a specific "weight" at $\rho_l = 0$. This situation is possible in liquids with a homogeneous space charge in an external electric field (if $\rho_l \ll \rho_h$). In this case, $\mu$ is calculated from the ratio of the specific gravities. The weight of the bottom liquid is controlled electrically. The dashed curve is plotted along the rising section of the dependence $\ddot{y}_j$, numerically calculated for $\mu = 0.1$. The solid curve for $\mu = 0.1$ is numerically calculated. The solid curves for $\mu = 0.03$ and $\mu = 0.2$ are plotted in the following way. The rising sections of both are the same. The onset of the decrease in acceleration depends on $\mu$. It can be calculated from the analytical model given below. The corresponding values are shown by the arrows in Fig. 3.7(a). The linear section of the curve in Fig. 3.7(a) is expressed by the following formula:

$$\ddot{y}_b = -\ddot{y}_j = (A_\upsilon \gamma \sinh \gamma t + A_\eta \gamma^2 \cosh \gamma t) \cos kx, \tag{3.10}$$

where $A_\upsilon$ and $A_\eta$ are the initial amplitudes of the velocity and boundary perturbations, respectively. The calculation for $\mu = 0.1$ was made for

$$A_\eta = 0, \quad A_\upsilon = 0.195\,(0.23\sqrt{(1-\mu)g\lambda}).$$

Let us calculate $y_{\mathrm{cr}}$. At $\mu \ll 1$, we get $y_{\sum}(t_{\mathrm{cr}}) \ll R_q(t_{\mathrm{cr}})$. The changes in $\dot{y}_j$ at $t \cong t_{\mathrm{cr}}$ occur at scale $\cong y_{\mathrm{cr}}$ and during the time $\cong y_{\mathrm{cr}}/\dot{y}_{\mathrm{cr}}$. These space and time scales are much greater than the respective dimensions of the head and the time of streamlining by the external flow. That is why a quasi-stationary approximation is sufficient for describing the flow in this region. Besides, we may assume with sufficient accuracy that the heavy liquid in the vicinity of point $T_j$ is at rest in the reference frame moving with $T_j$ (this holds till the moment of mushroom formation).

We shall describe the flow of the light medium in the following way. Let its potential in the reference frame moving with point $T_j$ be

$$f = -iUz + \frac{\dot{M}}{2\pi\rho_l}\ln\left(1 + \frac{iz}{d}\right). \tag{3.11}$$

This is the potential of a source, located at point $S$ with the coordinate $z =$ id, placed into the homogeneous flow (Fig. 3.7(c)). The velocity of this flow is

$$U = -\dot{y}_j. \tag{3.12}$$

Such a description is possible since $R_j \ll \lambda$. At the same time, the periodic structure may be neglected in the vicinity of $T_j$, arid we can limit ourselves to the simpler case of a solitary jet.

It is necessary that $z = 0$ be the stagnation point. This is the point $T_j$ in Fig. 3.7(c). In this case, the linear term in expansion of $f$ in power series of $z$ should drop out. From this condition we obtain

$$\dot{M} = 2\pi\rho_l U d. \tag{3.13}$$

Expanding (3.11) in $z$ to the third power inclusively, we find an equation for the stagnation streamline $\Psi = 0$ (see Fig. 3.7(c)), which passes through

$T_j$. The equation is $\Delta y = (\Delta x)^2/(3d) + O(x^4)$, where $z = \Delta x + i(\Delta y)$. Hence, for the radius of curvature at $\approx 0$, we obtain

$$R = \frac{3d}{2}. \tag{3.14}$$

The respective arc $G_c$, the radius $R$, and the center $C$ of the circle inscribed in $\psi = 0$ and having a point of contact with $\Psi = 0$ at $z = 0$ are all shown in Fig. 3.7(c).

Equating the mass flux from the source to the flux at infinity readily yields an asymptotic jet width $2\Delta$ in this model. We have

$$\Delta = \pi d. \tag{3.15}$$

In this way, solution (3.11) is found in accordance with (3.11)–(3.15) for the velocity $U$ and any of the three geometric parameters, i.e., $d$, $R$ or $\Delta$.

Formulas (3.11)–(3.15) describe the flow of the light liquid around the head of the jet. Now let us consider the jet of the heavy liquid. At $y_\Sigma > (0.3 \div 0.4)\lambda$, this jet can be described accurately by the following asymptotic formulae

$$|\dot{y}_j| = \sqrt{(1-\mu)\,g}\left(\sqrt{2y_\Sigma} - 0.23\sqrt{\lambda}\right) = F_1\left(y_\Sigma\right), \tag{3.16}$$

$$\Delta_j = \frac{\frac{0.23}{2\sqrt{2}}\lambda^{3/2}}{\sqrt{y_\Sigma}} = F_2(y_\Sigma). \tag{3.17}$$

The first term in (3.16) is due to free fall, and the second to the rise of the bubbles at a velocity of $0.23\sqrt{(1-\mu)g\lambda}$.[19–25] Formula (3.17) can easily be derived from the condition that the mass flux must be conserved if we write this condition in the reference frame moving with $T_b$.

Now let us connect the velocity and the geometric parameters in solutions (3.11), (3.16), and (3.17). The velocity is derived by means of (3.11), (3.12), and (3.16). Now let us combine the geometric parameters $\Delta_j$ of jet (3.17) and of approximate solution (3.11). It is natural to assume that

$$R = c\Delta_j, \tag{3.18}$$

where $c$ is a fixed constant. We obtain $R = \Delta_j$ at $c = 1$, and $\Delta = \Delta_j$ at $c = 3/(2\pi)$. Consequently, $c$ is in the range from $\approx 1/2$ to 1. Leaving this parameter uncalculated, let us find the answer and then analyze the effect of this parameter on the answer.

The procedure is as follows. We set $y_\Sigma$, and then derive $u_j$ and $R$ with the help of it, using (3.16), (3.17), and (3.18), respectively. Thus, we obtain the potential in (3.11). Next, we calculate the stagnation streamline

$\Psi(\Delta x, \Delta y) = 0$. Finally, we find the pressure drop, caused by the dynamic head, along the stagnation streamline $\psi = 0$ as a function of $\Delta_y$ (it should be recalled that $\Delta x$ and $\Delta y$ are measured from point $T_j$). We are interested in the $dp/dy$ gradient of the pressure. By comparing it with the gradient caused by the weight of the heavy liquid, we obtain a criterion for the initiation of mushroom flow. When these two gradients are of the same order of magnitude, mushroom flow is initiated.

To calculate the gradient, we determine the velocity from (3.11) by means of Bernoulli's integral

$$|v| \approx \frac{\dot{M}}{2\pi\rho_l\, d^2}(\Delta x), \quad \Delta x \approx \sqrt{3d\Delta y}, \quad \frac{dp}{dy} \approx \frac{(-3/2)\,\rho_l\, U^2}{d}. \tag{3.19}$$

To avoid misunderstanding, we should stress that this is the gradient of the dynamic head taken along the line $\Psi = 0$. That is why it is not equal to zero. An ordinary gradient of this head is equal to zero at the stagnation point $\mathrm{T}_j$ since the maximum head is reached at this point.

The heavy liquid is at rest in the vicinity of point $T_j$. That is why the pressure gradient at this point is equal to the hydrostatic gradient

$$\frac{dp}{dy} = -\rho_h g. \tag{3.20}$$

Taking into account the specific weight of the light liquid from (3.20) and (3.19), we obtain the following condition for the effect of the dynamic head, caused by the transport of the light liquid, on the dynamics of the heavy liquid:

$$|\dot{y}_j| \ll \frac{2}{3}\sqrt{(\mu^{-1} - 1)gR}.$$

In this inequality, $\dot{y}_j$ and $R$ are functions of $y_{\sum}$, therefore, it defines the critical amplitude $(y_{\sum})_{\text{cr}}$. The defining equation proves to be cubic:

$$\left(\frac{2.05}{\sqrt{\mu}}\right)\xi^3 + \xi^2 - c = 0, \quad \xi^2 = \frac{R_{\text{cr}}}{\lambda/2}. \tag{3.21}$$

The amplitude $(y_{\sum})_{\text{cr}}$ is related to $(\Delta_j)_{\text{cr}}$ by Eq. (3.17), and $(\Delta_j)_{\text{cr}}$ is in turn related to $R_{\text{cr}}$ by Eq. (3.18).

If we neglect the low (as compared to the velocity of the jet) velocity of the bubble in Eq. (3.16), which is equivalent to neglecting the second term at the right-hand side of Eq. (3.16) and the square of the $\xi$ term in (3.21), the

**Table 3.1.**

| $\mu$ | 0.001 | 0.01 | 0.03 | 0.1 | 0.2 |
|---|---|---|---|---|---|
| $\frac{R_{cr}}{\lambda/2}$ | 0.06 | 0.12 | 0.17 | 0.25 | 0.29 |
| $\frac{(y_\Sigma)_{cr}}{\lambda/2}$ | 15 | 3.7 | 1.8 | 0.9 | 0.6 |

approximate solution of (3.21) can easily be found. It has the form

$$\frac{R_{\rm cr}}{\lambda/2} \approx 0.62c^{2/3}\mu^{1/3}, \quad \frac{(y_\Sigma)_{\rm cr}}{\lambda/2} \approx \frac{0.14c^{2/3}}{\mu^{2/3}}.$$

Taking into account the rate at which the bubble rises slightly reduces the jet velocity. As a result, the mushroom structure forms later. This is why the exact solutions of (3.21) have somewhat greater values of $(y_\sum)_{\rm cr}$ than the approximations given above.

The accurate solutions of (3.21) are given in Table 3.1 for $c = 1$ in (3.18) and (3.21). Critical depths and radii of curvature are given. For $c = 0.5$, the value of $(y_\sum)_{\rm cr}$ is approximately 40% less.

It follows from the numerical simulation that if $(y_\sum)_{\rm cr}$ is determined from the maximum acceleration, then $(y_\sum)_{\rm cr} \approx 0.75(\lambda/2)$ at $\mu = 0.1$. Comparing these results with the analytical ones, we see that they are consistent. The best values of $c$ vary from 1/2 to 1.

The mushroom initiation is described above. The moment of its origination divides the process of the perturbation evolution into two stages. During the first stage, the density of the light liquid is insignificant. During the second stage, the flow region starts subdividing into two parts: in one region the density $\rho_{\rm l}$ is still insignificant, while in the other one it is very significant.

In turn, these two regions have four characteristic zones (Fig. 3.8). The first region, which arises before mushroom formation, consists of zones $\rm B^+$ and J. They are separated by the dashed lines in the figure. In zone $\rm B^+$ the bubble of the light liquid is pushed around by the concentrated flow of the heavy liquid. This zone is located above bubble B, and therefore it has the plus sign. Zone J is that of the jet. The liquid is in a near free-fall condition with an acceleration equal to $(1 - \mu)g$.

The second characteristic region, which arises after the mushroom structure is formed, consists of zones M and S. Zone M is characterized by strong deceleration of jet J with a force greater than $(1 - \mu)g$. This is the zone in which the mushroom has a "blunt" form. Zone B is the bubble

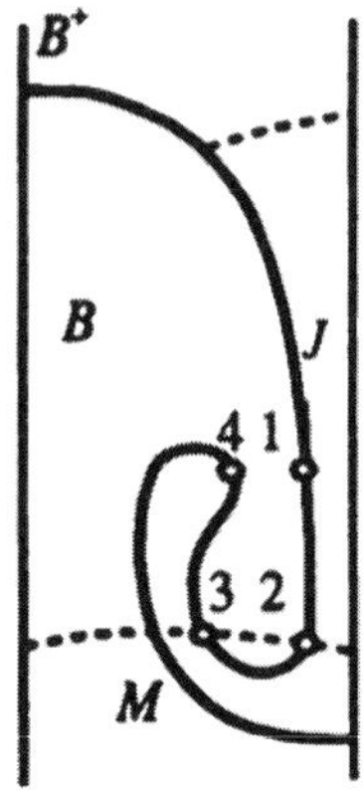


**Fig. 3.8.** Characteristic flow zones after the development of mushroom.

zone. Intricate motions of the light and heavy liquids characterize its lower portion; the heavy liquid has the form of an inverted decelerating jet.

Zones $B^+$ and J are described in Refs. 19–25, while a description of zone M is the substance of this part of chapter.

### *3.1.8. The mushroom flow structure*

The initiation of the mushroom flow structure is described above: it occurs when the hydrodynamic force becomes equal to the gravitational force. The former increases with the velocity of the jet, while the latter is fixed. It should be noted that the hydrodynamic force has its greatest value at the jet stagnation point $T_j$.

Now let us see what happens to the flow after the mushroom has been formed. It turns out that the jet velocity at point $T_j$ continues to increase in the laboratory reference frame. The hydrodynamic force increases accordingly. Small wonder that at $(y_\sum) > (y_\sum)_{\rm cr}$ in the vicinity of point $T_j$, the gradient of the hydrodynamic head appears to be much greater than that of the hydrostatic one. In this case, weight may be neglected in that region indicated.

In fact, the square of the Froude number (*Fr*) in zone M, which characterizes the dynamic contribution of gravitation and is calculated from the characteristic scale (here $A_j$), is

$$\mathrm{Fr}^2 = \frac{u_j^2}{(1-\mu)\, g\Delta_j} = \frac{2y_\Sigma}{\Delta_j}.$$

For $\mu < 0.2$ and $(y_{\sum}) > (y_{\sum})_{\rm cr}$, the square of the Froude number is very large. This is because the gravitational acceleration and deceleration occur at the distance $\cong y_{\sum}$. It is evident that over distance $\cong \Delta_j$ the velocity cannot change substantially.

The pressure variation in Secs. 3.1.3 and 3.1.4 of the turbulent pocket 1–2–3–4 may be neglected (see Fig. 3.8), first, because the liquid is light, and, second, because the circulation flow in this liquid has no stagnation points. Therefore, the change in the dynamic contribution to the pressure is small.

We have thus come to a problem of the theory of jets. It is necessary to consider the interaction of the periodic chain of jets and a thick target. It should be noted that the case of a solitary jet is discussed in. Ref. 31 This theory has not been used to describe the mushroom dynamics during Rayleigh–Taylor instabilities.

The solution on the physical plane $z$ and in the hodograph plane $\xi = \dot{z}^* = df/dz$ is shown in Fig. 3.9. The flow region is conformally mapped onto the interior of a circle with a cut. The edges of cut, $a \to b$ and $a \to d$, in plane $\xi$ (Fig. 3.9(b)) are referred to the respective rigid boundaries in plane $z$ (Fig. 3.9(a)). The arrows mark the direction of travel of the Lagrangian particles. In a periodic solution these boundaries are the lines of symmetry. The upper and the lower semicircles correspond to the lower, $c \to d$, and the upper, $c \to b$, branches of the free streamline, respectively. The centers $\xi = 0$ and $z = 0$ should be mapped one onto the other. This is the stagnation

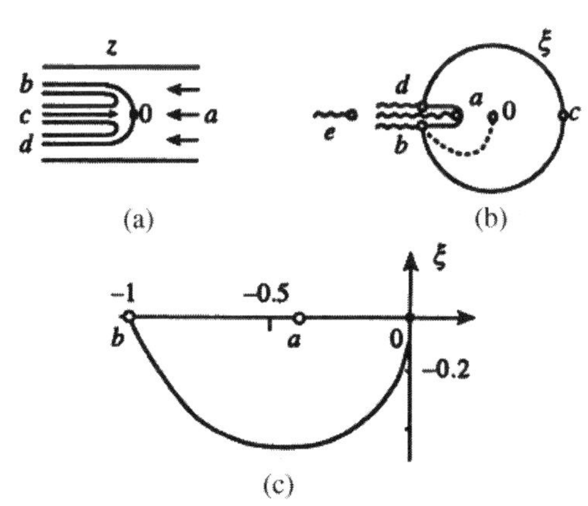


**Fig. 3.9.** (a) Physical plane; (b) hodograph; (c) hodograph of the stagnation streamline $a \to b$.

point of the jet. The velocities are given in the reference frame moving with point $z = 0$.

There are singularities, i.e., sources $(a, c, e)$ and sinks $(b, d)$, at points $a$, $b$, $c$, $d$, and $e$. The generation of the singularity at point $e$ is concerned with the necessity to reflect source $a$ relative to the circle, which should be the streamline. A source remains a source upon reflection. It is possible to show that

$$\frac{df}{d\xi} = F(\xi) = -\frac{2h}{\xi+1} + \frac{2h_b}{\xi-1} - \frac{a}{\xi-1} - \frac{a}{\xi-1/a}. \tag{3.22}$$

When deriving (3.22), it was assumed that $|\xi| = 1$ at the free streamlines. Here, $|\xi| = |\xi_a| = a$ is the absolute velocity as $Re\ z \to \infty$, and $\pi h_b$, $2\pi h$, and $2\pi$ are the widths of jets $b, c$, and $a$, respectively (see Fig. 3.9).

On the other hand, $\xi = df/dz$. Hence, we have

$$dz = \frac{df}{\xi} = \frac{df}{d\xi}\cdot\frac{d\xi}{\xi} = \frac{F(\xi)d\xi}{\xi}.$$

Resolving the fractions into simple ones and integrating them, we obtain

$$\begin{aligned} z = (1-a^2)\ln a + \frac{i\pi(1-a)^2}{2} + 2h\ \ln(\xi{+}1) \\ + 2h_b\ \ln(\xi{-}1) - \ln(\xi - a) - a^2\ln(\xi - 1/a), \end{aligned}$$

$$\begin{aligned} Rez = (1-a^2)\ \ln a + 2h\ln R + 2h_b\ln R_b - \ln R_a - a^2\ln R_e, \\ R = |\xi+1|, R_b = |\xi-1|, \quad R_a = |\xi - a|, R_e = |\xi - 1/a|, \end{aligned} \tag{3.23}$$

$$\begin{aligned} \operatorname{Im} z = \frac{\pi(1-a^2)}{2} + 2h\tan^{-1}\frac{-\operatorname{Im}\xi}{1-Re\,\xi} + 2h_b\left(\pi - \tan^{-1}\frac{-\operatorname{Im}\xi}{1+Re\,\xi}\right) \\ - \left\{\begin{array}{ll} \tan^{-1}\dfrac{\operatorname{Im}\xi}{Re\,\xi - a}, & -Re\,\xi > a \\ \pi - \tan^{-1}\dfrac{-\operatorname{Im}\xi}{a + Re\,\xi}, & -Re\,\xi < a \end{array}\right\} \\ - a^2\left(\pi - \tan^{-1}\frac{-\operatorname{Im}\xi}{1/a + Re\,\xi}\right). \end{aligned}$$

It should be noted that the term $\ln\xi$ appearing during the integration drops out. The factor it is multiplied by is proportional to the increase in the parallel $Re\ z$ component of the momentum in the flow region, and therefore is zero due to the stationary system and the momentum conservation law.

From the mass and momentum conservation laws and Bernoulli's integral, we obtain

$$a = 1 - 2\sqrt{h}, \quad h_b = \frac{(1+a)^2}{4}. \tag{3.24}$$

Consequently, all the geometric characteristics of the flow are determined by one parameter, that is, $h$.

The integration constant in (3.23) is determined on the basis that point $z = 0$ corresponds to point $\xi = 0$. The cuts of the complex plane of branching points are shown in Fig. 3.9(b) by wavy lines.

Equation (3.23) makes it possible to plot the free streamline. To do this, we parameterize the semicircle on plane $\xi$, using the angle $\alpha = \arg \xi$. Then, we obtain $\xi = \xi_f(\alpha)$, $Re\, \xi_f = \cos\alpha$, and Im $\xi_f = \sin\alpha$. As $\alpha$ varies from 0 to $\pi$, the function $\xi_f(\alpha)$ sweeps the upper semicircle. At the same time, $z = z_f(\alpha) = z(\xi_f(\alpha))$ where $z(\xi)$ is found from (3.23), sweeps the lower branch of the free streamline.

To find the other lines, a more accurate analysis is required. Imagine $\xi$ as a "physical" coordinate. Then in this "coordinate" plane, sources and sinks $a, b, c, d$, and $e$ set up a field of "velocities" $d\xi/d\tau$. By analogy with ordinary variables and formulae for $z$ and $dz/dt = (f_z)^*$, we find the "velocity" $d\xi/dt = (f_\xi)^*$. It should be noted that here we have $f_z \equiv df/dz$, or component-by-component,

$$X_\tau = Re\, f_\xi, \quad Y_\tau = -\text{Im}\, f_\xi, \quad X = Re\, \xi, \quad Y = \text{Im}\, \xi. \tag{3.25}$$

The values $Re\, f_\xi$ and Im $f_\xi$ are functions of $X$and $Y$. Thus, we come to an autonomous system of two ordinary differential equations (3.25).

Starting from the initial point and integrating either $dY/dX = -\text{Im}\, f_\xi/Re\, f_\xi$ or (3.25), say, by the Runge–Kutta method, we find the streamline from this initial point. This is the streamline in plane $\xi$. Due to conformity, when mapped $\xi \to z$ according to (3.23), this streamline will become the physical streamline.

To avoid misunderstanding, we note that the variable $\tau$ in (3.25) is not the physical time $t$. To find the path of a Lagrangian particle along a streamline in physical time, we must integrate another system of ordinary differential equations. This system is derived in the following way. We have $z = z(\xi)$. Let us differentiate this equation with respect to time $t$ and obtain

$$\frac{dz}{dt} = \frac{dz}{d\xi} \cdot \frac{d\xi}{dt}, \quad \dot{z} = \frac{(df/d\xi)\dot{\xi}}{df/dz}.$$

Since $\dot{z} = \xi^*$ and $df/dz = \xi$, we get

$$\dot{\xi} = \frac{\xi\xi^*}{F(\xi)}.$$

Now we use system (3.25) to find the stagnation streamline. Its $0 \to b$ branch is shown by a dashed line in Fig. 3.9(b). Point 0 is a singular saddle point with a regular crossed configuration of the incoming and outgoing separatrices. We expand (3.25) in the vicinity of this point to the squares of the terms inclusively. This makes it possible to find the curvature of the stagnation streamline at the stagnation point. Calculations show that

$$Y = \pm\sqrt{\frac{3A}{B}}\sqrt{-X}, \tag{3.26}$$

$$A = 2h - 2h_b + \frac{1}{a} + a^3, \tag{3.27}$$

$$B = -2h - 2h_b + \frac{1}{a^2} + a^4. \tag{3.28}$$

Expanding (3.23) in the vicinity of $z = 0$ in power of $z$ and substituting (3.26)–(3.28) into this expansion, we obtain the curvature radius $R$

$$R - \frac{(3/4)A^2}{B}. \tag{3.29}$$

It should be noted that, as $h \to 0$, we have $R \to 6h$. Hence, it follows that at very small A, the diameter of the inscribed circle is $12h \ll 2\sqrt{h}$, i.e., the distance by which the jet is deflected after hitting the target. Therefore, at these values of $h$, the jet head is not rounded.

The description given above deals with the asymptotics of the stagnation streamline near the stagnation point. Now let us define the asymptotics near sink $b$ (see Fig. 3.9(b)). From the balance of the mass fluxes in the run-off, it is easy to find the angle between the actual axis $\xi$ and the tangent to the dashed line in Fig. 3.9(b) in the vicinity of point $b$. This angle is equal to

$$\alpha_b = \frac{a\pi}{2(a+h)}. \tag{3.30}$$

Either (3.26) or (3.30) may be used to determine the initial point and the emergence from singular points 0 or $b$ in system (3.25) while integrating the stagnation streamline. For the sake of the stability of calculation, it is better to go along the separatrix of saddle 0 by means of (3.30) to check the integration error. The numerical integration of (3.25) for $h = 0.0914$ is shown in Fig. 3.9(c). We now present the computed hodograph of the stagnation streamline on plane $z$ with the help of (3.23). The stagnation and free streamlines for $A = 0.0914$ are shown in Fig. 3.10(a).

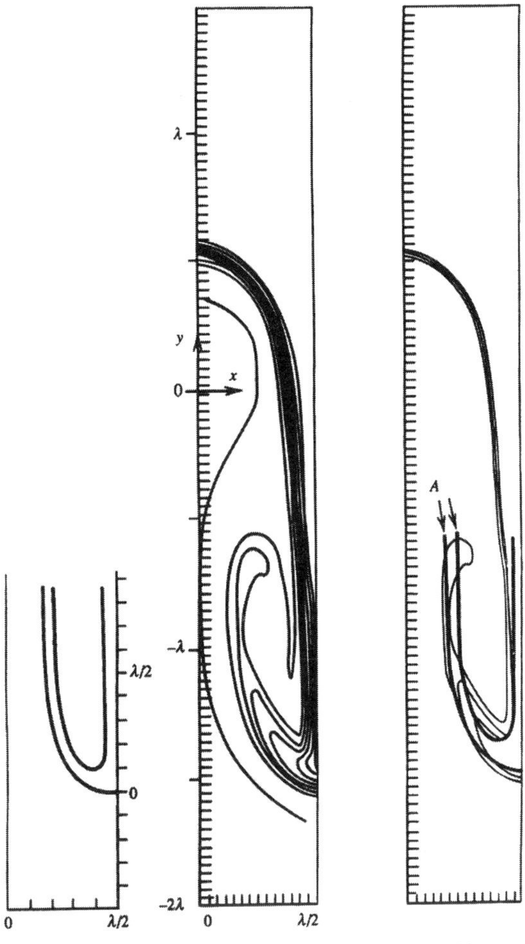


**Fig. 3.10.** (a) Analytical stagnation and free streamlines; (b) family of isochores; (c) comparison between numerical and analytical results.

### *3.1.9. Numerical simulation*

The numerical simulation was done by the pseudo-compressibility method. The method, usually used to investigate the dynamics of uniform media, has been adapted by Chekhlov to a nonuniform medium.

Figure 3.10(b) shows the density distribution at time

$$t = \frac{7.91}{\sqrt{Agk}}. \tag{3.31}$$

A model version shown was calculated for

$$\mu = 0.1, \quad \rho_l = 0.1, \quad \rho_h = 1, \quad A_\eta = 0, \quad A_\upsilon = 0.195(0.23\sqrt{(1-\mu)g\lambda}).$$

For an initially sinusoidal perturbation in an incompressible liquid, the motion is completely determined by $\mu$, $A_\eta$, and $A_\upsilon$ (3.10). The calculations were made on a square grid with dimensions $N_x = 15$ and $N_y = 105$. The laboratory reference frame is shown in Fig. 3.7(b). The symmetric half of the period was calculated. The lower and the upper boundary were at $y = -2\lambda$ and $1.5\,\lambda$, respectively (see Fig. 3.10(b)). Rigid boundary conditions, at which the speed component normal to the boundary is zero, were chosen at the bottom and at the top.

The boundaries were set at a considerable distance so that they did not affect the dynamics of the bubble and the jet. For instance, the increments for an unbounded liquid and for the liquid whose behavior was evaluated under the chosen boundary conditions differed in the third decimal digit. The boundaries begin affecting the dynamic behavior when the bubble or the jet approaches them more closely, but they were separated at time intervals of up to the value of $t$ given by Eq. (3.31) and their effect remained insignificant. This will become clearer later in the discussion from the consistency with which the analytical and numerical velocity distributions decline with increasing distance between the stagnation points and the boundaries. The point that must be borne in mind is that the analytical distributions were obtained under the assumption of an unbounded liquid.

Figure 3.10(b) shows the family of isochores with $\rho = 0.1, 0.2, 0.3, 0.4, 0.5, 0.6, 0.7, 0.8$, and $0.9$. Figure 3.10(c) illustrates isochores with $\rho = 0.3$, $0.4$, and $0.5$. The analytical contact boundaries from Fig. 3.10(a) are plotted on the same graph by thick lines $A$. As we can see, the shapes of the boundaries in the mushroom zone M and even above it agree well.

To compute the analytical boundaries, only one parameter, viz., the half-width $h_d$ of the neck of jet J, was taken from the numerically simulated fields. It is equal to the half-width of jet J at its narrowest point (see Figs. 3.8 and 3.10(b)). The isochore with $\rho = (\rho_l + \rho_h)/2$ was taken as the jet boundary. Here, the hydrodynamic head still does not affect jet J. At time $t$ (see Eq. (3.31)), $h = 0.0914$, where $h = h_d/(\lambda/2)$. Here, $h$ is a dimensionless value. It is included among the analytical expressions in Sec. 3.1.8. The other geometric parameters are calculated from Eq. (3.24). At this time, the analytical asymptotic distance between the external contact boundary of the mushroom and the right-hand boundary (see Figs. 3.10(a) and 3.10(c))

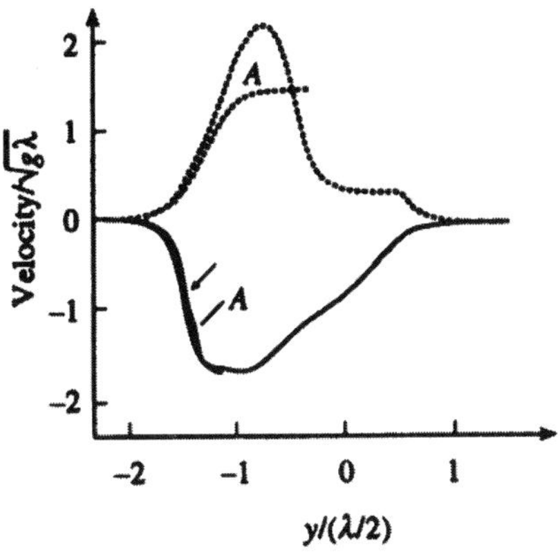


**Fig. 3.11.** Distributions of $u(x, y, t)$ against $y$. Dashed curves along the bubble axis $x = 0$, solid cirves along the jet axis $x = \lambda/2$ (see Fig. 3.10(b)).

is equal to $2\sqrt{h}(\lambda/2) = 0.60(\lambda/2)$, and the curvature radius at point $T_j$ is equal to $R = 0.46(\lambda/2)$, in accordance with (3.27)–(3.29).

Figure 3.11 shows several velocity distributions at $t$ (from Eq. (3.31)) for the left and right walls (see Figs. 3.7(b) and 3.10(b)). The dotted and solid curves pertain respectively to the verticals $x = 0$ and $x = \lambda/2$, i.e. to those lines along which the bubble rises and to those along which the jet descends. The analytical distributions are labeled "A". The arrow shows the position of the stagnation point. This point coincides with the isochore $\rho = (\rho_l + \rho_h)/2$. It should be noted that this point is characterized by a break in the analytical function $u(y)$. To normalize the analytical functions the velocity at the neck of the jet was taken as the unit velocity. We can see that the analytical and numerical distributions agree. The analytical solutions were found in the following way. Section $ab$ on the plane $\xi$, in Fig. 3.9(b) corresponds to side $ab$ in Fig. 3.9(a). We have taken 1000 points equally spaced on section $ab$ of the plane $\xi$ and computed the corresponding $z$-coordinate on the side $ab$. Plotting the inverse function $\xi(z)$ to the resultant function $z(\xi)$, we arrive at the desired distribution.

The model distribution was plotted in a similar way, the only difference being that here we have two sections: one from $a$ to 0 and the other from 0 to $c$ (see Figs. 3.9(a) and 3.9(b)). These two sections correspond to the two velocity distributions on opposite sides of the stagnation point. Since the liquids have different densities, their velocities should be normalized so that a continuous pressure is ensured at the interface between them. At the same time, the velocity of the heavy liquid in the reference frame moving with the stagnation point is $\sqrt{\mu}$ times smaller than that of the light liquid. It is then necessary to carry out a Galilean transformation to return to the laboratory reference frame. As a result, we obtain the distributions in Fig. 3.11.

## 3.2. Development of the Rayleigh–Taylor instability: numerical simulations

### *3.2.1. Introduction*

The problem of development of the RTI, apart from its indubitable theoretical significance, is of a great meaning for numerous important practical problems, like, for example, for the investigation of a stability of shell's compression in connection with problems of laser thermonuclear synthesis, by the creation of superstrong magnetic fields, etc.

An important contribution to the study of the problem under discussion was brought about by the works of Fermi.[34] In these works, the development of Taylor's instability is considered, as it proceeds at the boundary *fluid-vacuum*, both for linear and nonlinear cases, and at the boundary between two fluids: *light* one and *heavy* one. In doing so, for nonlinear cases the formulations are given, which are geometrically close to the numerical ones. In particular, the smooth surface partition is approximated by the piecewise smooth step function. There are some clearly pronounced stages which may be traced in development of RTI: *linear* one, *intermediate* one, *regular asymptotic* one, and *turbulent* one.[35,36] The linear stage is characterized by the smallness of amplitude $a$ in comparison with a length of perturbation wave $L$, and with an exponential rate of growth. When the value of a perturbation amplitude $a$ becomes equal to $0.4L$, the process enters the stage, which is intermediate between the linear one and regular asymptotic one. At the stage that begins at $a \approx 0.75\,L$, the "peaks" of heavy fluids are completely formed, being collapsed with a constant acceleration, as well as the "bubbles" of light fluid rising to the surface with a constant velocity. This stage of RTI is an unstable one[35,36] and is followed by the turbulent stage, in the course of which takes place an intensive interaction of perturbation of various wavelengths and fluid's mix-up.

The RTI phenomenon was most profoundly studied for the case of a planar partition surface, and for the ratio of densities of a heavy fluid and a light one, tending to infinity. The linear stage was minutely investigated in classical works of Rayleigh, Taylor, and Lewis,[37–39] the regular asymptotic stage — in works of Birkhoff,[40] while in Refs. 9 and 41, the phenomenological theory of the turbulent stage was developed, and in Ref. 36, some ideas were put forward concerning the mechanism of formation of the last stage.

Nevertheless, the analytical mathematical apparatus seems to be insufficient for the global analyses of such a complicated phenomenon as RTI; from the other side its experimental studies are rather labor-consuming. The most complete information might be obtained on the basis of numerical computations. Thus, the case of free surface was studied in Ref. 42, while the case of two incompressible fluids was treated in Ref. 43 and that of two compressible media — in Ref. 44.

From the point of the view of practical application it seems to be rather actual to determine the conditions of development of RTI, which would lead to a *stabilization*, or, at least, to a *deceleration* of the instability's evolution at the contact surface (the effects of the RTI's development would be minimal). Another problem consists in the investigation of the *mechanism of RTI's development during the turbulent stage and analysis of three-dimensional case*. Up to the present moment, the extensive series of theoretical and experimental studies of the RTI's development was conducted, but in the most of those only one-mode perturbation for the initial data was considered. Thus, it would be of a great interest to consider the *multi-mode interactions* and to study this process of *transition from the nonlinear stage of RTI's development to the turbulent one* on the basis of the complete dynamical equations.[32,33,45–47]

### *3.2.2. Numerical simulation of RTI development by the method of large particles*

Within the frame of a numerical experiment, the RTI problem was considered by Davydov and Panteleev.[47] The formulation of this problem was quite traditional: within the field of gravitation $g$, the contact surface separating the heavy (at the top, with density $\rho_h$) fluid from the light one (at the bottom, with density $\rho_l$), is subject to an initial perturbation, which might have a form of the velocity field. After that, it is necessary to analyze the dynamics of the behavior of the form of partition surface. Using the method of large particles[32,33] one is able to solve the complete *vortical system of Euler's equations for the motion of a compressible medium, taking into account the effect of gravitational field*[47]:

$$
\begin{aligned}
\frac{\partial \rho}{\partial t} + \operatorname{div}(\rho \mathbf{V}) &= 0,\\
\frac{\partial \rho u}{\partial t} + \operatorname{div}(\rho u \mathbf{V}) + \frac{\partial P}{\partial x} &= 0,\\
\frac{\partial \rho v}{\partial t} + \operatorname{div}(\rho v \mathbf{V}) + \frac{\partial P}{\partial y} + \rho g &= 0,\\
\frac{\partial \rho w}{\partial t} + \operatorname{div}(\rho w \mathbf{V}) + \frac{\partial P}{\partial z} &= 0,\\
\frac{\partial \rho E}{\partial t} + \operatorname{div}(\rho E \mathbf{V}) + \operatorname{div}(P \mathbf{V}) + \rho g v &= 0,
\end{aligned}
\tag{3.32}
$$

where $\mathbf{V} = \{u, v, w\}$ is velocity vector, $\rho$ is medium's density, $\mathrm{E} = e + \mathbf{V}^2/2$ is a specific total energy, $e$ is specific internal energy, $P$ is pressure, and gravity acceleration $g$ is directed along the $y$-axis.

Note that by the prescription of an initial perturbation for the computation with compressible media, it is important to observe everywhere (with exception of a partition surface) the condition $\operatorname{div}\mathbf{V} = 0$. Otherwise, the perturbation of density will arise, which would be able to distort a picture of RTI's development.

The two-dimensional computations of RTI, conducted with the help of the method of large particles, have demonstrated a good agreement with the results of Ref. 43. Shown in Fig. 3.12(a) are the forms of contact surface for $\rho_h/\rho_l = 10$, at the various time moments: 0.25, 0.49, 0.73, 0.96, 1.21; the solid lines correspond to the calculation by the method of large particles while dotted lines correspond to the results of Ref. 43. The graph of the velocity of peak's motion is also rather close to that given in Ref. 43, while the mean acceleration of a peak amounted to, like in Ref. 43, approximately the half of the gravitational acceleration. The ratio of radius of curvature of the bubble to a perturbations wavelength, $R/(2L)$, was equal to 0.40, which is closer to the 0.39 value given in Ref. 42, and to the theoretical value of 0.35, than the 0.48 value of Ref. 43. The velocity of bubble's ascent was found to be somewhat greater than that of Ref. 43. In Ref. 47, the value of $U_B\left(\frac{\rho_h-\rho_l}{\rho_h}gR\right)^{-1/2} = 0.44$ was obtained, which was equal to 0.32 in Ref. 43 whereas the theoretical value of that quantity, obtained by Birkhoff, was equal, approximately, to 0.5.

Note that in spite of a smearing of the contact surface, the large particles method permits to follow also the development of the Kelvin–Helmholtz's

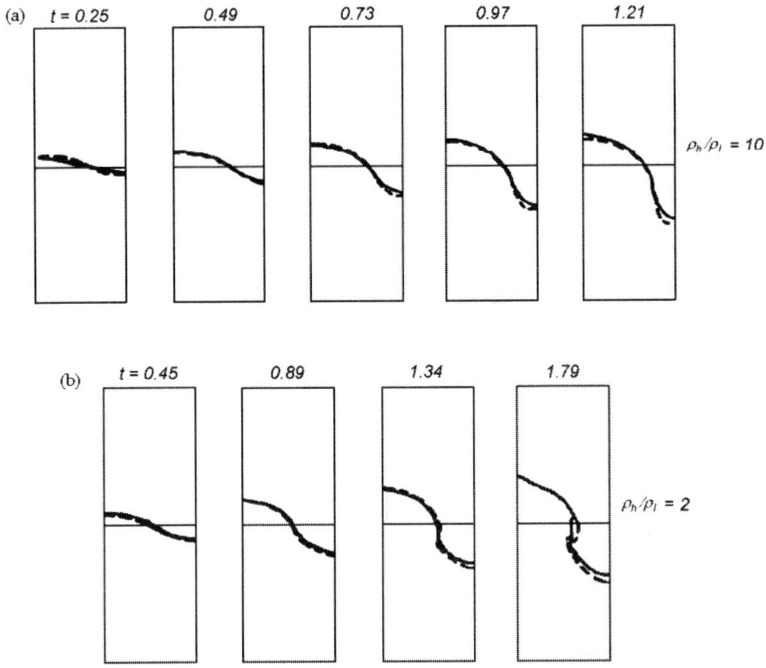


**Fig. 3.12.** The form of a contact surface at various moments of time, by the two-dimensional RTI: (—) calculations by the large particles' method, (- - -) data of paper by Daly.[43]

instability with small values of $\rho_h/\rho_l$. Given in Fig. 3.12(b) are the forms of a contact surface for $\rho_h/\rho_l = 2$, at the time moments: 0.45; 0.89; 1.34 ; 1.79.

Presently, one should turn to the survey of results of the numerical simulation of the three-dimensional RTI with the help of the large particles' method.[47] The computational area's boundaries consisted of $20 \times 60 \times 20$ cells along the axes $x$, $y$, and $z$, correspondingly, with $\Delta x = \Delta y = \Delta z$. In a three-dimensional case under consideration, the perturbation of a partition surface might form either hexagonal, or square lattice. In the present study, the last version was chosen (though there is no principal difficulty in simulation of a hexagonal lattice). For its realization (using kinematic formulation) the initial velocities perturbation was prescribed in the form[a]

$$u = \mathrm{A} \sin \frac{\pi x}{L} \cos \frac{\pi z}{L} [2H(y) - 1] \exp \left\{ \frac{-2\pi |y|}{L} \right\},$$

[a]In dynamical formulation the partition boundary within the fluid, which is initially at rest, has a perturbed form.

$$v = \mathrm{A}\cos\frac{\pi x}{L}\cos\frac{\pi z}{\exp}\left\{\frac{-2\pi|y|}{L}\right\},$$

$$w = \mathrm{A}\cos\frac{\pi x}{L}\sin\frac{\pi z}{L}[2H(y)-1]\exp\left\{\frac{-2\pi|y|}{L}\right\}. \tag{3.33}$$

The rest of initial and boundary conditions, as well as the problem parameters, were just the same as for two-dimensional case.

Shown in Fig. 3.13 are the partition surfaces in isometric projection, formed at successive moments of time $t = 0.6;\ 0.9;\ 1.2;\ 1.5$. One clearly

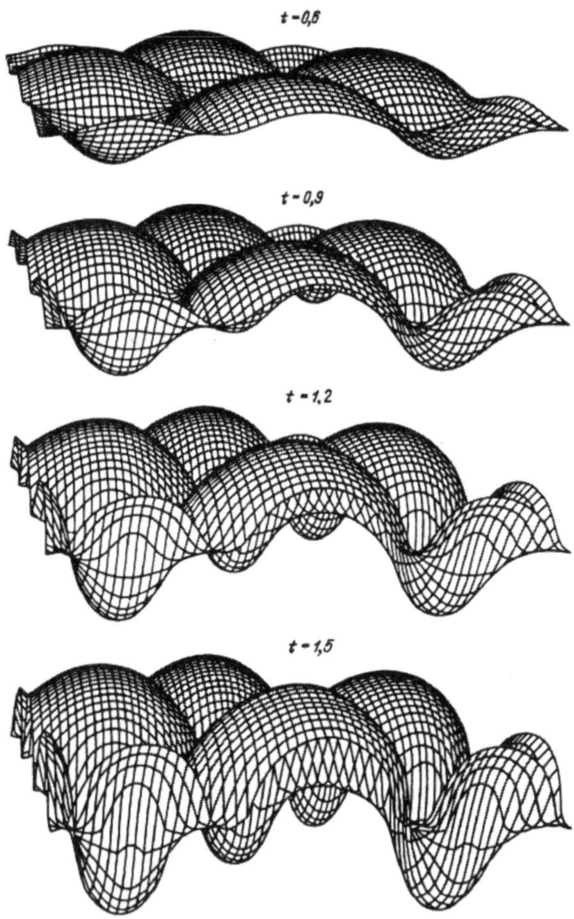


**Fig. 3.13.** Dynamics of the form of partition surface by the three-dimensional RTI, in isometrical projection ($\rho_h/\rho_l = 10$).

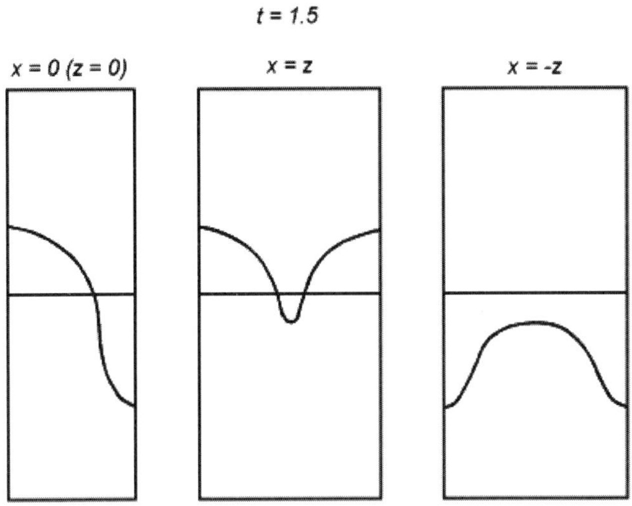


**Fig. 3.14.** The form of a surface in cross-section by the planes by the three-dimensional RTI ($t = 1.5$, $\rho_h/\rho_l = 10$).

sees the "ascent" of wide bubbles and the "collapse" of the sufficiently narrow peaks.[47] The main problem consists in the necessity of "passing" the large time intervals, aiming to determine the moments of a *break-up* (or *stabilization*) of the process.

Note that the "heights" (shifts along the $y$-axis) of the bubble and of the peak in Fig. 3.13 are not, actually, the same. This might be explained by the fact that the point at the partition surface having the coordinates $x = z = \pi/(2L)$, $y = 0$, in which it was initially $u = v = w$, moves downward with the acceleration, approximately equal to the half of the peak's acceleration. More distinct impression on the character of a process might be obtained from Fig. 3.14, in which are shown the cross-sections of a contact surface by the planes $x = 0$ ($z = 0$), $x = z$, $x = -z$, with $\rho = 0.5$ $(\rho_l + \rho_h)$, at the moment $t = 1.5$.

The ratio of the radius of curvature of a bubble to the perturbation's wavelength $R/(2L)$ is equal to 0.34 in the cross-section $x = 0$ ($z = 0$), and equal to 0.31 for $x = z$. The theoretical and experimental values of that quantity, which were obtained by investigation of the ascent of air bubbles in vertical tubes filled by a liquid, are equal to 0.35 according to Birkhoff.[31] The dimensionless speed of the bubble's ascent, $U_B(2Lg)^{1/2}$, was equal to $0.31 \pm 0.02$, while according to the data of Ref. 31, it was equal to 0.32 for cylindrical tubes, while for those of rectangular cross-section (with a rate of rectangle's sides 1:4), it was equal to 0.29. The peak's acceleration was in approximately 20% larger, then for a two-dimensional version, while the peak itself was somewhat wider.

It is quite possible that the computations presented here require more precise definition, but it is quite clear that they are of a considerable interest.

### *3.2.3. Intermode interaction in RTI*

Following Ref. 45, one will consider the influence of interaction of two or more waves on the evolution of contact surface for two-dimensional RTI.

Let, over the long wave mode, having the wavelength $\lambda$ and amplitude $A$, be superimposed some modes with shorter wavelengths $\lambda_1 < \lambda$ and with the amplitudes $A_i$. As the initial perturbation of a velocity fields one chooses the superposition of waves

$$u = [2H(y) - 1]\left[A\sin\left(\frac{2\pi}{\lambda}x\right)\exp\left(-\frac{2\pi}{\lambda}|y|\right) + \sum_{i=1}^{N} A_i \sin\left(\frac{2\pi}{\lambda_i}x\right)\exp\left(-\frac{2\pi}{\lambda_i}|y|\right)\right],$$

$$v = A\cos\left(\frac{2\pi}{\lambda}x\right)\exp\left(-\frac{2\pi}{\lambda}|y|\right) + \sum_{i=1}^{N} A_i\cos\left(\frac{2\pi}{\lambda_i}x\right)\exp\left(-\frac{2\pi}{\lambda_i}|y|\right), \tag{3.34}$$

where $H(y)$ is Heaviside function,

$$H(y) = \begin{cases} 1, & y \geq 0, \\ 0, & y < 0. \end{cases}$$

Schematically presented in Fig. 3.15 are four numerically investigated cases of the pairwise interaction of different waves, having the form of velocity's $v$ perturbation. It is evident that all the versions of interaction considered here and periodical, having the period $\lambda$. Therefore, just as for the case of a single wave, here, at the edges of a segment $0 \leq x \leq \lambda/2$, one can prescribe the conditions of symmetry.

Note that if the present problem is analyzed in linear approximation, then the perturbation with shorter wavelength "suppresses" one with longer wavelength.[34] But at the nonlinear stage of perturbation's development

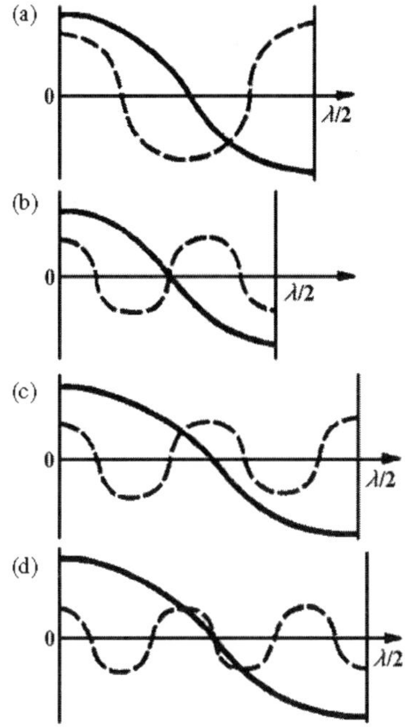


**Fig. 3.15.** The field velocities of the initial perturbation, by the pairwise interaction of waves of different lengths: (a) $\lambda/\lambda_1 = 2$; (b) $\lambda/\lambda_2 = 3$; (c) $\lambda/\lambda_3 = 4$; (d) $\lambda/\lambda_4 = 5$.

it was found, that the nonstationary process "remembers" the perturbation of a long wavelength, which in the course of time begins to play a decisive role.

In the paper by Inogamov,[10] on the basis of estimations it was shown that there exists a mechanism of transition to a turbulent stage, which leads to the stirring-up of the long-wave perturbations and to the formation of yet greater hydrodynamical scale (such an effect is observed experimentally, too). Development of the instability of a phenomenon results in the increase of cavern's (bubbles) sizes, in their drawing together, in their unification ("redoubling"), and in the formation of new (larger) caverns; however, such a flow is, once again, unstable, the process of a new confluence begins, which results in appearance of yet larger scale, etc. An analysis of the field of perturbation of the velocity and pressure, and of the resulting field of acceleration, has shown that there appears a force directed along the displacement, and this leads to the warp and to the unification of caverns. Naturally, it would be very interesting to check both these estimations and

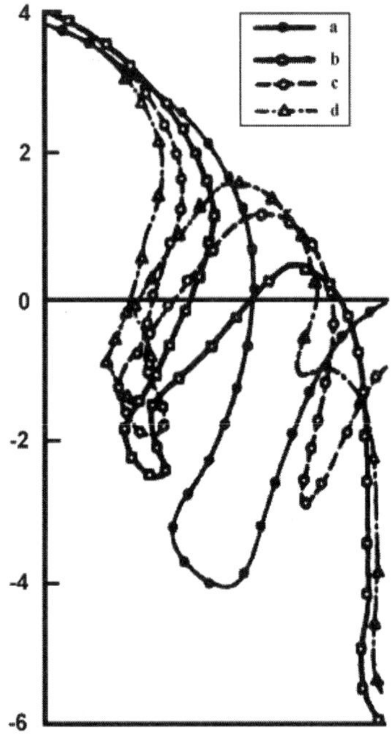


**Fig. 3.16.** Position of the contact surface for $g = 1$, by the pairwise interaction of the models with equal amplitudes (the same nomenclature as in Fig. 3.15).

the hypothesis of the redoubling of hydrodynamical scale, by way of their comparison with computational data.

Presented in Fig. 3.16 are the contact surface's positions at time moment $t = 4.5$, which were obtained as a result of numerical experiments with the help of the large particles method in application to the interaction of two perturbation modes for each of the cases shown in Fig. 3.15. The determining parameters were assumed to have the following values: the initial perturbation amplitudes $A = A_i = 0.3$, density ratio $\rho_h/\rho_l = 10$, wavelength $\lambda = 10$, acceleration of the free fall $g = 1$. It is seen that in case (a) $(\lambda/\lambda_2 = 3)$, the jet of a heavy component is "coming down" into a light medium and has a horizontal velocity component: an asymmetry of flow is initiated. There begins a *unification* of bubbles and a formation of a bubble with long-wave mode. In case (b) $(\lambda/\lambda_2 = 3)$, it is seen that side by side with a jet located near $x = \lambda/2$, one another internal jet is formed in the vicinity of $x \approx \lambda/4$, which, similarly to the case (a), has a tendency of deflation in horizontal direction. Also evident is a unification of two bubbles, and the tendency toward this unification is a result of the development process of thinning of the jet and of its clapping-in. In case (c) ($\lambda/\lambda_3 = 4$), two sloping jets are observed, one of which (located closer to the axis $x = 0$) begins to become thinner. At $x = \lambda/2$, the perturbation of wave $\lambda$ seems to be dominating, as compared with case (a). In case (d) ($\lambda/\lambda_4 = 5$), at large values of time (see Fig. 3.16), one might observe the vanishing of previously developed jet in the vicinity of $x = 0.3\lambda \div 0.4\lambda$, and the thinning of jet near the ray $x = 0.15\lambda \div 0.20\lambda$.

It is evident that the last case with $\lambda/\lambda_4 = 5$ of the series of computations considered is the most characteristic for the dynamics of *the bubbles' unification*, as well as of the *vanishing of jets and their clapping-in* (similar effects of the instability's development are observed in the cases $\lambda/\lambda_1 = 2$ and $\lambda/\lambda_2 = 3$, too, but for much larger values of time).

Shown in Fig. 3.17 are positions of the contact surface at $t = 4.5$ for double value of the gravity force ($g = 2$); all the other parameters remained the same. The RTI process is developing faster. With $\lambda/\lambda_1 = 2$, the typical deflection of the jet is seen, which might be explained by the increase of the acceleration of free fall $g$, and by the development of Kelvin–Helmholtz's instability (which leads to the even faster turbulent mixing). With $\lambda/\lambda_1 = 2$, the process is accompanied by the localization (and subsequent separation) of the bubble of a light component; equally noticeable is

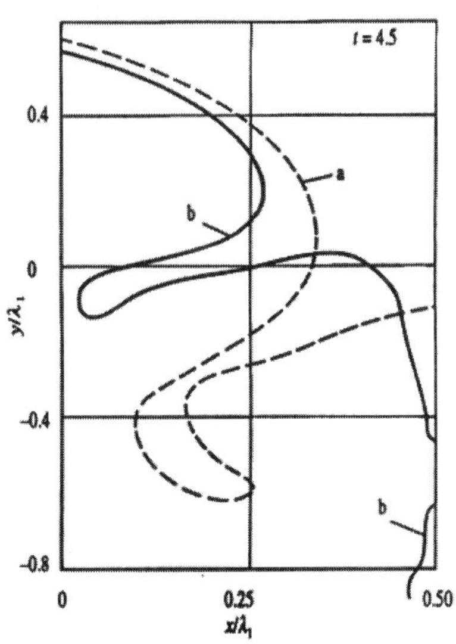


**Fig. 3.17.** Interface for Rayleigh–Taylor instability for two-mode disturbance and $\rho_h/\rho_t = 10$, $t = 4.5$: (a) $\lambda_2/\lambda_t = 2$; (b) $\lambda_2/\lambda_t = 3$.

the separation of a drop from a jet of heavy medium (in the vicinity of the ray $x = \lambda/2$).

Finally, one might consider the RTI process after the prescription of several perturbation's modes at the initial moment $t = 0$. Shown in Figs. 3.18 and 3.19 is the dynamics of RTI development: presented are the successive positions of a contact surface at various moment of time, in the case of interaction of five modes: $\lambda = 10$, $\lambda_1 = 5$, $\lambda_2 = 10/3$, $\lambda_3 = 5/2$, $\lambda_4 = 2$. The corresponding initial values of perturbation's amplitude were $A = 0.2$, $A_i = 0.1$, $i = 1, 2, 3, 4$, the density's overfall $\rho_h/\rho_l = 10$, the free fall's acceleration $g = 2$; one could observe, also, the process of the growth of the noticeable asymmetry of a flow.

It is evident that at the first moment of time the short-wave perturbations are developing faster than the long-wave ones (see Fig. 3.18). However, during the nonlinear stage ($t > 3$, see Fig. 3.19), the long-wave perturbations begin to "suppress" the short-wave ones, and the processes of vanishing of jets for small wavelength take place. At the same time it is observed the prevailing of a long-wave mode, and the formation of a larger spatial

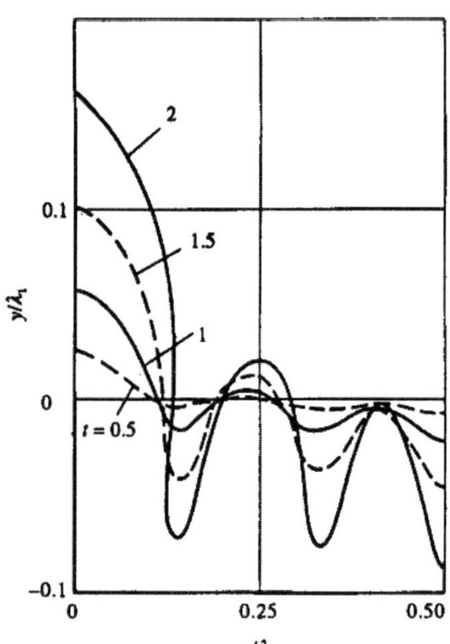


**Fig. 3.18.** Dynamics of the Rayleigh–Taylor instability for the five-mode disturbance at the initial stage.

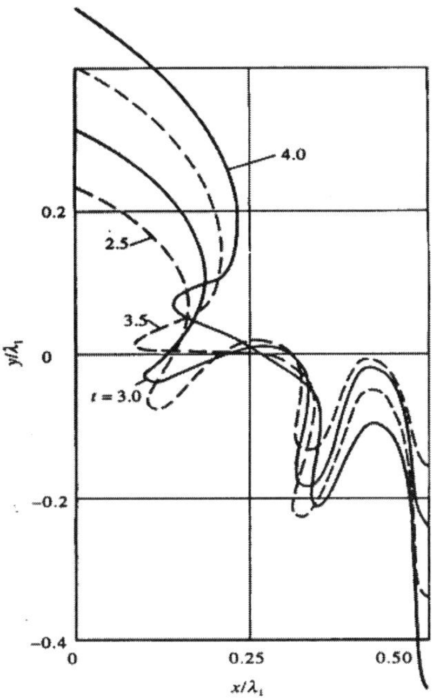


**Fig. 3.19.** Dynamics of the Rayleigh–Taylor instability for the five-mode disturbance at the nonlinear stage.

scale (in accordance with the predictions of Ref. 10). The described version of the dynamics of nonlinear interaction of the harmonics in RTI seems to be typical and reminds the Feigenbaum's scenario of the turbulence.

### *3.2.4. RTI simulation by the method of pseudo-compressibility*

A. V. Chekhlov has studied the RTI's dynamics at sufficiently large times of development — *up to the achievement* of the turbulent stage. In the frame of the *multi-mode* version the diverse-density two-dimensional motion of incompressible fluids having a common planar contact surface was considered. The computational approach used was based on the idea of the method of "artificial compressibility" (or on the introduction into the initial system of the $\varepsilon$-perturbation, where $\varepsilon > 0$ is a small number). Such an approach permits to avoid those complexes, which are connected with a satisfaction of the condition $(\nabla \mathbf{V}) = 0$. Some important aspects of that study will be analyzed here in more detailed form.

Thus, by the development of a *single-mode* perturbation with large ratio of densities of upper and lower fluids in the flow arising (which consists of bubbles and jets), the bubbles are ascending with a *constant speed*, while the jets are falling down with acceleration. With close values of the fluid's densities the main part of energy is wasted on the "rolling-up" of the appearing quasi-symmetrical vertical structures, created by the Kelvin–Helmholtz instability; these structures are moving upward and downward with *constant velocities*.

The process of development of the single-mode perturbations was studied simultaneously with their stability, up to the essentially nonlinear (regular asymptotic and turbulent) stage. In this case, superimposed over the main periodical perturbation was "the ripple" consisting of the short-wave mode with the ratio of wavelength equal to 1:20. Proceeding by just the single tendency of the flow's scale (Inogamov's hypothesis), by sufficiently large times of development the main mode "should be purified" and reveal the tendency to a single-mode solution. However, the computational experiment has demonstrated that even at large times there remain the *non vanishing short-wave perturbations*. Thus, by the RTI's development two competing tendencies are observed: the enlargement of the scale of flow and the generation of short-wave perturbations (a new effect!).

The matters discussed will be illustrated by the sample numerical results. Shown in Fig. 3.20 is an example of the typical flow (line of the

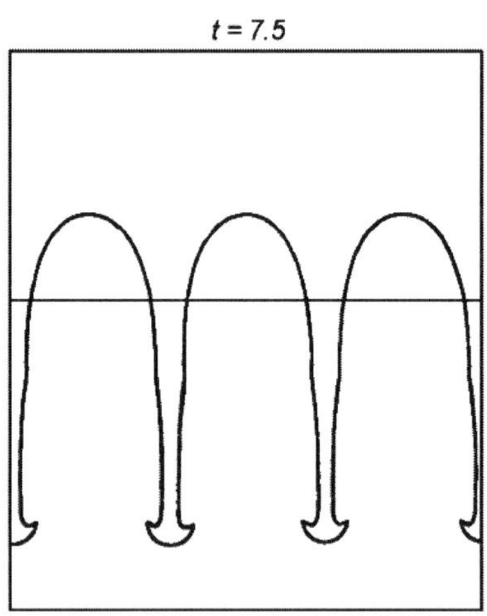


**Fig. 3.20.** The contact surface by the RTI (single-mode perturbation, $\rho_h/\rho_l = 10$, $t = 7.5$, regular asymptotic stage).

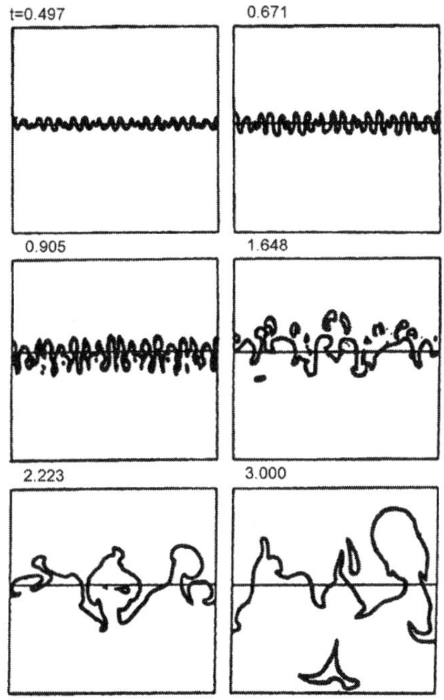


**Fig. 3.21.** The evolution of the contact surface by the RTI with a multi-mode initial perturbation (the scale's enlargement is observed).

contact) in the form of jets and bubbles, induced by a single-mode initial perturbation ($\rho_h/\rho_l = 10$), at sufficiently large value of time ($t = 7.5$).

At the regular asymptotical stage, presented here, the bubbles have a regular "large" form (which is already invariable in time) and ascend with a constant speed $v \approx 0.23\sqrt{g\lambda}$. The flow of such a type plays *the principal* role in evolution toward the turbulent stage and in enlargement of the scale by the multi-mode perturbation. Presented in Fig. 3.21 is the contact surface's form at the various moments of time, in the processes of RTI with the multi-mode initial perturbations. Quite clearly seen is the picture of "the scale's enlargement" — beginning with 23 bubbles in the first picture at $t = 0.5$, and ending with three bubbles in the last one, where $t = 3.0$ (the computational mesh is $120 \times 120$). Illustrated in Fig. 3.22 is the evolution of the mean density profile,

$$\bar{\rho}(y,t) = \frac{2}{\lambda}\int_0^{\lambda/2} \rho(x,y,t)dx,$$

at the same moment of time and by the same multi-mode computation.

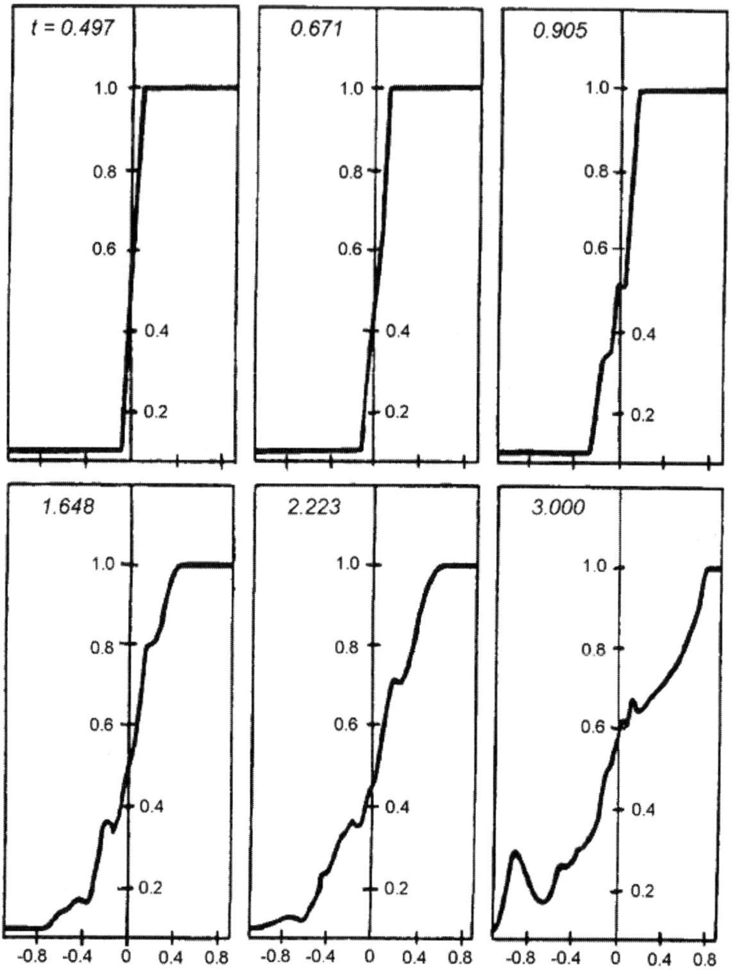


**Fig. 3.22.** The evolution of average profile of density (the moments of time and the initial perturbation are the same as in Fig. 3.21).

The results presented are used in the determination of the coefficient of turbulent mixing.

Presented in Fig. 3.23 (with the same data) is the evolution of Fourier's spectrum of the density function $f(x)$, where

$$f(x, y, t)\big|_{y=0} = \sum_{i=0}^{\infty} \mathrm{A}_i \cos(2\pi\, ix/\lambda),$$

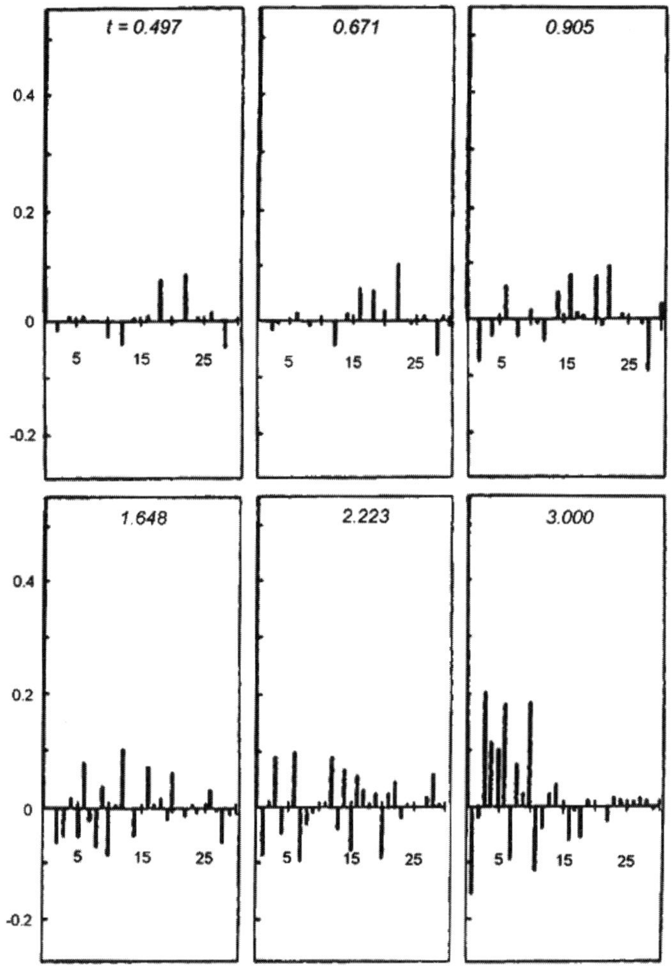


**Fig. 3.23.** The evolution of Fourier's spectrum of the function of density at $y = 0$ (the moments of time and the initial perturbation are the same as in Fig. 3.21).

$$A_n = (4/\lambda) \int_0^{\lambda/2} f(x) \cos(2\pi n x/\lambda)\, dx\,.$$

Illustrated here is the dependence of the amplitude of $n$th mode $A_n$ on its number $n$. One might clearly see the temporal enlargement of the characteristic scale (modes with maximal Fourier's amplitude) from $n = 23$ up to $n = 3$.

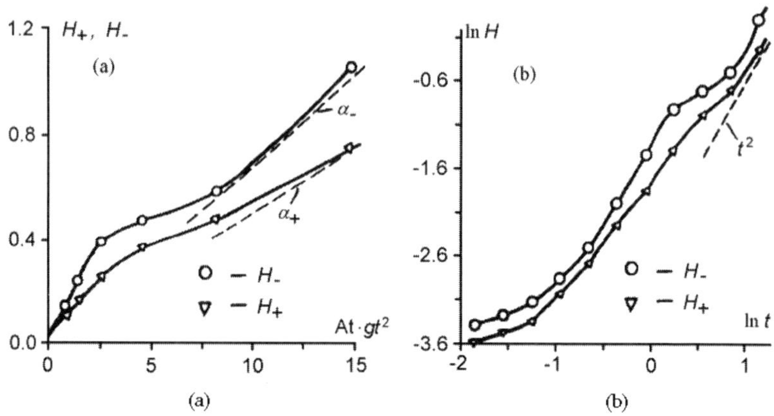


**Fig. 3.24.** The graphs for the determination of the coefficient of turbulent mixing: $H_+$ ($H_-$) is the thickness of perturbation of the light (heavy) fluid into the heavy (light) one, $\alpha_+ = 0.053 \pm 0.03$, $\alpha_- = (2 \pm 0.05)\alpha_+$ (see also Fig. 3.37).

Finally, pictures in Fig. 3.24 permit to determine *the coefficient of turbulent mixing*. Here, $H_+$ ($H_-$) is the thickness of penetration of the light (heavy) fluid into the heavy (light) one; $t$ is the time; $g$ is acceleration of gravity; $At = (\rho_h \rho_l)/(\rho_h + \rho_l)$ is the Atwood number. Dotted lines correspond to the law of development of the turbulent area (determined from the calculations):

$$H_{\pm} = \alpha_{\pm} \cdot At \cdot g \cdot t^2. \tag{3.35}$$

Therewith, from the first graph (Fig. 3.24(b)), it is deducted that the appearing in Eq. (3.35) of quadratic law of the temporal development (dotted line) is correct. The results by A. V. Chekhlov are in good agreement with experiment[16] and with two-dimensional computations[15] by other authors. The investigations of such a kind are available with a computational experiment only.

### *3.2.5. Numerical simulation of the RTI development by means of high-resolution Euler hydrocode*

Now one is going to discuss some results concerning the numerical simulation of the Rayleigh–Taylor's instability (RTI), which were obtained using quasimonotonous grid-characteristic approach of the second-order approximation. A. M. Oparin developed this approach and performed all simulations that are presented in this section. The mathematical model is based on the full system of the Euler's equations for a multicomponent compressible inviscid gas, and for that reason the potential resources of the hydrocode referred to are sufficiently vast.

Here, one has applied the quasimonotonous grid-characteristical method having the second-order of approximation. The hybrid monotonous method similar to that was previously developed and applied to the numerical simulation of the flows of incompressible fluid.[30] For the case of model linear convection equation such an approach is reduced to a combination of schemes with oriented and central differences. For that scheme, in application to the convection equation, the monotony condition is fulfilled strictly, that is, any arbitrary monotonous set of the values of function at the grid's nodes preserves its monotony after the temporal step. Resorting to the grid-characteristical formalism,[32,63] one has extended the scheme to the case of a compressible gas. As a result, the explicit computational scheme obtained possesses such useful properties as conservativeness, monotony, the second-order of approximation, but at the time it does not use neither artificial viscosity, nor smoothing-out, nor flux-limiting procedure, which are frequently used in the modern schemes of computational fluid dynamics. Following the present methodics, the switching between schemes with central differences and with oriented ones (which just determines the quasimonotony with preservation of the second-order approximation) is carried out separately for each of the characteristics and depends on the sign of the corresponding characteristic and on the sign of one additional parameter.

As an one-dimensional test of the hydrocode chosen here, one might consider a familiar problem of the collapse of discontinuity, for which an analytical solution is known. The integration area representing an unit section, is divided into two parts, and in each of these subsections at the initial time moment are prescribed their own values of density, pressure, and velocity. For the case considered, the initial values of the parameters of a gas with adiabatic index $\gamma = 1.4$ are the following: $\rho_L = 1$, $\rho_R = 0.125$, $P_L = 1000$, $P_R = 0.1$, and $u_L = u_R = 0$. An exact solution consists of the contact discontinuity and of the shock moving to the right, and the rarefaction wave moving to the left side. This analytical solution at the time moment $t = 0.008$ is represented by the solid line in (Fig. 3.25 (a) density, (b) velocity of flow and velocity of sound, (c) pressure). The exemplary version was particularly complicated due to the existence of a pressure drop of the fourth-order, and due to the fact that as a result of the discontinuity's collapse the area of supersonic flow is formed ($|u| > c$, $u$ and $c$ are local values of the velocity of flow and that of sound) within the rarefaction wave. The line of transition $|u| = c$ within the rarefaction wave is a dangerous

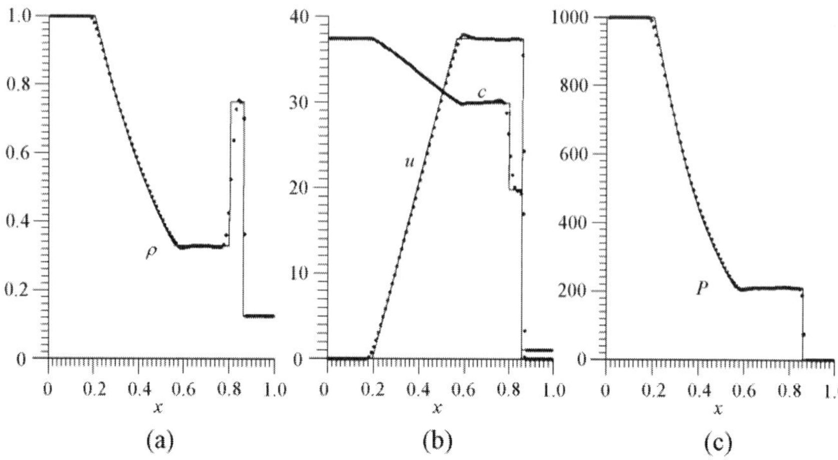


**Fig. 3.25.** Analytical (solid line) and numerical (marker) solutions for the one-dimensional problem of a discontinuity's collapse: (a) density, (b) velocity of flow and speed of sound, (c) pressure.

one — near this line some numerical methods are accumulating the error, which leads to a nonphysical rarefaction shock.

The circles in the same Fig. 3.25 represent the calculated values for this problem. The number of cells covering the integration area was equal to 100. The current value of a temporal step was determined from the stability condition with Courant number 0.4. The result obtained is quite satisfactory, since the fronts of both the shock and the contact discontinuity are rather steep, and there are no appreciable oscillations near both of them. The calculated rarefaction wave, as well as the other smooth sections, is, practically, coinciding with analytical results.

The important argument in favor of some or another numerical method or against it would be a conduction of comparative calculations for one and the same problem using the sequence of permanently refining grids. Such a series of calculations was conducted for the two-dimensional problem of development of RTI with a single-mode perturbation of velocity, the comparison being made for a rather distant time moment corresponding to the highly nonlinear stage of flow. The height of an integration area was $h = 7\lambda/2$, and its width was $l = \lambda/2$, where $\lambda = 2\pi$ is the length of a perturbation wave. The gravitational force is directed downward and $g = |\mathbf{g}| = 1$.

At the initial time moment the horizontal flat contact boundary divides the integration area in the ratio 4:3. Within the upper and smaller part of the area the heavy gas is located, which has the density $\rho_h = 1$, while within the lower part is the light gas with the density $\rho_l = 0.1$. The adiabatic index for both the heavy and the light perfect gases is $\gamma = 1.4$. The pressure in the area corresponds to the condition $P = P_{bt} - \rho g(y - y_{bt})$, where $P_{bt} = 100$ is the pressure at the lower boundary of the area, $y = y_{bt}$, if only the axis $y$ is directed upward. At the sidelong boundaries of the integration area the conditions of symmetry were prescribed, while at the upper and lower boundaries — those of nonflowthrough. By means of the potential a small initial perturbation of velocity is prescribed, which is equivalent to a small perturbation of the contact surface itself:

$$\mathbf{V} = \nabla\varphi \quad \varphi = -(A_k/k)\cos(kx)\exp(-k|y|)\,\mathrm{sign}(y), \tag{3.36}$$

where $k = 2\pi/\lambda = 1$ for the main mode, and the amplitude of perturbation is $A_1 = 0.05$.

Three versions of calculations of the RTI development were conducted with subsequently refining grids $15 \times 105$, $30 \times 210$, and $60 \times 420$. By virtue of the symmetry the calculations were made for only one half of the perturbation's wavelength, but by the visualization of results the pattern obtained was supplemented up to the total wavelength. Presented in Fig. 3.26 are the families of isochores $(0.15,\ 0.25, \ldots, 0.95)$ for the time moment $t = 8$ (the dimensionless time $T = t\sqrt{At \cdot g \cdot 2\pi/\lambda} \approx 7.236$, where $At = (1-\mu)/(1+\mu)$ is Atwood number, $\mu = \rho_l/\rho_h$ is the density ratio, $k = 2\pi/\lambda$ is the wave number) for each of the grids indicated. At this figure the coordinates are normalized with respect to the perturbation's wavelength $Y = y/\lambda$, $X = x/\lambda$. The initial position of a contact boundary corresponds to the line $Y = 0$.

It is to be noted that the main features of a pattern and the position of a contact boundary are adequately described even with the coarsest grid. The more refined grids permit to find the more precise position of the contact boundary (the isolines of the family are drawn closer together), and with these grids the new details of a flow are revealed, which could not be resolved on coarser grids due to their smallness.

The functional dependencies on time for the vertical coordinate of the bubble's upper point and of the jet's lower point are shown in Fig. 3.27. The vertical coordinate is normalized with respect to a perturbation's wavelength. The curves corresponding to the coarsest grid $15 \times 105$ are marked with circles; to the grid $30 \times 210$, with squares; and to the grid $60 \times 420$, with triangles.

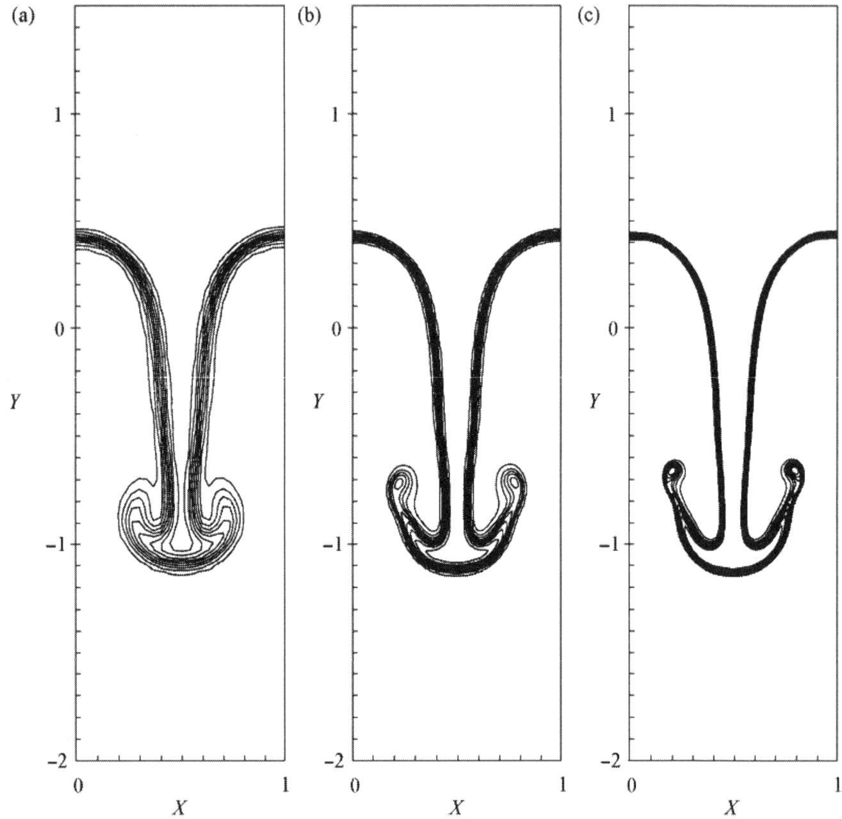


**Fig. 3.26.** Isochores for the single-mode two-dimensional problem of the RTI's development, for the time moment $t = 8$, obtained on three successively refined grids: (a) $15 \times 105$, (b) $30 \times 210$, and (c) $60 \times 420$.

As it is known, at the asymptotic stage the bubbles are rising with a constant velocity and have a fixed, nonchanging in time form. The ascent of bubble in Fig. 3.27, which is linear after the time moment $t = 5$, confirms this fact. The velocity of bubble's ascent, when determined by the way of drawing of the best straight line (in the sense of the method of least squares) for the interval of time from 6 to 10, is equal to

$$\nu_b \approx 0.23\sqrt{(1-\mu)g\lambda} \approx 0.57\sqrt{(1-\mu)g/k}.$$

From the results of numerical simulation one can extract important geometrical information like, for instance, the radius of curvature at the bubble's apex. To determine this radius we searched for a circumference

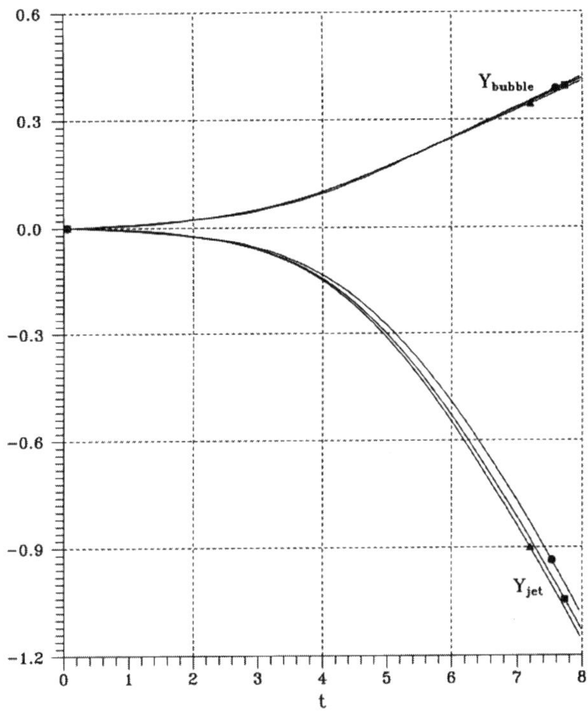


**Fig. 3.27.** Dependencies on time of the vertical coordinate of the bubble's upper point and of the jet's lower point. For the grid 15 × 105 (circles), 30 × 210 (squares), 60 × 420 (triangles).

with a center on the axis of symmetry, which would be most close to the contact boundary in the vicinity of the bubble's apex. Only the points located not farther then $\lambda/8$ from the axis of symmetry were taken into account. Coordinates of these points were calculated with the help of linear interpolation. For that we assumed that the contact surface coincides with the average isochore $(\rho_l + \rho_h)/2$. By analogy one can compute the radius of jet's curvature in its top. There are oscillations of these functions, which result from the passage of grid cells by the contact boundary. At the end of computation conducted the radius of the bubble's curvature (more accurately, its average value for oscillation period) approached the asymptotic value. Computed values at the end of simulation ($t = 10$) are equal to

$$R_b \approx 0.3\lambda \approx 1.9/k \quad \text{and} \quad R_j \approx 0.36\lambda \approx 2.3/k,$$

for bubble and jet, correspondingly.

The computational run to obtain the velocity of bubble's ascent and curvature radii for bubble and jet was conducted with a grid $30 \times 210$.

To be considered later on is the two-dimensional problem of the RTI's development with prescription at the initial time moment of perturbation mode having small wavelength. The computational experiment is described as follows. The integration area has the form of a rectangle stretched along the vertical direction $y$. The force of gravity is directed downward and $g = |\mathbf{g}| = 1$. The area's height is $h = 4\pi$, and its width is $l = 2\pi$. The flat horizontal contact boundary divides the integration area in two parts according to the ratio of 5 to 3. Within the smaller upper part is the heavy gas with density $\rho_h = 1$, while below is the light one, with density $\rho_l = 0.1$. The adiabatic index both of the heavy and the light gas is $\gamma = 1.4$. The pressure within the area is determined by the condition $P = P_{bt} - \rho g(y - y_{bt})$, where $P_{bt} = 100$ is the pressure at the lower boundary of the area $y = y_{bt}$. At the initial time moment the velocity perturbation ($\operatorname{div} \mathbf{v} = 0$) is prescribed as corresponding to the 23rd harmonics with the initial amplitude in (3.36) $A_{23} = 0.2$.

The computational grid $240 \times 480$ permits to resolve these harmonics with a sufficient degree of accuracy. At the lateral boundaries of the area the conditions of symmetry are prescribed, while at the upper and lower ones — those of nonflow through.

Presented in Fig. 3.28 are snapshots of the density distribution, obtained for successive time moments $t_n = 3, 4, \ldots, 14$. Even the first snap-shot already corresponds to the highly nonlinear stage of the RTI's development. As the time goes on, the topology of the contact boundary is permanently changed and becomes more and more complicated. There is an interaction between bubbles. The larger bubbles grow faster. The total number of bubbles reduces in time due to disappearing of smaller bubbles. At the same time, the collisions of jets occur both between themselves and with internal surfaces of the bubbles, the new jets are made up, the drops of heavy medium begin to separate, etc.

The evolution of the Fourier spectrum is of special interest for us. Presented in Fig. 3.29 are the Fourier coefficients for expansion of the density at the horizontal cross-section coinciding with the initial position of contact boundary. Shown is each moment as compared with Fig. 3.28. The results of calculation demonstrated here confirm the hypothesis concerning the enlargement of the scale of flow during the RTI's development. Qualitatively one can clearly follow this process in Fig. 3.28. And quantitatively, the

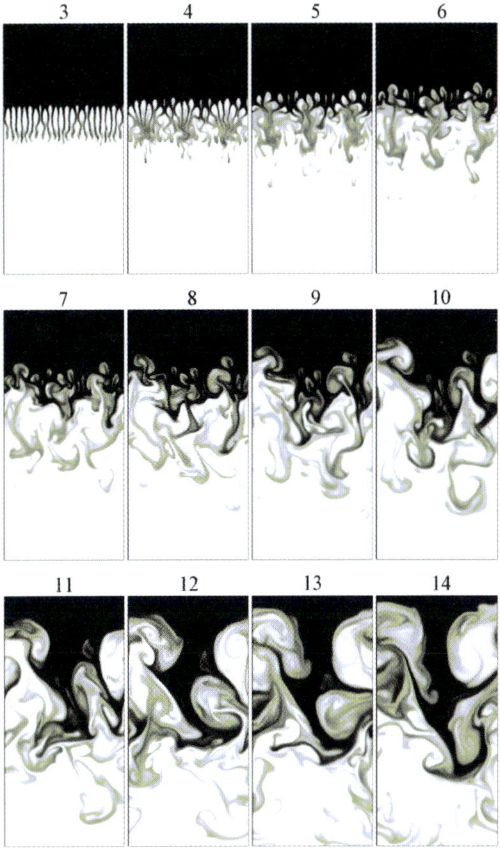


**Fig. 3.28.** Calculated distributions of density during the initiation of the spontaneous turbulence for a two-dimensional problem of the RTI's development, at the time moments $t_n = 3, 4, \ldots, 14$.

intensification of the main harmonics is shown in Fig. 3.29. At the same time, the short-wave perturbations do not vanish.

The great amount of publications having computational aspect was devoted to the two-dimensional numerical simulation of RTI.[15,42–45,57–60,64] In a number of occasions, these results give a chance for sufficiently good interpretation of a natural experiment, but they, however, do not explain many important details. The increase of the phenomenon's dimensionality (i.e. transition from two-dimensional flows to three-dimensional ones) is accompanied by the physical effects, which in the problems of lesser dimensionality are either absent, or reveal themselves in

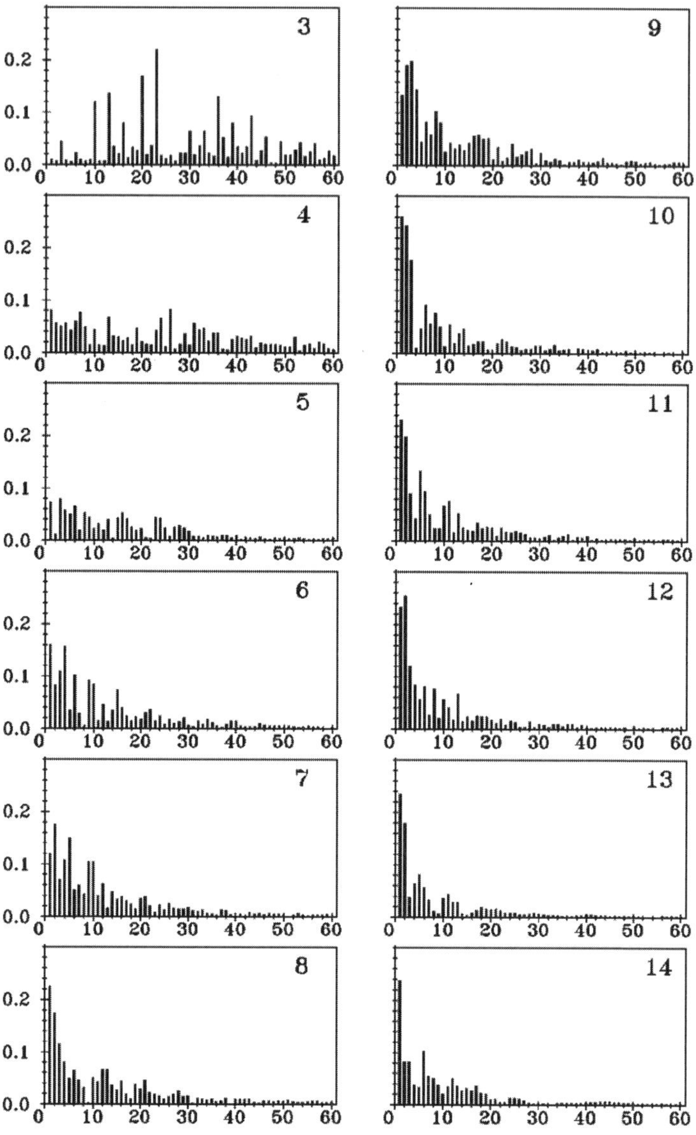

**Fig. 3.29.** The Fourier coefficients for the expansion of the density in horizontal cross-section coinciding with the initial position of contact boundary at the time moments $t_n = 3, 4, \ldots, 14$.

the quantitatively different degree. At the same time, the total number of computational publications on RTI problems in three-dimensional formulation is still rather limited.[47,51–53,61,62] So, our hydrocode was extended to a three-dimensional case to investigate the specific features of a spatial progress of RTI.

At first, one can find an information on calculation for the single-mode three-dimensional problems. In this case, the integration area has the form of a rectilinear parallelepiped having the height $h = 7\lambda/2$, with the width and thickness equal to each other, namely $\lambda/2$, where $\lambda = 2\pi$ is the perturbation's wavelength. Thus, due to the existence of two planes of symmetry, only one quarter was calculated, as referred to the area of the main perturbation. All the parameters needed for calculations (the value of gravity force $g = 1$, the ratio of densities $\mu = 0.1$, initial position of contact boundary, etc.) correspond here to the described earlier single-mode two-dimensional calculations.

The initial perturbation of velocity one finds from the potential function:

$$V = \nabla\varphi,$$

$$\varphi = -A_x/2 \cdot \cos(k_x x)\exp(-k_x\,|y|)\mathrm{sign}(y) \\ - A_z/2 \cdot \cos(k_z z)\exp(-k_z\,|y|)\mathrm{sign}(y), \tag{3.37}$$

where $k_x = 2\pi/\lambda_x$ and $k_z = 2\pi/\lambda_z$ with $\lambda_x = \lambda_z = \lambda = 2\pi$, i.e. $k_x = k_z = k = 1$. The both amplitudes $A_x$ and $A_z$ were equal to 0.05. The axis $y$ is directed upward. This calculation of the three-dimensional RTI development was conducted with two grids $15 \times 105 \times 15$, and $30 \times 210 \times 30$.

The numerical computation stably proceeds till sufficiently remote times, up to the complete collapse of the jet and even to further moments. Given in Fig. 3.30 are the projections of the family of isochores (0.2, 0.55, 0.9), obtained for three characteristic longitudinal planes (the plane $x = 0$ and two diagonal cuts $x = -z$ and $x = z - \pi$; all of them are the symmetry planes), and for four transverse cuts (the size of each is $\pi \times \pi$), the totality of which permits to visualize the rather complicated form of a contact surface. The patterns demonstrated here correspond to the time $t = 8$. Shown at right is the pattern for finer grid. For the density ratio indicated above, the jets have the narrow elongated form, and their "mushroom-like" exterior is revealed quite weakly in comparison with two-dimensional case. The calculated curvature's radius of the bubble (the average value for the interval of time from 8 to 10) is equal to

$$R_b \approx 0.57\lambda \approx 3.6/k,$$

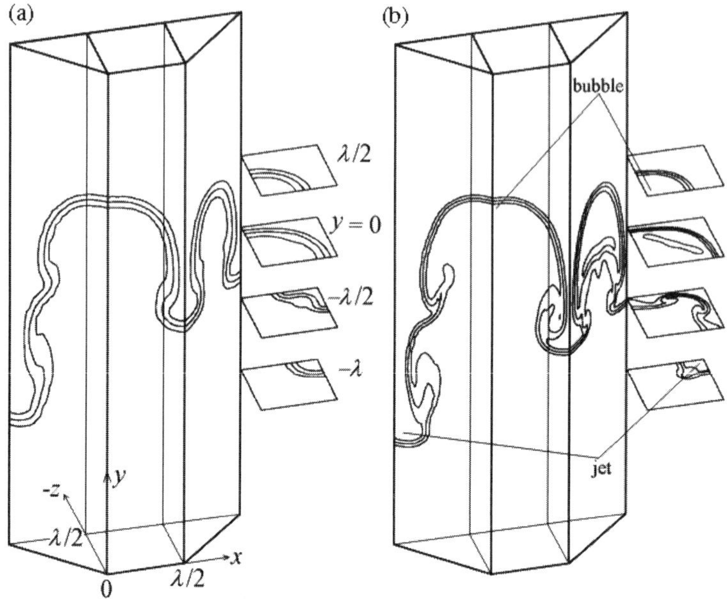


**Fig. 3.30.** Projections of the family of isochores (0.2, 0.55, 0.9) for three longitudinal planes and for four transversal cuts at the time $t = 8$, obtained for a single-mode three-dimensional problem of the RTI's development for grid: (a) $15\times105\times15$, (b) $30\times210\times30$.

which exceeds significantly the corresponding asymptotic value for two-dimensional case. The speed of bubble's ascent approaches at time $t = 8$ the constant value, which is equal to

$$v_b \approx 0.35\sqrt{(1-\mu)_g\lambda}.$$

The topology of the contact surface can be seen in detail in Fig. 3.31 for time moments $t = 5, 7, 9$ for the current case of RTI's development. The evolution of bubble is shown in the upper part of the figure and the evolution of jet is in the lower one.

In Fig. 3.32, one sees features of development of instability without gravity force, but with large initial perturbation. The time moments $t = 2, 8, 16$ are shown. The initial data for this run were the same as for above-mentioned case of RTI's development besides gravity force $g = 0$ and coefficients in (3.37) $A_x = A_z = 1$. The topology of the contact surface in current case differs somewhat from RTI's one.

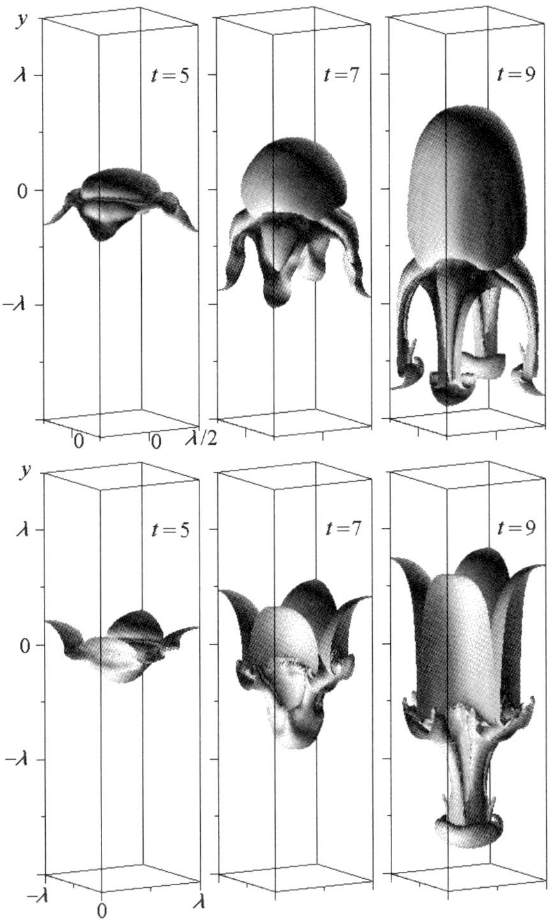


**Fig. 3.31.** Evolution of a bubble (above) and a jet (below) in the RTI's development. Numerical simulation performed for grid 30 × 210 × 30.

The calculated average value of the curvature's radius of the bubble for the interval of time from 16 to 20 is equal to

$$R_b \approx 0.45\lambda \approx 2.8k.$$

Some aspects of present numerical simulations (three-dimensional, single-mode perturbation) have to be studied more. Circumstantial comparison with latest three-dimensional analytical works, numerical simulations, and experiments[48–54] lies ahead of us. Here we give the only figure, which contains a comparison of calculated dependence of bubble ascent

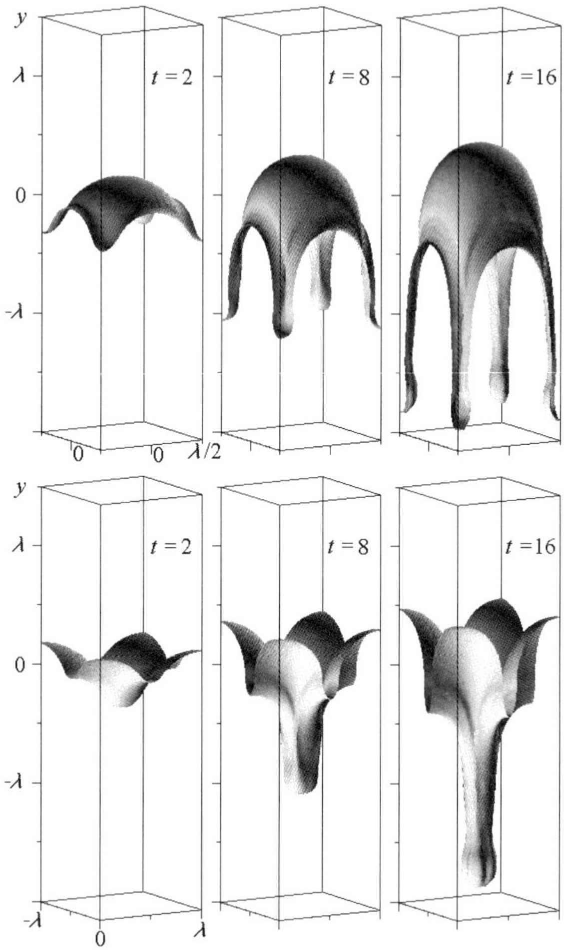


**Fig. 3.32.** Evolution of a bubble (above) and a jet (below) in the development of instability without gravity force, but with large initial perturbation. Numerical simulation performed for grid 30 × 210 × 30.

in the last case of instability development (without gravity force) with analytical solution obtained in the Layzer approximation.[19] Following by this approximation, Inogamov got the equation for the velocity of bubble's ascent in its apex $v(t)$

$$\frac{1}{v} - 1 + \frac{1}{2\sqrt{2}} \ln \left( \frac{\sqrt{2}+1}{\sqrt{2}-1} \cdot \frac{\sqrt{2}-v}{\sqrt{2}+v} \right) = t.$$

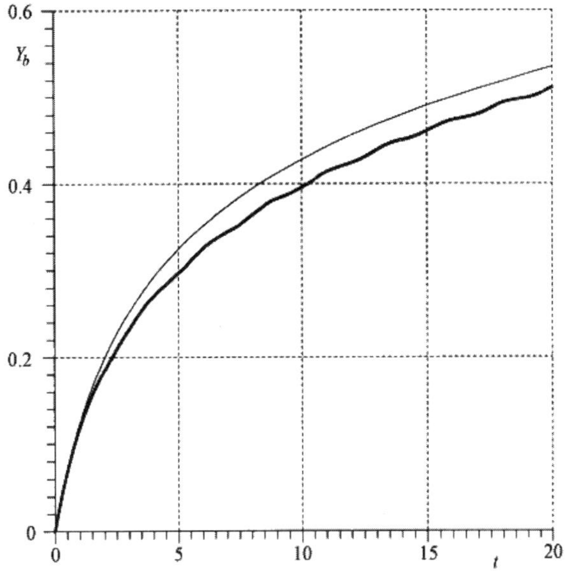


**Fig. 3.33.** Dependence on time of the vertical coordinate of the bubble's upper point in case of development of instability without gravity force, but with large initial perturbation. Thin line — theory and thick line — numerical simulation for grid $30 \times 210 \times 30$.

In Fig. 3.33, the thin curve was derived by integration of this dependence $v(t)$ and the thick line corresponds to our numerical simulation. One can see good agreement.

Let's go to the problem of evolution of three-dimensional structures from the initially two-dimensional perturbation. The area's dimensions for this problem are $2\pi \times 2\pi \times 4\pi$. In the present case, the small two-dimensional perturbation of a contact surface was prescribed, having the amplitude of 0.013. The contact boundary remains to be invariable along the $y$-axis, while along the $x$-axis it has a form of sinusoid (three periods). The contact boundary divides the area according to the ratio of 5 to 3. Within the larger lower part the light fluid was located, having the density $\rho_l = 0.5$, while the density within the upper part was $\rho_h = 1$. The density ratio for this case was $\mu = 0.5$, and Atwood number was $At = 0.333$. The initial pressure distribution was prescribed similarly to that of a single-mode problem. For the three dimensionality of flow to be emerged, one had to set a "noise" both within the light and heavy media, which means that all velocity components within each cell have had the random value, uniformly distributed along the interval $[-\varepsilon\colon \varepsilon]$. In the present example of calculation, the value $\varepsilon = 0.0005$

is negligibly small as compared with speed of sound. The adiabatic index was set to $\gamma = 5/3$. The computations were carried out with the grid $60 \times 60 \times 120$. At the side boundaries of the computational area are set the boundary conditions of periodicity, while at the upper and lower boundaries the nonflow through conditions are prescribed.

The calculations have shown that the development of a contact boundary remains to be two-dimensional for sufficiently long time, up to $t \approx 10$ (see Fig. 3.34(a)), but after that the accelerated formation of three-dimensional bubbles is initiated (see Fig. 3.34(b)), and these bubbles move much faster than two-dimensional ones. This fact is illustrated also at Fig. 3.35, where the graph of ascent of the upper boundary of bubbles is shown both for the present three-dimensional computation and for the corresponding two-dimensional one, which is carried out on a grid $60 \times 120$.

The last problem to be considered in this section concerns the growth of thickness of the three-dimensional turbulent mixing zone. The computational area had the sizes $2\pi \times 2\pi \times 2\pi$ and was divided into two parts by the contact boundary. Staying within the upper part was heavy fluid with density $\rho_h = 1$, while the lower part contained the light fluid with density $\rho_l = 0.5$, which means that in the present example the density ratio is $\mu = 0.5$ and the Atwood number is $At = 0.333$. The small initial three-dimensional perturbation of a contact surface was set in the form of random combination of the first $20 \times 20$ Fourier harmonics. The modes' amplitudes $a_{mn}$ were scaled in such a way that those would give $\left(\sum\sum a_{mn}^2\right)^{1/2} = 0.002\lambda$, where $\lambda = 2\pi$ is the area's size. The computations were carried out with a grid $100 \times 100 \times 100$. The boundary conditions are the same as for the previous problem considered.

The typical form of a contact surface at the sufficiently remote stage ($t = 10$) is shown at Fig. 3.36. As demonstrated by experiments,[16,55,56] in the case when instability develops itself due to the initial small random perturbation, the width of a zone of turbulent mixing grows in time proportionally to $gt^2$. Thus, for example, the height of ascent of the light fluid within the heavy one (see Refs. 16, 55, 56) is equal to

$$z = \alpha\, Atgt^2,$$

where the coefficient of turbulent mixing $\alpha$ is insensitive to the value of density ratio. Presented in Fig. 3.37 is the dependence of coordinate $z$ on the combination $At\, gt^2$. This dependence contains a considerable linear part with $\alpha = 0.073$, but this behavior reveals itself not from the very beginning of evolution. It is evident that certain temporal interval is needed for turning

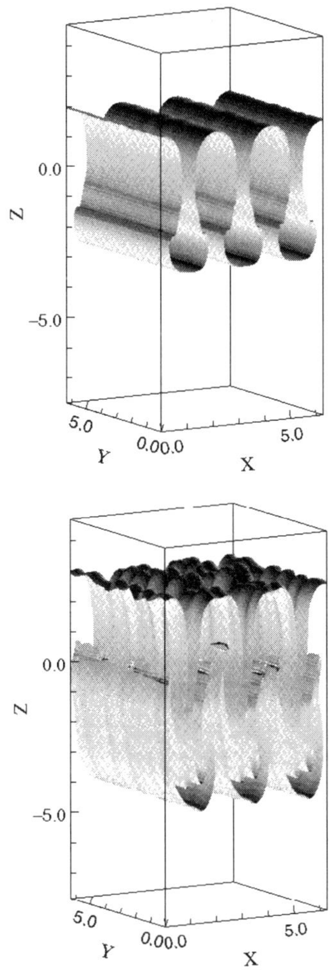


**Fig. 3.34.** The contact boundary in the problem of evolution of three-dimensional structures from the initially two-dimensional perturbation in RTI: for $t = 10$ (above) and after the formation of three-dimensional bubbles (below).

out to a self-similar regime with respect to $gt^2$. The reason for that is that given initially are the first $20 \times 20$ harmonics, and at the initial stage just high-frequency harmonics are the most fastly growing. Also presented in Fig. 3.37 is the similar dependence for the corresponding two-dimensional computation, which was carried out for the random perturbation consisted of 20 modes, on the grid $100 \times 100$. As it will be seen from comparison,

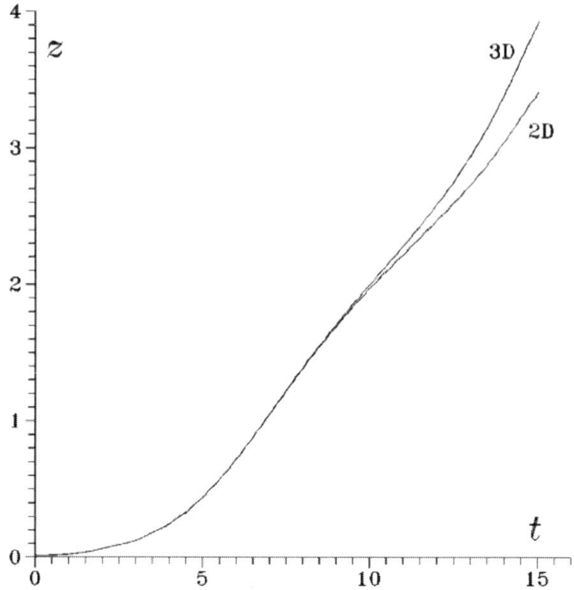


**Fig. 3.35.** The graph of ascent of the upper boundary of bubbles for three-dimensional computation and for corresponding two-dimensional one.

three-dimensional zone of turbulent mixing grows much more quickly than two-dimensional one. Moreover, as soon as one has carried out the two-dimensional computation on the grid $200 \times 200$, with 40 harmonics, one makes oneself certain that the coefficient of turbulent mixing $\alpha$ practically does not depend on the number of nodes in the grid. As a matter of fact, the difference is reduced just to that, that on more detailed grid the coming out to the linear (in coordinates $z$ and $At\,gt^2$) regime of evolution occurs somewhat earlier.

I am going to the final **discussion** of obtained numerical results. Let us note, that one has not specially used the front tracking, how it is frequently made for two-dimensional RTI problems.[15,59,60] Such a tracking for three-dimensional problems on RTI becomes extremely cumbersome and sufficiently expensive, especially for multi-mode problems. In the present case, the contact surface is implicitly controlled by the variations of density (concentration) at each of the grid's nodes.

The mathematical model presented here does not take into account neither real viscosity nor surface tension. Nevertheless, the structure of a scheme itself meeting the requirement of monotony provides a certain

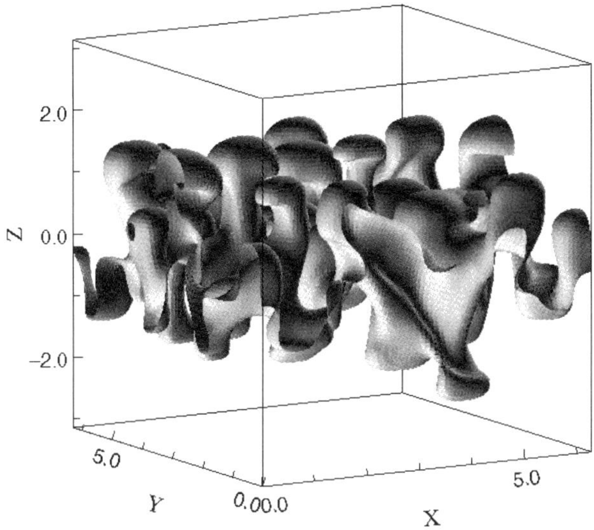


**Fig. 3.36.** The topology of contact boundary at the remote stage ($t = 10$) for the problem of growth of thickness of the three-dimensional turbulent mixing zone.

nonlinear dissipative mechanism, which leads to the damping of short-wave harmonics. In other words, the harmonics with the wavelength lesser than some effective one $\lambda_c$, are damped. This fact is confirmed by the presented calculations. It is evident that $\lambda_c$ is approximately equal to several steps of a computational finite-difference grid. In his paper on turbulent mixing[62] Youngs carries out the similar reasoning. Wishing to damp the short-wave harmonics and to provide the possibility for stable computation, Li[51–53] was compelled to introduce into his three-dimensional computations a certain finite viscosity.

Several three-dimensional problems of the RTI's progress were numerically studied presently. One problem concerned the analysis of development of a three-dimensional single-mode perturbation. The speed of a bubble's ascent, evaluated in computational experiment, is found to be in very good agreement with data of analytical works.[48–50] The calculated radius of curvature of a bubble and the same quantity, as it is expected from papers,[48–50] are also very close to each other $(4/k)$.

The numerical simulation of the next spatial problem has quite clearly shown that the two-dimensional perturbation is unstable, and even if the deviations from two dimensionality remain to be very small, the progress of instability will acquire a three-dimensional character sooner or later.

The last problem consisted in the study of instability produced by the three-dimensional random perturbation, and the numerical experiment permitted to measure the coefficient of turbulent mixing. One should bring attention to some peculiarities of that computational experiment. The first of those consists in the existence of a clearly pronounced portion of the development of the turbulent mixing zone, which has a self-similar character. The coefficient of turbulent mixing corresponding to that regime is considerably higher than the same coefficient resulting from two-dimensional formulation. Second, the calculations made for this problem on different two-dimensional grids show that the coefficient of turbulent mixing does not depend on the number of grid's nodes. This fact testifies to the authenticity of the results presented and to the adequacy of the computational experiment, which was set up here. And final and most important deduction consists in the fact that the measured coefficient of turbulent mixing proves to be in excellent agreement with experimental data.[16]

Thus, one can conclude that presented in Sec. 3.2.5 hydrocode is a worthwhile numerical tool that allows getting sufficiently accurate results in computational experiments on RTI for quite moderate grids.

## References

1. S. I. Anisimov, A. V. Chekhlov, A. Yu. Dem'yanov and N. A. Inogamov, The theory of Rayleigh–Taylor instability: modulatory perturbations and mushroom-flow dynamics, *Russ J. Comput. Mech.* **1**(2), 5–31 (1993).
2. A. M. Prokhorov, S. I. Anisimov and P. P. Pashinin, Laser-confinement fusion, *Uspekhi Fiz. Nauk.* **119**(3), 401–424 (1976) (in Russian).
3. V. B. Rozanov, I. G. Lebo, S. G. Zaitsev *et al.*, Experimental investigations of gravitational instability and turbulent mixing of stratified flows in acceleration field in connection with problems of inertial-confinement fusion, Preprint No. 56, Institute of Physics of Academy of Sciences, Moscow (1990) p. 63 (in Russian).
4. E. G. Gamaly, A. P. Favorsky, A. O. Fedyanin *et al.*, Nonlinear stage in the development of hydrodynamic instability in laser targets, *Laser Particle Beams* **8**(3), 399–407 (1990).
5. S. A. Grebenev and R. A. Syunyayev, Expected X-ray radiation from Supernova 1987A. Computations by Monte Carlo method, *Pis'ma v Astronom. Zh.* **13**(11), 945–963 (1987) (in Russian).
6. A. B. Bud'ko, A. L. Velikovich, M. A. Liberman *et al.*, Growth of Rayleigh–Taylor and volumetric convective instabilities in the dynamics of plasma liners and pinches, *Zh. Exper. i Teoret. Fiz.* **96**(1), 140–162 (1989) (in Russian).

7. R. A. Volkova, V. M. Goloviznin, F. R. Ulinich and A. P. Favorsky, Numerical simulation of compressing the magnetic field by cumulative liner, Preprint No. 111, Institute of Applied Mathematics, Moscow (1976), p. 66 (in Russian).
8. S. I. Anisimov, Ya. B. Zel'dovich, N. A. Inogamov and M. F. Ivanov, The Taylor instability of contact boundary between expanding detonation products and a surrounding gas, in *Shock Waves, Explosions and Detonation*, AIAA Progress in Astronautics and Aeronautics Series, Vol. 87 (AIAA, New York, 1983), pp. 218–227.
9. V. A. Andronov, S. M. Bakhrakh, E. E. Meshkov *et al.*, Turbulent mixing on contact surface accelerated by shock waves, *Zh. Exper. i Teoret. Fiz.* **71**(28), 806–811 (1976) (in Russian).
10. N. A. Inogamov, Turbulent stage of Taylor instability, *Soviet Tech. Phys. Lett.* **4**(6), 299–300 (1978).
11. E. E. Meshkov, Some results of experimental investigations of gravitational instability of interface separating heterodensity media, Collected Articles, Keldysh Institute of Applied Mathematics, USSR Academy of Sciences (1981) (in Russian).
12. V. A. Andronov, S. M. Bakhrakh, E. E. Meshkov *et al.*, Experimental investigations and numerical simulation of turbulent mixing in one-dimensional flows, *Dokl. Akad. Nauk SSSR.* **264**, 76–81 (1982) (in Russian).
13. V. E. Neuvazhayev, Properties of model of turbulent mixing of interface separating accelerated heterodensity liquids, *Zh. Prikl. Mekh. i Tekhn. Fiz.* (5), 81–88 (1983) (in Russian).
14. N. N. Anuchina and V. N. Ogibina, Numerical simulation of gravitational instability of interface between heterodensity media, *Physical Mechanics of Nonuniform Media*, Collected Articles, ed T. V. Gadiyaka, Institute of Theoretical and Applied Mechanics, USSR Academy of Sciences, Siberian Branch, Novosibirsk (1984) (in Russian).
15. D. L. Youngs, Numerical simulation of turbulent mixing by Rayleigh–Taylor instability, *Physica D* **12**, 32–44 (1984).
16. K. I. Read, Experimental investigation of turbulent mixing by Rayleigh–Taylor instability, *Physica D* **12**, 45–58 (1984).
17. V. V. Nikiforov, Turbulent mixing at contact boundary of heterodensity media, *Voprosy Atomnoi Nauki i Tekhniki* **1**, 3–10 (1985) (in Russian).
18. R. l. Ardasheva, S. l. Balabin, N. P. Voloshin *et al.*, Experimental studies of perturbation evolution in gravitationally unstable system with continuous density distribution, *Voprosy Atomnoi Nauki i Tekhniki.* (1), 20–27 (1988) (in Russian).
19. D. Layzer, The instability of superposed fluids in a gravitational field, *Astroph. J.* **122**(1), 1–12 (1955).
20. G. Birkhoff and D. Carter, Rising plane bubbles, *J. Math. Mech.* **6**(6), 769–779 (1957).
21. P. R. Garabedian, On steady state bubbles generated by Taylor instability, *Proc. Roy. Soc. London. Ser. A* **241**, 423–431 (1957).

22. N. A. Inogamov, On steady-state flow connected with Rayleigh–Taylor instability, *Problems of Plasma Dynamics and Stability*, Collected Articles, RSFSR State Committee for Science and Universities, Moscow Institute of Physics and Technology, Moscow (1990), pp. 115–124.
23. N. A. Inogamov, Higher-order Fourier approximations and exact algebraic solutions in the theory of the hydrodynamic RTI, *JETP Lett.* **55**(9), 521–525 (1992).
24. N. A. Inogamov and A. V. Chekhlov, Stationary solutions in the theory of RTI, Preprint, L. D. Landau Institute of Theoretical Physics (ITF) of Academy of Sciences, Chernogolovka (1992) p. 140 (in Russian).
25. N. A. Inogamov and A. V. Chekhlov, Multiplicity and uniqueness in the theory of RTI: a set of the theoretically possible solutions and the option between them, *Dokl. Akad. Nauk.* **328**(3), 39–42 (1993) (in Russian).
26. O. M. Belotserkovskii and Yu. M. Davydov, *Method of Large Particles in Gas Dynamics*, (Nauka, Moscow, 1982), pp. 392 (in Russian).
27. R. Temam, Navier–Stokes equations, *Theory and Numerical Analysis*, (Mir, Moscow, 1981), pp. 408 (Russian translation).
28. D. Kwak, J. L. C. Change, S. P. Shank and S. R. Chakravarty, A three-dimensional incompressible Navier–Stokes flow solver using primitive variables, *AIAA J.* (3), 390–396 (1986).
29. J. L. C. Chang and D. Kwak, On the method of pseudo-compressibility for numerically solving incompressible flows, *AIAA Paper* **84**(1), 252–260 (1984).
30. O. M. Belotserkovskii, V. A. Gushchin and V. N. Kon'shin, Decomposition method for investigating stratified liquid flows with free surface, *USSR Comput. Math. Math. Phys.* **27**(4), 181–191 (1987).
31. G. Birkhoff and E. Sarantonello, *Jets, Wakes, and Cavities*, (Mir Publishers, Moscow, 1964) pp. 466 (Russian translation).
32. O. M. Belotserkovskii, *Numerical Modeling on the Mechanics of Continuous Media*, 2nd edn. (Physmatlit, Moscow, 1994) (in Russian).
33. O. M. Belotserkovskii and Yu. M. Davydov, *Method of Large Particles in Gas Dynamics: Numerical Experiment* (Nauka, Moscow, 1982) (in Russian).
34. E. Fermi, Collection book, Vol. 2 (Nauka, Moscow, 1972) (in Russian).
35. G. Birkhoff, *Instability of Helmholtz and Taylor. Hydrodynamical Instability*, (Mir, Moscow), pp. 68–94 (in Russian).
36. N. A. Inogamov, Turbulent phase of the Rayleigh–Taylor instability, Preprint, L.D. Landau Institute for Theoretical Physics, Chernogolovka (1978).
37. Lord Rayleigh, *Theory of Sound*, Vol. 2 (Dover Publications Inc., New York, 1894).
38. G. Taylor, The instability of liquid surfaces when accelerated in a direction perpendicular to their planes, I, *Proc. Roy. Soc. Sec. A* **201**(1065), 192–196 (1950).
39. D. J. Lewis, The instability of liquid surfaces when accelerated in a direction perpendicular to their planes, II, *Proc. Roy. Soc.* **202**(1068), 81–96 (1950).
40. G. Birkhoff, Los Alamos Scientific Lab. Rep. No. LA–1862, Los-Alamos (1955).

41. S. Z. Belenkii and E. S., Fradkin, The theory of turbulent mixing, Works of Physical Institute of Academy of Sciences, **29**, 743–747 (1965).
42. F. H. Harlow and J. E. Welch, Numerical study of large amplitude free surface motion, *Phys. Fluids* **9**(5), 842–851 (1966).
43. B. J. Daly, Numerical study of two-fluid Rayleigh–Taylor instability, *Phys. Fluids* **10**(2), 297–307 (1967).
44. Yu. M. Davydov, Numerical investigation of Taylor instability in non-linear approximation, *Numerical Method in Mechanics of Continuous Media*, (Novosibirsk, Nauka, 1978), Vol. 9, No. 3, pp. 67–69 (in Russian).
45. O. M. Belotserkovskii, Yu. M. Davydov and A. Yu. Dem'yanov, Intermode interaction in the case of the Rayleigh–Taylor instability, *Dokl. Akad. Nauk SSSR.* **288**(5), 1071–1074 (1986) (in Russian).
46. Yu. M. Davydov, A. Yu. Dem'yanov and G. A. Zvetkov, Numerical simulation of harmonics stabilization and unification in Rayleigh–Taylor instability by method of large particles, Preprint Computer Center of Academy of Science USSR, Moscow (1987) (in Russian).
47. Yu. M. Davydov and M. S. Panteleev, Evolution of three-dimensional perturbations in Rayleigh–Taylor instability, *Zh. Prikl. Mekh. i Tekhn. Fiz.* (1), 117–122 (1981) (in Russian).
48. N. A. Inogamov and S. I. Abarzhi, Dynamics of fluid surface in multi-dimension, *Physica D* **87**, 339–341 (1995).
49. S. I. Abarzhi and N. A. Inogamov, Steady state solutions of three-dimensional Rayleigh–Taylor instability, *JETP* **80**, 132–143 (1995).
50. S. I. Abarzhi, Stable steady flows in Rayleigh–Taylor instability, *Phys. Rev. Lett.* **89**, 1332 (1998).
51. X. L. Li, Study of three-dimensional Rayleigh–Taylor instability in incompressible fluids through level set method and parallel computation, *Phys. Fluids A* **5**, 1904–1913 (1993).
52. X. L. Li, A numerical study of three-dimensional bubble merger in the Rayleigh–Taylor instability, *Phys. Fluids* **8**, 336–343 (1996).
53. X. L. Li, B. X. Jin and J. Glimm, Numerical study for the three-dimensional Rayleigh–Taylor instability through the TVD/AC scheme and parallel computation, *J. Comput. Phys.* **126**, 343–355 (1996).
54. G. Jourdan and L. Houas (eds.), *Proc. 6th Int. Workshop on the Physics of Compressible Turbulent Mixing*, Marseille, France, 1997.
55. P. Linden, J. Redondo and D. Youngs, Molecular mixing in Rayleigh–Taylor instability, *J. Fluid Mech.* **265**, 97 (1994).
56. M. M. Marinak, S. G. Glendinning, *et al.*, Non-linear evolution of a three-dimensional multimode perturbation, *Phys. Rev. Lett.* **80**, 4426 (1998).
57. R. Menikoff and C. Zemach, Rayleigh–Taylor, instability and the use of conformal maps for ideal fluid flow, *J. Comput. Phys.* **51**, 28 (1983).
58. G. R. Baker, D. I. Meiron and S. A. Orszag, Vortex simulations of the Rayleigh–Taylor instability, *Phys. Fluids* **23**, 1485–1490 (1980).
59. G. Gardner, J. Glimm, *et al.*, The dynamics of bubble growth for Rayleigh–Taylor instability, *Phys. Fluids* **31**, 447–465 (1988).

60. J. Glimm, X. L. Li, R. Menikoff, D. H. Sharp and Q. Zhang, A numerical study of bubble interactions in Rayleigh–Taylor instability for compressible fluids, *Phys. Fluids* **2**, 2046–2054 (1990).
61. T. Yabe, H. Hoshino and T. Tsuchiya, Two- and three-dimensional behavior of Rayleigh–Taylor and Kelvin–Helmholtz instabilities, *Phys. Rev. A* **44**, 2756–2758 (1991).
62. D. L. Youngs, Three-dimensional numerical simulation of turbulent mixing by Rayleigh–Taylor instability, *Phys. Fluids A* **3**, 1312–1320 (1991).
63. K. M. Magomedov and A. S. Kholodov, On construction of differences schemes for the equations of hyperbolic type on the basis of characteristic relationships, *USSR Comput. Math. Math. Phys.* **9**(2), 158–176 (1969).
64. O. M. Belotserkovskii, *Numerical Experiment in Turbulence: From Order to Chaos*, (Nauka, Moscow, 1994) (in Russian).

# Chapter 4

# Direct Statistical Approach for Aerohydrodynamic Problems

The chapter deals with the statistical particle-in-cell method,[1] widely used in rarefied gas dynamics as an effective numerical method for solving problems based on the Boltzmann equation. The method concerned has a number of variations. In many countries Bird's schemes are applied.[2] Problems in aerodynamics, vacuum technologies, and other fields are solved in one-dimensional, two-dimensional, stationary and nonstationary formulations.

When speaking of one- or two-dimensional problems, we mean a space of physical coordinates $\{X, Y, Z\}$. The advantage of the method involved is that the space of molecular velocities is regarded here as a three-dimensional one. The space of pulsations in a turbulent flow is known to be always three-dimensional, and it was interesting to examine the possibility of using the statistical particles-in-cell method to simulate turbulence. Yanitskii was the first to carry out such numerical experiments, this chapter being oriented to them.

The first section describes the main principles of the method for the problems in rarefied gas dynamics, to which it was originally intended. Sections 4.2 and 4.3 contain the results of using the statistical approach for turbulence problems in different formulations.

## 4.1. Statistical modeling in rarefied gas-dynamics

### 4.1.1. Introduction

The emergence of powerful computers stimulated the development of numerical methods for solving problems in the kinetic theory of gases, formulated on the basis of the Boltzmann equation. Suppose that the flow of a Boltzmann gas consisting of spherically symmetric molecules is considered in a domain $\Omega$ with a boundary $\tilde{\mathrm{A}}$. Each molecule, whose position and velocity are denoted by $r$ and $c$, collides with other molecules and with the boundary $\tilde{\mathrm{A}}$. The laws of elastic collisions of molecules labeled $i$ and $j$

are given by the differential ($d\sigma_{ij}$) and total ($\sigma_{ij}$) cross-sections, $\sigma = \int d\sigma$. Let $f = f(t, x, c)$ be the concentration of molecules at the instant $t$ in the vicinity of the point $(x, c)$ in the six-dimensional phase space. The Boltzmann equation has the following form:

$$\frac{\partial f}{\partial t} + c\frac{\partial f}{\partial x} + \frac{G}{m}\frac{\partial f}{\partial c} = \int (f' f_1' - f f_1) g \, d\sigma \, dc_1, \tag{4.1}$$

where $G$ is the external force acting on a molecule, $m$ is the mass of the molecule, $f_1 = f(t, x, c_1)$, $g = |c - c_1|$, and the prime over $f$ indicates that the values of $f$ and $f_1$ are computed from the velocities $c'$ and $c_1'$ of the molecules after a collision between them. The problem involves the solution of Eq. (4.1) satisfying a given initial data and the boundary conditions

$$f(t, x_\Gamma, c) = \int K(c, c_1) f(t, x_\Gamma, c_1) dc_1.$$

Monte-Carlo methods are found to be the most effective for a numerical solution of this problem. Various approaches to the construction of such methods have been described in Refs. 1–5. The approach considered in Refs. 1–3 is based on the imitation of the Boltzmann gas by a system of a finite number $N$ of particles ($N = 10^3$–$10^4$).

We shall consider one of the possible models of a gas of particles,[1,3] whose realization makes it possible to compute the kinetics of spatially inhomogeneous gas flows. A fine grid is introduced in the physical space $\Omega$, dividing it into cells of volume $V$. The continuous time $t$ is replaced by the discrete time $t_\alpha = \alpha \Delta t$. At the initial moment of time, the space $\Omega$ is filled by particles whose coordinates $r_i$ and velocities $c_i$ are drawn by Monte-Carlo methods in accordance with the densities $n_0(x) = \int f_0(x, c) dc$ and $\varphi_0(c/x) = f_0(x, c)/n_0(x)$ of the initial coordinate and velocity distributions. The computation of the evolution of this system of particles over a small time step $\Delta t$ is divided into the following two, physically meaningful stages. At the first stage, binary collisions of particles are simulated in cells, only velocities being changed, while at the second stage the free flight of particles from one cell to another and their interaction with the boundary $\Gamma$ are simulated.

Suppose that at the instant of time $t_\alpha (\alpha = 0, 1, \ldots)$, the cell with its center at $x_j (j = 1, 2, \ldots, J)$ contains $N(\alpha, j)$ particles with velocities $c_1, \ldots, c_{N(\alpha,j)}$. The state $\{N(\alpha); C(\alpha)\}$ can be defined if a sequence of $J$ points having the form $\{N(\alpha, j); c_1, \ldots, c_{N(\alpha,j)}\}$ is known. Alternation of the above-mentioned stages of collisions and displacements of particles gives the trajectory $\{N(\alpha), C(\alpha)\}; \alpha = 0, 1, \ldots$. Once the set of realized trajectories has been defined, any macroparameter $\Psi$ of the gas can be computed at the instant $t_\alpha$ at the point $x_j$ by using the consistent estimate[1] of the appropriate integrals:

$$\Psi(t,x) = \frac{1}{n(t,x)} \int \psi(c) f dc, \quad n(t,x) = \int f dc.$$

Such a method was first realized by Bird.[2] The main features of his method lie in the collision-modeling schemes and in the approach to the construction of such schemes. The algorithms described in Bird's work were used to compute the structure of a shock wave and heat transfer in a rarefied gas between two planes. The results were compared to the direct numerical solution of the Boltzmann equation[1,3] and were found to be in good agreement with the latter. A refinement of the method made it possible to compute flows past bodies with two- and three-dimensional geometry.[6]

The introduction of weight factors in the collision-modeling schemes made it possible to extend the method for studying the kinetics of neutral and reactive gas mixtures as well.[7–10]

The theoretical aspects of the method considered below are important for understanding its connection with the Boltzmann equation which is of a prime importance in the dynamics of rarefied gases, and also for a further refinement of this numerical technique.

### *4.1.2. Stochastic analogue of the Boltzmann equation*

The most comprehensive statistical description of a system of $N$ identical molecules, which occupies a volume $\Omega$ with a boundary $\Gamma$, is based on the Liouville equation.[11] Let $F_N(t, R, C)$ denote the probability distribution density for various positions of the point $(R, C) = \{r_1, c_1, \ldots, r_N, c_N\}$ in the $6N$-space. Molecules with mass $m$ are assumed to be the force-field centers and $\zeta_{ij}$ denotes the potential of molecular interaction between pairs of particles $i$ and $j$. The external field force acting on the molecule $i$, including that exerted by the boundary $\Gamma$, is given by $G_i$. The Liouville equation for such a system has the form

$$\frac{\partial F_N}{\partial t} + L_N F_N = 0. \tag{4.2}$$

Within a factor equal to the imaginary unit, the operator

$$L_N = \sum_{i=1}^{N} \left( c_i \frac{\partial}{\partial r_i} + \frac{G_i}{m} \frac{\partial}{\partial c_i} \right) - \sum_{1 \le i \le j \le N} \Theta_{ij}, \tag{4.3}$$

is the Liouville operator, while $\Theta_{ij}$ is the molecular interaction operator defined by the formula

$$\Theta_{ij} = \frac{1}{m} \left( \frac{\partial \pi_{ij}}{\partial r_i} \frac{\partial}{\partial c_i} + \frac{\partial \pi_{ij}}{\partial r_j} \frac{\partial}{\partial c_j} \right).$$

The numerical model of the Boltzmann gas of $N$ particles is constructed by proceeding from Eq. (4.2). At a certain stage in the analysis, the prerequisites of the Boltzmann gas are introduced into the structure model. Suppose that the solution $F_N(t, R, C)$ of the Liouville equation (4.2) is known at the instant $t$. The numerical solution of this equation at the instant $t + \Delta t$ can be obtained according to the following splitting scheme:

Stage I

$$\frac{\partial \Phi_N}{\partial \tau} = \left[ \sum_{1 \le i \le j \le N} \Theta_{ij} \right] \Phi_N,$$

$$t < \tau \le t + \Delta t, \quad \Phi_N^*(t + 0, R, C) = \Phi_N(t + 0, R, C). \tag{4.4}$$

Stage II

$$\frac{\partial \Phi_N^*}{\partial \tau} + \sum_{i=1}^{N} \left( c_i \frac{\partial}{\partial r_i} + \frac{G_i}{m} \frac{\partial}{\partial c_i} \right) \Phi_N^* = 0,$$

$$t < \tau \le t + \Delta t, \quad \Phi_N^*(t + 0, R, C) = \Phi_N(t + \Delta t, R, C). \tag{4.5}$$

Such a scheme is known to give an approximate solution of transport equations of the type (4.2):

$$F_N(t + \Delta t, R, C) = \Phi_N^*(t + \Delta t, R, C) + \mathrm{O}(\Delta t^2).$$

By realizing the collisions of pairs of particles in exact accordance with the operator $\Theta_{ij}$, one can carry out a direct modeling, instead of a successive integration of Eqs. (4.4) and (4.5), according to the basic scheme described in Sec. 4.1.1. However, such a procedure would reconstruct the situation inherent in elementary processes typical for a real gas, while the collision process in the Boltzmann gas has a more imitative character.

Let us consider the peculiarities of the Boltzmann model. Passing to the $s$-particle probability distribution densities $F$, for low values of $s$

equal to 1 and 2 attains the simplification of statistical description associated with this model:

$$F_S(t, r_1, c_1, \dots, r_s, c_s) = \int F_N(t, R, C) \prod_{i=s+1}^{N} dr_i dc_i.$$

Applying the operation $\int (\cdots) dr_s + 1 dc_{s+1} \cdots dr_N dc_N$ to Eq. (4.2) and making well-known transformations,[11] we obtain the BBGKY (Bogolyubov–Born–Green–Kirkwood–Yvon) chain of equations

$$\frac{\partial F_s}{\partial t} + L_s F_s - (N - s) \sum_{i=1}^{s} \int \Theta_{i,s+1} F_{s+1} dr_{s+1} dc_{s+1} = 0.$$

The last link in this chain corresponds to $s = 1$ and has the form

$$\frac{\partial F_1}{\partial t} + c_1 \frac{\partial F_1}{\partial r_1} + \frac{G_1}{m} \frac{\partial F_1}{\partial c_1} - (N - 1) \int \Theta_{12} F_2 dr_2 dc_2 = 0. \tag{4.6}$$

This relation is reduced to the Boltzmann equation (4.1) under certain additional assumptions, which will be formulated below.

Rewriting the Boltzmann equation in a form analogous to (4.6) and using the relation $f = NF_1$ between the kinetic distribution function and the density $F_1$, one obtains

$$\frac{\partial F_1}{\partial t} + c_1 \frac{\partial F_1}{\partial r_1} + \frac{G_1}{m} \frac{\partial F_1}{\partial c_1} - N \int B_{12} F_2^* dc_2 = 0. \tag{4.7}$$

Here,

$$F_2^* = F_1(t, r_1, c_1) F_1(t, r_1, c_2),$$
$$B_{ij} F_2 = g_{ij} \sigma_{ij} (T_{ij} - I) F_2. \tag{4.8}$$

Operators $T_{ij}$ and $I$ are defined as follows:

$$IF \equiv F,$$
$$T_{ij} F \equiv \int F(c_i', c_j') \frac{d\sigma_{ij}}{\sigma_{ij}}. \tag{4.9}$$

As usual, the quantities $c_i'$ and $c_j'$ correspond to the velocities of particles $i$ and $j$ after their collision. A comparison of the collision terms in formulae (4.6) and (4.7) makes it possible to formulate the main hypotheses, which distinguish the Boltzmann gas from a real one. The Boltzmann collision operator acts on $F_2$ only as on a function of velocities, and the coordinates of the particles are assumed to be identical: $r_1 = r_2$. Thus, a collision in the Boltzmann model is an act of random change in velocity

$(c_i, c_j) \to (c_i', c_j')$ localized at a point. Segments of particle trajectories on the scale $r_0$ of the molecular interaction radius are neglected and hence the collision process is assumed to be instantaneous. It follows from Eq. (4.8) that the function $F_2(t, r_1, c_1, r_2 = r_1, c_2)$ can be presented in the form of the product $F_1(t, r_1, c_1)F_1(t, r_1, c_2)$. This assumption, called the "local hypothesis of molecular chaos", is usually satisfied under the assumption $N \gg 1$. Thus, in order to make the direct modeling scheme with a splitting of elementary processes according to the physical principle applicable to the Boltzmann gas, the basic equation (4.4) of stage I must be replaced by some approximation taking into account the prerequisites of the Boltzmann model.

The total number $N$ of particles in a model gas must be quite large though finite ($10^3$–$10^4$) for the condition $N \gg 1$ to be satisfied. Hence, the idea of a strict fulfillment of the hypothesis of molecular chaos must be rejected. Numerical experiments show that when simulating a rarefied gas flow with the Boltzmann collision statistics, the effect of disturbing the molecular chaos is not pronounced, at least in fluid-dynamical parameters.[1,3]

The peculiarities of the Boltzmann statistics (localization) of collision can be taken into account in the model of particles in cells, when the physical space $\Omega$ is divided into cells of a small volume $V$. The collisions are assumed to be possible only between particles in the same cell, the probability and consequence of the collision being independent of the position of particles in a cell and depending only on their velocity. In this respect, each subsystem of $N$ particles in a cell with velocities $c_1, \ldots, c_N$ must be treated as spatially homogeneous. For a statistical description of such a system at the collision modeling stage, it is sufficient to know the probability distribution density $\varphi_N(\tau, c_1, \ldots, c_N)$ for the velocities of these $N$ particles in the $3N$-space. Thus, in the numerical model of the Boltzmann gas of $N$ particles based on the splitting principle, Eq. (4.4) in the collision stage is replaced by a series of independent equations in the form

$$\frac{\partial \varphi_N}{\partial \tau} = \sum_{1 \le i \le j \le N} \bar{\Theta}_{ij} \varphi_N, \tag{4.10}$$

one equation for each cell. Obviously, the difference between the operator $\bar{\Theta}_{ij}$ and Boltzmann operator $B_{ij}$ may be associated only with the difference in the normalization of functions $\varphi_s(t, c_1, \ldots, c_s)$ and $f_s(t, r_1, c_1, \ldots, r_s, c_s)$ (or $F_s(t, r_1, c_1, \ldots, r_s, c_s)$). In the spatially homogeneous case and under the validity of the molecular chaos hypothesis, one should have $f_s = n^s \varphi_s$.

In our model, the role of the concentration $n$ is played by the ratio $\langle N\rangle/V$, where $\langle N\rangle$ is the number of particles in a given cell averaged over the ensemble of realizations. In the operator form, the Boltzmann equation is analogous to (4.7) without convective derivatives and in terms of the densities $\varphi$:

$$\frac{\partial\varphi_1}{\partial\tau} = \frac{\langle N\rangle}{V}\int B_{12}\varphi_2 dc_2, \quad \varphi_2 = \varphi_1(t,c_1)\varphi_1(t,c_2).$$

A comparison with (4.6) shows that we must put $\bar{\Theta}_{12} = B_{12}/V$. In this case, the basic equation of the collision relaxation process in a cell can be presented in the form

$$\frac{\partial\varphi_N}{\partial\tau} = \sum_{1\le i\le j\le N}[\omega_{ij}(T_{ij}-I)]\varphi_N, \quad \omega_{ij} = \frac{g_{ij}\sigma_{ij}}{V} \tag{4.11}$$

or

$$\frac{\partial\varphi_N(\tau,C)}{\partial\tau} = \frac{1}{V}\sum_{1\le i\le j\le N} g_{ij}\int[\varphi_N(\tau,C'_{ij}) - \varphi_N(\tau,C')]d\sigma_{ij}, \tag{4.12}$$

where

$$C'_{ij} = \{c_1,\ldots,c_{i-1},c'_i,c_{i+1},\ldots,c_{j-1},c'_j,c_{j+1},\ldots,c_N\}.$$

Equation (4.12) was considered earlier in Refs. 12 and 14, in connection with the stochastic treatment of the Boltzmann equation and the investigation of the molecular chaos hypothesis.

The above model of particles in a cell is stochastic. At any instant $t$, its state $\{r_1(t),c_1(t),\ldots,r_N(t),c_N(t)\}$ cannot be defined uniquely from the state at the preceding instant $t-\Delta t$, for an arbitrarily small interval $\Delta t$. The trajectory of the process in the $6N$-space is not reversible, contrary to the trajectory in the $N$-particle model of a real gas. The model of the Boltzmann gas becomes stochastic owing to the prerequisites of local homogeneity (spatial homogeneity within the cell volume $V$).

### *4.1.3. Probabilistic approach to the basic equation of the collision stage*

The methods of direct statistical modeling of the collision process (computational stage I over a step $\Delta t$) are based on the probabilistic approach to Eq. (4.12). The random process involving the basic equation (4.12) can be constructed by using the following definition and postulates.

Let $\{c_1(\tau), \ldots, c_N(\tau)\} = C(\tau)$ be a $3N$-vector whose components are the velocities of $N$ particles in a homogeneous cell volume $V$.

**Definition 4.1.** A collision is an instantaneous random transition of the system $\{c_1, \ldots, c_N\}$ from the state $C$ to the state $C'$, satisfying the following properties:

(i) the collision may change the components of only one pair[a] of vectors $(c_i, c_j)$; this pair is referred to as that undergoing the collision;
(ii) the new components of the vectors $(c'_i, c'_j)$ for the pair that has undergone a collision are random quantities, but such that $G_{ij} = \frac{1}{2}(c_i + c_j)$ and $g_{ij} = |c_i - c_j|$ do not change as a result of the collisions;
(iii) the probability that the new vector $g'_{ij} = g_{ij}e$ proves to be within the solid angle $de$ after the collision of the pair $(c_i, c_j)$ is

$$B(g_{ij}, \theta)de = \frac{g_{ij}\sigma(g_{ij}, \theta)}{g_{ij}\sigma(g_{ij})} de,$$
$$e = \{\sin\theta\cos\chi,\ \sin\theta\sin\chi,\ \cos\theta\}, \tag{4.13}$$

where the given function $d\sigma = \sigma(g, \theta)de$ is called the differential elastic cross-section and satisfies the condition

$$0 < \int_{4\pi} g\sigma(g, \theta)de = g\sigma(g) < +\infty, 0 \leq g \leq +\infty.$$

*Postulates.* The evolution of a $3N$-point $C(\tau)$ is defined by the sequence of collisions separated by random intervals of time $T$, and

(1) the probability that a pair $(c_i, c_j)$ of particles collides at the instant $\tau$ (given that one of the pairs does collide at the instant $\tau$) is

$$W_{ij} = \frac{\omega_{ij}}{\lambda}, \tag{4.14}$$

where

$$\lambda = \sum_{1 \leq i \leq j \leq N} \omega_{ij};$$

[a]In the above definition, a "pair" means an unordered set of elements $c_i$ and $c_j$, i.e., $(c_i, c_j)$ and $(c_j, c_i)$ represent the same pair, and the number $k = N(N-1)/2$ of pairs is equal to the number of combinations of $N$ elements taken in pairs at a time.

(2) the waiting time of the next collision has the distribution[b]

$$F(\tau) = P\{T \leq \tau\} = 1 - \exp(-\lambda\tau), \tag{4.15}$$

which is independent of the choice of the zero time reference, as well as of the pair $(c_i, c_j)$ participating in the collision, and is defined by the state $C$ of the system as a whole before the collision.

The evolution of the state $C(\tau)$ as the model described above is a uniform jump-like strictly Markovian process concentrated at the hypersphere of constant energies $E$ and momentum $P$ of the entire system:

$$E(C) = \sum_{i=1}^{N} c_i^2, \quad P(C) = \sum_{i=1}^{N} c_i.$$

The second postulate defines the necessary condition for a process to be strictly Markovian.

It can be proved that the basic equation of the model defined in this way coincides with Eq. (4.12). For deriving the basic equation, it should be noted that the state of a model at the instant $\tau + d\tau$ lies in the vicinity of the given point $C$ only as a result of the following two incompatible events:

(a) at the instant $\tau$, the state of the model was in the vicinity of the point C and not a single collision occurred during the time $d\tau$;
(b) at the instant $\tau$, the state of the model was in the vicinity of one of the points $C'_{ij} (1 \leq i \leq j \leq N)$:

$$C'_{ij} = \{c_1, \ldots, c_{i-1}, c'_i, c_{i+1}, \ldots, c_{j-1}, c'_j, c_{j+1}, \ldots, c_N\},$$

whose coordinates $(c'_i, c'_j)$ form a pair of the possible velocities acquired by the particles $c_i$ and $c_j$ after their collision, and the "reverse" collision $(c'_i, c'_j) \to (c_i, c_j)$ takes place during the time $d\tau$.

The total probability that the state of the model is found in the vicinity of the point $C$ at the instant $\tau + d\tau$ is the sum of the probabilities of the events (a) and (b). It follows from (4.15) that the probability of a collision in the time interval $d\tau$ is equal to $\lambda d\tau$. The probability that a pair $(c_i, c_j)$ undergoes a collision is $\omega_{ij}/\lambda$. Thus, the collision probability for a given pair in time $d\tau$ is

$$P_{ij} = \frac{\omega_{ij}}{\lambda}\lambda d\tau = \omega_{ij} d\tau = \frac{d\tau}{V} g_{ij}\sigma(g_{ij}).$$

[b]Henceforth, the probability of an event $A$ will be denoted by $P\{A\}$.

The probability of event (a) is

$$\left[1 - \sum_{1\le i\le j\le N} P_{ij}\right]\varphi_N(\tau, C) = \left[1 - \frac{d\tau}{V}\sum_{1\le i\le j\le N} g_{ij}\sigma(g_{ij})\right]\varphi_N(\tau, C). \tag{4.16}$$

The quantity $dP_{ij} = d\sigma(g_{ij},\theta)/\sigma(g_{ij})$ describes the probability that the direction of the relative velocity $g'_{ij}$ lies within the solid angle $de$ covering a given direction $e$. For elastic collisions, $g_{ij} = g'_{ij}$, consequently, $\omega_{ij} = \omega'_{ij}$ and $dP_{ij}$ assumes the form $dP_{ij} = g_{ij}d\sigma(g_{ij},\theta)/(V\omega'_{ij})$. Taking this distribution into account, the probability of event (b) is given by

$$\sum_{1\le i<j\le N} d\tau\omega'_{ij}\int \varphi_N(\tau, C'_{ij})dP_{ij} = \frac{d\tau}{V}\sum_{1\le i<j\le N} g_{ij}\int \varphi_N(\tau, C'_{ij})d\sigma_{ij}. \tag{4.17}$$

The density $\varphi_N(\tau + d\tau, C)$ is equal to the sum of the expressions (4.16) and (4.17). Consequently,

$$\begin{aligned}&\varphi_N(\tau + d\tau, C)\\&\quad = \varphi_N(\tau, C) + \frac{d\tau}{V}\sum_{1\le i<j\le N} g_{ij}\left[\int \varphi_N(\tau, C'_{ij})d\sigma_{ij} - \sigma(g_{ij})\varphi_N(\tau, C)\right].\end{aligned}$$

Transposing the term $\varphi_N(\tau, C)$ to the left-hand side, dividing the entire equality by $d\tau$, and using the definition of the total collision cross-section $\sigma(g)$

$$\sigma(g_{ij}) = \int d\sigma_{ij},$$

we arrive at Eq. (4.12).

### *4.1.4. Algorithms for modeling the collision relaxation*

The definition and postulates used for formulating the model described in Sec. 4.1.3. lead to the following algorithm for computing the state $C(t+\Delta t)$ from the known state $C(t)$.

Let $T_r$ be the time between the collisions with numbers $r$ and $r-1$. We introduce the quantity

$$S_n = \sum_{r=1}^{n} T_r(S_0 = 0),$$

as the total waiting time for the collision with number $n$. The terms $T_r$ are random quantities distributed according to the exponential law

$$P\{T_r \le \delta t\} = 1 - \exp(-\lambda_r \delta t). \tag{4.18}$$

Here, $\lambda_r$ is the value of $\lambda$ before the collision with number $r$. Collisions of particles can be simulated as follows.

Step 1. A random time $T_n$ with distribution (4.18) is drawn, and the quantity $S_n = S_{n-1} + T_n$ is calculated. A collision is assumed to have occurred if $S_n \le \Delta t$, otherwise the process of computation of collisions in the subsystem $\{c_1, \ldots, c_N\}$ is terminated.

Step 2. A pair of particles with velocities $c_i$ and $c_j$ participating in the collision is chosen in accordance with the conditional collision probability

$$W_{ij} = \frac{g_{ij}\sigma(g_{ij})}{\sum_{1\le i\le j\le N} g_{ij}\sigma(g_{ij})}.$$

Step 3. The velocities after the collision are computed by the formulae

$$c_i' = \frac{1}{2}[(c_i + c_j) + g_{ij}e], \quad c_j' = \frac{1}{2}[(c_i + c_j) - g_{ij}e], \quad g_{ij} = |c_i - c_j|.$$

The unit vector $e$ is generated as a random vector distributed in all directions with a probability density $B(g_{ij},\theta)$. After step 3, computations proceed to step 1 for the $(n+1)$th collision.

The above algorithm was first presented in Ref. 15 in the form equivalent to that presented here. Though cumbersome, this algorithm correctly reflects the collisional statistics in a perfect gas over any time interval $\Delta t$, the total number of operations required for an exact simulation of collisions is proportional to $N^3$, and real problems can hardly be solved by using this algorithm.

The only particular case in which the above algorithm turns out to be the simplest and fastest is the modeling of collision relaxation in a subsystem of $N$ "pseudo-Maxwellian" particles. The elastic cross-section $\sigma\ (g)$ of pseudo-Maxwellian molecules is inversely proportional to $g$. Hence, $\omega = g\sigma/V$ is independent of $g$ and the frequency $\lambda = N(N-1)\omega/2 = \text{const.}$ may be computed beforehand just once for all the collisions. The number of collisions $s(\Delta t)$, defined in the general case by the inequality

$$\sum_{r=1}^{s} T_r \le \Delta t < \sum_{r=1}^{s+1} T_r,$$

given that $\lambda = \text{const.}$, is distributed according to the Poisson law:

$$P\{s(\Delta t) = n\} = \frac{(\lambda \Delta t)^n}{n!} \exp(-\lambda \Delta t). \tag{4.19}$$

The number $s$ of colliding pairs of particles is chosen with equal probabilities from the subsystem $\{c_1, \ldots, c_N\}$ and their velocities are recalculated.[16]

The density $\varphi_N^{(1)}(C)$ after the first collision in the system of $N$ pseudo-Maxwellian particles is equal to $T_{\varphi_N}(t, .)$, where $T$ is the operator

$$T = \frac{2}{N(N-1)} \sum_{1 \leq i \leq j \leq N} T_{ij}.$$

After the collision with number $s$, the velocity distribution will have the density

$$\varphi_N^{(S)}(.) = T^s \varphi_N(t, .).$$

The number $s(\Delta t)$ of the collisions over the time interval $(t, t + \Delta t)$ is distributed according to the law (4.19), hence $\varphi_N(t + \Delta t, .) = \sum_{n=0}^{\infty} P\{s(\Delta t) = n\} \varphi_N^{(s)}(.)$ is defined by the following transformation:

$$\varphi_N(t + \Delta t, .) = G(\Delta t) \varphi_N(t, .), \tag{4.20}$$

where

$$G(\Delta t) = e^{-\lambda \Delta t} \sum_{n=0}^{\infty} \frac{(\lambda \Delta t)^n}{n!} T^n. \tag{4.21}$$

Formula (4.21) can be presented in the exponential form

$$G(\Delta t) = \exp[\Delta t \lambda (T - I)]. \tag{4.22}$$

This representation for the "transition operator" $G(\tau)$ of a random process $C(\tau)$ with the basic equation (4.11) remains valid for an arbitrary dependence of $\sigma$ on $g$. In this case, the operator $T$ is defined by the equality[17]

$$T = \sum_{1 \leq i \leq j \leq N} \frac{\omega_{ij}}{\lambda} T_{ij}.$$

We shall now consider the schemes for approximate modeling of collisions in the $N$-particle model.

*Modeling by the Bernoulli trials.* This algorithm is a modification of the scheme of $k = N(N-1)/2$ trials of all pairs of particles for collisions with various probabilities of success. It generates the process $C(\alpha)$ with a discrete time $t_\alpha = \alpha \Delta t$. In what follows wherever possible, a single subscript will be used for numbering pairs: $(i, j) \sim m$; $m = 1, 2, \ldots, k$. Let us consider a Markovian chain $\{C_m\}_{m=1}^{k}$, generated by the state $C$ after trials for collisions of pairs with numbers $m = 1, 2, \ldots, k$ respectively. The density of distribution of the state $C_m$ is denoted by $\mu_\alpha^{(m)}$.

The exact realization of the model on $(t_\alpha, t_{\alpha+1})$ corresponds to the following density transformation: $\varphi_N(t_{\alpha+1}, .) = G(\Delta t)\varphi_N(t_\alpha, .)$. The transition operator approximating the algorithm $\bar{G}(\Delta t)$ is obtained as follows. For the half-interval $(t_\alpha, t_{\alpha+1}]$, Eq. (4.22) can be written in the form

$$G(\Delta t) = \prod_{m=1}^{k} \exp(\Delta t \omega_m (T_m - 1)),$$

and each cofactor in this representation is replaced by an approximation linear in $\Delta t$. This gives

$$\bar{G}(\Delta t) = \prod_{m=1}^{k} [(1 - \Delta t \omega_m) I + \Delta t \omega_m T_m]. \tag{4.23}$$

The density transformation $\varphi_N(t_{\alpha+1}, .) = \bar{G}(\Delta t)\varphi_N(t_\alpha, .)$ is realized according to the following iterative scheme:

$$\begin{aligned} \mu_\alpha^{(m)}(C) &= (1 - \Delta t \omega_m)\mu_\alpha^{(m-1)}(C) \\ &\quad + \Delta t \omega_m \int_{4\pi} \mu_\alpha^{(m-1)}(C'_m) B(g_m, \theta) de, \quad m = 1, \ldots, k; \end{aligned} \tag{4.24}$$

$$\mu_\alpha^{(0)}(C) = \varphi_N(t_\alpha, C), \varphi_N(t_{\alpha+1}, C) = \mu_\alpha^{(k)}(C).$$

Here, $C'_m$ is obtained from $C$ by replacing the velocities $(c_i, c_j)$ of the pair with number $m$ by their values after the collision.

The iterative scheme (4.24) has a probability interpretation if the time step $\Delta t$ is chosen in such a way that

$$\omega_m \Delta t \leq 1, \tag{4.25}$$

for all admissible values of $C$. The process corresponding to (4.24) is realized as follows.

For each pair $m$ of particles $(c_i,\ c_j)$ in the system $(c_1, \ldots, c_N)$, the following procedure is adopted.

Step 1. The fact of the collision of the pair $(c_i, c_j)$ is drawn with the probability

$$p_{ij} = \Delta t \omega_{ij}.$$

If this trial is successful, the next step is taken.

Step 2. The velocities $c_i$, $c_j$ are replaced by $c'_i, c'_j$:

$$c'_i = \frac{1}{2}[(c_i + c_j) + g_{ij}e], \quad c'_j = \frac{1}{2}[(c_i + c_j) - g_{ij}e].$$

*"Ballot-box scheme"*. If the trial at step 1 is unsuccessful, the values of $c_i$ and $c_j$ are not changed. In the transitional operator (4.23), the term linear in $\Delta t$ is isolated:

$$\tilde{G}(\Delta t) = I + \Delta t \sum_{m=1}^{k} \omega_m (T_m - I).$$

The term $\Delta t$ is inserted under the summation sign and each term $\Delta t \omega_m$ is replaced by $k^{-1}(k\Delta t \omega_m)$. The transformation $\varphi_N(t_{\alpha+1}, .) = \tilde{G}(\Delta t)\varphi_N(t_{\alpha+1}, .)$ will correspond to the following relation:

$$\varphi_N(t_{\alpha+1}, C) = \left[1 - \sum_{m=1}^{k} \frac{1}{k}(k\omega_m \Delta t)\right] \varphi_N(t_\alpha, C)$$
$$+ \sum_{m=1}^{k} \frac{1}{k}(k\omega_m \Delta t) \int_{4\pi} \varphi_N(t_\alpha, C'_m) B(g_m, \theta) de. \qquad (4.26)$$

The time step $\Delta t$ is then put equal to an indefinitely small quantity $\delta t$ such that the following inequality is valid for all admissible values of $C$:

$$k\omega_m \delta t \leq 1. \qquad (4.27)$$

In this case, the relation (4.26) is the basic equation for the random process $C(\alpha)$ with the following constructive description.

Step 0. From the vectors $(c_1, \ldots, c_N)$ a random pair $(c_i, c_j)$ is chosen with the probability $1/k$.

Step 1. The collision of particles of the chosen pair is drawn with the probability

$$p_{ij} = k\omega_{ij}\delta t.$$

Step 2. If this trial is successful, the velocities $c_i$ and $c_j$ are replaced by their postcollisional values $c'_i$ and $c'_j$, respectively. The modeling of collisions of all the particles is terminated at this step and other possible pairs are not considered irrespective of the results of the trial for the first pair chosen.

It should be noted that in computational practice the conditions (4.25) and (4.27) are replaced by a less stringent condition

$$P\{\omega(g)\Delta t > 1\} \ll 1 \quad \text{or} \quad P\{k\omega(g)\delta t > 1\} \ll 1.$$

Both the above schemes of approximate modeling of collisions of particles in a cell are used for computing nonequilibrium kinetic processes in gas flows, when a molecule cannot be treated as a pseudo-Maxwellian.[1,3,7–10] The number of operations realized in the scheme of modelling by the Bernoulli trials increases with $N$ as $N^2$. However, this scheme does not introduce any errors in the collision frequency as, for example, the Bird's algorithm.[2] Therefore, computations can be carried out for a comparatively small $N$. Numerical experiments[1,3,9] carried out for computing the structure of a shock wave showed that the average number $N_0$ of particles in the cells corresponding to the flow upstream of the shock may be of the order of unity. In such a situation, the above-mentioned quadratic dependence for the number of arithmetic operations of collision simulation on $N$ is not manifested significantly.

The advantages of the ballot-box scheme over the Bernoulli trial scheme are manifested when the requirements on computational error associated with the splitting schemes (4.4) and (4.5) make it necessary to use a very small time step $\Delta t$. In this case, $N$ can be increased.[1,9] In the shock-wave computation,[1] the ballot-box scheme considerably reduces the computer time for $\Delta t \approx 10^{-3}$ and $N_0$ ranging from 12 to 18.

## 4.2. Direct statistical modeling of the shock wave in gaseous flow with velocity pulsations

On the basis of Boltzmann equation the numerical study of shock waves in a flow with gas-dynamical nonuniformities is carried out. There revealed is the effect of a temperature growth behind the shock wave's front, as well as the essential changes in the profiles of the macroparameters within the forward part of the shock's front.

### *4.2.1. Introduction*

The propagation of a shock wave in the medium with statistical nonuniformities gives an example of nonlinear interaction, which simultaneously with the pressure changes the fluctuations of vorticity and entropy are present.

Of a special interest are the studies of the interaction of shock waves with turbulent domain of the flow. The shock waves affect the fluctuations within a medium, while the medium's nonuniformities, in their turn, affect the propagation of shock waves.

The propagation of weak and initially planar shock waves within a gas with random nonuniformities was earlier studied experimentally and numerically on the basis of Euler and Navier–Stokes equations.[19,20] The results of modeling described here refer to the shock waves of greater intensity ($M \approx 1.7$–$3$). The calculations were carried out on the basis of Boltzmann equation, and such an approach renders the most correct way of solution of the problem of shock wave propagating in viscous and heat conducting gas. The problem was solved in one-dimensional, nonstationary formulation. The turbulent flow was modeled with the help of determinated and random pulse functions, which characterize the velocity perturbations, and that gave rise to the development of corresponding pulsations of density, temperature, and pressure.

### *4.2.2. Problem formulation*

For a numerical modeling of the shock wave's passing through domain of nonuniform fields of the hydrodynamical parameters was applied statistical method of a direct simulation, used in molecular gas-dynamics (statistical method of particles in cell). The details of this method in application to the problem of shock wave's structure in the one-component gas and in mixture of neutral and chemically active gas are exposed in Refs. 1 and 18.

One should dwell a little more on the peculiarities of realization of the above-mentioned problem. Let $f(t, x, y, z, u, v, w)$ be a function of distribution of the molecules near the point $(x, y, z, u, v, w)$ of the phase space at the moment of time $t$. We restrict ourselves to a one-dimensional in coordinate space and non stationary in time formulation; this means that $f = f(t, x, u, v, w)$ and all hydrodynamical parameters $(\rho, T, U = \langle u \rangle, \ldots)$ depend only on $t$ and $x$. Denote by $f_1(x, u, v, w)$ a solution of a stationary kinetic problem of the shock wave structure. When using the statistical method of particles in cells, this function is found numerically, by application of the scheme of splitting and steadification to the problem formulated for the Boltzmann equation,

$$\frac{\partial f}{\partial t} + u\frac{\partial f}{\partial x} = I(f_1 f) = \int (f' f_1') g d\sigma du_1 dv_1 dw_1, \tag{4.28}$$

with boundary conditions corresponding to the uniform gas in the thermodynamical equilibrium both before and after the shock,

$$f \to f_{\mathrm{eq}}^{(2)} = \frac{n_\alpha}{(2\pi k T_\alpha)^{3/2}} \exp\left[-\frac{(u-U_\alpha)^2 + v^2 + w^2}{2kT_\alpha}\right], \tag{4.29}$$

for $x \to \pm\infty$. Index $\alpha$ takes up value 1 for conditions before the shock, $x \to +\infty$, while $\alpha = 2$ for conditions after the shock, $x = -\infty$. Symbol $n$ stands for number density. The initial data correspond to gas-dynamical shock wave (jump of macroparameters) at the preliminary chosen point $x_S$. The values of density, temperature and velocity of flow before the shock and after it are related by the Rankine–Hugoniot relations, for which the case of monatomic gas $\gamma = 5/3$ look as

$$\frac{\rho_2}{\rho_1} = \frac{U_1}{U_2} = \frac{4M_S^2}{M_S^2+3}, \quad \frac{T_2}{T_1} = \frac{(M_S^2+3)(5M_S^2-1)}{16M_S^2}, \quad M_S = \frac{U_1}{a_1},$$

where $M_S$ and $a_1$ are Mach number and velocity of sound for the flow running into the wave. Such a formulation is made under assumption that the reference system is connected with a shock wave. As it was already mentioned, solution of the problem formulated is obtained by means of steadification method,

$$f_S(x,u,v,w) = \lim_{t\to\infty} f(t,x,u,v,w).$$

Practically, it is assumed that the steady-state solution is obtained as soon as $t = t_0 \gg 1$.

Presented in Fig. 4.1 are the results of modeling conducted in accordance to the methodology described, for the case of a shock wave with $M_S = 1.7$ in a monatomic gas. The model of molecules corresponds to that of elastic spheres with diameter $d$. The main units of measurement (and normalizing factors by the quantitative presentation of results) are mean

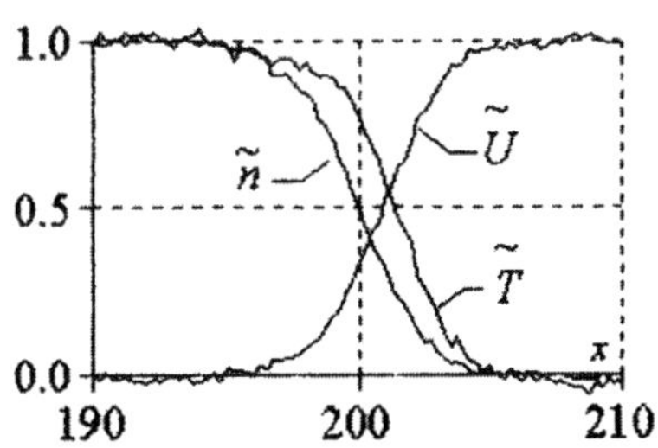


**Fig. 4.1.** Calculated profiles of shock wave for $M_S = 1.7$.

free path and the most probable thermal velocity in the oncoming flow, and mass of molecules. The size of a cell is $\Delta x = 0.25$, and is equal to the temporal step of splitting $\Delta t$. The mean number of particles within each cell of the oncoming flow is $N_0 = 2$–$4$. The initial length of a computational domain corresponds to the value $X = 300$. The mean location of the shock wave is $X_f = 200$. The time of a settlement of the stationary structure of a wave was taken equal to $t_0 = 10$. The shock wave's thickness, as determined according to the density profile, is $d_S = 5$. All the quantities shown in the graph are reduced to the range of variation (0, 1), according to a formula

$$\tilde{A}(x) = \frac{A(x) - A_1}{A_2 - A_1},$$

where $A_1$ and $A_2$ are the values of a macroparameter $A$ before and after the shock, correspondingly; for a transversal velocity one has $\tilde{V} = V/(U_1 - U_2)$.

To model the interaction of a shock wave with fields of nonuniform distribution of hydrodynamical parameters, at the moment $t_0 = 10$, when the stationary structure of a shock wave is already formed, in the flow running into the wave, at a certain upstream distance from the wave's center was introduced a region of the perturbed macroparameters having a length $L$.

After that the nonstationary problem (4.28), (4.29) was solved with the following initial data:

$$f(t_0, x, u, v, w) = f_S(x, u, v, w)(1 - I_\omega(x)) + I_\omega(x)\frac{n_*}{(2\pi kT)^{3/2}}$$
$$\times \exp\left[-\frac{(u - U_*)^2 + (v - V_*)^2 + (w - W_*)^2}{2kT_*}\right]. \quad (4.30)$$

Here, $n_*(x), T_*(x), U_*(x), V_*(x)$, and $W_*(x)$ are the functions describing the perturbed values of hydrodynamic parameters in a domain $\omega = [x_*, x_* + L]$ on the axis $OX$, and $I_\omega(x)$ is an indicator of the domain $\omega : I_\omega(x) = 1$ for $x \in \omega$, and $I_\omega(x) = 0$ for $x \notin \omega$.

Following the first domain of perturbations, at the moment of time $t_1$ is introduced another domain, similar to the first one, and within the temporal interval $[t_1, t_2)$ the problem (4.28)–(4.30) for Boltzmann equation is solved, the function $f_S$ in initial data (4.30) being substituted for a solution $f(t_1, x, u, v, w)$ of the same problem for the preceding interval $[t_0, t_1)$. In a similar way might be introduced the third domain of perturbation, and any desirable subsequent one, and the solution for the corresponding intervals of time will be found. Just in such a way the passing of a shock wave through a

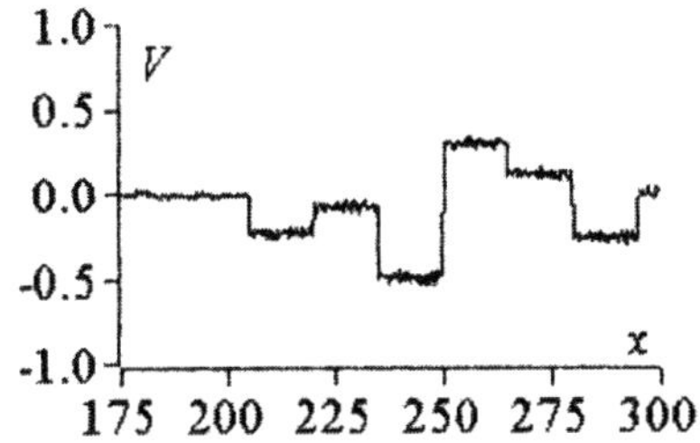


**Fig. 4.2.** Computed passing of shock wave through a sequence of domains.

sequence of domains with nonuniform distribution of hydrodynamic parameters way was modeled. These domains might follow each other immediately, if $(t_n - t_{n-1})U_1 = L$, or with a delay in time, if $(t_n - t_{n-1})U > L$. As an example, presented in Fig. 4.2 is the domain of perturbation of macroscopic velocity $V$, normal to the direction $OX$ of the wave's propagation. This domain appears as a package of six rectangular pulses, the amplitudes of which form a sequence of six independent random quantities, distributed according to a normal law with the density of probabilities

$$f(\xi) = \frac{1}{\sqrt{2\pi\sigma^2}} \exp\left[-\frac{\xi^2}{2\sigma^2}\right], \tag{4.31}$$

where $\sigma$ is the mean square value of pulsations $V$. In this figure and in subsequent figures, along the axis of abscissa are measured the distances along the axis $OX$, normalized with respect to the mean free path $\lambda_1$ of molecules in the flow running into a wave.

### *4.2.3. Results of the numerical modeling*

In the numerical experiments on modeling of the passing of shock waves through the nonuniform hydrodynamic fields, the wide spectrum of domains of perturbations of density, temperature, and macroscopic velocities $U$, $V$, and $W$ were considered. By means of the analysis of the results obtained it proved to be possible to isolate the domain of perturbations, the interaction of which with a shock wave generates all the main effects revealed in the series of calculations indicated above. As such a domain was chosen the degenerating package of velocity perturbations. It represents itself as a file of rectangular pulses following each other.

As it is known, one of the main indications of turbulence in the flow is a nonzero rate of dissipation $\varepsilon = \langle \mu \frac{\partial u_i' \partial u_j'}{\partial x_j \partial x_i} \rangle$ of the energy of velocity's pulsations $q' = \langle u'_i u'_j \rangle$. The degenerating package of velocity perturbations possesses the property indicated, and for this reason such a model is considered as some approximation to the simulation of turbulized regions of gaseous flow. The degeneration manifests itself in the decrease of amplitude and in the modification of the form of pulses in the course of time. This phenomenon is connected with the influence of viscosity, heat conductivity, and diffusion. All these effects are taken into account by the present scheme of numerical modeling because of its being based on the Boltzmann equation. Moreover, it is to be noted that the initial perturbations of velocity generate the corresponding perturbation of the fields of density, temperature, and pressure. The characteristic linear scale of such perturbations $l$ was chosen on the grounds of a condition that $l$ should be essentially larger than the thickness of a shock wave $d_S$.

Presented in Fig. 4.3 are the profiles of macroscopic velocities — longitudinal $U$ and parallel to the plane of wave $V$, at the moment before the interaction of a shock wave with the perturbation's domain having the form of eight rectangular, varying in sign pulses of transversal velocity $V$. The wave's Mach number is $M_S = 3$. Pulses have the same amplitudes equal to $\max V = a_1$, velocity of sound in the oncoming flow, and the length of each pulse is $l = 15$, while the shock wave's thickness is $d_S = 3$ (in Fig. 4.3 left and right vertical scales refer to the variables $U$ and $V$, correspondingly).

Figures 4.4(a)–4.4(d) demonstrate the results of modeling of the interaction of a shock wave and a package of pulses of transversal velocity $V$, indicated above (see Fig. 4.3). Presented are the profiles of density, Fig. 4.4(a), pressure Fig. 4.4(b), temperature Fig. 4.4(c), longitudinal $U$ and transversal $V$ components of velocity Fig. 4.4(d), at the moment when the major part of the perturbed domain has already passed through the shock layer. The small-scale pulsations in the graphs of Fig. 4.4 are connected with

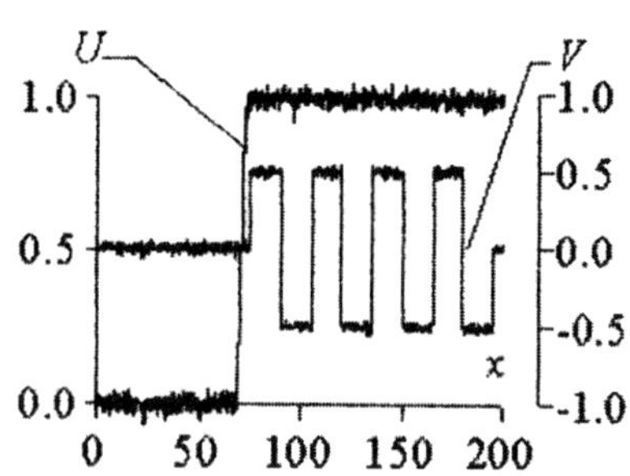


**Fig. 4.3.** Profiles of macroscopic velocities at the moment before interaction.

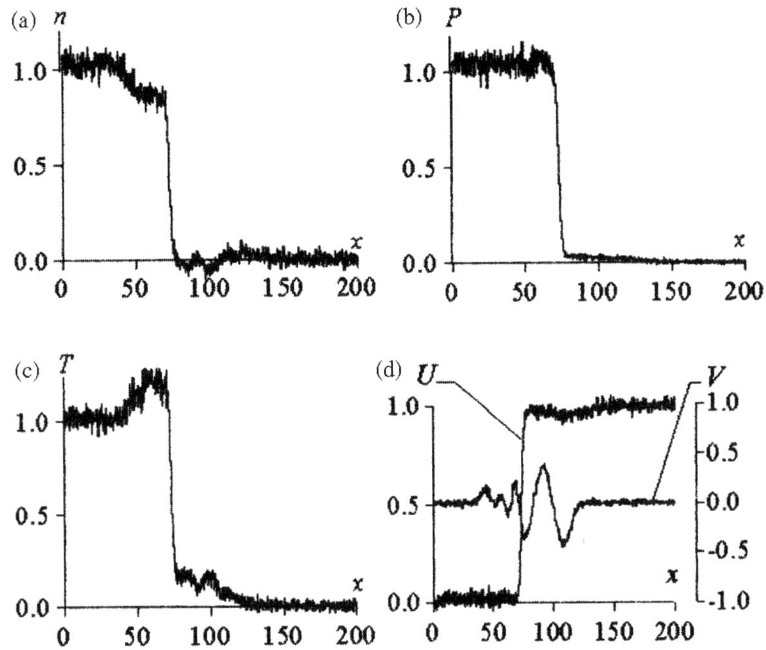


**Fig. 4.4.** The results of modeling of the interaction of a shock wave and a package of pulses of transversal velocity (see Fig. 4.3): (a) density, (b) pressure, (c) temperature, (d) components of velocity.

a statistical nature of the numerical method applied. As for the large-scale pulsations and for the changes of the profiles of all the parameters, those are the macroeffects induced by the interaction of wave and perturbation's domain. The dissipation of pulsation of the transversal velocity $V$ leads to a heating of gas. In Fig. 4.4(c), one might see the temperature's rise, most pronounced immediately behind the shock wave's front. To that rise corresponds the tendency toward the decrease of a mean of density behind the shock's front (Fig. 4.4(a)). The dissipation of a package of velocity's pulses goes on more intensively after its passing through the shock layer, and in the course of that one is able to observe the decrease of a linear scale of these pulsations (Fig. 4.4(d)).

Were conducted also the calculations of a shock wave with Mach number $M_S = 1.7$ passing through the stochastically nonuniform domain. This domain was modeled by a series of four degenerating pulse packages, which followed immediately one after another. Each package consisted of six rectangular pulses of all the components of macroscopic velocity, which had

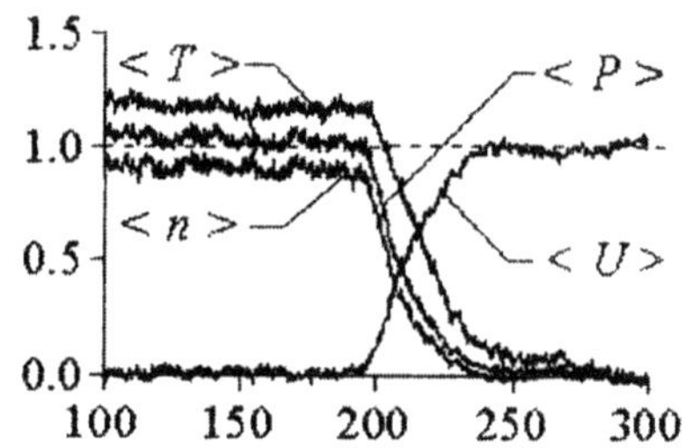


**Fig. 4.5.** The average profiles of density $n$, temperature $T$, pressure $P$, and longitudinal component of mean velocity $U$ after the shock wave's passing through the series of pulse's packages (see Fig. 4.3).

random amplitudes $\{A_u, A_v, A_w\}$, statistically independent and distributed identically according to a normal law (4.31) with a mean square deviation $\sigma = 0.18U_1$. Perturbations of this kind, in their physical contents, are much closer to the turbulized domain in a flow.

Presented in Fig. 4.5 are the average profiles of density $n$, temperature $T$, pressure $P$, and longitudinal component of mean velocity $U$ after the shock wave's passing through the above-mentioned series of pulse's packages. These results were obtained by way of the averaging of all the macroparameters over the period of passing of shock wave through the fourth package, and effects, which were accumulated during the time of shock wave's passing through the series of perturbations.

The rise in a mean level of temperature behind the shock wave's front (see Fig. 4.5) is connected with the heating of gas due to a dissipation of the velocity's pulsations. As this takes place, the level of pressure behind the front stays practically unchanged, which leads to a decrease of the mean level of density behind the wave in accordance to equation of state $P = nkT$ for the mean parameters of a gas. The heating of gas in the flow running into a wave, which is caused by dissipation, leads to a blurring of profiles of all the main macroparameters in the forward part of shock wave's front, as well as to their asymmetry. The characteristic thickness of a shock layer, which is approximately equal to 2.5, is essentially larger than the thickness of an unperturbed shock wave, $d \approx 5$, and becomes comparable to the linear scale of perturbations introduced.

The decrease of a density level behind the wave, together with the unchanged level of a mean velocity (see Fig. 4.5), does not imply that the mean flow rate of a substance is decreased, because for fluctuating quantities the mean flow is equal to $\langle nu \rangle = \langle n \rangle \langle u \rangle + \langle n'u' \rangle$. The numerical scheme chosen guarantees the conservation of flow rates and their equality to the initial value,[18] which means that one always has $\langle nu \rangle = n_2 u_2 = n_1 u_1$. As it is implied by the last equality, in the fluctuating part of flow behind the wave the correlator of pulsations of density and velocity is not vanishing, $\langle n'u' \rangle \approx 0.08 n_2 u_2$.

### 4.2.4. Conclusion

By means of a method of direct statistical modeling, the interaction of shock wave and turbulent-like perturbations of flow was numerically studied. The Mach number was equal to 1.7 and 3, the magnitude of pulsations was, correspondingly, 20% and 100% of the speed of sound in the flow running into a wave.

There were detected the effects of temperature rise behind the shock wave's front, of the decrease of intensity of density's jump, as well as of the blurring of profiles of macroparameters in the forward part of a shock wave's front and of their asymmetry.

O.A. Azarova carried out similar numerical experiments, but the problem was formulated on the basis of the Euler equations.[21] The fields of density pulsations, the coefficient of the increase, correlational and spectral functions were closely investigated.

## 4.3. Direct statistical simulation for some problems of turbulence

The possibilities of a statistical simulation of turbulence are studied by means of two exemplary problems of the free turbulence. Two different approaches are considered. The first of these is similar to the method of direct Monte-Carlo simulation in rarefied gas dynamics, while the other one is based on the analogy of the turbulent mixing of a fluid with the Brownian motion of fluid particles. Some of the results of simulation are shown.

### 4.3.1. Introduction

The incompressible fluid is considered as an ensemble of fluid particles — "moles". The flow is determined by chaotic motion of these moles, each of them possessing its own velocity and coordinate. The change of the character of flow as a whole, for example — of the field of mean velocities, occurs due to the turbulent mixing of moles with different velocities of their own. Generally speaking, any characteristical feature of the flow appears as a result of averaging of the similar features of moles contributing to the flow under consideration. J. Boussinesque and L. Prandtl already used an analogy between the molar mixing in a turbulent flow and the molecular transfer in gases for derivation of the well-known formulae of the turbulent friction. The Boussinesque's formula looks as

$$\tau = \rho\nu_T \frac{\partial u}{\partial y}, \quad \nu_T = \langle l\upsilon \rangle;$$

here, the random quantities $l$ and $\upsilon$ are, correspondingly, mixing length and pulsation velocity of a liquid particle.

The Prandtl's formula has a form

$$\tau = \rho\nu_T \frac{\partial u}{\partial y}, \quad \nu_T = L^2 \left| \frac{\partial u}{\partial y} \right|,$$

where the empirical quantity $L$ is a so-called "displacement path".

Essentially, the method chosen here is nothing more as a direct numerical computer-aided simulation (imitation) of the physical processes forming the basis of the phenomenon under consideration.

### *4.3.2. An application of the statistical method of particles in cell for simulation of the momentumless wake*

For further development of the above-mentioned analogy it will be desirable to have kinetic equations for a density $\varphi(t, x, c)$ of distribution of probabilities of the instantaneous values of velocity $c$ of one mole at the given moment of time $t$, at the point $x$. Unfortunately, in the theory of turbulence there is no universal closed kinetic equation for a one-point density $\varphi(t, x, c)$ of the distribution of probabilities, similar to the Boltzmann equation in rarefied gas dynamics. There were numerous publications (e.g., Refs. 22–26) describing the efforts to find such an equation. However, the theory of turbulence is still a science of semiempirical models even at the kinetic level of description. It would be preferable for the purpose of construction of the imitation scheme to adopt those mathematical models in which elementary physical features of the phenomenon are not "buried" under a great number of additional assumptions.

Thus, for our purpose the role of a basis, qualitatively presenting an analogy between molar mixing in fluids and molecular one in gases, might be played by the following, comparatively simple relaxational kinetic equation[25,26]:

$$\frac{\partial \varphi}{\partial t} + c\frac{\partial \varphi}{\partial x} + \frac{\partial}{\partial c}\left[\left(-\frac{v}{2\tau_1} + F\right)\varphi\right] = \frac{\varphi_M - \varphi}{\tau_2}. \tag{4.32}$$

Here, $v = c - u$ is a pulsational velocity, while $u = \langle c\rangle$ is a mean velocity of flow. The function

$$\varphi_M = \left(\frac{3}{4\pi q^2}\right)^{3/2} \exp\left(-\frac{3v^2}{2q^2}\right) \tag{4.33}$$

is a density of probability for a normal law of distribution of the velocity pulsations. Furthermore, $q^2 = \langle v_x^2 + v_y^2 + v_z^2\rangle$ is a doubled mean value of a specific kinetic energy of these pulsations. The vector field $F$ is that of mass forces acting on a fluid particle. For the case of incompressible fluid $\rho = 1$ and force $F = g - \nabla P$, where $P$ is a mean pressure in the turbulized fluid and $g$ is a vector of the acceleration of gravity.

Formally, relaxational parameters $\tau_i$ are expressed in terms of the scales of turbulence $L_i$ and the specific energy of turbulence $q^2$ as $\tau_i = L_i/q$. The scales of turbulence $L_i$ are not functionals of the one-point distribution $\varphi$, but are determined by the two-point correlation function. In semiempirical kinetic theories of turbulence the laws of variation of $L_i$ are chosen following different considerations; for example, those might be found from the problem's geometry or from the approximate equations, which relate evolution of $L_i$ to the variation of the mean characteristics of a turbulent medium.

Here, it will be assumed that the functional dependence $L_i(q)$ is prescribed. This dependence might be actually determined for a number of practical cases. For example, in a turbulent flow behind the grating the quantities $L_1$ and $q$ are connected by means of Loitsianskii's invariant $L_1^{5/2}q = \text{const}$. The first of the approaches mentioned earlier assumes the existence of the following power laws connecting $L_i$ and $q$:

$$L_1^{-1} \sim q^{(2\gamma_1 - 1)} \quad \text{and} \quad L_2^{-1} \sim q^{(2\gamma_2 - 1)}. \tag{4.34}$$

It is to be noted that for $\gamma_1 = 0.7$ from Eq. (4.34) will follow the Loitsianskii's invariant for the dissipation scale $L_1$. The main empirical constants of the above model appear in the form of the exponents $\gamma_1$ and $\gamma_2$ and the initial values of integral scales $L_1^0$ and $L_2^0$.

If one assumes that the field of force $F$ is a prescribed function of $t$ and $x$, then Eq. (4.32) will differ from BGK model (P. L. Bhatnagar, E. P. Gross, M. Krook), well known in rarefied gas-dynamics, by the additional term

$$-\frac{\partial}{\partial c}\left(\frac{v}{2\tau_1},\varphi\right).$$

For the better understanding of the role of such a term in the kinetic equation, let us consider the Cauchy problem for a distribution function independent of the spatial variables, that is, governed by the equation

$$\frac{\partial \varphi}{\partial t}-\frac{\partial}{\partial c}\left(\frac{v}{2\tau_1}\varphi\right)=\frac{\varphi_M-\varphi}{\tau_2}, \tag{4.35}$$

$$\varphi(t=+0,c)=\varphi_0(c). \tag{4.36}$$

If Eqs. (4.32) and (4.33) are multiplied, successively, by $c$ and $v^2$, after integrating each time over the whole velocity space $\{c\}$ one obtains the equations for $u(t)$ and $q^2(t)$ in the form

$$\frac{du}{dt}=0, \qquad \frac{dq^2}{dt}=-\frac{q^2}{\tau_1} \tag{4.37}$$

$$u(t=+0)=u_0, \quad q^2(t=+0)=q_0^2. \tag{4.38}$$

As it is evident from the resulting equations, within the system described by Eq. (4.35) the momentum $u$ is conserved, while the energy $q^2$ is a monotonously decreasing function of time. The dependence of the form of Eq. (4.34) leads to the following form of the function $\tau_1(q^2)=\tau_1^0(q_0^2/q^2)^{\gamma_1}$. Then, the last of Eq. (4.37) assumes the following form:

$$\frac{d\tilde{q}^2}{dt}+\frac{\tilde{q}^{2(1+\gamma_1)}}{\tau_0}=0, \quad \tilde{q}=q/q_0.$$

After the analytical solution of that equation, one obtains for $\gamma_1>0$

$$\tilde{q}^2=\frac{1}{(1+\gamma_1 t/\tau_1^0)^{1/\gamma_1}}.$$

This equation gives an approximate expression for a decrease of the energy of turbulence within the wake.

The meaning of the relaxational term in a right-hand side of Eq. (4.32) is easily revealed for a spatially uniform case. Without the dissipation and the force $F$ one obtains the familiar Bhatnagar–Gross–Krook equation

$$\frac{\partial \varphi}{\partial t}=\frac{\varphi_M-\varphi}{\tau_2},$$

of which the solution has the form

$$\varphi(t,c) = e^{-t/\tau_2}\varphi_0(c) + (1 - e^{-t/\tau_2})\varphi_M(c), \tag{4.39}$$

where $q^2 = q_0^2 = \text{const.}$ and $\tau_2(q^2) = \tau_0^2 = \text{const.}$

As it follows from Eq. (4.39), the quantity $\tau_2$ characterizes the time interval during which any initial distribution $\varphi_0(c)$ evolves toward the normal law $\varphi_M(C)$.

### *4.3.3. An application of the statistical method of particles in cell to the problem of a turbulent spot*

As an example of using the imitation as a method of the numerical simulation of turbulence, an experiment consisting in a solution of the two-dimensional unsteady problem of a "turbulent spot"[27,28] was conducted. At the moment of time $t = 0$ within a stationary fluid, for $u = 0$, is in some way created an initial distribution $q^2(0, Y, Z)$ of the energy of turbulent pulsations. The problem consists in a determination of the spatial distribution of the characteristic parameters of turbulence at the arbitrary moment of time $t$. This problem arises from the consideration of the momentumless wake behind the generator of pulsations having diameter $D$ and placed within a flow with velocity $u_0$. Such an axisymmetric wake extended along the axis $OX$ was experimentally realized and investigated in Ref. 29. Its property of zero momentum, that is, in average over the cross-section, the absence of a mass transfer along the wake's axis $OX$, permits at the sufficiently large distance from source to neglect the nonuniformity of the profiles of mean velocity and friction stress. Within this domain a spatial distribution of energy $q^2$ is self-similar, and an evolution of the turbulent wake proceeds in such a way as if it was initiated by the point source of pulsations placed within a concurrent flow having a velocity $u$; thus during the calculation of fields it is possible to assume that at the distances $x \geq 4D$ one has $t = x/u_0$.[29] It means that in terms of the corresponding temporal variable $t = x/u_0$, for $t \geq t_0 = 4D/u_0$, the function to be measured, $q^2(t, Y, Z)$, assumes the Gaussian form,

$$q^2(t,r) = q^2(t,Y,Z) = q_m^2(e)\exp(-0.69r^2/r_q^2(t)),$$
$$r = (Y^2 + Z^2)^{1/2}, \quad q^2(t, r_q) = q_m^2(t)/2. \tag{4.40}$$

The energy $q_m^2(t)$ at the center of a spot, and the typical radius of it, $r_{q(t)}$, are characterized by the following relations[30]:

$$q_m^2(t) = q_0^2\left(1+\frac{t-t_0}{\tau_0}\right)^{-n_1}, \quad r_q(t) = r_0\left(1+\frac{t-t_0}{\tau_0}\right)^{n_2}, \tag{4.41}$$

where $n_1 \cong 1.58, n_2 \cong 0.35, \tau_0 = 2D/u_0$ for $50D/u_0 < t < 130D/u_0$; for $t_0 \le t \le 50D/u_0$ the exponents $n_1$ and $n_2$ are equal to 1.80 and 0.25 correspondingly.

The mathematical formulation of the problem of a turbulent spot on the kinetic level of description may be described as follows. The Cauchy problem for a kinetic equation, Eq. (4.32), is to be solved within an unbounded five-dimensional phase space $(Y, Z, c_x, c_y, c_z)$. As an initial distribution function $\varphi_0(Y, Z, c)$ was accepted the density of a normal law for the distribution of probabilities of the instantaneous values of hydrodynamic velocity $c$, that is,

$$\varphi_0(r,c) = \left(\frac{3}{2\pi q_0^2(r)}\right)^{3/2}\exp\left(-\frac{3}{2q_0^2(r)}\right), \tag{4.42}$$

where $q_0^2(r) = q_0^2\exp(-0.69r^2/r_0^2)$.

The scheme of simulation is constructed in accordance with the principles described in a preceding paragraph, which are the same as in rarefied gas-dynamics. Each particle is a model of a small element of the turbulized fluid and is characterized by its position $\{Y(t), Z(t)\}$ and by its instantaneous hydrodynamic velocity $c(t)$. The computational domain within which the evolution of a system of $N$ particles is traced through is nothing more than a rectangle having size $F_y \times F_z = 2\times 2$ and located within a first quadrant of the coordinate plane; thus it is implicitly assumed that the problem possesses the specular symmetry with respect to the axes $OY$ and $OZ$. The external boundaries $Y = F_y$ and $Z = F_z$ were considered as open for the particles escaping from the computational domain. The uniform computational grid was used, which had the total number of cells $20\times 20$, the size of each cell being equal to $L_z = L_y = 0.1$. At each particular time moment $t = 0$, the distribution of particles within the computational domain was uniform, the mean number of particles within each call being equal to 10. The velocities of particles $c_i$ were raffled in the correspondence with the density of probability distribution, Eq. (4.42), where $r = r_i$ is a distance between the $i$th particle and the coordinate's origin. In correspondence to physical processes stored in Eq. (4.32), the scheme of simulation of the model's evolution during the small temporal interval, which in our calculations was equal to 0.085, consists of the three following stages.

The main equation of the stage I (transfer of the turbulence),

$$\frac{\partial f}{\partial t} + c\frac{\partial f}{\partial x} + F\frac{\partial f}{\partial c} = 0,$$

describes the motion of a system of $N$ particles, noninteracting between themselves, within a field of force $F$. The equations of motion for each particle have the following form:

$$\frac{dr_i}{dt} = c_i, \quad \frac{dc_i}{dt} = F; \quad i = 1, \ldots, N.$$

Stage II simulates the dissipation of the turbulent energy — each of the subsystems of particles in cells evolves in accordance to the main equation of a process

$$\frac{\partial f}{\partial t} - \frac{\partial}{\partial c}\left(\frac{v}{2\tau_1}f\right) = 0.$$

At this stage only velocities $c_1, \ldots, c_N$ of the totality of $N$ particles located within a given cell are changed. These velocities vary according to the law

$$\frac{dc_i}{dt} = -\frac{1}{2\tau_1}v_i; \quad i = 1, \ldots, N;$$

$$v_i = c_i - \langle c\rangle_N^{(1)}; \quad \langle c\rangle_N^{(1)} = \frac{1}{N}\sum_{j=1}^{N} c_j^{(1)}. \tag{4.43}$$

The system of Eq. (4.43) allows for an analytical solution, with results in that $L_1(q)$, and hence $\tau_1(q)$, is a power function. The direct recount of the velocities in correspondence to the formula given below just represented the realization of the stage of energy dissipation,

$$c_i^{(2)} = \langle c\rangle_N^{(1)} + \frac{c_i^1 - \langle c\rangle_N^{(1)}}{(1+\gamma_1\Delta t/\tau_1)^{\gamma_1/2}};$$

here the upper indices symbolize the number of the stage, at which the values considered are obtained.

In the capacity of the main equation of stage III ( redistribution of the pulsations over the degrees of freedom) there appears the spatially uniform BGK-equation

$$\frac{\partial \varphi}{\partial t} = \frac{\varphi_M - \varphi}{\tau_2}, \tag{4.44}$$

which is modeled for each of the cells separately. Strictly speaking, the pulsational velocities of each of the $N$ particles within a cell, in accordance to Eq. (4.44), should be calculated in such a way:

$$v_i^{(3)} = v_i^{(2)}\chi + (1 - \chi)\eta. \tag{4.45}$$

Here, $\chi$ is a random quantity assuming the values 1 or 0 with probabilities $\exp(-t/\tau_2)$ and $1 - \exp(-t/\tau_2)$, correspondingly, while $\eta$ is a random quantity distribution with a probability density $\varphi_N$. However, in such a situation the computational algorithm equation (4.45) does not conserve the energy of the subsystem as a whole. The random oscillations $\langle v^2 \rangle_N$ are, on the average, of the order of $N^{-1/2}$.

For the realization of the present stage of imitation, an alternative method which provides an exact conservation of the energy $E = c_1^2 + \cdots + c_N^2$ and of the momentum $P = c_1 + \cdots + c_N$ of the system as a whole was proposed[31] and used. This method is reduced to the following transformation:

$$C^{(3)} = C^{(2)}\chi + (1 - \chi)TC^{(2)}, \quad C = \{c_1, \ldots, c_N\}.$$

Here $T$ is such a random matrix of the rotation of a $3N$-dimensional vector $C$, that the point $TC$ is distributed over the hypersphere of constant energy $E$ and constant momentum $P$ of the system of $N$ particles within the given cell with a uniform probability. Each of the pulsational velocities $v_i^{(3)}$ obtained in such a way is asymptotically (for $N \gg 1$) equivalent to the quantity $v_i^{(3)}$ in Eq. (13), the speed of convergence being of the order of $N^{-1}$. The degeneration of the turbulent spot within a uniform (nonstratified) medium occurs due to the dissipation of the energy of pulsations and its diffusion. Within the frame of certain approximation it is possible to neglect the influence of the gradient of mean pressure $P$ for these processes, the absence of stratification signifying that $g = 0$. Hence, the force acting on each mole was equal to zero. For the zero values of the initial velocity field, $u(t = +0, r) \equiv 0$, and for the condition $P = \text{const.}$, after some time appears a mean velocity $u_r \neq 0$ of the turbulized fluid in radial direction. Beginning with a certain moment of time, the derivative $\frac{\partial q^2}{\partial t}$ becomes to be comparable with a convective term $u_r \frac{\partial q^2}{\partial r}$, and after that one should take into account the field of mean pressure. The initial parameters $q_0^2$ and $r_0^2$, equal, correspondingly, to 0.0104 and 0.612, were chosen in such a way that within the temporal interval $t \leq 20$ were simultaneously small both effects of the zero $\nabla P$ and of the finite dimensions of a computational domain. The point $x_0 = 4D$ of the experiment[29] was put into correspondence to the momentum $t = 0$, and this permits to determine the quantities $\tau_0 = 2D/U_0$, $n_1 = 1.58, n_2 = 0.35$ in the approximations (4.41) of the physical experiment. The initial kinetic model possesses four empirical constants, namely, the initial values of integral scales $L_1^0$ and $L_2^0$ and the exponents $\gamma_1$ and $\gamma_2$ entering the power-law relations for $L_i(q)$. As it was shown by means of a special analysis,[30] these constants are uniquely determined through the initial data, if only for the criterions of agreement with experiment one takes the coincidence of the functions calculated, $q_m^2(t)$ and $r_q(t)$, with the approximations of the results of Ref. 29 (see Figs. 4.6 and 4.7); that was made with assumption of the self-similarity of a profile $q_m^2(t, r)$. In the calculations, the values $\gamma_1 = 0.63, \gamma_2 = 0.80, L_1^0 = L_2^0 = 0.14$ $(= 0.23r_0)$ were chosen; those correspond to the following values of the relaxational parameters: $\tau_1^0 = \tau_2^0 = 1.31$. In the rarefied gas-dynamics the flows with similar values of parameters $L$ and $\tau$ are classified as belonging to the "transitional" regime, which means that they are considered on the kinetic level of description.

Figures 4.6–4.8 show the comparison with experimental data of Naudasher.[29] The spatial distribution of energy obtained in calculations in the course of time assumes a self-similar form, which, however, differs from the Naudasher's profile (see Fig. 4.8), and within the interval $r/r_q \leq 1.3$ is well approximated by the Bessel function $J_0(1.52r/r_q)$. A number of calculations was made with the following initial distributions of the pulsational

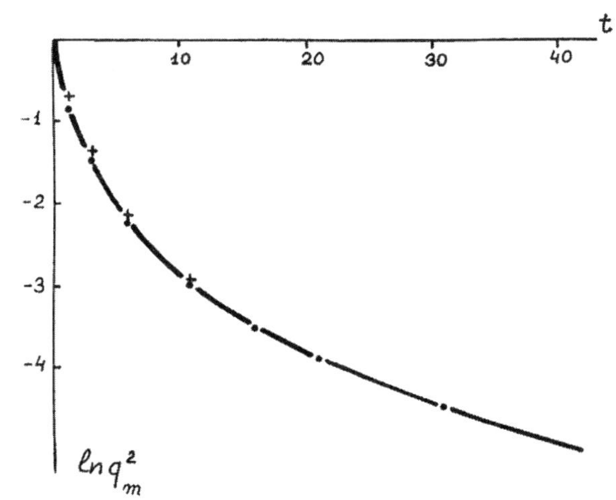


**Fig. 4.6.** The dependence of energy in the center of a spot on time.

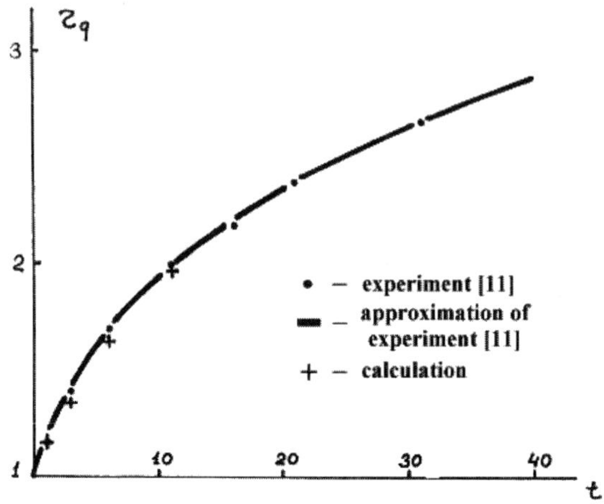


**Fig. 4.7.** The dependence of the typical size of a spot on time.

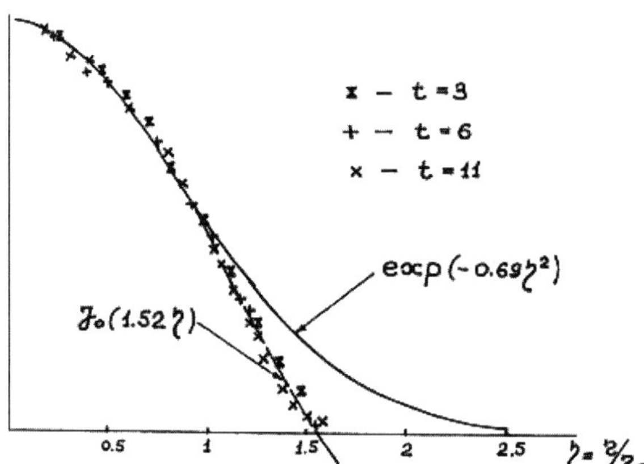


**Fig. 4.8.** Self-similar distribution of the energy of pulsations along the radius of a spot. (Solid curve) experiment[10]; (signs) results of the calculation.

energy over the radius:

$$q_0^2(r) = q_0^2 J_0(1.52r/r_0); \quad q_0^2(r) = \begin{cases} r \leq r_0, & q_0^2 = \text{const.}, \\ r > r_0, & 0. \end{cases}$$

These calculations have confirmed the same character of an eventual similarity as it was presented in Fig. 4.8. The similar discrepancy between the calculated results and those of experiment was detected by numerous investigators, which tried to solve the problem numerically on the basis of various models of turbulence. As it was shown in Ref. 30, the appearance of a "tail" in the experimental curve $q^2(t,r)$ and, correspondingly, the absence of a definite edge in a spot, is explained by the intermittence effect, that is, by the irregular alternation of the laminar and turbulent phases. Hence, in the above-mentioned discrepancy an inadequacy to the physics of the phenomenon is revealed, which is common to a number of numerical models and consists of neglecting the intermittence effect.

One of the advantages of the kinetic approach to the study of turbulence consists in a possibility to obtain the function of distribution of the pulsations and thus to check the hypotheses connected with it. The Millionshchikov's hypothesis is well known concerning the cumulants of the fourth- and higher-orders being equal to zero. This assumption permits to express the high-order moments of the distribution function in terms of the lower-order moments, namely, of the first-order (mean velocity), second-order (Reynolds stresses), and third-order (fluxes of energy). Upon doing that it will be possible to write down the closed set of Reynolds equations. For the experimental check of the above assumptions one needs to measure the indices of asymmetry and of excess, which are the cumulants of the distribution of velocity $v$ having the third- and fourth-orders. The presented, here, analysis of the characteristic function of the distribution of pulsations in the turbulent spot proves to be the natural generalization of the approach described.

Let $v_r = (c_r - u_r)/\sqrt{D(c_r)}$ be the value of a radial pulsation of velocity, normalized with respect to a standard deviation. The density of distribution of probabilities for this quantity is $f(t,r,v_r)$. By definition,[13] the characteristic function corresponding to this distribution is

$$\chi_{t,r}(z) = \int e^{izv_r} f(t,r,v_r)dv_r.$$

This function might be presented in exponential form

$$\chi_{t,r}(z) = e^{\rho(z)+i\theta(z)}.$$

Functions $\rho(z)$ and $\theta(z)$ are called the producing functions of the cumulants $æ_n$ of the even and uneven orders, correspondingly; they are expressed as

$$\rho(z) = -\frac{z^2}{2} + \sum_{k=2}^{\infty}(-1)^k æ_{2k}\frac{z^{2k}}{(2k)!}, \quad æ_2 = D(v_r) = 1,$$

$$\theta(z) = -\alpha\frac{z^3}{6} + \sum_{k=2}^{\infty}(-1)^k æ_{2k+1}\frac{z^{2k+1}}{(2k+1)!}, \quad æ_3 = \alpha.$$

Here, $\alpha$ is an index of asymmetry, and $æ_4$ is an excess of the distribution of pulsations of a radial velocity $v_r$.

From the hypothesis of normality follow the relations $\theta(z) \equiv 0$ and $\rho(z) = -z^2/2$, or, equivalently,

$$U(z) \equiv Re\chi(z) = e^{-z^2/2} \quad \text{and} \quad V(z) = \text{Im}\chi(z) \equiv 0.$$

From the Millionshchikov's hypothesis, the assumption that $\rho(z) = -z^2/2$ and $\theta(z) = -\alpha z^3/6$ follows, that

$$U(z) = e^{-z^2/2}\cos(\alpha z^3/6), \quad V(z) = e^{-z^2/2}\sin(\alpha z^3/6). \tag{4.46}$$

The statistical method of particles permits to realize a direct evaluation of $U(z)$ and $V(z)$ as of the mean values of quantities $\cos(zv_r)$ and $\sin(zv_r)$ and in accordance with the simulation to get the approximate values $\hat{U}(z)$ and $\hat{V}(z)$; the mean square scattering (standard deviation) of these quantities is equal to $S_U$ and $S_V$, correspondingly. If the mean number of particles within the domain of measurement is $N_*$, while the number of realizations in the process of determination of the mean values over the ensemble is $L$, then one has

$$S_U = \sqrt{\frac{D(\cos(zv_r))}{N_* L}}, \quad S_V = \sqrt{\frac{D(\sin(zv_r))}{N_* L}}.$$

The characteristic function $\chi_{t,r}(z)$ of the distribution of radial pulsations $v_r$ was calculated for the moments of time $t = 0, 1, 3, 6,$ and 11. The measurements were conducted within six annular domains $\{\underline{x}r_q(t) \leq r \leq \bar{x}r_q(t)\}$, of which the boundary values for the domains of measurement of $\chi_{t,r}(z)$ are in Table 4.1.

The areas of domains I–VI (see Fig. 4.9) are chosen in such a way that on the average they would contain the same number of particles $N_*$. The domains I and II form "the nucleus of a spot", the domains III and IV are directly attached to a conditional boundary of a spot, $r = r_q$, while the domains V and VI represent "the external domain". Presented in Fig. 4.10 are the results for the time moment $t = 6$. The solid lines depict the values

**Table 4.1.**

| No of domain | I | II | III | IV | V | VI |
|---|---|---|---|---|---|---|
| $\underline{x}$ | 0.00 | 0.40 | 0.92 | 1.00 | 1.17 | 1.23 |
| $\bar{x}$ | 0.40 | 0.57 | 1.00 | 1.08 | 1.23 | 1.30 |

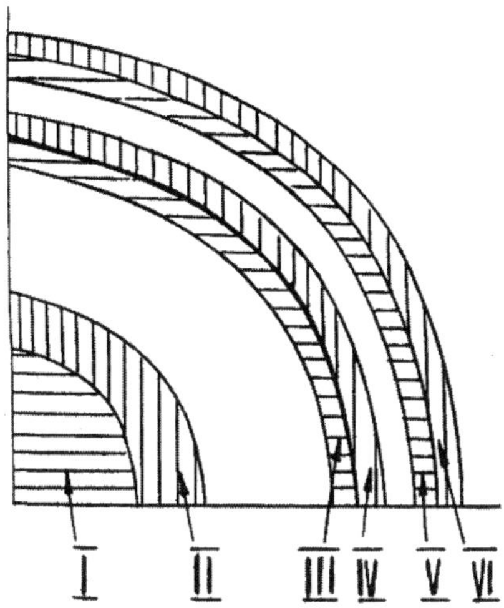


**Fig. 4.9.** The domains for measurements.

$\hat{U}(z)$ and $\hat{V}(z)$, calculated within the domains indicated by means of imitation (the numbers of curves in Fig. 4.10 correspond to the numbers of domains).

The circles correspond to the exact values of $U(z)$ and $V(z)$ for a normal distribution. The vertical segments indicate the band of a statistical scatter $\pm 3S$. One can readily see that only curves 1 are found within this band. This means that only near the center of a spot one can adopt a hypothesis on a normal law for the distribution of pulsational velocities.

The straight crosses in Fig. 4.10, which practically coincide with curves 2, depict the values of $U(z)$ and $V(z)$ calculated with the help of formulae of Eq. (4.46) which correspond to the Millionshchikov's hypothesis for $\alpha = 0.25$. The oblique crosses, coinciding with curves 3 at $z < 1.4$, also depict $U(z)$ and $V(z)$ after Millionshchikov, but with $\alpha = 0.65$. As it is evident, for $z \geq 2$ curve 3 representing $V(z)$ goes outside of the band $\pm 3S$ of the trustworthy interval for Millionshchikov's hypothesis. The curves 5 and 6 are found to be in a qualitative disagreement with this hypothesis.

From the analysis exposed above one could make the following conclusion. Only at the center of spot the distribution of pulsations is isotropic and is subject to a normal law. As soon as one goes further away from the center to the periphery, the anisotropy tends to increase; hence near the spot's core it is possible to accept the Millionshchikov's hypothesis, and deviations from it use to begin upon crossing of the conventional boundary

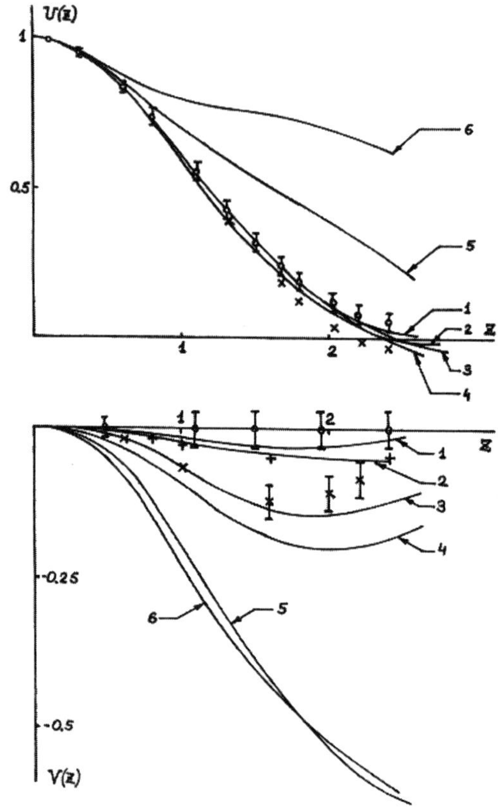


**Fig. 4.10.** Real part $U(z)$ and imaginary part $V(z)$ of the characteristical function of distribution of the pulsations.

$r = r_q$. The distribution of pulsations within the external domain cannot be made agreeable to the hypothesis indicated.

It is possible, however, that taking into account the intermittence phenomenon extends the domain of applicability of the Millionshchikov's hypothesis.

### *4.3.4. The direct statistical modeling of the turbulence within a wake behind the cylinder*

An alternative approach to the construction of the physically sensible models of turbulence consisted, in consideration, of the free turbulence flow as an ensemble of liquid particles, everyone of which is in its motion similar to the Brownian particle. The long-range wake behind the cylinder was modeled as a two-dimensional, stationary problem with self-similar (due to the remoteness from a cylinder) profiles of the mean velocity of flow and its viscosity. The Reynolds equation for mean velocities, corresponding to this problem, might be reduced to a form[32]:

$$u(x,y)\frac{\partial u(x,y)}{\partial x} = -\frac{\partial \overline{u'v'}}{\partial y}, \quad \bar{v}(y) = 0, \tag{4.47}$$

where $u(x,y)$ is the velocity of flow along the $x$-axis, of which the direction coincides with that of a wake; $v(x,y)$ is the velocity normal to that direction, while symbols $u' = u - \bar{u}$ and $v' = v - \bar{v}$ correspond to the pulsational velocities, the line over the quantities being the symbol of a mean value.

Upon the introduction of the deficiency of instantaneous velocity $u^* = U_\infty - u$, where $U_\infty$ is the velocity of flow running against the cylindrical body normal to its axis, Reynolds equation will reduce to a form

$$U_\infty \frac{\partial \bar{u}^*}{\partial x} = -\frac{\partial \overline{u^{*\prime}v'}}{\partial y}. \tag{4.48}$$

As it is known, the deficiency of instantaneous velocity in a long-range wake, under condition of a self-similarity of the flow, is fairly described by the Goertler's model. This model corresponds to the following approximation of the friction stress:

$$\overline{u^{*\prime}v'} = -\nu_* \frac{\partial \overline{u^*}}{\partial y},$$

where $\nu_* = \text{const}$.

The main equation of this model has a form

$$U_\infty \frac{\partial \overline{u^*}}{\partial x} - \nu_* \frac{\partial^2 \overline{u^*}}{\partial y^2} = 0. \tag{4.49}$$

The self-similar solution of Eq. (4.49) under the condition of conservation of momentum, with $\nu_*$ value equal to 0.0225,[14] has a form

$$\overline{u^*}(x,y) = 2U_\infty \sqrt{\frac{\ln 2}{\pi}} \sqrt{\frac{C_W d}{x}} \exp\left\{-16 \ln 2 \frac{y^2}{C_W dx}\right\} \tag{4.50}$$

here and further on $C_W$ is a coefficient of the body's drag and $d$ is cylinder's diameter. The variable most convenient for statistical modeling is $t = (x - x_0)/U_\infty$, where $x_0$ is a coordinate along the wake's axis which corresponds to a conventional boundary for self-similar solution ($x_0 \gg d$).

It might be noted that the basis of the method of direct statistical modeling is formed by the idea of L. Prandtl concerning the turbulent mixing of the liquid particles–moles.

To describe the mole's velocity for the present planar problem, it will be convenient to introduce the longitudinal component $u$ along the wake's axis and the transversal (or vertical) one, $v$, which is normal to the axis. One should keep in mind that the vertical velocity $v$ is purely pulsational; this property is determined by the fact that due to the character of the flow within a long-range wake the mean value $\bar{v}$ of the transversal velocity vanishes.

The distinctive feature of the present approach is the assumption that each mole, in its transversal motion, behaves like a Brownian particle. Moreover, the moles conserve the $x$-component of their momentum. Thus, being invariable in their volumes, they conserve their own constant velocity $u$ during the whole period of observation. All the moles are of one and the same size and the mole's position is characterized by the coordinate of its center. At the initial moment of time the moles are continuously distributed along the $y$-axis. Depending on the mole's position $y$ each of them is put in accordance with the longitudinal velocity $\bar{u}(0, y)$. Later, for the determination of the characteristics of flow it would be sufficient to follow the evolution of this particular and, generally, infinite "string" of moles distributed along the $y$-axis. During the time interval $\Delta t$ the moles will be shifted along the $x$-axis, and the step of this shift is expressed in terms of the velocity of the oncoming stream: $\Delta x = U_\infty \Delta t$. Within the same interval the mole's coordinates along the $y$-axis vary in accordance with the law of motion of the Brownian particles. The diffusion coefficient $\nu_*$ serves as a parameter of this motion, it determines the intensity of the mole's shift and of their mixing. Note that all the moles possess equal properties at any moment during the modeling process, and keep to move isotropically along the $y$-axis, but that fact does not spoil the uniformity of their distribution along the axis.

As a result of the Brownian motion, at the end of the observation interval $\Delta t$, in the vicinity of the point on the axis $y$, there will be a mole, which is generally different from the one located at this point at the initial moment of observation. Due to the fact that the flow's velocity is that of the moles, the flow's velocity at the point considered is changing, too. To calculate the mean velocity at the point $y$, it is sufficient to determine the mean values of the velocities of those moles that are found to be located in the vicinity of that point.

It is important that the present model permits to calculate even such quantities like the friction stress. As it is known, the friction stress at the point $y_*$ is nothing more as a transfer of the corresponding momentum over this point. For this reason, such a modeling is sufficient to record all the moles crossing the point $y_*$ during the time $\Delta t$, taking into account their moments. Each of the moles brings its momentum equal to the product of its longitudinal velocity on its size and on fluid's density. The sum of these moments with consideration of their signs (plus if the mole is moving in the direction of the increase of $y$, and minus if in the inverse direction) is just the total momentum's transfer over the point $y_*$, within the time $\Delta t$.

The constant $\nu_*$, which is taking part in the process of modeling, is a free parameter of the model. The numerical experiments, conducted with computer, have demonstrated the complete coincidence between the model profile of a mean velocity and the Goertlerian one, if only the value $\nu_*$ is chosen to be the same as in Goertler's approximation, that is, in Eqs. (4.49) and (4.50). Moreover, the viscosity coefficient obtained in modeling also coincides with the Goertler's constant $\nu_*$. Thus with the help of a methodology presented here one obtains the results which are in complete agreement with the Goertler's model of the long-range wake, as concerns the mean velocity and the friction drag. It is to be noted once again that putting into our model only the initial profile of mean velocity one is able to obtain, practically, complete information on the flow resulting, including that on viscosity which in the Goertler's equation was selected *a priori*. The coincidence of the results of the model considered with the Goertler's solution should be looked at only as a fair physical interpretation for the equation chosen on the basis of empirical consideration. From the other side, this coincidence just stresses the fact of model viscosity being in disagreement with experimental data. Thus, according to the results of Ref. 35, the turbulent viscosity in a long-range wake, being calculated in self-similar variables, proves to be constant along $OY$ direction (see Fig. 4.15, dashed line). For this reasons, there arises a necessity in expanding the model in such a way that the disagreement mentioned would be overcome.

The direct modeling might be looked upon, moreover, as a way of construction of the model having a real physical significance; thus it would be quite natural to turn to the empirical physical reasons. First of all, it seems to be necessary to define the boundaries of the domain of turbulence. As it is known, the concurrent flow under consideration consists, actually, of two parts: a strongly turbulized core, and the laminar fluid surrounding it. The distinction between these two subdomains is determined by the index of intermittence $\gamma$, which reflects the multi-scaled character of turbulence. Following Landau,[33] let us acknowledge the existence of two scales — a fine one connected with pulsations within the turbulent part of flow, and a coarse one connected with oscillations of the form of the closed domain occupied by the turbulized fluid. In this sense, the model described above is a single-scaled one, and it considers the flow of a fluid as uniformly perturbed within the whole space. This means that, actually, the flow modeled in such a way corresponds only to the turbulent core of a wake. The restriction of a turbulent domain by introduction of the boundary dividing the flow into two logical parts is, actually, just a presentation of a second scale of turbulence. Wishing to define this boundary one can use the well known experimental expression for the index of intermittence.[34] As soon as the flow is divided into subdomains, the character of the motion of moles will be changed, too. It might be contemplated that those moles which prove to be found within the zone of turbulence, i.e. those, whose coordinates occurred within the coordinates of a boundary, move like Brownian particles, normally to the axis of a wake. If in the process of its motion the mole reaches the boundary, it will reflect from it and will move as a Brownian particle, once again. Those moles which form the laminar part of the flow, i.e. those whose coordinates are located outside the boundary of turbulence, do not have any vertical velocities at all. It is to be noted that the boundary's position is a random quantity, and thus the turbulent domain might later capture the mole that is presently located within the laminar zone.

Such a way of modeling of the turbulent mixing will be called "modeling with a boundary". The character of the flow resulted differs significantly from that in the case of a model which does not take into account intermittence effects. The mean velocity's profile becomes to be steeper and does not posses any "tail" dragging on, which is characteristic for the Goertler's model. The viscosity is not constant and is proportional to the intermittence index $\gamma(\dots, y)$.

Thus, starting with the natural preconditions and constructing the physical model of a real significance one obtains the results that are close to the experimental data.

As concerns the statistical modeling "without the boundary", it will correspond to the following Cauchy problem:

$$\frac{\partial \bar{u}(t,y)}{\partial t} - \nu_* \frac{\partial^2 \bar{u}(t,y)}{\partial y^2} = 0,$$
$$t = (x - x_0)/U_\infty, \quad \bar{u}(0,y) = \bar{u}_0(y). \tag{4.51}$$

As a quantity $\bar{u}_0(y)$ chosen a certain initial profile of the mean velocity like, for example, the Goertler's solution (4.50), or the approximate experimental profile by Townsend[35] at the chosen point $x = x_0$; here $x_0$ is that coordinate at the wake's axis, from which the modeling process begins. As it is evident, Eq. (4.51) corresponds to Eq. (4.49) with accuracy of the change of variables.

Given below is a detailed description of all the computational procedures connected with the evolution of $N$ moles. Here, mole is understood as certain volume of fluid. It is characterized by its size $\Delta y_0$, by the coordinate of its center $\xi_i(t)$, and by the value of the actual longitudinal velocity $u_i$. The modeling is carried out until the moment of time $T$, by means of proceeding in discrete steps $\Delta t$. The space is bounded in the direction of $y$-axis by the quantity $Iy_m$, so that only the interval $[-y_m, y_m]$ is considered. At the initial moment the moles are distributed along the interval $[-y_m, y_m]$ as follows:

$$\xi_i(0) = -y_m + i\Delta y_0 - \Delta y_0/2, \quad i = 1, \ldots, N.$$

When thus located, the moles are not only uniformly distributed over the $y$-axis, but the number of moles within each cell of the grid is one and the same and equal to $N_0 = \Delta y/\Delta y_0$. In correspondence to the mole's coordinate, the value of its longitudinal velocity is defined as $u_i = \bar{u}_0(\xi_i(0))$. The mole conserves this velocity during all the modeling process.

The calculation of the model's evolution within the small interval $\Delta t$ is reduced to a realization of $N$ Brownian trajectories. As this takes place, the variation of the model's coordinates at the axis $y$ proceeds according to the law:

$$\xi_i(t + \Delta t) = \xi_i(t) + S_i, \quad i = \overline{1, N}, \quad S_i = \sqrt{2v_* \Delta t}\zeta;$$

here and further on $\zeta$ is a normally distributed random quantity having the density of probability $n(0, 1)$.

The calculation of macroparameters proceeds in the following way. The mean longitudinal velocity of flow is determined at the end of the current step $\Delta t$ at the cell's center, i.e. at the points $y_j^* = -y_m + \Delta y(j - 1) + \Delta y/2, j = \overline{1, N}$, as a mean value of the longitudinal velocity of those moles, whose coordinates belong to the cell considered, that is,

$$\bar{u}(t + \Delta t, y_j^+) = \frac{1}{K}\sum_{k=1}^{K} u_{i_k}, \quad \text{where } i_k : \xi_{i_k}(t + \Delta t) \in [y_j, y_{j+1}].$$

At the moment $t$, the turbulent friction $\overline{u'v'}$ is considered as the transfer of the longitudinal momentum through this point, per unit time. Let the fluid's density be equal to $\rho$ , and then each mole will have the longitudinal momentum $u_i\rho\Delta y_0$, which does not vary with time. Prior to each step $\Delta t$ the quantities $\overline{u'v'}|_{y=y_j}, j = \overline{1, J}$ are given the zero values. If in the process of the Brownian motion, the $i$th mole at the present step is crossing the cell's boundary $y_j$, then to the quantity $\overline{u'v'}|_{y=y_j}$ is added the quantity $\text{sign}(S_i)u_i\rho\Delta y_0$. At the end of a step the resulting sum $\overline{u'v'}|_{y=y_j}$ is divided by $\Delta t$. Thus, the formula for the calculation of the momentum's transfer is written in a form

$$\overline{u'v'}\Big|_{t,y=y_i} = \frac{1}{\Delta t}\sum_{i_k} \text{sign}(S_{i_k})u_{i_k}\rho\Delta y_0,$$

where $i_k$ are the numbers of those moles, whose trajectories have crossed the point $y_j$.

The macroparameters might be determined at each particular step $\Delta t$. After the calculation of the flow's characteristics at the final moment $T$ all the results obtained at each step $\Delta t$ are put into memory, and the modeling begins anew, starting from $t = 0$, which is repeated $N_S$ times. This kind of action is necessary for the statistical accumulation and for finding the mean values on the basis of obtained ones, using $N_S$ realizations.

The model described is easily yielded to the theoretical analysis. The law of the Brownian motion of the moles might be presented as

$$\dot{\xi} = \sqrt{2\nu_*}V(t), \tag{4.52}$$

where $V(t)$ is a standard white noise.

As it is known, the transitional probabilities for the random process with a statistical equation

$$\dot{\xi} = f(\xi, t) + \bar{\sigma}(\xi, t)V(t), \tag{4.53}$$

satisfy the Fokker–Planck equation (4.51):

$$\begin{aligned}\frac{\partial p(t_0, y_0; t, y)}{\partial t} &= \frac{\partial}{\partial y}\{f(y, t)p(t_0, y_0; t, y)\} \\ &\quad + \frac{1}{2}\frac{\partial^2}{\partial y^2}\{\tilde{\sigma}^2(y, t)p(t_0, y_0; t, y)\}\end{aligned} \tag{4.54}$$

and the initial data: $p(t_0, y_0; t, y) = \delta(y - y_0)$. For our particular case $f(\xi, t) = 0, \quad \tilde{\sigma}^2(\xi, t) = 2\nu_*$. Thus, the transitional probability $p(t_0, y_0; t, y)$

is a solution of the following problem:

$$\frac{\partial p(t_0, y; t, y)}{\partial t} + \frac{\partial}{\partial y}\left(-\nu_* \frac{\partial p(t_0, y_0; t, y)}{\partial y}\right) = 0,$$

$$p(t_0, y_0; t, y) = \delta(y - y_0).$$

If $\bar{u}_0(y)$ is an initial profile of velocity, then, the distribution of mean velocity at the arbitrary moment $t$ is expressed by

$$\bar{u}(t, y) = \int_{-\infty}^{+\infty} \bar{u}_0(y_0) p_0(t_0, y_0; t, y) dy_0, \tag{4.55}$$

and proves to be a solution of the problem formulated as Eq. (4.51).

According to the method of modeling described, one obtains the value of a turbulent viscosity $v_T = \overline{u'v'}/\frac{\partial \bar{u}}{\partial y} = v_*$, constant over the wake's cross-section; this, however, does not correspond to the experimental data (see Fig. 4.15). From there it follows that the model chosen inaccurately describes the mechanism of turbulence. Actually, it does not take into account the intermittence effect and the division of a flow into turbulent and laminar domains. For the model "with a boundary", the alternative way of modeling is considered, which is based on the introduction of a random boundary separating the turbulized core of a wake from its laminar surrounding. As it might be conjectured from experimental data, the intermittence coefficient characterizing such a boundary is properly approximated by the following formula:

$$\gamma(x, y) = \int_{y}^{+\infty} \frac{1}{\sqrt{2\pi\sigma^2(x)}} \exp\left\{\frac{-(\xi - b(x))^2}{2\sigma^2(x)}\right\} d\xi, \quad y > 0. \tag{4.56}$$

Here $\gamma(x, y)$ might be interpreted as a probability of the given point $y$ being captured by the turbulent domain: $\gamma(x, y) = P\{y < \xi_{\gamma+}(x)\}$, where $\xi_{\gamma+}$ is a positive quantity taken from a solution of normally distributed random quantities with a probability density $n(b(x), \sigma^2(x))$. The determination of a boundary of turbulence in the upper half-plane of a wake, $y > 0$, is reduced to the drawing of a random quantity $\xi_{\gamma+}$. For $y < 0$, the flow is asymmetrical, and the boundary's location is determined by the quantity $\xi_{\gamma-}$, i.e. by the negative value taken from a selection of normally distributed quantities with a probability density $n(-b(x), \sigma^2(x))$.

The major part of the modeling process coincides with that of a model "without boundary". The only modification concerns the block of the moles' displacement, and that is constructed as follows.

Let the current coordinate of the $i$th mole be $\xi_i(t)$. The boundaries of a turbulent domain are drawn as

$$\xi_{\gamma+} = b(t) + \sigma(t)\varsigma, \quad \xi_{\gamma-} = -b(t) + \sigma(t)\varsigma.$$

If the mole considered is not located within this zone, i.e. if $\xi_i(t) < \xi_{\gamma-}$ or $\xi_i(t) > \xi_{\gamma+}$, then during the time $\Delta t$ its coordinate along the axis $y$ does not vary, and $\xi_i(t + \Delta t) = \xi_i(t)$. If $\xi_{\gamma-} \le \xi_i(t) \le \xi_{\gamma+}$, then the Brownian displacement of the mole is drawn, $S_i = \sqrt{2v_* \Delta t}\varsigma$. If one finds that $\xi_{\gamma-} \le \xi_i(t) + S_i \le \xi_{\gamma+}$, i.e. the moving mole does not reach the boundary, then one sets $\xi_i(t + \Delta t) = \xi_i(t) + S_i$. In those cases when $\xi_i(t) + S_i > \xi_{\gamma+}$ or $\xi_i(t) + S_i < \xi_{\gamma-}$, it is necessary to find the time needed for the mole to reach the boundary of

$$\text{Positive domain: } \tau = \frac{\xi_{\gamma+} - \xi_i(t)}{(S_i/\Delta t)}, \quad \text{or}$$

$$\text{Negative domain: } \tau = \frac{\xi_{\gamma-} - \xi_i(t)}{(S_i/\Delta t)}.$$

Then, the mole is reflected from the boundary, and in the quality of a Brownian particle continues to move in reverse direction,

$$\xi_i(t + \Delta t) = \xi_{\gamma+} - \sqrt{2v_*(\Delta t - \tau)}\varsigma_+,$$

if the reflection is realized from $\xi_{\gamma+}$ boundary, and

$$\xi_i(t + \Delta t) = \xi_{\gamma-} + \sqrt{2v_*(\Delta t - \tau)}\varsigma_+,$$

if the reflection occurs from $\xi_{\gamma-}$ boundary, where $\varsigma_+$ is a positive quantity taken from a selection of the normally distributed random quantities with the probability density $n(0, 1)$. The prescription of initial data and the determination of macroparameters of the flow are conducted in a complete accordance with the above-described method of modeling without boundaries. This means that the distinction of the present model from the preceding one consists in that along the $y$-axis only those moles are moving, which happen to be in the zone of turbulence $[\xi_{\gamma-}(t), \xi_{\gamma+}(t)]$ and cannot go out of this zone. Due to the introduction of a boundary, the present model will contain, in addition to a constant $v_*$, two empirical functions, $b(x)$ and $\sigma(x)$.

By means of such an approach, and using the direct modeling "with a boundary", one is able to solve the following problem for the field of a mean velocity:

$$\begin{cases} \dfrac{\partial \bar{u}(t,y)}{\partial t} - \dfrac{\partial}{\partial y}\left\{ v_* \gamma(t,y) \dfrac{\partial \bar{u}(t,y)}{\partial y} \right\} = 0, \\ \bar{u}(0,y) = \bar{u}_0(y), \end{cases} \tag{4.57}$$

where the intermittence coefficient $\gamma(t,y)$ is determined by Eq. (4.56) with the corresponding change of the variable $x$ for $t$. In distinction from the problem defined by Eq. (4.51), the friction stress is expressed as

$$\langle u'v' \rangle = v_* \gamma(t,y) \frac{\partial \bar{u}(t,y)}{\partial y}.$$

Following the reasoning similar to those which were made by the analysis of the model "without boundary", one can put into a correspondence to the problem of Eq. (25), the following problem formulated for the transitional probabilities of the corresponding random process:

$$\frac{\partial p(t_0,y_0;t,y)}{\partial t} - \frac{\partial}{\partial y}\left\{ v_* \gamma(t,y) \frac{\partial p(t_0,y_0;t,y)}{\partial y} \right\} = 0,$$

$$p(t_0,y_0;t,y) = \delta(y-y_0). \tag{4.58}$$

The first of Eq. (4.58) might be rewritten in a form

$$\frac{\partial p(t_0,y;t,y)}{\partial t} - \frac{\partial}{\partial y}\left\{ \frac{\partial v_* \gamma(t,y)}{\partial y} p(t_0,y_0;t,y) \right\}$$
$$- \frac{\partial^2}{\partial y^2} \{ v_* \gamma(t,y) p(t_0,y_0;t,y) \} = 0,$$

which coincides with that of Fokker–Planck equation (Eq. (4.54)) if one takes

$$f(y,t) = \frac{\partial v_* \gamma(t,y)}{\partial y} \quad \text{and} \quad \tilde{\sigma}^2(y,t) = 2v_* \gamma(t,y).$$

Hence from the problem of Eq. (4.57), one can able to pass to a random process of Eq. (4.53) with $\xi = f(\xi,t) + \tilde{\sigma}(\xi,t)V(t)$. For the present case, taking into account the expression of Eq. (4.56) for the intermittence coefficient, one gets

$$f(y,t) = -v_* (2\pi\sigma^2)^{-1/2} \exp\left\{ \frac{-(\xi-b)^2}{2\sigma^2} \right\}.$$

Prior to the mole's displacement one should fix the randomly drawn boundary $\xi_\gamma$, and that might be interpreted as a limiting transition $\sigma/b \to 0$. Then, $f(y,t)$ tends to become a delta-function $v_*\delta(\xi - \xi_\gamma)$, and $\gamma(y,t)$ tends to $\theta(\xi - \xi_\gamma)$ — Heaviside function, which might be put out of the radical sign. Thus, the equation of the mole's motion might be presented in the form

$$\dot{\xi} = -v_*\delta(\xi - \xi_\gamma) + \sqrt{2v_*}\theta(\xi - \xi_\gamma)V(t).$$

The first term of the right-hand side describes the reflection from the boundary of turbulence $\xi_\gamma$, if the mole reaches it as a result of the Brownian motion characterized by the second term of the same expression. Modeling of the random process corresponding to this equation was, actually, described above.

### 4.3.5. *Simulation results*

To solve the problems concerning the long-range turbulent wake calculations of the evolution of the model at a small time interval $\Delta t$ were made in such a way as shown above.

As the characteristic parameters the following values may be taken $y_m = 8.1; \Delta y = 0.2; \Delta y_0 = 0.004; N_S = 1000$. The cylinder's diameter $d$ and the velocity of the oncoming flow $U_\infty$ were taken as the units of measurement.

Calculations were made with the number of empirical constants. The results described were calculated with the following parameters: $\nu_*/U_\infty C_W d = 0.017, b/\sqrt{C_W x d} = 0.39, \sigma/b = 0.38$. Simulation also gives rise to experimental data at any initial profile of average velocity when the following conditions are satisfied: $\Delta t < 5; \Delta y < S = \sqrt{2\nu_\gamma \Delta t}$, and when the values of empirical parameters are chosen as shown above.

Figures 4.11–4.15 show the results of simulation in comparison with the experimental and calculated data. The solid lines are Townsend's[35] experiments; the squares are the simulation results; the dashed lines are the results in accordance with Goertler. Everywhere there is coincidence of our results with Townsend's data. Comparison of Figs. 4.11 and 4.12 demonstrates that the velocity according to Goertler is valid only in the core of the flow. More important is the difference in the profiles of the friction strain (Fig. 4.13). Turbulent viscosity fits well in the intermittence curve (Fig. 4.14), except the regions of small velocity gradients. The turbulent viscosity (Fig. 4.15) according to Goertler is a constant value.

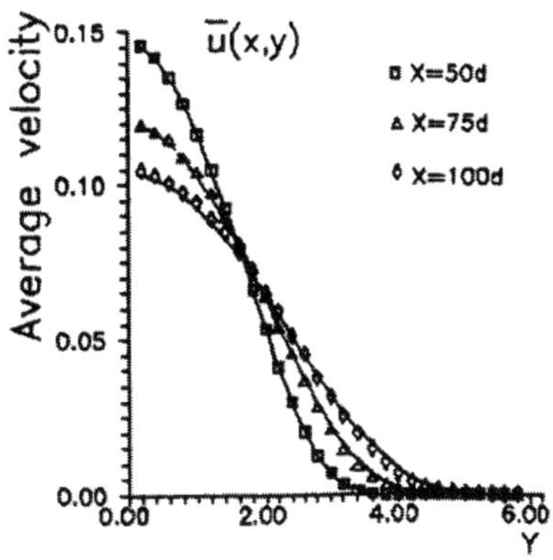


**Fig. 4.11.** Profiles of average velocity.

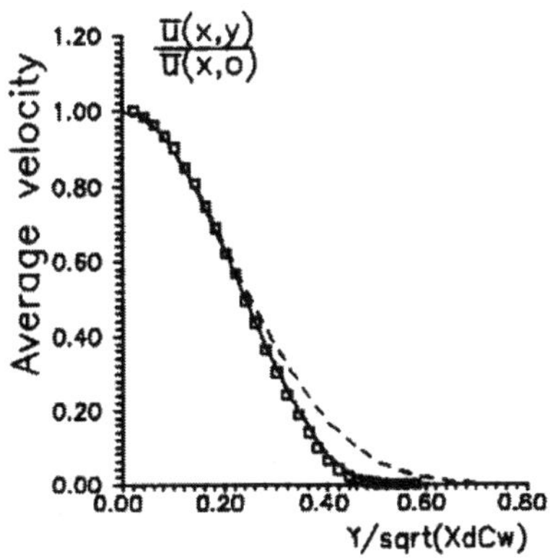


**Fig. 4.12.** Normalized profiles of average velocity.

### *4.3.6. Conclusion*

Thus, in the present chapter, two different approaches to the statistical modeling of a free turbulence are presented within a long-range wake of the incompressible fluid. Both of these approaches are based on the assumption that the medium to be modeled is considered as a discrete totality of a finite number of mobile points — liquid moles.

The first approach is a modification of the statistical method of particles, widely used in rarefied gas-dynamics. Its mathematical basis is made up of the kinetic equation for the one-point function of the distribution of probabilities for the pulsation of the hydrodynamic parameters. This

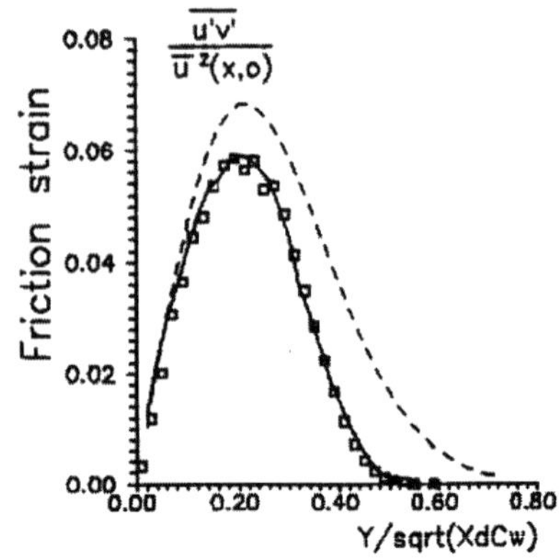


**Fig. 4.13.** Profiles of friction strain.

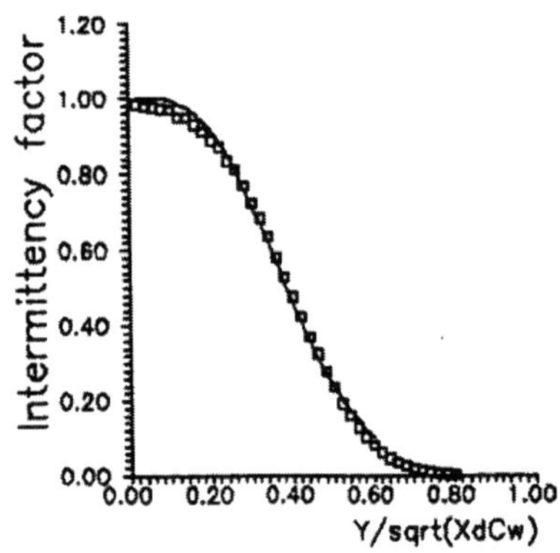


**Fig. 4.14.** Profiles of intermittency factor.

permits to realize the numerical determination of the function indicated itself, or of its arbitrary moments (one-point correlators of the pulsations of corresponding parameters). In this sense, the present approach is more informative than the approaches based on the truncated chains of Reynolds equations. The main obstacle on a way to the general inculcation of the method considered consists in the lack of the universal kinetic equations for turbulence. It seems, however, that in many cases it would be possible to limit oneself by some simple models like, for example, those of the relaxational type; just that was done here.

The second approach is essentially different from the first one. Its mathematical basis is formed by the stochastic equation describing the motion

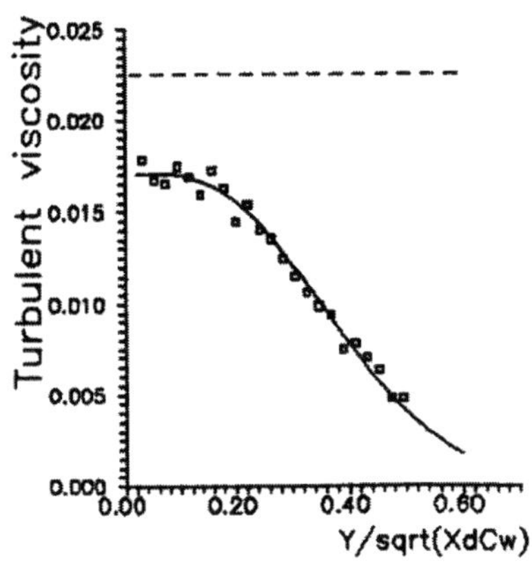


**Fig. 4.15.** Profiles of turbulent viscosity.

of moles as Brownian particles. From the point of view of the numerical realization such a method is simpler than the first one, though it is less informative. This method permits to calculate the main characteristic features of turbulent flows, such as the fields of mean velocities and Reynolds' friction stresses.

## References


1. O. M. Belotserkovskii and V. E. Yanitskii, Statistical "particle-in-cell" method for the solution of the problems of rarefied gas dynamic, *USSR Comput. Math. Math. Phys.* **15**(5/6), 184 (1975).
2. G. A. Bird, *Molecular Gas Dynamics and the Direct Simulation of Gas Flows* (Clarendon Press, Oxford, 1976).
3. V. E. Yanitskii, Chisl. metody mekh. splosh. sredy, *Novosibirsk.* **6**(4), 139 (1975) (in Russian).
4. S. M. Ermakov, V. V. Nekrutkin and A. S. Sipin, *Random Processes for Solving Classical Equations of Mathematical Physics* (Nauka, Moscow, 1984) (in Russian).
5. A. A. Arsen'ev, *USSR Comput. Math. Math. Phys.* **27**(2), 51 (1987).
6. O. M. Belotserkovskii, A. I. Erofeev and V. E. Yanitskii, *USSR Comput. Math. Math. Phys.* **20**(5), 82 (1980).
7. A. E. Korolev and V. E. Yanitskii, *USSR Comput. Math. Math. Phys.* **25**(2), 70 (1985).
8. S. P. Radev, S. K. Stefanov and V. E. Yanitskii, *Proc. 14th Int. RGD Symp.*, Tsukuba, Japan (1984).
9. A. P. Genich, S. V. Kulikov, G. B. Manelis, V. V. Serikov and V. E. Yanitskii, *USSR Comput. Math. Math. Phys.* **26**(6), 153 (1986).

10. N. A. Nurlybaev, *Development of Weighting Schemes for Modeling Nonequilibrium Chemical Reactions in Gases*, Computing Center of the USSR Academy of Sciences Moscow (1986) (in Russian).
11. J. M. Ferziger and H. G. Kaper, *Mathematical Theory of Transport Processes in Gases* (North Holland Publ. Co., Amsterdam, 1972).
12. M. L. Leontovich, *Zh. Eksp. Teor. Fiz.* **5**, 211 (1935).
13. M. Kac, *Probability and Related Topics in Physical Sciences* (Interscience Publishers Ltd., New York, 1957).
14. I. R. Prigogine, *Nonequilibrium Statistical Mechanics* (Interscience Publishers, New York, 1962).
15. S. A. Denisik, S. N. Lebedev, Yu. G. Malama and A. I. Osipov, *Fizika Goreniya i Vzryva* **8**, 331 (1972).
16. V. E. Yanitskii, *USSR Comput. Math. Math. Phys.* **13**(2), 310 (1973).
17. W. Feller, *An Introduction to the Probability Theory and Its Applications* (Wiley, New York, 1971).
18. A. P. Genich, S. V. Kulikov, G. B. Manelis, V. V. Serikov and V. E. Yanitskii, Application of the weighted schemes of statistical modeling of the flows of multicomponent gas to the calculation, of the structure of shock wave, *USSR Comput. Math. Math. Phys.* **26**(12) (1986).
19. L. Hesselink and B. Sturtevant, Propagation of weak shocks through a random medium, *J. Fluid. Mech.* **196**, 513–533 (1988).
20. D. Rotman, Shock wave effects on a turbulent flow, *Phys. Fluids A* **3**, 1792–1806 (1991).
21. O. A. Azarova, E. A. Bratinkova, L. S. Stemenko, F. V. Shugaev and V. E. Yanitskii, Influence of a shock wave on density pulsations in the flow, *Vestnik Moscow University* **3**(5) (1997).
22. A. S. Monin, Equation for the finite-dimensional distributions of the field of turbulence, *USSR Acad. Sci. Dokl.* **177**(5), 1036–1038 (1967) (in Russian).
23. E. A. Novikov, Kinetic equation for the field of vortex, *USSR Acad. Sci. Dokl.* **177**(2), 299–305 (1967) (in Russian).
24. V. V. Struminskii, On the possibility of the application of dynamical method for description of the turbulent flows, in *Turbulent Flows* (Nauka, Moscow, 1974), pp. 19–33 (in Russian).
25. T. S. Lundgren, Model equation for nonhomogeneous turbulence, *Phys. Fluids* **12**(3), 969–975 (1969).
26. A. T. Onufriev, On the equation of semiempirical theory of turbulent transfer, *Appl. Mech. Phys.* (2), 66–71 (1970) (in Russian).
27. O. M. Belotserkovskii, A. I. Erofeev and V. E. Yanitskii, Direct Monte-Carlo simulation of aerohydrodynamics problems, in *Proc. 13th Int. RGD Symp.*, Novosibirsk, RSFSR, 1982 (Plenum Press, New York, 1985), Vol. 1, pp. 313–332.
28. O. M. Belotserkovskii, *Numerical Modeling in Continuum Mechanics* (Nauka, Moscow, 1984) (in Russian).
29. E. Naudasher, Flow in the wake of self-propelled body and related sources, *J. Fluid Mech.* **22**(4), 625–656 (1965).

30. O. V. Troshkin and V. E. Yanitskii, *The Intermittence in One Problem of the Free Turbulence*, Computing Center of the USSR Academy of Sciences, Moscow (1988) (in Russian).
31. V. E. Yanitskii, *Stochastic Models of Perfect Gas Consisting of the Finite Number of Particles*, Computing Center of the USSR Academy of Sciences, Moscow (1988) (in Russian).
32. H. Schlichting, *Boundary Layer Theory* (McGraw–Hill, New York, 1960).
33. L. D. Landau and E. M. Lifshitz, *Hydrodynamics* (Nauka, Moscow, 1986) (in Russian).
34. J. C. LaRue and P. A. Libby, Statistical properties of the interface in the turbulent wake of a heated cylinder *Phys. Fluids* (12), 1864–1875 (1976).
35. A. A. Townsend, *The Structure of Turbulent Shear Flow* (Cambridge University Press, 1956).

# Appendix A

# Computational Experiment: Direct Numerical Simulation of Complex Gas-Dynamical Flows on the Basis of Euler, Navier–Stokes, and Boltzmann Models*

## A.1. Introduction

### *A.1.1. The use of numerical methods*

At present, specialists in the applied sciences are confronted with various kinds of practical problems whose successful and accurate solution, in most cases, may be attained only by numerical methods with the aid of computers. Certainly, this does not mean that analytical methods that permit us to find the solution in "closed" form will not be developed. Nevertheless, it is absolutely clear that the range of problems permitting such a solution is rather narrow; therefore, the development of general numerical algorithms for the investigation of problems of mathematical physics is important. It is especially urgent in continuous mechanics (gas-dynamics, theory of elasticity, etc.) for at least two basic reasons:

(1) *The difficulty of carrying out the experiment.* In studying the phenomena taking place, for example, at hypersonic flight velocities, the resulting high temperatures give rise to the effects of dissociation, ionization in the flow, and, in a number of cases, even to "luminescence" of a gas. In these cases it is enormously difficult to simulate the experiment in the laboratory, since, for similarity between the natural environment

*O. M. Belotserkovskii, Computational experiment: direct numerical simulation of complex gas-dynamics flows on the basis of Euler, Navier–Stokes, and Boltzmann models, Karman's Lecture, von Karman Institute for Fluid Dynamics, March 15–19, 1976, in *Numerical Methods in Fluid Dynamics*, eds. H. J. Wirz and J. J. Smolderen (Hemisphere, Washington, London, 1978), pp. 339–387.

and the conditions of the experiment, it is not sufficient to satisfy the classical criteria of similarity, i.e. the equality of the Mach and Reynolds numbers. The equality of absolute pressures and absolute temperatures is also required, and this is possible only if the sizes of the model and the real object are equal. All this involves numerous technical difficulties, which lead to high experimental costs. In addition, in many cases, the results of the tests are rather scarce. However, the importance of experimentation, in principle, must not be underestimated. Experimentation is always the basis of investigations confirming (or rejecting) the scheme and the solution according to some theoretical approach.

(2) *The complexity of the equations considered.* Deep penetration of numerical methods into the mechanics of continua is also explained by the fact that the equations of aerodynamics, gas-dynamics, and the theory of elasticity represent the most complicated systems of partial differential equations (as compared to other branches of mathematics). In a general case, this is a nonlinear system of mixed type with the unknown form of the conversion surface (where the equations change their type) and with "movable boundaries", i.e. the boundary conditions are given on surfaces or lines that, in turn, are determined by calculations. Moreover, the range of the unknown functions is so wide that ordinary methods of analytical investigations (linearization of equations, series expansion, separation of a small parameter, etc.) cannot be used to derive the full solution of the problem for this general case. It should be noted that, in solving complicated problems on electronic computers, the preliminary analytical investigation of a problem might be of great aid; sometimes, this investigation is decisive in the successful realization of the numerical algorithm.

Let us dwell on one more peculiarity of the algorithms used for solving concrete problems of the mechanics of continua. As is known, numerical methods have found wide practical application in design offices and research institutes. Substantial progress in the exploration of the cosmos, the optimum control of vehicles, the choice of rational configurations of vehicles, and so on, is, to a considerable extent, due to serial calculations and the use of scientific information obtained in this way. The volume of information obtained by means of the calculation is far more complete and substantially cheaper than the corresponding experimental investigations if the problem is correctly formulated, well simulated, and algorithmically rational. However, wide application of numerical methods for practical purposes requires sufficient simplicity and reliability.

Thus, on one hand, one has to deal with rather complicated mathematical problems. On the other hand, it is necessary to develop rather simple and reliable numerical methods permitting us to carry out serial calculations at project institutes and design offices.

Note that, for most problems in gas-dynamics, the mathematical theorems of existence and uniqueness have not been proved, and very often there is no confidence that such theorems can be derived. As a rule, the very mathematical formulation of the problem is not strictly given, and only the physical treatment is presented — which are far from being one and the same thing. The mathematical difficulties of investigating such types of problems are related to the nonlinearity of the equations, as well as to the great number of independent variables.

The state of affairs with regard to the methods of solving gas-dynamics equations is no better. So far, investigations related to the possibility of realization of the algorithm, its convergence to the unknown solution, and its stability have been rigorously performed only for linear systems, and, in a number of cases, only for equations with constant coefficients. When confronted with the necessity of solving a problem, the mathematician has to use the known algorithms and to develop new methods without rigorous mathematical basis for their applicability.

This does not mean that the situation is very much different from that in any other new field. In science, as well as in mathematics, one can find many examples of new ideas and concepts that were originated and successfully used without a solid basis, which appeared later. Of course, this is not to imply that, when developing new numerical algorithms, one must act at random, without thinking about the accurate formulation of the problem or without comprehending its physical meaning. That path inevitably leads to numerous mistakes and to the waste of time — and the experience, without being theoretically interpreted, does not lay the foundation for further development of the method.

We draw your attention to this rather clear question only because there is still an opinion that the main thing is to write down differential equations, and all the rest reduces to a trivial substitution of differences for derivatives and to programming (to which too much importance is sometimes attached). In this connection, it is reasonable to formulate the main stages of the numerical solution of a mechanical or physical problem with the aid of computers in the following way:

1. Construction of a physical model and the mathematical statement of a problem.

2. Development of a numerical algorithm and its theoretical interpretation.
3. Programming (manual or automatic) and the formal adjustment of the program.
4. Methodical adjustment of the algorithm, i.e. a test of its operation with concrete problems, elimination of the drawbacks found, and experimental investigation of the algorithm.
5. Serial calculations, accumulation of experience, and estimation of the effectiveness and the range of applicability of the algorithm.

At all stages, the mathematical theory and physical and numerical experimentation (with the aid of computers) are used jointly and consistently. The realization at each stage may be illustrated by solving concrete problems, and this will be done later in this chapter. At this point, we make only some common observations.

The main principle in using mathematical results is that the conditions providing the solution of a problem for simple and special cases must be fulfilled for more common and more complicated cases. Parallel to this, consideration of the physical meaning of the phenomenon gives a qualitative picture with the help of which the statement of the problem is checked and defined more exactly. Ultimately, the final experimental test allows us to judge the correctness of the assumptions and to estimate the accuracy of the algorithm and the derived solution. It should be noted that the accuracy of the numerical solution of the formulated differential problem must be estimated purely mathematically, without using the results of the physical experiment. The latter may be used for qualitative comparison, but quantitative comparison between the calculation and the experiment must provide information on how closely the physical model approaches the natural environment.

### *A.1.2. Numerical methods applicable to gas-dynamical problems*

In the exact sciences there may arise many important problems whose investigation is tied to the solution of a system of nonlinear partial differential equations. Gas-dynamics is one of these sciences, and, furthermore, it includes many problems with discontinuous solutions.

The construction of reasonably accurate solutions of the exact equations of gas-dynamics in the general case has become possible only with the aid of numerical methods that exploit the advantages of high-speed electronic digital computing machines. Technology has called for an intensive development of numerical methods and their application to the solution of a wide variety of gas-dynamical problems. Scientists and research engineers in the area of gas-dynamics have contributed significantly to the development of modern numerical methods for solving systems of nonlinear partial differential equations.

Four universal numerical methods are applicable to the solution of the nonlinear partial differential equations for gas-dynamical problems.

#### *A.1.2.1. Method of finite differences*

This method is the most highly developed of the four at the present time and is widely applied to the solution of both linear and nonlinear equations of the hyperbolic, elliptic, and parabolic types. The region of integration is subdivided into a network of computational cells by a generally fixed orthogonal mesh. Derivatives of functions in the various directions are replaced by finite differences of one form or another; usually, a so-called implicit difference scheme is applied to the integration of the equations. This results in the solution, at each step of the procedure, of a system of linear algebraic equations involving perhaps several hundred unknowns.

Finite-difference schemes are often used for solving steady and unsteady gas-dynamical equations. Lagrangian and Eulerian approaches are widely used here. In the first case, where the coordinate network is related to the liquid particles, the structure of the flow is better defined and one succeeds in constructing rather accurate numerical schemes for flows with comparatively small relative displacements. In the second case, when the calculational network is fixed over space, the schemes are used for constructing flows with large deformation. Recently, the approaches mentioned here have also found wide application in the calculation of steady flows.

#### *A.1.2.2. Method of integral relations*

In this method, which is a generalization of the well-known method of straight lines, the region of integration is subdivided into strips by a series of curves whose shape is determined by the form of the boundaries of the region. The system of partial differential equations written in divergence form is integrated across these strips, the functions occurring in the integrands being replaced by known interpolation functions. The resulting approximate system of ordinary differential equations is integrated numerically. The method of integral relations, like the method of finite differences, is applicable to equations of various types.

### *A.1.2.3. Method of characteristics*

This method is applied only to the solution of equations of the hyperbolic type. The solution, in this case, is computed with the aid of a grid of characteristic lines, which is constructed in the course of the computation. Actually, the method of characteristics is a difference method for integrating systems of hyperbolic equations on the characteristic calculational network, and it is used mainly for the detailed description of flows. Its distinguishing feature, as compared to other difference methods, is the minimum utilization of interpolation operators and associated maximum proximity of the region of influence of the difference scheme as well as the region of influence of the system of differential equations. The smoothing of the profiles inherent in the difference schemes with fixed network is minimal here, since the calculational network used in the method of characteristics is constructed exactly with the region of influence of the system taken into account.

Irregularity (nonconservativeness) of the calculational network should be mentioned as a drawback of the method of characteristics. It is possible to develop a technique, based on this method, in which the calculations are carried out in layers bounded by fixed lines. The method of characteristics permits one to determine accurately the point of origin of secondary shock waves within the field of flow, as the result of the intersection of the characteristics of one family. However, if a large number of such shock waves is developed, difficulties are encountered in their calculation. Accordingly, the method of characteristics is most expediently applied to hyperbolic problems in which the number of discontinuities is small (for example, problems concerning steady supersonic gas flow).

### *A.1.2.4. Particle-in-cell (PIC) method*

In certain respects, the PIC method incorporates the advantages of the Lagrangian and Eulerian approaches. The range of solution here is separated by the fixed (Eulerian) calculation network. However, the continuous medium is interpreted by a discrete model, i.e. the population of particles of fixed mass (Lagrangian network of particles) that moves across the

Eulerian network of cells is considered. The particles are used for determining the parameters of the liquid itself (mass, energy, velocity), whereas the Eulerian network is employed for determining parameters of the field (pressure, density, temperature).

The PIC method allows the investigation of complex phenomena of multicomponent media in dynamics, because particles carefully "watch" free surfaces, lines of separation of the media, and so on. However, because of the discrete representation of the continuous medium (the finite number of particles in a cell), calculational instability (fluctuations) often arises, the calculation of rarefied regions is specifically difficult, and there are other problems. Limitations in the power of modern computers do not permit a considerable increase in the number of particles.

For problems in gas-dynamics in the presence of a uniform medium, it seems more reasonable to use the concept of continuity by considering the mass flow across the boundaries of Eulerian cells rather than "particles".

Only numerical methods using high-speed computers and careful experiments allow us to find the complete solution to a complex gas-dynamical problem and to determine the necessary flow characteristics. Thus, the elaboration of numerical schemes, the calculation of different gas-dynamical problcms, and the study of the analytical properties of the solutions and their asymptotic behavior are of significant interest at present.

### *A.1.3. Development of numerical algorithms*

This chapter is, in essence, a review of the numerical methods used for the determination of the aerodynamic characteristics of high-speed vehicles with transonic and supersonic velocities. The numerical schemes were developed under our supervision and collaboration in the Moscow Physical Technical Institute and the Computing Center of the Academy of Sciences of the USSR. Here we discuss problems in the development and use of numerical algorithms for carrying out serial calculations in solving modern engineering problems arising in practice.

#### *A.1.3.1. Steady-state schemes*

In determining the steady aerodynamic characteristics of bodies (especially when electronic computers of average power were employed) we made wide use of the following methods for solving steady gas-dynamical equations: the method of integral relations (MIR), the method of characteristics

(MCh), and some finite difference schemes (e.g. schemes with "artificial viscosity" and others). We wish to consider problems in which discontinuities and singularities are given beforehand, together with some associated boundary conditions; the solutions are to be carried out in regions where functions vary continuously.

As is known, three different MIR schemes were developed for the determination of flow in the region of a blunt nose; namely, using an approximation for the initial functions across the shock layer (scheme I), along it (scheme II), and in both directions (scheme III). As a result, the boundary-value problem was solved for an approximate system of ordinary differential (algebraic) equations. To solve the three-dimensional problem, some additional trigonometric approximations in the circumferential coordinate were introduced. For different flow conditions and different body shapes, one or another of the MIR schemes has been found applicable; they are widely used in our country as well as abroad.[1–6]

The main advantage of these schemes is that, by means of different transformations, one succeeds eventually in approximating functions (or groups of functions) with comparatively weak variations. They allow us to obtain reliable results and a high degree of accuracy with a comparatively small number of interpolation nodes (usually three or four calculated points were used).

The choice of the independent variables, the form of the initial system of equations of motion (that is, the introduction of the integrals into the initial system, the use of the divergent form of the laws of conservation, and others), the use of conservation schemes, the approximation of the integrals, etc., are all of great importance in writing the numerical algorithm using MIR, and hence in producing results.

The main difficulty in carrying out MIR schemes is the solution of multiparameter boundary problems for the approximating system of equations. This was overcome by means of different iteration schemes. Moreover, these schemes were used in transonic regions mainly for bodies of comparatively simple form, but for a supersonic zone one had to adopt another algorithm.

In calculating supersonic flow, the two- and three-dimensional MCh schemes of Chushkin, Magomedov, and their coworkers were used.[7,8] As is known, once the initial form of the system is written in terms of characteristic variables, one requires the approximation of simple derivatives only. Using a fixed linear computational network, we get a system of finite-difference equations with its several advantages.

With the help of the above-mentioned approaches, a large number of gas-dynamical problems have been solved, namely, ideal gas flows with chemical reactions and radiation, transonic and three-dimensional flows, and viscous flows. In most cases sufficiently steady and reliable results were obtained, which were in perfect agreement with experiment.[6] However, these approaches to the solution of the steady-state equation may be successfully used only for problems in which there are no singularities, discontinuities, intersections, and interactions. The application of these approaches is difficult for bodies of complex form with a large number of discontinuities. Besides, a single algorithm for the calculation of different types of flow is preferable.

### *A.1.3.2. Unsteady-state schemes*

The next step in the evolution of numerical methods, which was motivated by urgent practical needs and aided by the availability of electronic computers, was the development of nonsteady schemes and the use of the stability method for the solution of steady-state aerodynamic problems. We tried to keep to the general principles and ideas of the MIR and MCh in approximating the nonsteady equations with respect to space variables. The divergent or characteristic forms of the initial equations were used, and the same calculational networks were employed.

In this way the nonsteady schemes II and III of the method of integral relations and the grid-characteristical method were developed.[9,10] These allow us to consider rather complicated types of flow with a single algorithm. It is natural that the problems of computational stability and the attainment of steady-state solutions should become crucial. They require some specific technique such as the introduction of artificial viscosity into the initial system, and of dissipation terms into the difference equations. In a number of cases, the accuracy of the results obtained is less than that in the steady-state methods, but these approaches enabled us to consider new classes of problems, for example, the determination of the aerodynamic characteristics of three-dimensional flow for specific configurations, and the calculation of viscous transonic flows.[10]

### *A.1.3.3. Large-particle method*

Finally, in the third stage of development, it seemed reasonable and advantageous to introduce elements of the Harlow particle-in-cell method[11–13] into the algorithms. At first, only the equation of continuity is represented as the mass flow across the Euler cell, using the simplest finite-difference or integral approximation along the coordinates.

Thus, the modified method of large particles[14–15] came into existence, which (again by means of the stability process) allowed us to consider from one point of view such complicated tasks as, for example, the subsonic, transonic, and supersonic flow past a flat-nosed body in two dimensions or with axial symmetry. Such an approach is used in calculating viscous flows, and it may permit us to study the characteristics of separated flows.

It should be stressed that the development of the numerical schemes mentioned above benefits from the improvement and extension of the methods of solving the boundary-value problems for the corresponding approximating equations; the consideration of a new, wider class of problems; and the development and improvement of electronic computers, machine languages, input and output arrangements, and so on.

### *A.1.4. Computational experiments*

In recent years the introduction of large computers has aroused a considerably greater interest in various numerical methods and algorithms whose realization enlarges the effectivity of computational experiment. The need for such an approach to the solution of problems of mathematical physics is the result of ever-growing practical demands; in addition, it is connected to an attempt to construct more rational general theoretical models for the investigation of complex physical phenomena.

Let us outline the principal steps of a computational experiment. At first, one chooses a mathematical model of a physical object based on analytical study. Then one works out a tool for the investigation of the phenomenon in question, namely, a difference scheme that permits the experiment itself to be carried out, i.e. the computational process. The next step comprises a detailed analysis of the results, leading to improvements and corrections in the mathematical model. This feedback procedure leads to modification and perfection of the methodology of numerical experiments.

The close analogy to physical experiments comprising similar steps is evident: analysis of a phenomenon under study; development of an experimental scheme; modification of design elements of the experimental installation; and measurements and their analysis.

In recent years the Computing Center of the USSR Academy of Sciences carried out a number of experiments associated with the studies of complex gas-dynamical flows using a nonstationary method of large particles.[14,15] Characteristic features of flows past bodies of different shapes were studied over a wide range of velocities, from subsonic through transonic and up to hypersonic. The results of a number of such experiments are presented in this chapter, but without the details of the computations.

Our approaches are based on the splitting of physical processes by a time step and on the stabilization of a process for solution of stationary problems. The main purpose of this research is to consider mathematical models for more complex and general gas flows in the presence of large deformations. This approach finds application in the solution of both Euler[14,15] and Navier–Stokes equations.[16,18]

With the help of the large-particles method we succeeded in investigating complex gas-dynamical problems, including transonic overcritical phenomena, injected flows, and separation zones. It is important to note that the above class of problems was regarded from a single viewpoint: subsonic, transonic, and supersonic flows, transition through the sound velocity and the critical regime. The calculation for plane and axisymmetrical bodies was carried out by a single numerical algorithm as well.[10,14,15,19,20]

It also seems promising to apply the main principles of the approach in question to the simulation of rarefied gas flows. The application of a statistical variant of such an approach to the solution of the Boltzmann equation is studied in Refs. 21–23. We shall not dwell on the description of the techniques (it is given in detail in the references); instead, only the characteristic features of each approach will be given.

The basis of the technique of splitting a pattern into physical processes is as follows: the simulated medium may be replaced by a system of $N$ particles (fluid particles for a continuous medium, and molecules for a discrete one) which, at the initial instant of time, are distributed in cells of the Eulerian mesh in a coordinate space in accordance with the initial data.

The evolution of such a system in time $\Delta t$ may be split into two stages: (1) changes in the internal state of subsystems in cells that are assumed to be "frozen" or stable (Eulerian stage for a continuous medium, and collision relaxation for a discrete one) and (2) subsequent displacement of all the particles proportionally to their velocity and $\Delta t$, without changing the internal state (Lagrangian stage for a continuous medium, and free motion of molecules for a discrete one).

The stationary distribution of all the medium parameters is calculated after the process is stabilized in time.

## A.2. "Large-particles" method for the study of complex gas flows

In numerical models constructed by Davydov[14,15] on the basis of Eulerian equations, the mass of a whole fluid (Eulerian) cell, i.e. a "large particle" (from which comes the name of the method), is considered instead of the ensemble of particles in cells. Furthermore, nonstationary (and continuous) flows of these "large particles" across the Eulerian mesh are studied by means of finite-difference or integral representations of conservation laws.

Actually, conservation laws are used in the form of balance equations for a cell of finite dimensions (which is a usual procedure in deriving gas-dynamical equations without further limit transition from cell to point). As a result,we obtain divergent-conservative and dissipative-steady numerical schemes that allow us to study a wide class of complex gas-dynamical problems (overcritical regimes, turbulent flows in the wake of a body, diffraction problems, transition through the sound velocity, etc.).[10,14,15,20]

### *A.2.1. Calculations*

Consider the motion of an ideal compressible gas. Our starting point is provided by the Euler differential equations in the divergence form (the equations of continuity, momentum, and energy):

$$\begin{gathered}
\frac{\partial \rho}{\partial t} + \operatorname{div}(\rho \mathbf{v}) = 0, \\
\frac{\partial \rho U}{\partial t} + \operatorname{div}(\rho U \mathbf{v}) + \frac{\partial P}{\partial x} = 0, \\
\frac{\partial \rho V}{\partial t} + \operatorname{div}(\rho V \mathbf{v}) + \frac{\partial P}{\partial y} = 0, \\
\frac{\partial \rho E}{\partial t} + \operatorname{div}(\rho E \mathbf{v}) + \operatorname{div}(P \mathbf{v}) = 0.
\end{gathered} \tag{A.1}$$

It was shown in Ref. 14 that, in the large-particle method, the set of equations of gas-dynamics, written as laws of conservation in integral form, might be used instead of (A.1). The important point is that the difference scheme approximating the initial set of equations should be homogeneous, so that "through" computation may be performed without isolating singularities.

Equations (A.1) are completed by the equation of state

$$P = P(\rho, E, \mathbf{v}). \tag{A.2}$$

The various stages of the computational cycle will be considered separately. Let us briefly describe the main principles of the large-particle method. The region of integration is covered by a fixed (over space) Eulerian mesh composed of rectangular cells with sides $\Delta x$ and $\Delta y$ ($\Delta z$, $\Delta r$ are along a cylindrical coordinate system).

In the first (Eulerian) stage of the calculations, only those quantities change which are related to a cell as a whole, and the fluid is supposed to be instantaneously decelerated. Hence, the convective terms of the form $\mathrm{div}(\varphi\rho\mathbf{v})$, where $\varphi = (1, U, V, E)$, corresponding to displacement effects, are eliminated from Eq. (A.1). Then it follows from the equation of continuity, in particular, that the density field will be "frozen" and the initial system of equations will be of the form

$$\rho\frac{\partial U}{\partial t} + \frac{\partial P}{\partial x} = 0, \quad \rho\frac{\partial V}{\partial t} + \frac{\partial P}{\partial y} = 0, \quad \rho\frac{\partial E}{\partial t} + \mathrm{div}\,(P\mathbf{v}) = 0. \tag{A.3}$$

Here we have used both the simplest finite-difference approximations and, to improve the calculation stability, the schemes of the method of integral relations, in which "sweeping-through" approximations of the integrands with respect to rays ($N = 3$, 4, 5) are used.

In the second (Lagrangian) stage we find mass flows across the cell boundaries. In this case we assume the total mass to be transferred only by a velocity component normal to the boundary. Thus, for instance,

$$\Delta M^n_{i+1/2,j} = \langle\rho^n_{i+1/2,j}\rangle\langle U^n_{i+1/2,j}\rangle\Delta y\Delta t. \tag{A.4}$$

The symbol $\langle\ \rangle$ denotes the value (of $\rho$ and $U$) across the cell boundary. The choice of these values is of great importance, since they substantially influence the stability and accuracy of the calculations. The consideration of the flow direction is characteristic of all possible ways of writing $\Delta M^n$.

Here, different kinds of representations for $\Delta M^n$, of the first- and second-orders of accuracy, are considered. These are based on central differences, without account being taken of the flow direction and other conditions, as well as on the discrete model of a continuous medium comprising a combination of particles of a fixed mass in a cell.[14,15]

In the third (final) stage, we estimate the final fields of the flow parameters at the instant of time $t^{n+1} = t^n + \Delta t$ (all the errors in the solution of the equations are "*removed*"). As was pointed out, the equations at this stage are laws of conservation of mass $M$, impulse $\mathbf{P}$, and total energy written for a particular cell in the difference form

$$F^{n+1} = F^n + \Sigma \Delta F_b^n, \tag{A.5}$$

where $F = (M, \mathbf{P}, E)$.

According to these equations, inside the flow field there are no sources or sinks of $M$ and $\mathbf{P}$, and their variations in time $\Delta t$ are caused by interaction across the external boundary of the flow region.

### *A.2.2. Boundary conditions*

To retain the unified nature of the computations and to avoid special expressions for the boundary cells, layers of fictitious cells are introduced along all the boundaries, into which the parameters from the neighboring flow cells are written. The number of such layers depends on the order of the difference scheme (one layer for the first-order of accuracy, etc.). Two kinds of boundaries then have to be distinguished: the rigid boundary (or axis of symmetry) and the "open" boundary of the computational region.

In the first case, the velocity component normal to the boundary changes sign, while the remaining flow parameters are taken unchanged (non-penetration condition). The normal velocity component thus vanishes on the rigid walls, and the flow transmission conditions are thus realized. It will be shown below that another type of boundary condition is possible, namely, walls without slip (condition of sticking). In this case, both velocity components change sign, and the entire velocity vector vanishes at the wall (condition of sticking).

The fluid can flow in and out through the open boundaries of the region, and some conditions on the continuity of the movement are required. Let the fluid flow into the rectangular mesh from the left; then the parameters of the entering flow will be specified here. On the remaining open boundaries of the region, we extrapolate the parameters of the flow "from within", i.e. we transfer to the fictitious layer the parameter values of the layer nearest to the boundary (zero-order extrapolation). A more complicated statement of the conditions is possible, as is more accurate extrapolation (linear, quadratic, etc.).

It is natural that the outer boundary of the region should be fairly remote from the source of the disturbance, in which case methods of "outward" flow extrapolation are possible. This topic will be discussed in more concrete terms below. It is merely mentioned here that the basic principle underlying the statement of the conditions is that no substantial disturbances should penetrate through the open boundaries of the region into the computational region.

### *A.2.3. Viscosity effects*

It has already been remarked that our approach employs homogeneous difference schemes, whereby "through" computation by a unified algorithm is possible both in the smooth-flow regions and at discontinuities. This is achieved by using finite-difference schemes with a viscosity approximation. Let us dwell briefly on this topic.

While the equations of gas-dynamics for a nonviscous gas were taken as the initial equations, viscosity effects are in fact inherent in our difference scheme. They are produced, firstly, by the introduction into the scheme of an explicit term with artificial viscosity ("viscosity pressure") and secondly by the presence of an essentially schematic viscosity, dependent on the structure of the finite-difference equations.

The form of the approximation viscosity and estimates for the stability of the scheme can be obtained by writing, as Taylor series, the difference operators appearing in the equations in all three stages. The terms of zero (lowest)-order should then represent the initial differential equations, while the structure of the approximation viscosity can be determined by retaining higher-order terms in the expansions ("expansion errors"). The resulting differential equations will be termed the *differential approximation* of the finite-difference scheme, while an expansion up to second-order terms in time and space is termed the *first differential approximation.*[14,15,24]

The stability of the difference schemes may be investigated by means of the differential approximation. Such investigations were made by Yanenko and Shokin for one-dimensional quasilinear equations of the hyperbolic type.[24] While a strict mathematical groundwork has not yet been supplied for the case of nonlinear equations, the method of differential approximations has in fact been used here.[13]

Taking the one-dimensional case for simplicity, let us describe the first differential approximation of our difference scheme. Take, say, $U_{i+1}^n$ written as function $U(x + \Delta x, t)$, and expand each term of the finite-difference equations in Taylor series in the neighborhood of the point $(x, t)$.

For instance, in computations of $\Delta M^n$ from the expressions (A.4) of the second-order of accuracy, we obtain

$$\begin{aligned}
\frac{\partial \rho}{\partial t} + \frac{\partial \rho U}{\partial x} &= 0, \\
\frac{\partial \rho U}{\partial t} + \frac{\partial (P + \rho U^2)}{\partial x} &= -\frac{\partial q}{\partial x} + \frac{\partial}{\partial x}\left(\rho\varepsilon\frac{\partial U}{\partial x}\right), \\
\frac{\partial \rho E}{\partial t} + \frac{\partial}{\partial x}(U(P + \rho E)) &= -\frac{\partial qU}{\partial x} + \frac{\partial}{\partial x}\left(\rho\varepsilon\frac{\partial E}{\partial x}\right),
\end{aligned} \tag{A.6}$$

or, using expressions (A.4) of the first-order of accuracy,

$$\begin{aligned}
\frac{\partial \rho}{\partial t} + \frac{\partial \rho U}{\partial x} &= \frac{\partial}{\partial x}\left(\varepsilon\frac{\partial \rho}{\partial x}\right), \\
\frac{\partial \rho U}{\partial t} + \frac{\partial (P + \rho U^2)}{\partial x} &= -\frac{\partial q}{\partial x} + \frac{\partial}{\partial x}\left(\rho\varepsilon\frac{\partial U}{\partial x}\right) + \rho\varepsilon\frac{\partial^2 U}{\partial x^2}, \\
\frac{\partial \rho E}{\partial t} + \frac{\partial}{\partial x}(U(P + \rho E)) &= -\frac{\partial qU}{\partial x} + \varepsilon\frac{\partial^2 (\rho E)}{\partial x^2} + \frac{\partial \varepsilon}{\partial x}\frac{\partial \rho E}{\partial x},
\end{aligned} \tag{A.7}$$

where $\varepsilon = |U|\Delta x/2$.

The differential approximations may be written similarly in the case of two-dimensional problems.

On the left-hand sides of (A.6) and (A.7), the exact expressions for the initial differential equations have been obtained, while on the right we have the terms that are a consequence of the presence of "viscosity" effects in the difference equations. The terms involving result from the explicit introduction of an artificial viscosity, while the terms involving are due to the schematic viscosity, which appears when the exact differential equations are replaced by finite-difference equations ("expansion errors").

It may easily be seen that, when the mesh is refined ($\Delta x \to 0$), we have ($\Delta\varepsilon \to 0$) and the equations of the differential approximation convert into the exact set of initial equations. In concrete computations (owing to $\Delta x$, $\Delta t \ldots$ being finite), terms containing $\varepsilon$ always appear implicitly in the difference scheme even when $q = 0$; these terms are, in turn, analogous to the dissipative terms of the Navier–Stokes equations. The role of the coefficient of actual viscosity is here played by the coefficient of schematic viscosity, which depends on the local flow velocity and the size of the difference mesh.

In the two-dimensional case, it follows from the equations of momentum that the schematic viscosity (with $q = 0$) has the tensor form

$$\Delta = \frac{\rho}{2}\begin{vmatrix} U\Delta x \cdot \dfrac{\partial U}{\partial x} & V\Delta y \cdot \dfrac{\partial U}{\partial y} \\ U\Delta x \cdot \dfrac{\partial V}{\partial x} & V\Delta y \cdot \dfrac{\partial V}{\partial y} \end{vmatrix} = \frac{1}{2}\rho\mathbf{v}\Delta\mathbf{r} \cdot \nabla\mathbf{v}, \qquad \text{(A.6a)}$$

where $\Delta\mathrm{r} = \Delta x\mathbf{i} + \Delta y\mathrm{j}$.

It is clear from this that, owing to the presence of the vectors $\Delta\mathbf{r}$ and $\mathbf{v}$, the schematic viscosity does not possess invariance under Galileo transformations; in practice it only appears in zones where the gradient is large, e.g. in the shock wave, at the body surface, and at a breakaway of the flow. The coefficient of schematic viscosity $\varepsilon$ (and hence the width of the "smeared" shock wave obtained) then depends on the size of the local flow velocity and the cell size. In the regions of smooth flow, where the gradients of the flow parameters are relatively small, the influence of the schematic viscosity is negligible.

It will be shown below that in certain cases (when expressions (A.4) of the first-order of accuracy are used for computing $\Delta M^n$), the schematic viscosity ensures stable computation without introducing an explicit term involving the pseudo-viscosity, whereas when the second-order expressions (A.4) are used in regions where the local velocity is small compared with the velocity of sound, the introduction of a term with $q$ is necessary to obtain a stable solution.

### *A.2.4. Stability of the scheme*

While it is natural for different types of difference equations to be possible in different stages, the computations become strongly unstable on occasion, and rapidly increasing and oscillating solutions appear, which no longer reflect the behavior of the solutions of the initial differential equations.

The difference schemes quoted above are of the multilayer type, while the difference equations are strongly nonlinear with variable coefficients. This makes it impossible to employ Fourier's method, devised for linear equations with constant coefficients, for investigating the stability of the difference scheme as a whole. In essence, Fourier's method presupposes that the equations are linearized in the neighborhood of the flow with constant parameters, and it ignores nonlinear effects (influence of the flow gradients), which are sometimes the true sources of the instability.

A heuristic approach will therefore be employed here to analyze the stability of the difference schemes, based on the consideration of their differential approximations[13–15] and appropriate for nonlinear equations.

In this approach, we determine the signs of the coefficients $\alpha_i$, ("diffusion coefficients") in the dissipative terms of the differential approximation; these terms contain second partial derivatives in the space variables. For example, a linear equation can be indicated such that, when the value of the coefficient is negative, the equation of the differential approximation admits a solution that is exponentially increasing in time (unstable).[14]

In short, the necessary conditions for stability are obtained here from the condition $\alpha > 0$ (parabolicity condition). In the case of linear equations, the results of stability analyses obtained by means of the differential approximation, and obtained by Fourier's method, are exactly the same.

Let us examine how the different ways of writing the equation of continuity (second stage of the computations) contribute to the instability, assuming that the equations of momentum and energy are stable.

If $\Delta M^n$ is determined from expressions (A.4) of the second-order of accuracy, we find, on expanding the relevant differential equations in Taylor series and retaining terms containing $\partial^2\rho/\partial x^2$,

$$\frac{\partial \rho}{\partial t} + \frac{\partial \rho U}{\partial x} = \Delta_1 - \frac{\Delta t}{2}(U^2 + c^2)\frac{\partial^2 \rho}{\partial x^2}; \tag{A.8}$$

if $\Delta M^n$ is evaluated from the expressions (A.4) of the first-order of accuracy, we get

$$\frac{\partial \rho}{\partial t} + \frac{\partial \rho U}{\partial x} = \Delta_1^* + \left[\frac{\Delta x}{2}|U| - \frac{\Delta t}{2}(U^2 + c^2) - \frac{\Delta x^2}{4}\frac{\partial U}{\partial x}\right]\frac{\partial^2 \rho}{\partial x^2}, \tag{A.9}$$

where $\Delta_1$ and $\Delta_1^*$ are terms of the first differential approximation proportional to $\Delta x$ and containing the first derivatives. In our case,[14,15]

$$\Delta x \approx 0.0071, \quad \Delta t \approx 0.0071, \quad \rho_\infty = 1, \quad U_{-\infty} = 1. \tag{A.10}$$

In practical computations, when shock waves, contact discontinuities, and rarefaction waves appear,

$$\rho|U| \approx 1 \left|\frac{\partial U}{\partial x}\right| \Delta x < 0.3, \quad \left|\frac{\partial \rho}{\partial x}\right| \Delta x < 2. \tag{A.10a}$$

It follows that the coefficient of $\partial^2\rho/\partial x^2$ in (A.9) is positive, whereas it is negative in (A.8), i.e. scheme (A.8) has rapidly increasing solutions and is computationally unstable, while scheme (A.9) is stable.

### A.2.5. Advantages

It follows from the very character of the construction of the calculation scheme that a complete system of nonstationary gas-dynamical equations is essentially solved here, while each calculation cycle represents a completed process in calculating a given time interval. Besides all initial nonstationary equations, the boundary conditions of the problem are satisfied, and the real fluid flow at the time in question is determined.

Thus, the large-particles method allows us to obtain the characteristics of nonstationary gas flows and, by means of the stability process, their steady magnitudes as well. Such an approach is especially applicable to problems in which a complete or partial development of physical phenomena with respect to time takes place. For example, in a study of transonic gas flows, the flows around finite bodies, flow in local supersonic zones, and separation regions, develop comparatively slowly while the major part of the field develops rather rapidly.

In contrast to the FLIC method,[25] our investigation is wholly devoted to systematic calculations of a wide class of compressible flows in gas-dynamical problems (transonic regimes, discontinuity, separation, and "injected" flows).

The divergent forms of the initial and difference equations are considered in the large-particle method; the energy is used; different kinds of approximations are used in the first and second stages; and additional density calculations are introduced in the final stage, which helps us to remove fluctuations and makes it possible to obtain satisfactory results with a relatively small network (usually 1000–2500 cells are used). All this results in completely conservative schemes, i.e. laws of conservation for the whole mesh region are an algebraic consequence of difference equations. Fractional cells are introduced for the calculation of bodies with a curvature in the slope of the contour.

The investigation of the schemes obtained (approximation problems, viscosity, stability, etc.) was carried out by successively considering the zeroth, first, and second differential approximations.[13–15] The investigation showed that the large-particle method yields divergent-conservative and dissipative-steady schemes for sweeping-through calculations. These enable us to carry out stable calculations for a wide class of gas-dynamical problems without introducing explicit terms with artificial viscosity. This may be of particular significance in studying flows around bodies with a curvature in the slope of the contour, since the methods of introducing explicit terms with artificial viscosity are different for whole and fractional cells. Moreover, by varying only the second stage of the calculation procedure, we can arrive at the conservative "particle-in-cell" method so that the calculational algorithm is of general use.

As for discontinuities, the stability of the calculations is provided here by the presence of approximate viscosity in the schemes (dissipative terms in difference equations), which results in a "smearing" of shock waves into several calculating cells and the formation of a wide boundary layer near the body. It should be stressed that the magnitude of the approximate viscosity is proportional to a local flow velocity and to the dimension of the difference mesh. Therefore, its effect is practically evident only in zones with high gradients.

### A.2.6. Results

We now give some results of the calculations of transonic and overcritical flows around profiles and plane and axisymmetrical bodies obtained by the large-particle method.[20]

It is reasonable to characterize the overcritical regimes of transonic flows around bodies with the value of the critical Mach number of the oncoming flow $M_\infty^*$ (when a sonic point develops on the body), as well as with the extent of a local supersonic zone (as compared to a characteristic dimension of the body) and with its intensity (maximum supersonic velocity realized in the zone).

Figure A.1 presents flow-field patterns (lines $M$ = constant) for a 24% circular arc profile ($v = 0$) extending from purely subsonic ($M_\infty = 0.6$) to supersonic ($M_\infty = 1.5$) regimes. The dynamics of the formation and development of a local supersonic zone, transitions through the critical Mach number (here $M_\infty^* = 0.65$), sound velocity, and other details are shown.

Figures A.1(b)–A.1(g) illustrate a supercritical flow around a profile ($0.65 < M_\infty < 1$). One can distinctly see the position of the shock in the region of crowded lines $M$ = const. that bounds the local supersonic line together with the sonic line ($M = 1$). The region of low velocities is located behind the shock wave. When the velocity of the flow increases, it reaches the parameters of an undisturbed flow at a large distance from the body. With $M_\infty > 0.9$, the zone becomes considerable both in size and in intensity (supersonic velocities are attainable up to $M = 1.7$–$1.8$), and in the case of a sonic flow (Fig. A.1(g)), lines of the level $M = 1$ end at infinity.

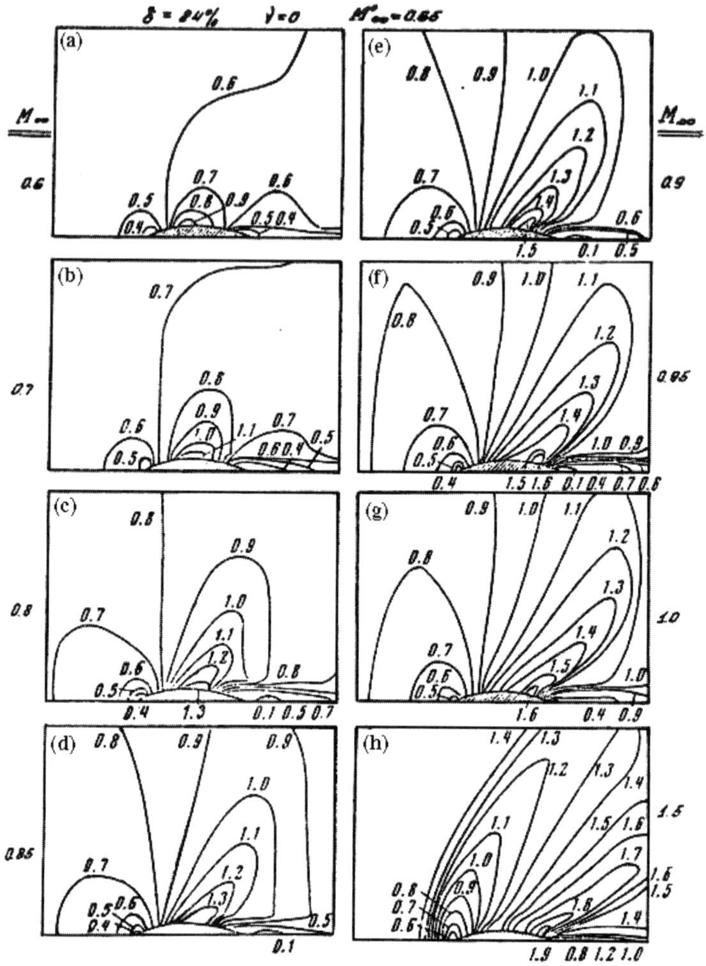


**Fig. A.1.** Flow-field parameter for a 24% circular are pattern.

The asymmetry of the whole flow pattern is noticeable (even at purely subsonic velocities in Fig. A.l(a)); it results from nonpotentiality of the flow (supercritical regimes) and from the presence of viscous effects as well (subsonic regimes, formation of a wake behind the body, etc.).

In the case of a supersonic flow around a profile (Fig. A.1(h), where $M_\infty = 1.5$), a shock wave ahead of the body develops and bounds the disturbed region. Behind the wave, in the vicinity of the axis of symmetry, a region of subsonic velocities is created. Afterward, the flow velocity along

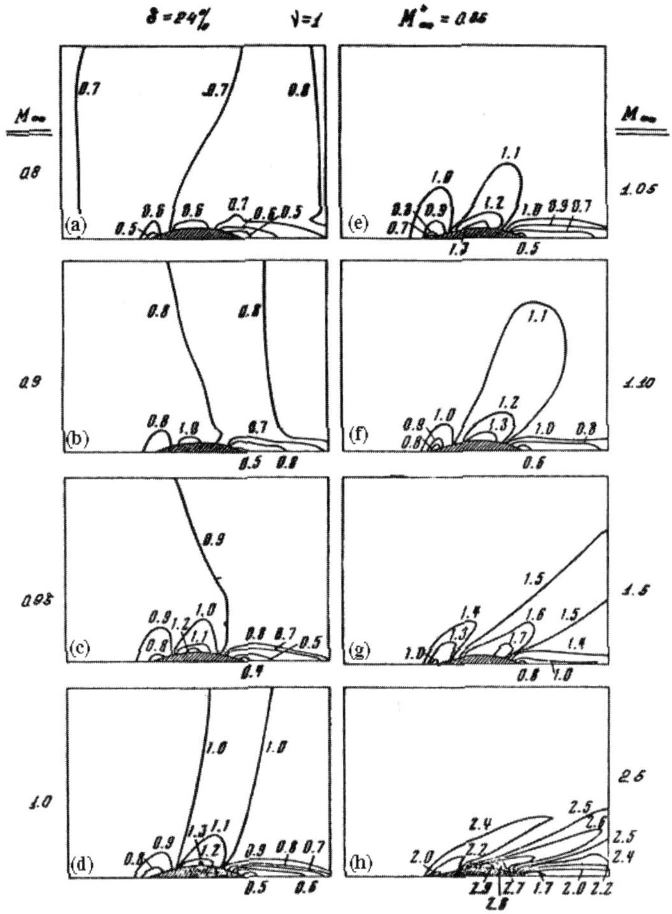

**Fig. A.2.** Flow-field parameter for a 24% axisymmetrical body.

the contour of the body increases, and, as a result, a terminal shock occurs near the stem of the body.

For comparison, the results of calculations by the above method of a flow around a 24% axisymmetrical spindle-like body ($v = 1$ and $0.8 \leq M_\infty \leq 2.5$) are given in Fig. A.2. Here a critical regime occurs at $M = 0.86$. Local supersonic zones are less developed as compared to the plane case, and are of weaker intensity (for example, values of $M = 1.3$–$1.4$ are realized), although, naturally, the main singularities of transonic flow are seen here too.

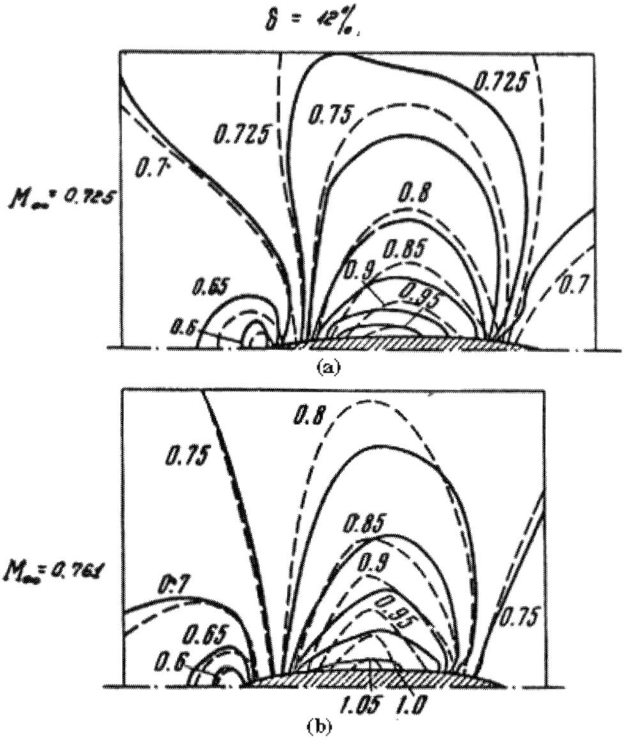


**Fig. A.3.** Calculated (solid) and experimental (dotted) flow fields around a 12% profile: (a) subcritical (b) supercritical.

Figure A.3 compares the flow fields calculated by the above method (solid line) and those of the Wood–Gooderum[26] experiment (dotted line). Subcritical (Fig. A.3(a), where $M = 0.725$) and supercritical (Fig. A.3(b), where $M_\infty = 0.761$) flows around a 12% profile are shown (in accordance with the calculations and the experiment, $M_\infty^* = 0.74$).

Analysis of the internal reference tests and the results of the comparisons reveal that the error in the calculations carried out by the large-particle method does not usually exceed several percent. The calculations were carried out using a Soviet BESM-6 computer; the time of the calculation in this case did not exceed an hour.

Figures A.4–A.6 show results of calculations for some complicated flows past bodies of different shapes in the presence of discontinuities in the wake as well as under the influence of a fluid injected upstream from the front

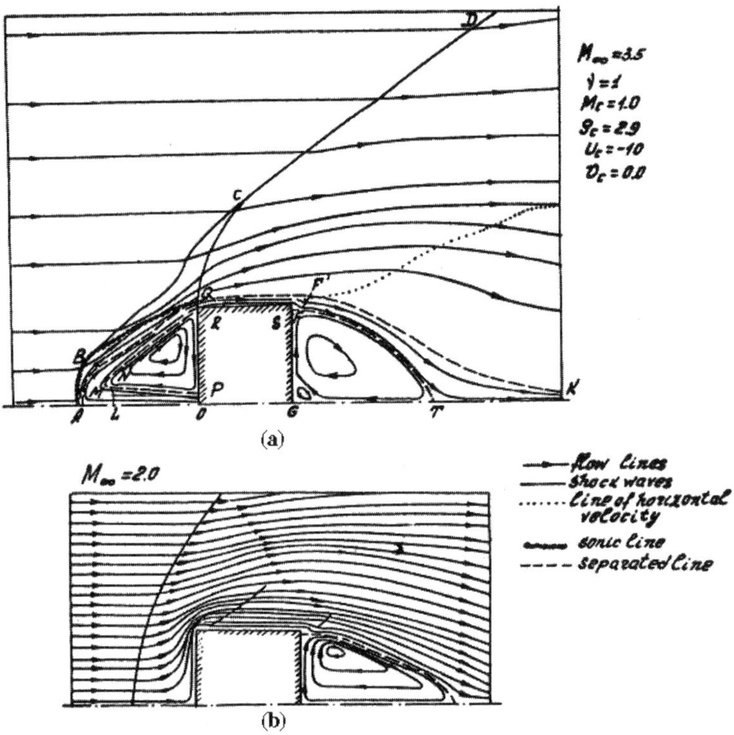


**Fig. A.4.** Calculated results: with (a) fluid injection and (b) without injection.

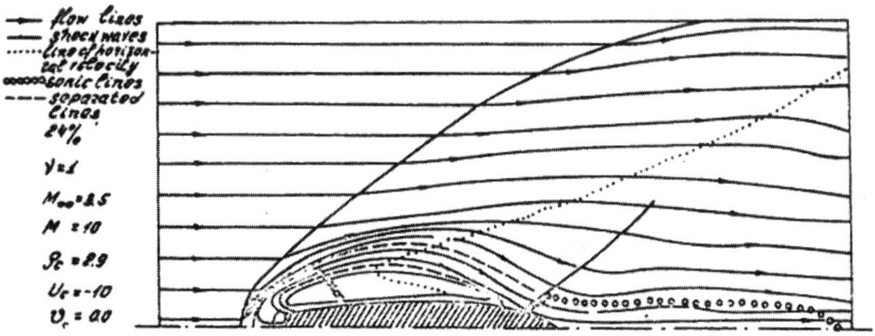


**Fig. A.5.** Calculated results with fluid injection.

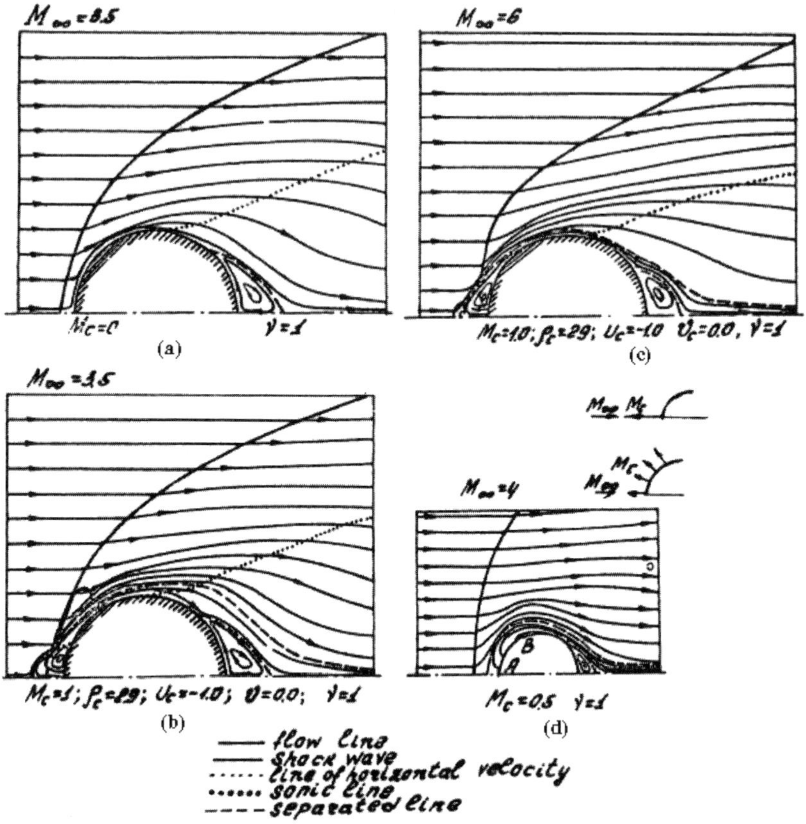


**Fig. A.6.** Calculated results: (a) without injection; (b, c) with injection; (d) with injection.

surface of the body. Such flows are of great practical interest in the study of wakes, turbulence, etc.

The results of the numerical experiments with fluid injection are given in Figs. A.4(a), A.5, and A.6(b) and A.6(c). They include the case of the interaction of a supersonic flow around a finite thick circular disk (Fig. A.4(a), where $M_\infty = 3.5$), a 24% body of revolution (Fig. A.5, where $M_\infty = 3.5$), and a sphere (Fig. A.6(b), where $M_\infty = 3.5$, and Fig. A.6(c), where $M_\infty = 6$) with a sonic axial stream (i.e. one where $M_c = 1.0$, $\rho_c = 2.0$, $U_c = 1.0$, and $v_c = 0$) issuing out of a nozzle situated at the axis of symmetry of the body. Figure A.6(d) presents results for the case in which distributed injection of the flow takes place at the surface of a sphere. In all the figures, streamlines, shock waves, horizontal velocity lines (dots), and sonic lines (circles) are indicated; dashes denote lines separating the main flow from the injected stream. In Figs. A.6(a) (sphere with $M_\infty = 3.5$) and A.4(b) (cylinder with $M_\infty = 2.0$), results obtained for flows past the same bodies without injection ($M_c = 0$) are presented for comparison.

The action of the jet markedly complicates the flow pattern. For instance, in the flow past a cylinder, the head shock wave ($ABCD$ in Fig. A.4(a)) is pushed toward the oncoming flow, and its distance from the body increases significantly. The jet issues out of the body in the direction of the axis of symmetry at a sonic velocity and expands, forming a local supersonic region $OLMNPO$ that is closed by a triple $\lambda$-shock intersection (normal front $ML$, oblique front $MN$, transverse front $MP$), having a common point $M$. In front of the body, a stagnation zone with a complicated vortex structure develops; sonic line $BQ$ is situated much lower than in a jetless flow. Behind the front stagnation zone, a secondary shock $QC$ is formed, and at some distance from the body it merges with the head shock wave $ABCD$ (at point $C$).

Behind the bodies in Figs. A.4–A.6, both with and without injection, one can observe the development of separated zones of recirculating backward flows. In the cases considered, these zones are closed, localized in the wake of the body, and separated from the external flow by a "nonflow" line, i.e. a contact surface indicated by dashes in the figures. In the vicinity of the separation (it is interesting to note that in Fig. A.4, the separation point is situated somewhat lower than the rear angular point of the body), a transverse shock wave $FF$ develops. Backward recirculation flows are essentially subsonic and rarefied (gas density and pressure are low here), so that viscosity effects are negligible.

The large-particle method has also been applied to the study of internal gas flows, diffraction problems, and other problems.

Figure A.7 presents some results of computations for flow through a straight channel ($\nu = 0$ in Fig. A.7(a)) and a straight tube ($\nu = 1$ in Fig. A.7(b)) in the presence of a central body ($M_\infty = 1.5$) for the case in which a triple shock intersection is formed as a result of the interaction of the flow with the upper wall. (This can be seen by examining the behavior of the lines $M =$ const. in Fig. A.7(a) and rot $\mathbf{W} =$ const. in Fig. A.7(b).).

In calculating separated flows using various mesh sizes, the cell dimensions of the large particles were changed several times so that, in the wave of the body of size $R$, from 4 to 30 computational intervals were used (Fig. A.8). In all cases there was an ample reserve of computational stability (over 100 Courant, where the Courant number represents the ratio of the time step to the space width of the cell).

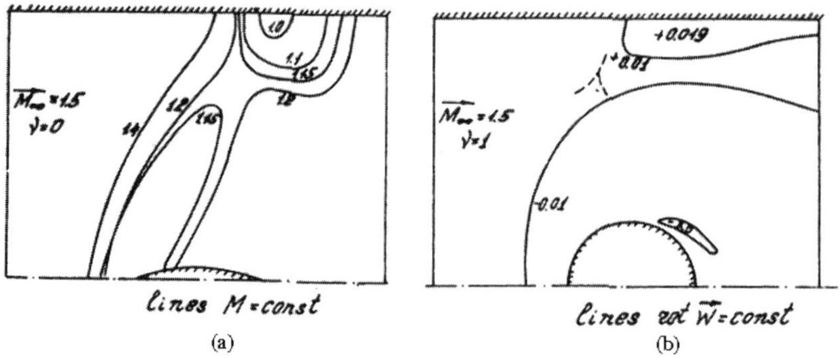

**Fig. A.7.** Calculated results for (a) a straight channel and (b) a straight tube in the presence of a central body.

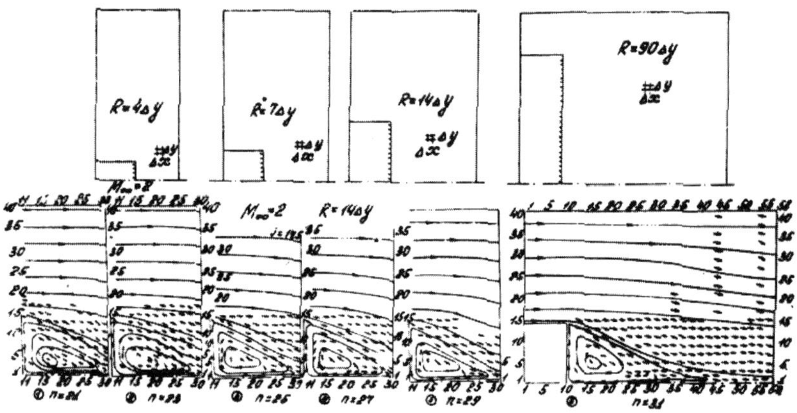

**Fig. A.8.** Base flows behind an axisymmetric cylinder ($M_\infty = 2.0$).

Figure A.8 shows base flows behind an axisymmetric cylinder ($M_\infty = 2.0$) for $R = 14\Delta y$. A gradual development of the flow in time is shown (in dimensionless units) from $t^n = 21$ to $t^n = 31$, when the zone is practically located. Streamlines are represented by solid lines; velocity vectors, by arrows. It follows from this diagram that at $t^n \approx 25$, the flow has already been formed but still continues to "breathe". It is interesting to note that similar flow patterns were obtained with denser meshes and (which is quite important) the zone of "breathing", i.e. changes in its dimensions, internal structure, and other features of the flow, occurred approximately at the same time intervals $t^n$ in various approximations.

The formation of breakaway zones in the case of strong interaction seems to be explained here by the fact that, as a result of viscosity effects and the treatment of the boundary, conditions close to sticking conditions are realized on the body itself. A fairly wide boundary layer forms around the body surface (comparable to the thickness of the body at its tail), and this layer then moves away from the body surface and forms a breakaway zone with complicated vertical structure behind the tail part. It must be emphasized here that, while the boundary layer is in fact the result of viscosity effects in the scheme, in the breakaway zone itself the influence of the approximation viscosity $\varepsilon$ (which is proportional to the local velocity and the size of the computational mesh; see above) is quite small, since in this zone only small values of the subsonic velocities are realized. Computations on different approximation meshes revealed only a slight change (within the limits of one step) in the zone contour.

The fact that the solution does not strongly depend on viscosity ($\varepsilon \approx \rho|U|\mathbf{r}$) shows, by the way, that flows corresponding to high Reynolds numbers can be treated by our methods of analysis. Thus, our calculations of separated zones might give quantitative information for the case of limiting flows ($Re \to \infty$) as, for example, the calculation of shock waves by a scheme including viscosity and other effects. Naturally, if necessary, the accuracy in determining the characteristic features of such zones can be further increased by using the results of preliminary calculations (e.g. the position of separation and closure points, or a zone contour) as initial data.

However, it should be pointed out that in the calculations, the flow parameters on the front part of the body under study are determined comparatively quickly, while local supersonic zones and separation regions continue, as mentioned above, to "breathe". This may be due to a physical (nonstationary) character of the phenomenon itself. The difference scheme prescribed by the nonstationary large-particle method appears to be especially well suited for such a case.

## A.3. Computation of incompressible viscous flows

### *A.3.1. The problem*

At present, many numerical methods are known for the solution of Navier–Stokes equations describing viscous incompressible flows. Most of them were developed for equations containing the flow function $\psi$ and vortex $\omega$.

A common disadvantage of these methods is the utilization in some form of a boundary condition (the Tom condition) for a vortex on a solid surface, which is omitted in the physical formulation of the problem. The rate of convergence of the numerical algorithms is limited by the presence of an additional iteration process due to this boundary condition for a solid-surface vortex.

Moreover, an apparent limitation of methods for the solution of the system $(\psi, \omega)$ is connected with their inapplicability in cases of spatial viscous flows and compressible gas flows. This accounts for the recent interest in the numerical solution of Navier–Stokes equations represented in natural variables:

$$\frac{\partial \mathbf{v}}{\partial t} + (\mathbf{v} \cdot \nabla)\mathbf{v} = -\nabla P + v\Delta\mathbf{v}, \quad \nabla\mathbf{v} = 0, \tag{A.11}$$

where $P$ is the pressure, $\mathbf{v}$ is the velocity vector, $\nu$ is the coefficient of kinematical viscosity.

Using the principles of the large-particle method, Gushchin and Shchennikov[16,17,29] studied viscous incompressible gaseous flows with velocity-pressure variables by means of a numerical scheme of splitting analogous to the SMAC method.[27]

### *A.3.2. The difference scheme*

The problem is to obtain, for Eqs. (A.11), a difference scheme with a high degree of accuracy that will enable us to carry out calculations for plane, axisymmetric, and spatial flows of a viscous incompressible fluid with a single algorithm.

Let us consider the following scheme for splitting a time cycle:

Stage I. Determination of the intermediate values of the velocities $\mathbf{v}(\tilde{U}, \tilde{V})$:

$$\frac{\tilde{\mathbf{v}} - \mathbf{v}^n}{\tau} = -(\mathbf{v}^n\nabla)\mathbf{v}^n + v\Delta\mathbf{v}^n, \quad \nabla\mathbf{v} = 0. \tag{A.12a}$$

Stage II. Calculation of a pressure field:

$$\Delta P = \frac{\Delta\mathbf{v}}{\tau}, \quad \nabla\mathbf{v}^{n+1} = \mathbf{D}^{n+1}. \tag{A.12b}$$

Stage III. Determination of the final values of the velocities:

$$\Delta P = \frac{\Delta\mathbf{v}}{\tau}, \quad \mathbf{v}^{n+1} = \tilde{\mathbf{v}} - \tau \cdot \nabla P. \tag{A.12c}$$

Stages I and III lead to the solution of the Navier–Stokes equations, and stages II and III are the conditions of solenoidality (second of Eqs. (A.11)). In view of this, at stage I, the transfer is accomplished only by convection and diffusion. The velocity field $\tilde{\mathbf{v}}$ so obtained does not satisfy a continuity equation ($\tilde{D} \neq 0$). Therefore, it is necessary to change ("correct") the field $v$ at the expense of a pressure gradient $P$, so that $D^{n+1} = 0$ (stage III); $P$ is found by solving the Poisson equation (stage II).

For a proportional calculational mesh, a two-dimensional difference scheme of the second-order of accuracy with respect to space is presented in Ref. 17. The main difficulties of the numerical realization of the scheme involve the calculation of a pressure field and the formulation of boundary conditions.

It should be noted that in some works the projection of an equation of motion on the normal to the surface at boundary points is used as a boundary condition on a solid surface. This reduces the efficiency of the numerical method, since such conditions are not part of the physical formulation of the problem.

Easton[28] proposes an original modification of the boundary conditions, in the MAC method, that allows homogeneous boundary conditions to be provided for pressure. Moreover, in the SMAC method[27] and in the modified MAC method,[28] owing to the difference schemes chosen, realization of the sticking condition necessarily results in the determination of a boundary vortex value on a solid surface satisfying the Tom condition of the first-order of accuracy. In addition, in the SMAC method, the sticking condition does not provide a balance of forces on a solid surface.

An essential point of the proposed method is the choice of boundary conditions. For the solution of problems concerned with viscous incompressible flow around bodies of finite dimensions, we can distinguish two basic types of boundary conditions: conditions on a solid surface and those on a line sufficiently remote from a body. Let us dwell on each of these conditions.

For boundary conditions on a solid surface, we have

$$\begin{aligned} V^n_{i,-1/2} &= 0 \quad \text{(nonpenetration condition)}, \\ U^n_{i+1/2,-1/2} &= 0 \quad \text{(sticking condition)}. \end{aligned} \tag{A.13}$$

From the latter, it follows that

$$\tilde{U}_{i+1/2,0} = \frac{U^n_{i+1/2,0}}{2} + \frac{U^n_{i+1/2,1}}{6} + O(h^3). \tag{A.14}$$

Condition (A.14) allows us to determine a boundary value for $\tilde{U}$ with the second-order of accuracy with respect to internal field points. We thus avoid the necessity of introducing a layer of fictitious cells (inside a solid body), which in schemes of the MAC, SMAC, and modified MAC types[28] gives rise to the inaccurate calculation of a vortex value on a solid surface with the first-order of accuracy. Note that, in the limits of the proposed approach, one need not calculate a vortex value on a solid surface. The latter can be determined from a calculated velocity field using some of the difference representations of the vortex expression

$$\omega = \frac{\partial U}{\partial y} - \frac{\partial V}{\partial x}$$

at the boundary points.

The boundary conditions on a line remote from a body are those for undisturbed flow; for the case $\bar{U}_\infty || OX$, they have the form

$$V^n_{i,N+1/2} = 0, \quad U^n_{i+1/2,N} = U_\infty.$$

In the calculation of a pressure field, homogeneous boundary conditions are attained with the help of an approach in Ref. 28 that consists in the following: supposing $V^{n+1}_{i,-1/2} = 0$ (for the case of a solid surface) and $V^{n+1}_{i,N+1/2} = 0$ (for the case of a line remote from a body), we have, from the finite-differences approximation (A.12c),

$$\tilde{V}_{i,-1/2} = \frac{\tau}{h}(P_{i,0} - P_{i,-1}), \tilde{V}_{i,N+1/2} = \frac{\tau}{h}(P_{i,N+1} - P_{i,N}). \tag{A.15}$$

By taking account of (A.15), it is now not difficult to write a difference equation to calculate the pressure in the boundary cells.[17]

The stationary solution of system of equations (A.12) is derived by repeating the above stages until the following criterion of establishment is fulfilled:

$$\max | U^{n+\lambda}_{i+1/2,j} - U^n_{i+1/2,j}| \leq \varepsilon^*.$$

Stability can be investigated in stages. The stability criterion at the first stage can be supplied by the first differential approximation (condition of a parabolicity). With regard to Eqs. (A.12a), the first differential approximation is[17]

$$\begin{aligned}
\frac{\partial U}{\partial t} + \frac{\partial U^2}{\partial x} + \frac{\partial UV}{\partial y} &= \left(v - \frac{\tau}{2}U^2\right)\frac{\partial^2 U}{\partial x^2} + \left(v - \frac{\tau}{2}V^2 - \frac{h^2}{4}\frac{\partial V}{\partial y}\right)\frac{\partial^2 U}{\partial y^2}, \\
\frac{\partial V}{\partial t} + \frac{\partial UV}{\partial x} + \frac{\partial V^2}{\partial y} &= \left(v - \frac{\tau}{2}U^2 - \frac{h^2}{4}\frac{\partial U}{\partial x}\right)\frac{\partial^2 V}{\partial x^2} + \left(v - \frac{\tau}{2}V^2\right)\frac{\partial^2 V}{\partial y^2}.
\end{aligned} \tag{A.16}$$

The stability criterion for the difference scheme employed follows from (A.16):

$$\tau \leq \frac{4v}{U^2 + V^2}.$$

By eliminating $P$ from (A.12b) and (A.12c), it is easy to show the absolute stability of the second and third stages by means of the Fourier method.

Thus, the difference scheme for the method enables us to calculate a flow without vortex and pressure values on a solid surface. This markedly increases the accuracy of the calculations, and the results attest to its effectiveness. The difference scheme (of the second-order of accuracy) provides us with a single algorithm for calculating viscous incompressible flows around plane, axisymmetric, and three-dimensional bodies of complex configuration, as well as internal flows with a wide range of Reynolds numbers.[17]

### *A.3.3. Results*

A great number of external hydrodynamical problems were solved with this method. In a wide range of Reynolds numbers ($1 \leq Re \leq 10^3$), viscous incompressible flows around different bodies of finite dimensions were studied: a rectangular slab and a cylinder of finite length with axis parallel to the velocity vector of the flow,[29] a sphere and a cylinder with axis perpendicular to $\bar{U}_\infty$, a rectangular parallelepiped (three-dimensional flow),[7] as well as bodies of more complex form.

Figure A.9 shows flow patterns around a cylinder (a plane problem) for Reynolds numbers 1, 10, 30, and 50 ($Re = 2RV_\infty/v$, where $R$ is the cylinder radius).

Figure A.10 presents flow patterns around a cylinder for $Re = 10^3$ at times $T_1 = 162$, $T_2 = 166$, and $T_3 = 170$. In the latter case, a nonsteady flow pattern is observed (the stagnation zone seems to grow, and at some instant of time there is a flopping and overshooting of the fluid from the stagnation zone). The result probably should be verified.

Figures A.11 and A.12 present calculated results for the full Navier–Stokes equations for nonstationary three-dimensional flows. There, the problem is that of viscous incompressible flow around a cube (with linear

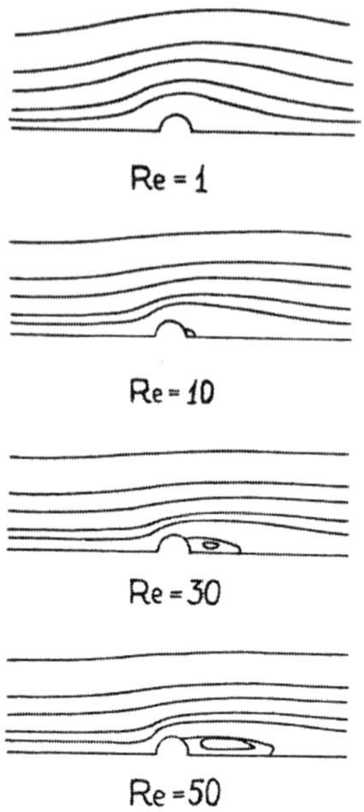


**Fig. A.9.** Flow patterns around a cylinder.

dimension 2a) when the oncoming flow velocity $\tilde{U}_\infty$ is parallel to an axis $Ox$. Owing to the presence of two planes of symmetry ($Oxy$ and $Ozx$), the calculation is carried out in the positive quadrant $Oxyz$ (Fig. A.11).

As is known, there are difficulties in presenting the results of studies of three-dimensional flows. We give here the velocity profiles $U$ (parallel to a vector $\bar{U}_\infty$) for various sections a with $x = \text{const}$.

Figure A.11 shows, for $Re = 1(Re = 2\alpha U_\infty/v)$, a velocity profile $U$ in an undisturbed flow ($x = -\infty$) and for a section $x = 3\text{a}$. Figure A.12(a) shows, for $Re = 1, 8, 40$, and 100, the dynamics of the change in the velocity profile $U$ along the cube (section a coincides with the cube face $x = 2\text{a}$; the distance between sections is a constant $\Delta x = 0.5(\text{a})$). Figure A.12(b) illustrates the change with time ($1.0 \leq t \leq 1.29$) of a velocity component at the section $x = 4\text{a}$. It follows from Fig. A.12 in particular, that with $Re = 40$ and 100, a reverse circular zone ($U < 0$) develops; after some time, a certain flow stabilization is observed. For further details concerning these results, the reader is referred to Refs. 16, 17, and 29.

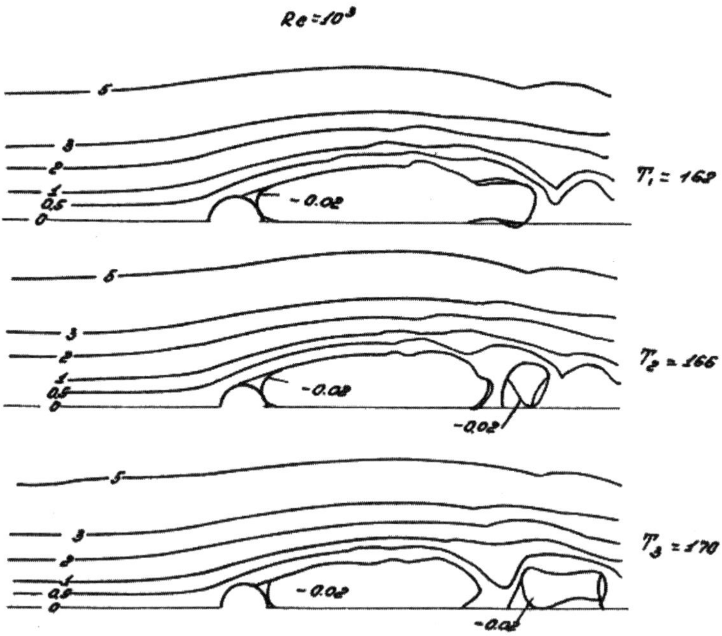


**Fig. A.10.** Flow patterns around a cylinder.

## A.4. Computation of viscous compressible gas flow (conservative flow method)

### *A.4.1. The method*

The calculation of viscous compressible gas flows was performed by Severinov and Babakov with the help of an approximation of the conservation laws in integral form for each cell of the calculation scheme ("flow" method).[18] The conservation laws for the mass, momentum, and energy of a finite volume have the form

$$\frac{\partial}{\partial t}\iiint_{\omega} F d\Omega = -\oint\int_{S_\Omega} \mathbf{Q}_F d\mathbf{s}, \quad F = \{M, X, Y, Z, E\}, \tag{A.17}$$

where

$S_\Omega$ — lateral surface of $\Omega$,
$M$ — mass,
$X, Y, Z$ — momentum components,
$E$ — energy terms in a cell volume,

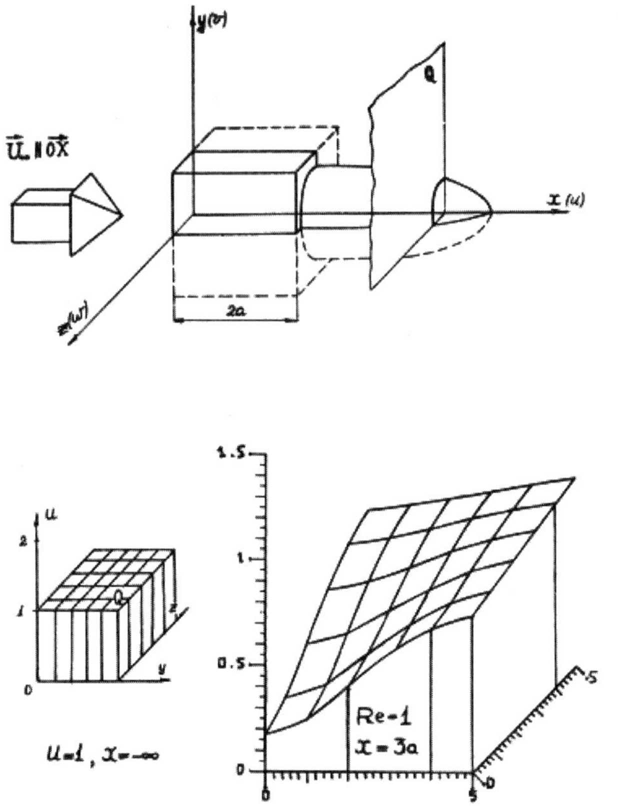


**Fig. A.11.** Calculated results for nonstationary three-dimensional flow.

$\mathbf{Q}_F$ — flow density vector for each of the quantities.

Equations (A.17) account for boundary conditions and are solved numerically for each cell of the calculation region.

If the values of $M^n = M(t^n)$, $X^n, Y^n$, $Z^n$, and $E^n$ are known at an instant $t^n = \tau n$, where $\tau$ is a time integration step, then at the $t^{n+1} = (n+1)\tau$, these quantities can be calculated with error $O(\tau^2)$ as follows[18]:

$$F^{n+1} = F^n - \tau \oint\!\!\int_{S_\Omega} \mathbf{Q}_F^n \cdot d\mathbf{s}. \tag{A.18}$$

In the finite-differences equations supplementary conditions, the form of which depends on the particular problem, must make it possible to determine the flow-density vectors on the boundary of the domain in which the solution is sought. The three-dimensional coordinate system, the shapes

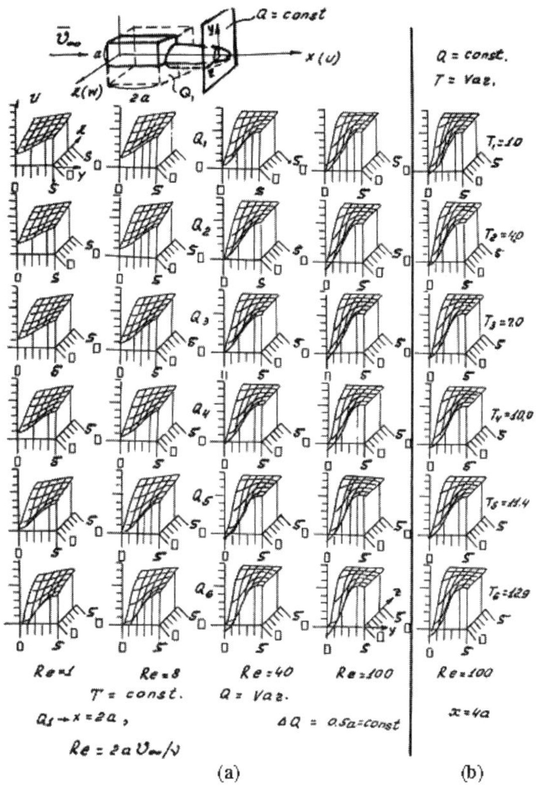


**Fig. A.12.** Calculated result for nonstationary three-dimensional flow.

of the elementary volumes $\Omega$, and the methods of determining the field variables and their first derivatives on the surfaces $S_\Omega$ must be chosen to ensure stability and monotonicity of the difference method and a fairly simple approximation of the integrals in the system (A.18).

In the solution of a specific problem, the integrals in (A.18) are calculated on separate segments of the surface which are the boundaries between two adjacent volumes $\Omega$. Depending on the directions of the flow density vectors, the values of $M$, $X$, $Y$, $Z$, and $E$ vary (they increase in some cells and decrease in others) by quantities determined by the flows of mass, momentum, and total energy through the corresponding segments of the boundary. Apart from rounding errors, this calculation method cannot lead to the loss or generation of the quantities $M$, $X$, $Y$, $Z$, and $E$ due to computational errors. Therefore, the flow method is conservative with respect to mass and momentum components and total energy.[18]

In Eq. (A.18), Stokes' assumption of the equality of the mean values of the principal tensions (with reversed sign) and pressure has been used. If the field variables are sufficiently smooth and the assumptions used in calculating the mass, momentum, and energy flow-density vectors are satisfied, the conservation laws (A.17) imply the complete Navier–Stokes equations for a compressible gas, with the volume $\Omega$ arbitrary.

### *A.4.2. Analysis*

We now consider some problems in the numerical investigation of Eq. (A.17).

Knowing the values of the quantities *M, X, Y, Z,* and *E*, we can calculate, for a given cell volume $\Omega$ fixed in space, the average values of the distribution densities of the quantities $\rho, \xi, \eta, \zeta$, and $\varepsilon$, where

$$\rho = \frac{M}{\Omega}, \quad \xi = \frac{X}{\Omega}, \quad \eta = \frac{Y}{\Omega}, \quad \zeta = \frac{L}{\Omega}, \quad \varepsilon = \frac{E}{\Omega}.$$

On the basis of these functions, it is easy to arrive at the generally accepted field variables — the components $U$, $V$, $W$ of the velocity $\mathbf{v}$ and the specific internal energy of the gas $e$:

$$U = \frac{\xi}{\rho}, \quad V = \frac{\eta}{\rho}, \quad W = \frac{\zeta}{\rho}, \quad e = \frac{\varepsilon}{\rho} - \frac{U^2 + V^2 + W^2}{2}, \quad \varepsilon = \frac{E}{\Omega}.$$

Using certain procedures of interpolation and numerical differentiation, we determine the values of the field variables and the first derivatives of $U$, $V$, $W$, and $e$ on the boundaries $S$ of the cells $\Omega$.[18]

In determining the values of all the functions (except the distribution densities $\rho, \xi, \eta, \zeta$, and $\varepsilon$) and the first derivatives present, we have used symmetric formulas; for example,

$$\begin{aligned}
U_{m+1/2,k} &= \frac{U_{m+1,k} + U_{m,k}}{2}, \\
\left(\frac{\partial U}{\partial x}\right)_{m+1/2,x} &= \frac{U_{m+1,k} + U_{m,k}}{h_1}, \\
\left(\frac{\partial U}{\partial y}\right)_{m+1/2,x} &= \frac{U_{m,k+1} - U_{m,k-1} + U_{m+1,k+1} - U_{m+1,k-1}}{4h_2}.
\end{aligned}$$

To calculate the density values of the distributions $\rho$, $\xi$, $\eta$, $\zeta$, and $\varepsilon$, we take asymmetric formulae:

$$\rho_{m+1/2,k} = \begin{cases} 1.5\rho_{m,k} - 0.5\rho_{m-1,k}, & \text{if } U_{m+1/2,n} > 0, \\ 1.5\rho_{m+1,k} - 0.5\rho_{m+2,k}, & \text{if } U_{m+1/2,n} < 0. \end{cases}$$

These equations ensure second-order approximation accuracy.

In approximating flow-density vectors $\mathbf{Q}_F$, an essential element of the method is that the distribution densities of additive characteristics such as densities $F$ are calculated on the boundary $S_\Omega$ of volume $\Omega$ in a nonsymmetrical way (extrapolation toward a gas flow). The other parameters, e.g. pressure, transfer velocities of additive characteristics, and derived values of $U$, $V$, and $W$ are calculated according to symmetry formulae in the viscous-stress tensor and in the thermal-conduction law. We believe it allows us to take account of influence regions, which form an important factor in the investigation of complex physical flow patterns.

The presence of a "convective" transfer renders space directions unequivalent, and it is desirable to take account of this fact in constructing difference schemes.

The transition to integral conservation laws essentially requires the approximation of derivatives whose order is one lower than those approximated in methods for the numerical solution of Navier–Stokes equations.

It is not hard to see that in essence the flow method is conservative with respect to mass, momentum, and total energy; conservativeness occurs both locally (for each cell of a difference mesh) and integrally, i.e. for the whole calculational region.[18] As follows from (A.17), the conservative property results because the approach is based upon the difference approximation of conservation laws written for each cell of the calculation mesh in terms of surface integrals of vectors of flow densities $\mathbf{Q}_F$; i.e. the conservation law is used in a form that holds true for an arbitrary gas volume.

Indeed, in the solution of an actual problem, the surface integrals in (A.17) are calculated on separate surface segments $S_\Omega$ that constitute boundaries between two adjacent volumes $\Omega$. Depending upon the direction of the flow vectors, the values of $F = \{M, X, Y, E)$ vary (they increase in some cells and decrease in others), and the new values are determined by flows of mass, momentum, and total energy across coincident boundary zones. Within the accuracy of rounded error, calculations of this kind cannot result in the loss or formation of quantities $F$ owing to the calculation procedures themselves, which testifies to its conservativeness.

A transfer (and, therefore, an approximation) of function "complexes" — distribution densities of the mass, momentum, and energy — is performed, which is consistent with the physics of the phenomenon. The approach is based upon the generality of a "transfer factor". From the viewpoint of obeying conservation laws, the scheme analysis seems significant, as it is known that a calculational scheme provides more accurate results when it rigorously preserves the quantities that occur in the physical process involved.

The flow method is essentially a development of the large-particles method. The difference formulae of the flow method can be deduced by using a scheme for splitting (A.3)–(A.5) for a "transfer" of the components of the quantities $F$.

### *A.4.3. Results*

A systematic study of the characteristics of viscous compressible gas flow around bodies of finite dimensions was carried out with the above approach, for a wide range of Reynolds numbers $Re$. The method formally "works" for large values of $Re$; however, the results are reliable when the boundary-layer thickness is much greater than a step of the calculational mesh. It should be emphasized that the division of the flow-density vector $\mathbf{Q}_F$ into convective and viscous components allows us to easily use the given algorithm for the calculation of ideal-gas flow.

The results given below were obtained by means of a method of establishment for the solution of a stationary boundary-value problem. According to the investigation of a linear model and the calculations, the difference scheme of the second-order of accuracy proves conventionally stable and conventionally monotonous.[18] The reliability of the results was experimentally examined for a general case by dividing a step of the calculation mesh, by using various forms of boundary conditions, and by comparing with the results of other calculations, with the results for an ideal gas, and with experiment.[18,30]

Figures A.13 and A.14 give the flow patterns for a sphere (separation zones of a reverse circular flow) with $M_\infty = 20$ and $550 \leq Re_\infty \leq 10^4$. Figure A.14 shows the behavior of lines $P =$ const. in a separation zone behind the sphere, with $Re_\infty = 10^4$ and 1500 ($Re_\infty = RV_\infty/v$).

Figure A.15 illustrates the density behavior across a shock layer in the region of the axis of symmetry ($x = 3°$) for $75 \leq Re_\infty \leq 10^4$. A density graph for an ideal gas ($M_\infty = 20, x = 1.4, Re_\infty = \infty$) is also plotted. The

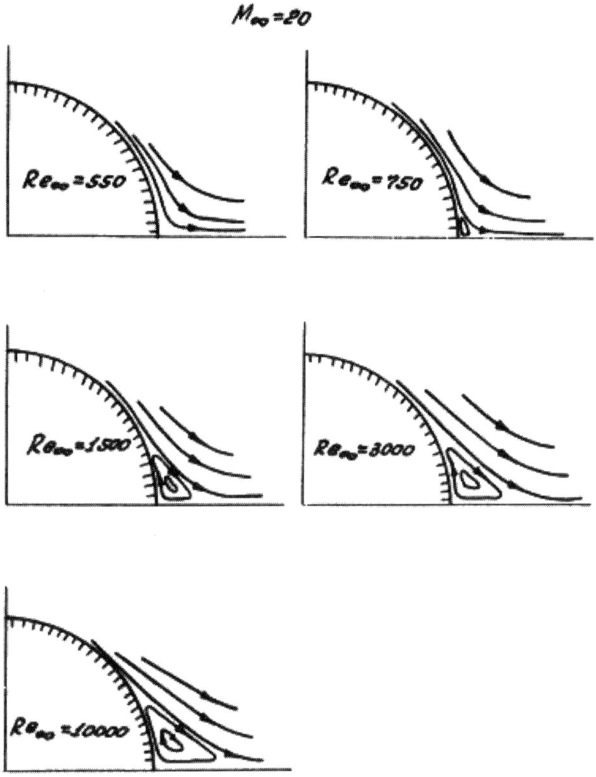


**Fig. A.13.** Calculated flow patterns for a sphere.

behavior of the curves shows that, for $Re_\infty \to \infty$, the gas density tends to acquire its limit value in a viscous nonthermoconducting gas; a tendency toward the formation of a shock wave is distinctly seen.

Figure A.16 shows a pressure distribution along a body (relative to the pressure at a critical point). Solid lines designate the results obtained by the flow method ($M_\infty = 6.05, Re_\infty = 6.43 \times 10^6$); crosses, experimental data (G. M. Riabinkova); and circles, the results for an ideal gas (Belotserkovskii[31]). There is very good agreement among the data. Thus, a limit transition from the viscous equations (A.17) to an ideal gas is obtained.

The above approach therefore allows the study of viscous compressible gas flows in a wide range of flow regimes (separation zones inclusive) up to large Reynolds numbers.

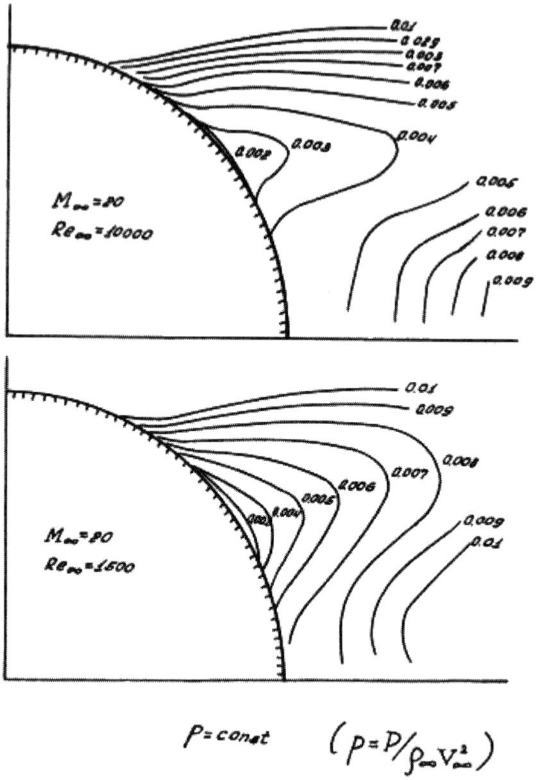


**Fig. A.14.** Calculated flow patterns for a sphere.

## A.5. Statistical model for the investigation of rarefied gas flows

### *A.5.1. The model*

The applicability of a statistical variant of our general approach was investigated by Yanitsky[21–23] for the solution of the Boltzmann equation. The main problem in this field is the development and investigation of a model for the behavior of a gas medium consisting of a finite number of particles. The model is based upon a combination of the idea of splitting the large-particles method in terms of Bird's statistical treatment[32,33] and Kats' ideas[34] about the existence of models asymptotically equivalent to the Boltzmann equation.

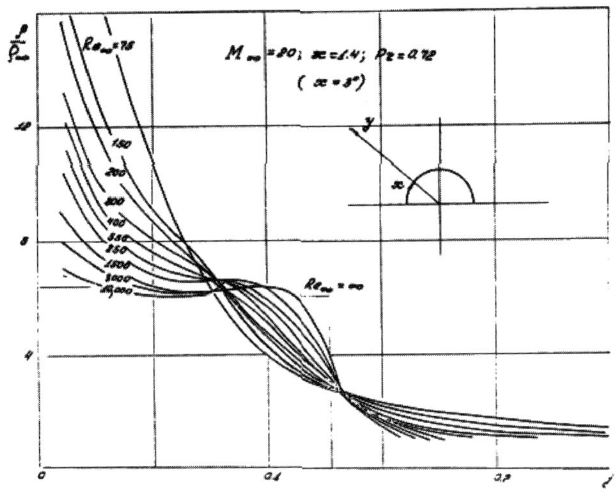

**Fig. A.15.** Density behavior across a shock layer near the axis of symmetry.

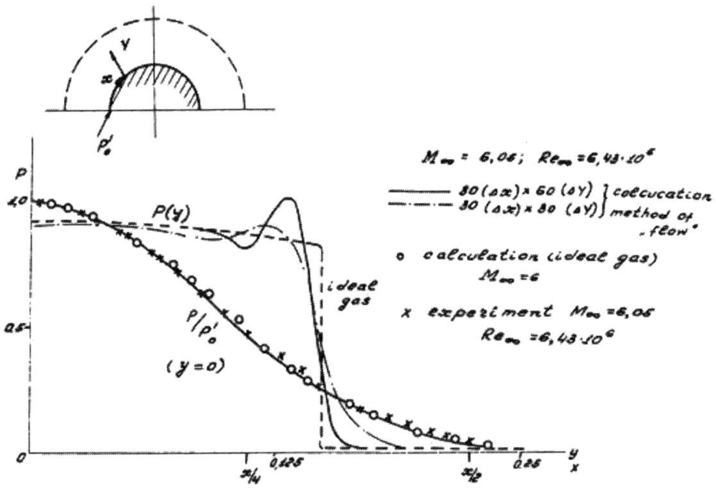


**Fig. A.16.** Flow-method results compared with experiment and ideal-gas behavior.

As is typical of particle-in-cell methods, the simulated medium is replaced by a system containing a finite number $N$ of particles of fixed mass. At a given instant of time $t_\alpha$ in each cell $j$ there are $N(\alpha, j)$ particles endowed with certain velocities. The main calculation cycle is comprised of two stages.

At the first stage, particles collide only with their counterparts in a cell (collision relaxation).

At the second stage, they are only displaced and interact with the boundary of a reference volume and with the surface of a body (collisionless relaxation).

The main distinction between the model suggested in Refs. 21–23 and Bird's model is that, at the first stage of the calculation, each group of $N$ particles in a cell is regarded as Kats' statistical model for an ideal monoatomic gas consisting of a finite number of particles in a homogeneous coordinate space. In simulating collisions, our approach makes use of Monte Carlo methods for the numerical solution of the main equation of Kats' model; this enables us to correctly determine the time between particle collisions in accordance with collision statistics for an ideal gas.

In contrast to the previously proposed Bird's methods,[32,33] the approach in Refs. 21–23 is a rigorously Markovian process. The main equation of this approach is linear (unlike the Boltzmann equation), which substantially simplifies the numerical realization of the algorithm. The feature of the propagation of molecular chaotic motion implies that Kats' model is asymptotically equivalent to the Boltzmann equation without convective derivative. The integration of the main equation of Kats' model results (with accuracy up to the realization of the assumption of molecular chaotic motion) in the Boltzmann equation.

For the realization of the second stage of the calculation of the evolution of a simulated gas, it is suggested in Refs. 21–23 that use should be made of the numerical algorithms for the displacement of particles, utilizing incomplete information about the positions of the particles in a coordinate space. This reduces the required volume of processor memory, which significantly increases the method's effectiveness. The method can be realized in a two- or three-dimensional coordinate space as well.

### *A.5.2. The method*

Let us dwell here upon the principal aspects of the suggested statistical particle-in-cell method.[21–23] We suppose that the problem of rarefied gas flow around a body can be solved by means of a distribution function and that the gas is monoatomic. Then any macroparameter of gas flow $\Psi(t, \mathbf{x})$ related to a molecular feature $\Psi(\mathbf{c})$ is a functional of the form

$$\Psi(t, \mathbf{x}) = \frac{1}{n(t, \mathbf{x})} \int \Psi(\mathbf{c}) \cdot f(t, \mathbf{x}, \mathbf{c}) \, d\mathbf{c},$$

where $f(t, x,\ \mathbf{c})$ is a molecular distribution function in a six-dimensional space $(\mathbf{x}, \mathbf{c})$ of the coordinates and velocities of the particles.

If $\Omega$ denotes the region of a control volume, and $\Gamma$ the boundary comprising a body surface as well, then the problem is reduced to obtaining the solution of the Boltzmann equation

$$\frac{\partial f}{\partial t} + \mathbf{c} \cdot \frac{\partial f}{\partial \mathbf{x}} = \int (f' \cdot f_1' - ff_1) g d\delta \cdot d\mathbf{c}_1, \tag{A.19}$$

satisfying the given initial conditions

$$f(t+0, \mathbf{x}, \mathbf{c}) = f_0(\mathbf{x}, \mathbf{c}), \quad \mathbf{x} \in \Omega, \quad -\infty < \mathbf{c}_{X,Y,Z} < +\infty, \tag{A.20}$$

and boundary conditions

$$f(t, \mathbf{x}_\Gamma, c) = \int k(\mathbf{c}, \mathbf{c}_1) f(t, \mathbf{x}_\Gamma, \mathbf{c}_1) d\mathbf{c}_1, \quad \mathbf{cn}(\mathbf{x}_\Gamma) > 0, \quad \mathbf{c}_1\mathbf{n}(\mathbf{x}_\Gamma) < 0. \tag{A.20a}$$

Here, $\mathbf{n}(\mathbf{x}_\Gamma)$ is normal to the surface $\Gamma$ at the point $\mathbf{x}_\Gamma \in \Gamma$ and directed into the volume $\Omega$; the nucleus shape $k$ is derived from the interaction law "gas-surface".

In deducing the Boltzmann equation, the following suppositions are made:

1. The mechanics of the collisions are described in the classical way.
2. The force fields of molecules are spherically symmetric.
3. Only binary collisions are considered (two molecules take part in any collision).
4. The molecules move randomly (the hypothesis of molecular chaos is valid; i.e. the distribution function of molecular pairs $f_2(t, \mathbf{x}, \mathbf{c}_1, \mathbf{c}_2) = f_1(t, \mathbf{x}, \mathbf{c}_1) \cdot f_1(t, \mathbf{x}, \mathbf{c}_2)$, which implies a statistical independence of particles).
5. The collision time is negligibly small.

The difficulty in constructing the solution of the Boltzmann equation in a nonlinear integrodifferential form results both from the great number of independent variables (there are seven of them in the general case: time, geometric coordinates, and molecular-velocity components) and from the complex structure of an integral of collisions.

Quadratic nonlinearity in the integrands, their dependence upon "dashed" functions of distribution (determined by the values of molecular velocities after a collision), the high level of multiplicity of integration (equal to five in a general case), and the complex formulation of boundary condition (A.20) are the main peculiarities which complicate the direct solution of the Boltzmann equation (A.19) and the application of ordinary numerical algorithms.

For an approximate solution of the problem formulated in this fashion, we shall construct a statistical model of an ideal monoatomic gas consisting of $N$ particles[a] with coordinates $r_i$, and velocities $c_i (i = 1, 2, \ldots, N)$ so that the equation of evolution of the model approximates Eq. (A.19). The only additional assumption is that of molecular chaos:

$$f_2(t, \mathbf{x}, \mathbf{c}_1, \mathbf{c}_2) = f_1(t, \mathbf{x}, \mathbf{c}_1) \cdot f_1(t, \mathbf{x}, \mathbf{c}_2), \qquad \text{(A.21)}$$

where

$$f_S(t, \mathbf{x}, \mathbf{c}_1, \ldots, \mathbf{c}_S) = \frac{N}{(N-S)!} \Im_S(t, \mathbf{r}_1, \ldots, \mathbf{r}_S, \mathbf{c}_1, \ldots, \mathbf{c}_S) \cdot f_1(t, \mathbf{x}, \mathbf{c}_2),$$

and with $\mathbf{r}_1 = \mathbf{r}_2 = \ldots = \mathbf{r}_S = \mathbf{x}, \Im_S$ being an $S$-partial function of distribution in a phase space of $6N$ dimensions.

If $\{\mathbf{R}(t), \mathbf{c}(t)\} = \{\mathbf{r}_1(t), \mathbf{c}_1(t); \ldots; \mathbf{r}_N(t), \mathbf{c}_N(t)\}$ designates the model state at time $t$, the solution of the problem is reduced to the numerical realization of a finite number of trajectories $\{\mathbf{R}(t), \mathbf{c}(t)\}$ with initial parameters corresponding to (A.20); the modeling of particle interaction with the boundary $\Gamma$ is accomplished in accordance with the given nucleus $k$ (Eq. (A.20a)). Once a number of trajectories are realized, one can calculate any macroparameter using adequate estimates of the Monte Carlo method for integrals.

The synthesis of the basic ideas of splitting, the particle method, and Kats' statistical model enable us to construct the desired model $\{\mathbf{R}(t), \mathbf{c}(t)\}$ for a space-inhomogeneous case when $\partial f / \partial \mathbf{x} \neq 0$.

Let us suppose that at time interval $t_\alpha (\alpha = 0, 1, \ldots)$ in a cell with center $x_j (j = 1, 2, \ldots, J)$ there are $N(\alpha, j)$ particles with velocities $\{\mathbf{c}_1, \ldots, \mathbf{c}_{N(\alpha,j)}\}$. The center $x_j$ of a cell in which a particular particle is situated is taken as a coordinate $\mathbf{r}_i$ of a particle $i$. The state of such a modeling gas $\{\mathbf{R}, \mathbf{c}\}$ is uniquely defined by a sequence of $J$ points of the form

$$\{\mathbf{R}(t), \mathbf{c}(t)\} \sim \{N(\alpha, j); \mathbf{c}_1, \ldots, \mathbf{c}_{N(\alpha,j)}\}, j = 1, 2, \ldots, J, N = \sum_{j=1}^{J} N(\alpha, j).$$

The principal cycle of calculation of the model evolution at time $\Delta t$ is split into two stages.

[a] A real gas is modeled by an ensemble of about 1000 rigid ball-like molecules that can be regarded as typical representatives of many trillions ($10^{12}$) of molecules, e.g. in the study of phenomena occurring in a real shock wave.[22]

At the first stage, for a gas at rest, we model the variation of the internal state of subsystems enclosed in the cells: collisions of particles (with their counterparts in a cell) in subsystems $\{\mathbf{c}_1, \ldots, \mathbf{c}_N\}$ are simulated independently in each cell, and thus the particles acquire new velocities.

Vector $\mathbf{c} = \{\mathbf{c}_1, \ldots, \mathbf{c}_N\}$ is regarded here as a state of Kats' model. Let $\phi(t, \mathbf{c})$ be the density of the probabilistic distribution of the state $\mathbf{c}(t)$; then the governing equation of this model (Kats' "master equation" [A.24]) has the form

$$\frac{\partial \varphi(t, \mathbf{c})}{\partial t} = \frac{1}{V} \sum_{1 \le l < m \le N} q_{lm} \int [\varphi(t, \mathbf{c}_{lm}) - \varphi(t, \mathbf{c})] d\sigma_{lm} \equiv K_{\varphi}(t, \mathbf{c}). \quad \text{(A.22)}$$

Here $K$ is Kats' operator of collisions; $q_{lm} = |\mathbf{c}_l - \mathbf{c}_m|$, where $\mathbf{c}_l$ and $\mathbf{c}_m$ denote the velocities of the $l$th and $m$th particles upon their collision; $d\sigma_{lm}$ is a differential section of elastic dissipation of a pair of particles ($\mathbf{c}_l$, $\mathbf{c}_m$); and a normalizing parameter $\mathbf{V}$ is determined by the choice of measurement units and can be interpreted as a cell volume.

If we introduce distribution functions

$$f_S(t, \mathbf{x}, \mathbf{c}_1, \ldots, \mathbf{c}_S) = \frac{N}{(N - S)! V^S} \int \varphi(t, \mathbf{c}) \prod_{i=S+1}^{N} d\mathbf{c}_i,$$

then by integrating (A.22), it is not difficult to obtain

$$\frac{\partial f_1(t, \mathbf{c}_1)}{\partial t} = \int [f_2(t, \mathbf{c}'_1, \mathbf{c}'_2) - f_2(t, \mathbf{c}_1, \mathbf{c})] q_{12}\, d\sigma_{12}\, d\mathbf{c}_2,$$

which coincides with the Boltzmann equation having a zero convective derivative when satisfying equality (A.21).

The algorithm for realization of the first calculation stage of the evolution of a spatially-inhomogeneous model corresponds to the Monte Carlo method of numerical solution of Kats' basic equation (A.22), which (unlike the Boltzmann equation) is linear.

At the second stage, we model a collisionless transfer of particles from a particular cell to any neighboring cells without changing the internal state of the subsystems; their interaction with a control-volume boundary and a body surface is also considered. This stage corresponds to the Monte Carlo method of numerical solution of the Boltzmann free molecular equation in the form

$$\frac{\partial f}{\partial t} + \mathbf{c} L f = 0. \quad \text{(A.23)}$$

Here $L$ is a finite-difference operator approximating a derivative $\partial/\partial x$; its introduction is closely related to an incomplete description of the system state in a coordinate space.

The simplest numerical algorithms for this approach[21–23] correspond to the solution of time-explicit, conventionally stable finite-difference schemes of the first-order of accuracy for Kats' equations and the Boltzmann free molecular equation, respectively. The equation of evolution of a modeling gas $\{\mathbf{R}(t_\alpha), \mathbf{c}(t_\alpha)\}$ with enough accuracy to satisfy the equality defining molecular chaos has the form (a one-dimensional flow)

$$\frac{\Delta_\alpha f}{\Delta t} + c_x \frac{\Delta_\alpha f}{\Delta x} = J[ff_1] - \Delta t c_x \frac{\Delta_\alpha J[ff_1]}{\Delta x}.$$

Here $\Delta_\alpha/\Delta t$ and $\Delta_\alpha/\Delta x$ are finite-difference first-order operators approximating derivatives $\partial/\partial t$ and $\partial/\partial x$, respectively. $J[ff_1]$ designates the right-hand side of the Boltzmann equation. The finite-difference scheme given is conventionally stable, and it approximates the Boltzmann equation within the accuracy of $O(\Delta t)$ and $O(\Delta x)$. As mentioned above, incomplete information concerning the spatial positions of the particles is used here.

The above approach can naturally be extended to the cases of two- and three-dimensional space. The extension to plane and space flows is trivial, and it consists of a sequence of one-dimensional displacements along coordinate axes. This corresponds to the splitting of a multi-dimensional transfer equation

$$\frac{\partial f}{\partial t} + \mathbf{c}\frac{\partial f}{\partial \mathbf{x}} = 0 \tag{A.23a}$$

into a sequence of one-dimensional finite-difference schemes.

The Boltzmann equation is known to imply a molecular chaos or a statistical independence of particles.[b] Our model includes the same premises as the Boltzmann equation, but without the molecular-chaos (or statistical independence) assumption. Consequently, in the model, there exists a statistical particle independence giving rise to a molecular-chaos disturbance. It should be noted that the inherent statistical independence is based upon theoretical and physical premises and does not depend upon the mesh dimension (it exists at $\Delta x \to 0$ as well).

[b]The molecular-chaos hypothesis implies that particle velocities are statistically independent (M. N. Kogan, *Rarefied Gas Dynamics* (M. Nauka, 1967)).

The results of the calculations of rarefied gas flows reveal that:

1. Results for various quantities of particles in a cell (e.g. with $N = 3$ and $N = 20$) practically coincide.
2. The results are in good agreement with the solution of the Boltzmann equation (Cheremisin's and Rykov's data). Therefore, the molecular-chaos disturbance is small in the problems involved here (though statistical particle independence exists, it is weakly manifested in rarefied gas problems and, apparently, it can be neglected here).

For consideration of turbulence, statistical independence is of crucial significance, and we can suppose that the statistical independence of the suggested method will become apparent when turbulent flows are considered.

### A.5.3. Results

The model was tested with a problem dealing with the structure of a direct shock in a gas consisting of elastic spheres in the Mach-number

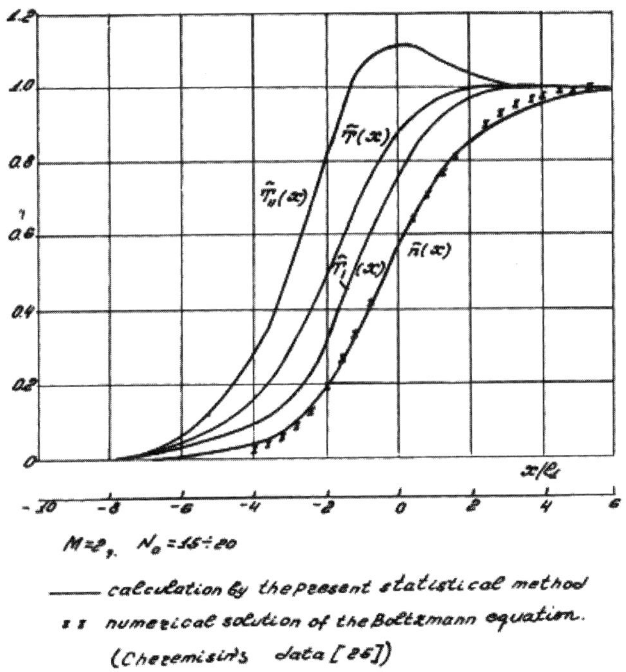


**Fig. A.17.** Statistical method versus direct numerical integration.

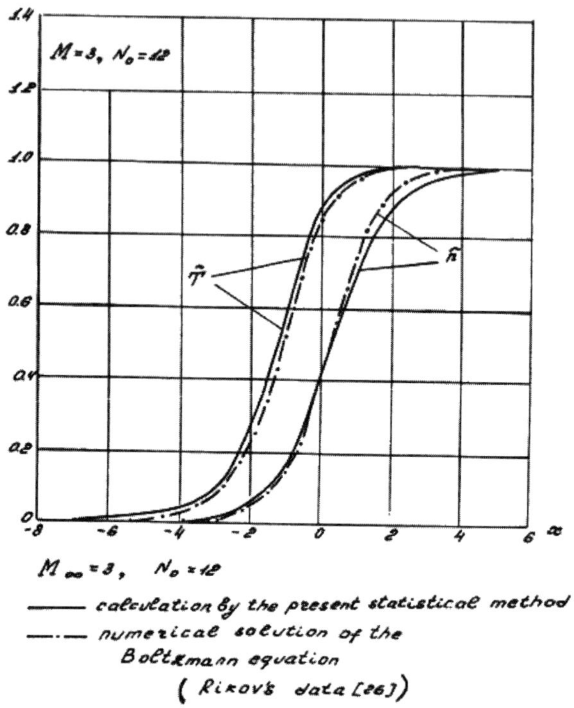


**Fig. A.18.** Statistical method versus direct numerical integration.

range $M_\infty = 1.25–4$. Figures A.17 and A.18 show graphs of the density $\tilde{n}(x)$, longitudinal temperature $\tilde{T}_{||}(x)$, transverse temperature $\tilde{T}_1(x)$, and total temperature $\tilde{T}(x)$ for Mach numbers of 2 and 3. The unit of length is a free mean path of the molecules in the flow. The relation $\Delta t/\Delta x$ is chosen to satisfy stability conditions. The average number of particles in the cells corresponding to the oncoming flow is $N_0 = 15–20(M = 2)$ and $N_0 = 12(M = 3)$. For comparison, the figures give the density $\tilde{n}(x)$ and temperature $\tilde{T}(x)$ obtained by direct numerical integration of the Boltzmann equation[35,36] on a network $\Delta x$ similar to the one used in our calculations ($\Delta x = 0.2–0.3$).

Figures A.19 and A.20 show the dependence of the results upon the average number $N_0$ of particles in the cells. In gas-dynamical problems concerned with a rarefied gas, this dependence is obviously rather weak.

This approach is probably also suitable for the investigation of turbulent gas flows.

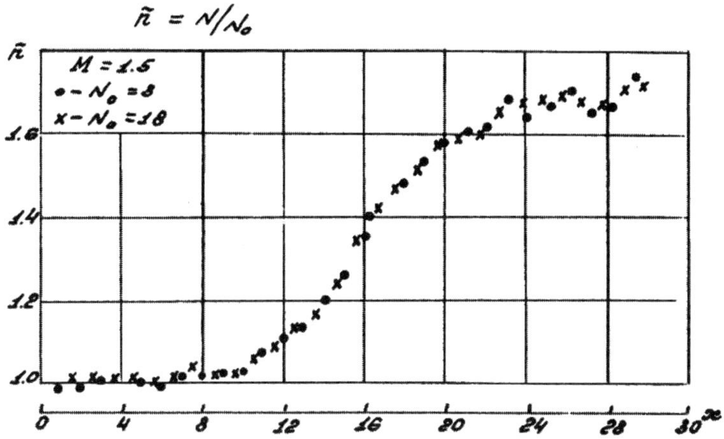


**Fig. A.19.** Comparison of results for $N_0 = 3$ and $N_0 = 18$.

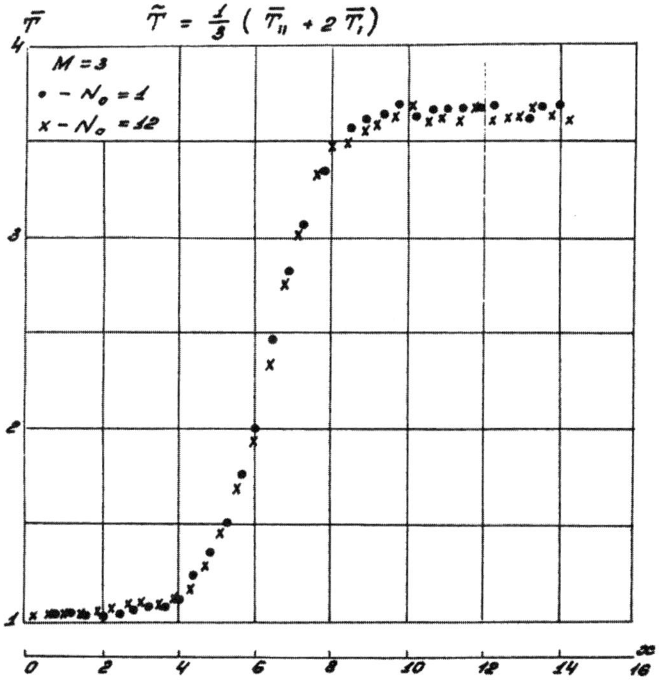


**Fig. A.20.** Comparison of results for $N_0 = 1$ and $N_0 = 12$.

## A.6. Conclusion

In conclusion, we should say that a series of numerical algorithms (the computational experiment) allows us to effectively investigate a wide class of complex gas-dynamical phenomena from a single viewpoint. These methods may also be used for obtaining the aerodynamic characteristics of bodies, vehicles, and aircraft.

While individual fine and local details of the flow may be missed by this approach, it hardly seems likely that the basic properties of a flow will not be significantly determined by an exact description of small structures. The computational experiment is the only way of obtaining the general characteristics of a complex phenomenon and a picture of the flow as a whole.

## References

1. A. A. Dorodnitsyn, On one method of the equation solution of a laminar boundary layer, *Zh. Prikl. Mekh. i Tekh. Fiz.* **1**(3), 111–118 (1960).
2. O. M. Belotserkovskii, A flow with a detached shock wave around a circular cylinder, *Dokl. Akad. Nauk SSSR* **113**(3), 509–512 (1975).
3. O. M. Belotserkovskii, A flow with a detached shock wave around a symmetrical profile, *Prikl. Mat. Mekh.* **22**(2), 206–219 (1952).
4. O. M. Belotserkovskii and P. I. Chushkin, A numerical method of integral relation, *USSR Comput. Math. Math. Phys.* **2**(5), 823–858 (1962).
5. O. M. Belotserkovskii, A. Bulekbayev and V. G. Grudnitskii, Algorithms for schemes of the method of integral relations applied to the calculations of mixed gas flows, *USSR Comput. Math. Math. Phys.* **6**(6), 162–184 (1966).
6. O. M. Belotserkovskii, (ed.), Flow past blunt bodies in supersonic flow: theoretical and experimental results, *Vychisl Tsentr. Akad. Nauk SSSR*, 1st edn. (Moscow, 1996); (2nd edn., revised and extended, 1967).
7. P. I. Chushkin, Blunt bodies of simple form in supersonic gas flow, *Prikl. Mat. Mekh.* **24**(5), 927–930 (1960).
8. P. I. Chushkin, Method of characteristics for three-dimensional supersonic flow, *Vychisl Tsentr. Akad. Nauk SSSR* (Moscow, 1968).
9. K. M. Magomedov and A. S. Kholodov, On the construction of difference schemes for equations of hyperbolic type based on characteristic coordinates, *USSR Comput. Math. Math. Phys.* **9**(2), 158–176 (1969).
10. O. M. Belotserkovskii, (ed.), *Numerical Investigation of Modern Problems in Gas Dynamics* (Nauka, Moscow, 1974).
11. F. K. Harlow, The particle-in-cell computing method for fluid dynamics, in *Methods in Computational Physics*, eds. B. Alder, S. Fernbach and M. Rotenberg, Vol. 3 (Academic, New York, 1964).

12. M. Rich, A method for Eulerian fluid dynamics, Report LANS-2826, Los Alamos Scientific Laboratory, New Mexico (1963).
13. C. W. Hirt, Heuristic stability theory for finite-difference equation, *J. Comput. Phys.* **2**(4), 339–355 (1968).
14. O. M. Belotserkovskii and Yu. M. Davidov, The use of unsteady methods of large particle for problems of external aerodynamics, Preprint, *Vychisl Tsentr. Akad. Nauk SSSR* (1970).
15. O. M. Belotserkovskii and Yu. M. Davidov, A non-stationary "coarse particle" method for gas dynamical computations, *USSR Comput. Math. Math. Phys.* **11**(1), 241–271 (1971).
16. V. A. Gushchin and V. V. Shchennikov, On one numerical method for the solution of the Navier–Stokes equation *USSR Comput. Math. Math. Phys.* **14**(2), 512–520 (1974).
17. O. M. Belotserkovskii, V. A. Gushchin and V. V. Shchennikov, Method of splitting applied to the solution of problems of viscous incompresslble fluid dynamics, *USSR Comput. Math. Math. Phys.* **15**(1), 190–199 (1975).
18. O. M. Belotserkovskii and L. I. Severinov, The conservative flux-method and the calculation of the flow of a viscous heat-conducting gas past a body of finite size, *USSR Comput. Math. Math. Phys.* **13**(2), 141–156 (1973).
19. O. M. Belotserkovskii and E. G. Shifrin, Transonic flow behind a detached shock wave, *USSR Comput. Math. Math. Phys.* **9**(4), 908–931 (1969).
20. O. M. Belotserkovskii and Yu. M. Davidov, Computation of transonic "super-critical" flows by the "coarse particle" method, *USSR Comput. Math. Math. Phys.* **13**(1), 187–216 (1973).
21. V. E. Yanitsky, Use of Poisson's stochastic process to calculate the collision relaxation of a non-equilibrium gas, *USSR Comput. Math. Math. Phys.* **13**(2), 310–317 (1973).
22. V. E. Yanitsky, Application of random motion processes for modeling free molecular gas motion, *USSR Comput. Math. Math. Phys.* **14**(1), 264–267 (1974).
23. O. M. Belotserkovskii and V. E. Yanitsky, Statistical "particle-in-cell" method for the solution of the problem of rarefied gas dynamics, *USSR Comput. Math. Math. Phys.* **15**(5)(Part I); (6)(Part II), 184–198.
24. N. N. Yanenko and Y. I. Shokin, On the first differential approximation of difference schemes for hyperbolic sets of equations, *Sib. Mat. Zh.* **10**(5), 1173–1187 (1969).
25. R. A. Gentry, R. E. Martin and J. Daly, An Eulerian differencing method for unsteady compressible flow problems *J. Comput. Phys.* **1**, 87–118 (1966).
26. C. Ferrari and F. G. Tricomi, *Transonic Aerodynamics* (Academic, New York, 1968).
27. A. A. Amsden and F. K. Harlow, The SMAC method, Report LA-4370, Los Alamos Scientific Laboratory, New Mexico (1970).
28. C. R. Easton, Homogeneous boundary conditions for pressure in MAC method, *J. Comput. Phys.* **9**(2), 375–379 (1972).

29. V. A. Gushchin and V. V. Shchennikov, Solution of problems of viscous incompressible fluid dynamics by the method of splitting, *USSR Comput. Math. Math. Phys.* **14**(2), (1974).
30. A. V. Babakov, O. M. Belotserkovskii and L. I. Severinov, Numerical investigation of a viscous heat-conducting gas flow past a blunt body of finite size, *Izv. Acad. Nauk SSSR. Mech. zhidkosti i gaza.* (3), 112–123 (1975).
31. O. M. Belotserkovskii, Calculation of the flow around axially symmetric bodies with a detached shock wave, Preprint, Computing Center AN SSSR (1961).
32. G. A. Bird, The velocity distribution function within a shock wave, *J. Fluid Mech.* **30**(3), 479–487 (1967).
33. G. A. Bird, Direct simulation and the Boltzmann equation, *Phys. Fluids* **13**(11), 2677–2681 (1970).
34. M. Kats, *Probability and Related Topics in Physical Sciences* (Mir, Moscow, 1965).
35. F. G. Cheremisin, Numerical solution of the Boltzmann kinetic equation for one-dimensional, stationary gas motion, *USSR Comput. Math. Math. Phys.* **10**(3), 125–137 (1970).
36. V. A. Rikov, On averaging the Boltzmann kinetic equation with respect to a transverse velocity for the case of one-dimensional gas motion, *Izv. Acad. Nauk SSSR, Mech. zhidkosti i gaza.* (4), 120–127 (1969).

# Appendix B

# Formation of Large-Scale Structures in the Gap Between Rotating Cylinders: the Rayleigh–Zeldovich Problem*

The methodology for direct numerical simulation of turbulence presented in the book by O. M. Belotserkovskii and A. M. Oparin is developed and illustrated. The methodology relies on "rational" approximations of integral conservation laws and rules out the use of any semi-empirical models or constants associated with grids or schemes. The Taylor–Couette flow is considered. The rotation velocities are varied. The large-scale structures evolving from an initially unstable flow preserve their characteristic scales. The flow profile develops through the formation of large-scale structures and is independent of numerical viscosity. The kinetic energy of turbulent motion is mainly carried by large-scale vortices. The high-frequency portion of the spectrum is generated in the course of nonlinear interactions of large-scale structures with one another and with the cylinder walls and weakly depends on numerical dissipation, which is clear from the fact that entropy is conserved in computations.

## B.1. Introduction

In recent years, great effort has been devoted to numerical simulation of turbulence development in various terrestrial and astrophysical flows (see Refs. 1–3). For example, studies of rotational motion are of key importance for understanding the physics of accretion disks.[4] Accretion of matter onto compact objects is controlled by the rate of loss of angular momentum in

*O. M. Belotserkovskii, A. M. Oparin and V. M. Chechetkin, *Comput. Math. Math. Phys.* **42**(11), 1661–1670 (2002).

accretion disks. It is generally believed that the loss of angular momentum is explained by turbulent viscosity, which exceeds molecular viscosity by orders of magnitude. This leads one to the following question: what is turbulent viscosity and what are the physics behind it in accretion disks? We believe that two physical processes should be distinguished here. One of them is the transfer (loss) of angular momentum dominated by large-scale structures. The other process is responsible for dissipation of turbulent kinetic energy into heat. The relative importance of these two processes depends on various factors: existence of a steady flow, turbulent spectrum associated with it, flow geometry, and possible influence of magnetic field and other factors that affect the turbulent spectrum. For this reason, analysis of turbulence (including turbulent convection and turbulent mixing) is crucial for understanding the processes taking place in astrophysical objects, as pointed out in Ref. 3. Development of turbulence is an unsteady process that is extremely difficult to investigate. However, some understanding has been gained in recent years by applying mathematical modeling methods.[1]

It is well known (see Ref. 1) that turbulence can be analyzed via mathematical modeling or direct simulation of unsteady motion of coherent and large-scale structures (by solving the Euler equations). Alternatively, various hypotheses about stochastic properties of velocity field and other physical variables can be invoked. In the latter approach, the physical origin of turbulence development (the source of energy of chaotic motion) is generally left outside the scope of analysis. Our research concerns the problem of generation of large-scale structures. The most representative illustration of the problem is provided by the far-field structure of the wake behind a cylinder. It mainly consists of coherent large-scale vortices, but also involves small-scale vortices. The principal goal of turbulence theory is to explain the development of the spectrum of scales exhibited by turbulent flow. It should be noted here that most studies dealt with specific problems characterized by narrow ranges of physical parameters, which made it impossible to gain sufficient insight into the problem.

However, the existence of isotropic free turbulence is currently being questioned in an increasing number of papers (e.g. see Ref. 5). It has been pointed out (e.g. see Ref. 6) that turbulence is intermittent, i.e. regions of almost laminar flow alternate with turbulent regions. For this reason, it was suggested in Ref. 5 that flow analysis should be focused on (linear or nonlinear) interaction rather than (laminar or turbulent) structure. In this context, the meaning of the Reynolds number introduced to characterize the problem becomes obscure, because the Reynolds number depends on local flow structure. Therefore, local dynamic viscosity should be introduced instead of turbulent viscosity, as proposed in Ref. 5.

## B.2. Background

We began our studies of turbulence development with a physically transparent problem: relaxation of the velocity profile in the gap between cylinders starting from a highly unstable shear flow. The problem of flow between cylinders has a history more than a century old. The basic results were obtained by Taylor in his classical studies (1923, 1936), where it was shown that convective cells develop at a Reynolds number $Re \sim 60$ provided that angular momentum decreases outward. (Recall that turbulence in a parallel flow develops at $Re \sim 2000$.) Moreover, it was shown in Ref. 6 that laminar flow remains stable if angular momentum increases outwards (the corresponding critical Reynolds number was estimated as $Re \sim 2 \times 10^5$ by extrapolating the results obtained by Taylor in 1936). Our research concerned the mechanism of transition to the steady velocity profile in the gap starting from an initially unstable shear flow. Our goal was to analyze the evolution of flow structure (including vortices) by using a model of the gap flow. Our approach was based on the hypothesis that viscous stresses are unimportant as compared to inertial forces at high Reynolds numbers and turbulent flow develops from the large-scale vortices generated by the shear flow through the Kelvin–Helmholtz instability. In other words, the energy of chaotic motion in turbulent flow is the energy of shear flow!

We focused on the flow between coaxial cylinders, which is the easiest to simulate computationally. It was pointed out by Rayleigh that a centrifugal force stabilizes the Taylor–Couette flow if the following condition is fulfilled (see Ref. 7):

$$d(Vr)^2/dr \leq 0. \tag{B.1}$$

The onset of turbulence plays a particularly important role in astrophysical problems. For example, it is generally assumed that flows in accretion disks can be characterized by the conventional turbulent viscosity, since the corresponding Reynolds numbers are very high. However, this assumption was questioned in a number of studies, where it was hypothesized that the Keplerian flow can be laminar. The main argument behind this hypothesis is that differential fluid rotation characterized by a radial increase in angular momentum is stable. Zeldovich performed several analytical studies of flows between rotating cylinders characterized by different laws of angular-momentum variation with radius. In particular, Zeldovich introduced the dimensionless Taylor number (see Ref. 6)

$$Te = \frac{d}{dr}(\omega r^2)^2 \left[ r^5 \left( \frac{d\omega}{dr} \right)^2 \right]^{-1}, \tag{B.2}$$

which can be used to derive a stability criterion for a uniform-density flow:

$$Te < 0. \tag{B.3}$$

For example, when the fluid rotation is described by the power law $\omega \sim r^n$, the Taylor number $Te = 8(n+2)/n^2$ is constant in the entire domain, and the flow is stable if $n > -2$. In the case of solid-body rotation, $Te = \infty$ and the flow is stable. The criterion introduced by Zeldovich is better suited for the analysis of transition to turbulence in astrophysical flows. Studies of this kind are important because the flow of a homogeneous fluid induced by a rotating outer cylinder is analogous to a gravity-stabilized shear flow. Criterion (B.3) can be derived by performing a simple analysis and invoking an analogy with the Richardson number (which determines the stability of laminar flow in a density-stratified atmosphere in the presence of gravity force).

## B.3. Direct numerical simulation methodology

We have developed a general methodology for direct numerical simulation of a broad class of nonlinear spatiotemporal phenomena of interest for modern fluid dynamics. These phenomena involve the development of free shear turbulence and/or hydrodynamic instabilities.[1,2] Our versatile approach to turbulence and instabilities relies on the construction of "rational" numerical models consistent with the phenomena under study and essentially differs from the approaches to turbulence simulation used by other authors. In structural analysis of turbulent flows, one should use the models best suited for describing the particular interaction mechanisms under study. For example, large-scale convection is simulated by invoking models of inviscid flow dynamics, models of laminar/turbulent flows allow for the viscous interaction mechanism, and stochastic processes are considered on a kinetic level. This approach to "rational" simulation makes it possible to reflect the interaction mechanisms contributing to "structural" turbulence and to substantially reduce the required computing resources as compared to other approaches.

The development of turbulent flow in the gap between cylinders is difficult to analyze because of the simultaneous action of the inertial force $VdV/dr$ and viscosity. Viscosity is responsible for the linear Couette flow profile developing when the Reynolds number is sufficiently low. However, the effective value of $Re$ and the relative contributions of the forces require further analysis, because a Kelvin–Helmholtz-type instability similar to that of the boundary (tangential discontinuity) between flows having different velocities can develop when the velocity gradient is sufficiently steep and the kinetic energy of the flow is sufficiently high. When this is the case, the viscous term becomes unimportant as compared to the convective term. The estimated rate of growth of the Kelvin–Helmholtz instability is independent of viscosity.[8] As pointed out in Ref. 1, such models can be analyzed within the framework of the Euler equations. The important role played by large-scale structures in turbulence was originally emphasized in Refs. 9 and 10. More detailed discussions of this problem and the structural approach to turbulent flow computations can be found in Refs. 1 and 2. It was shown that the energy of turbulent flow is mainly associated with large-scale motion, i.e. with inertial forces.

In the present formulation, inertia can act via couples of forces to create vortices (cyclones and anticyclones) that control the development of the flow profile in the gap. One important issue in this approach is the breakup of vortices. It is not clear whether viscous or inertial forces are responsible for the breakup, and the characteristic time of the process is yet to be determined. Our hypothesis is that the high-frequency portion of the spectrum is generated as a result of nonlinear interactions of large-scale structures with one another and with walls, while molecular viscosity does not play any role in this process (at sufficiently high $Re$).

## B.4. Statement of the problem and results

Following the ideas outlined above, we conducted the following numerical experiments. We considered coaxial cylinders with the relative gap width $\Delta R/R = 0.227$. Two-dimensional flows were computed by solving the Euler equations in a polar coordinate system $(r, \phi)$. We used a second-order accurate in space and time, quasi-monotone grid-characteristic scheme.[11]

Molecular viscosity was neglected, but the numerical technique involved an implicit nonlinear dissipation mechanism depending on the mesh size (dissipation decreased with the mesh size). Note, however, that the rate of energy dissipation in turbulent flows is determined by the rate of inertial energy transfer and is insensitive to the characteristics of the dissipation mechanism (molecular viscosity).

In a numerical experiment, the relative importance of inertia and dissipation can be estimated from the experiment itself by measuring the rate of kinetic-energy dissipation $\theta_{\text{num}}$ and calculating the dimensionless number $Re_{\text{num}}$ defined, by analogy with the Reynolds number, as

$$Re_{\text{num}} \sim \frac{(\Delta u)^2}{\theta_{\text{num}} \ell},$$

where $\Delta u$ is the change in velocity over a length scale $\ell$.

The grids employed in these computations had 100 radial points and 2500 angular points. We examined the following variants differing in initial and boundary conditions.

### *B.4.1. The inner cylinder is at rest and the outer cylinder is rotating*

In this variant, the inner cylinder remained permanently at rest ($V_0 = 0$), while the outer one rotated counterclockwise with a constant velocity ($V_1 = 1$). (In our computations, we used dimensionless units determined by a reference velocity and a reference length.) At the initial moment, the flow had the following structure. In the annular half-gap adjoining the inner cylinder, the fluid was quiescent. In the outer half-gap, the fluid rotated as a solid body with an angular velocity equal to that of the outer cylinder. The tangential discontinuity was perturbed by introducing a random small-amplitude high-frequency disturbance of the radial velocity component (its amplitude was less than 1% of $V_1$ and its size was similar to the mesh size).

The boundary conditions on the inner and outer cylinders were set by using an implicit conservative correction of the angular velocity $v_\varphi$ and internal energy in the cells that adjoin the cylinders:

$$\nu_\varphi^{n+1} = \frac{\tilde{\nu}_\varphi + \alpha \nu_{\text{cyl}}}{1+\alpha}, \quad \alpha = v_* \frac{2\tau}{h_r^2}, \tag{B.4}$$

where $\tilde{\nu}_\varphi$ is the velocity obtained at a current time step by solving the Euler equations subject to an impermeability condition, $v_{\text{cyl}}$ is the cylinder velocity, $\tau$ is the time step, $h_r$ is the radial mesh size, $\nu^*$ is an artificial viscosity effective only in boundary cells. A decrease in kinetic energy resulting from the velocity correction defined by (B.4) was interpreted as dissipation of kinetic energy, i.e. its conversion into heat, whereas an increase in kinetic energy was attributed to the work done by the external forces that sustained the cylinder rotation.

Figure B.1 shows the snapshots of a rapid development of the Kelvin–Helmholtz instability taken at $t = 0.05$, 0.1, and 0.2 (from top to bottom). In Fig. B.l(a), black and white areas correspond to the fluids initially located in the upper and lower layers, respectively. Figure B.1(b) shows the distribution of vorticity. The inverse increment of growth of the Kelvin–Helmholtz instability is proportional to $\Delta R/V_1$ (see Ref. 8), a short time interval corresponding to 1 /30 of a revolution of the outer cylinder. After this time interval has elapsed, further growth of vortices was suppressed because of the finite width of the gap between the cylinders. After that, the number of vortices decreased and their rotation accelerated as they attracted one another and merged under the action of the Joukowski force.

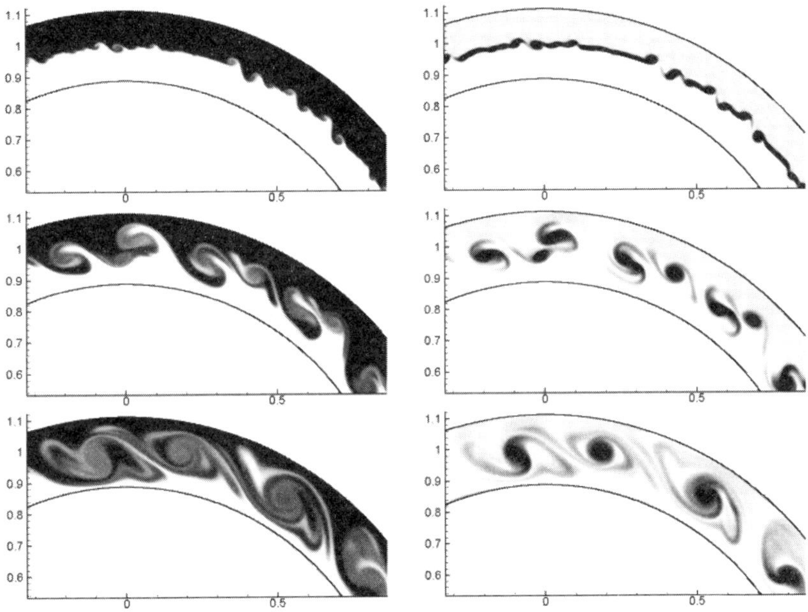


(a) (b)

Fig. B.1.

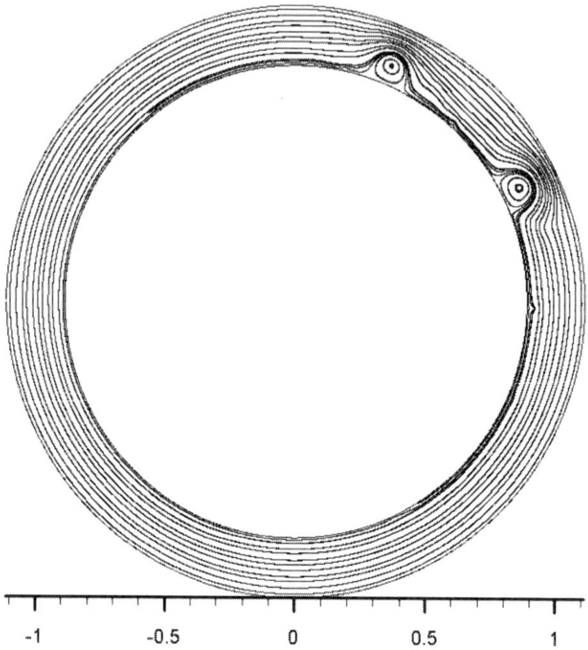


**Fig. B.2.**

When the computation of this variant was terminated at $t = 122$ (after approximately 20 revolutions), two vortices remained. The corresponding instantaneous streamline pattern is shown in Fig. B.2. A steady flow profile develops through redistribution of angular momentum in large vortices. Figure B.3 shows the steady profiles of angular momentum obtained in this computation (symbols) and in Ref. 7, where the steady state was approached through relaxation dominated by the viscous mechanism (line). According to the Taylor–Zeldovich theory, there is no turbulence after the steady flow profile has developed in the gap. However, the profiles obtained in our computations are inconsistent with the velocity $u = \text{const}.r^{6.5}$ estimated in Ref. 6. This discrepancy can be explained by the fact that the critical numerical Reynolds number obtained in the experiment is lower than the critical Reynolds number estimated by Zeldovich[6] (see estimate in Sec. B.2).

Figure B.4 shows the correlation functions of the radial velocity $u$ evaluated after one-third of a revolution (at $t = 2$) and after 20 revolutions

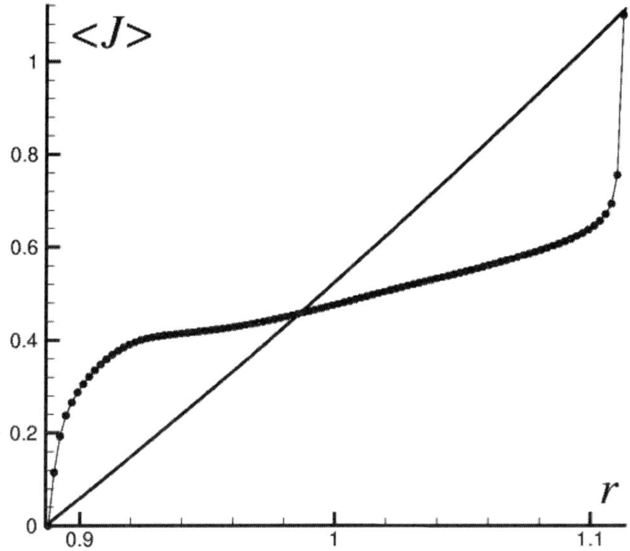


**Fig. B.3.**

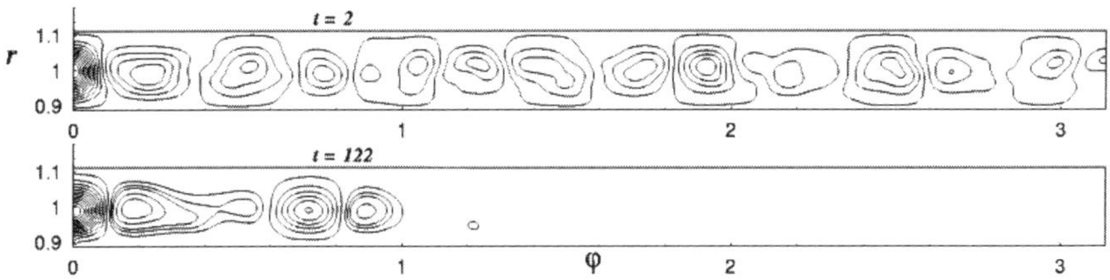


**Fig. B.4.**

(at $t = 22$). The correlation function $K_u$ is defined here as

$$K_u(r, \varphi) = \int u(r, \varphi + \xi) u(r, \xi) d\xi.$$

The figure demonstrates that the characteristic angular size (say, the distance between the minimum and maximum of the correlation function) corresponds to the gap between the cylinders, i.e. the sizes of the large-scale structures are equal at the initial and final stages of their evolution (when there are many and few well-developed structures, respectively).

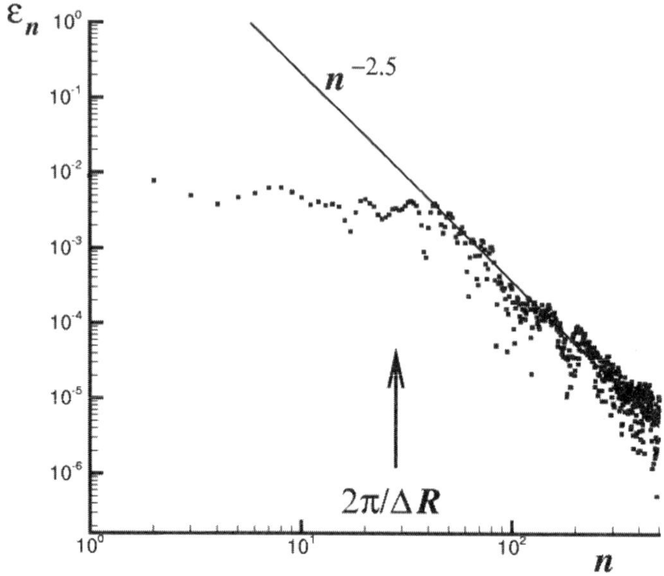


**Fig. B.5.**

Figure B.5 shows a typical spectrum of kinetic energy $\varepsilon_n$ (against wavenumber $n$) calculated as

$$\varepsilon_n = \sqrt{a_n^2 + b_n^2}, \quad a_n = \pi^{-1} \int_0^{2\pi} \left( \frac{u^2}{2} + \frac{\nu^2}{2} \right) \cos(n\varphi) d\varphi,$$

$$b_n = \pi^{-1} \int_0^{2\pi} \left( \frac{u^2}{2} + \frac{\nu^2}{2} \right) \sin(n\varphi) d\varphi.$$

The spectrum was computed at the radial location where the kinetic energy reaches its maximum, i.e. at $r \approx 1$. Two trends should be noted here: first, the "tail" of the spectrum does not obey the Kolmogorov law; second, the spectrum has a cusp at point $n_1 = 2\pi/\Delta R$ determined by the gap width.

### *B.4.2. The inner cylinder is at rest and the outer cylinder is brought to rest*

Two additional variants were computed to examine the development of turbulence under boundary conditions modified so as to violate condition (B.1) corresponding to an outward-increasing momentum. After approximately 20 revolutions, we stopped the rotation of the outer cylinder so that the angular momentum would drop in the outer region of the gap. The computed results showed that the characteristic vortex size decreased, and vorticity concentrated near the inner cylinder.

The system tends to rotate as a solid body. This is illustrated by Fig. B.6, which shows the angular momentum averaged over the angular coordinate at $t = 190$. The radial-velocity correlation functions depicted in Fig. B.7 also demonstrate the decrease in the characteristic size of the vortices and their tendency to "stick" to the inner cylinder.

The results of this numerical experiment can be used to estimate the rate of dissipation of kinetic energy and hence the effective turbulent viscosity.

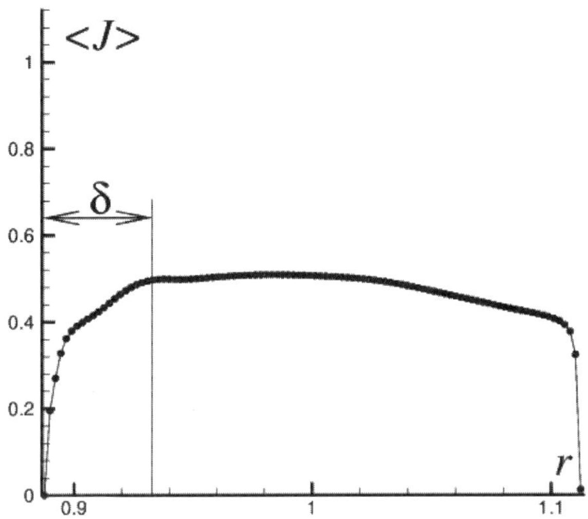


**Fig. B.6.**

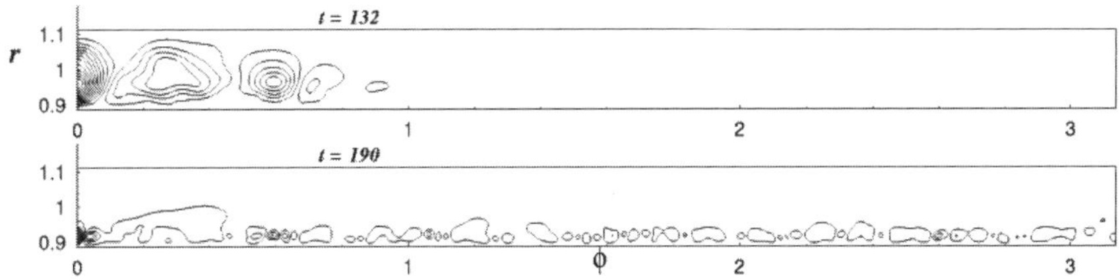


**Fig. B.7.**

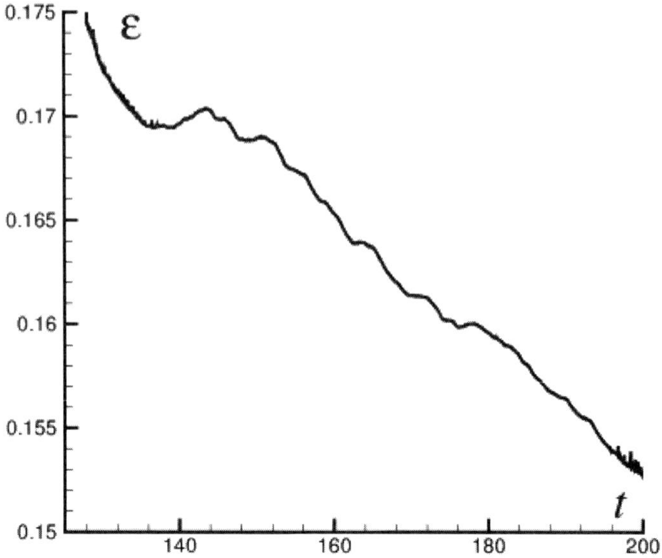


**Fig. B.8.**

Figure B.8 illustrates the change in the integral kinetic energy with time. According to (B.4), the corresponding numerical Reynolds number is of the order of 1000.

Turbulence does not develop, which should have been expected since the computed distribution of angular momentum corresponds to solid-body rotation (except for boundary layers).

### *B.4.3. The inner cylinder is rotating and the outer cylinder is at rest*

In the last variant, we examined a classical scenario of turbulence development in which either condition (B.1) or (B.3) is violated, i.e. the inner cylinder is rotating while the outer one is at rest. Figures B.9–B.12 show numerical results corresponding to this case. Figure B.9 shows the instantaneous streamline pattern at $t = 840$; Fig. B.10 — the velocity correlation functions at $t = 700$ and $t = 840$; Fig. B.11 — the spectrum of kinetic energy at $t = 840$. The distribution of energy over wavenumbers demonstrates that the high-frequency "tail" of the spectrum presented here, in contrast to the classical Kolmogorov spectrum, carries an insignificant fraction of the total energy. Accordingly, the rate of conversion of kinetic energy into

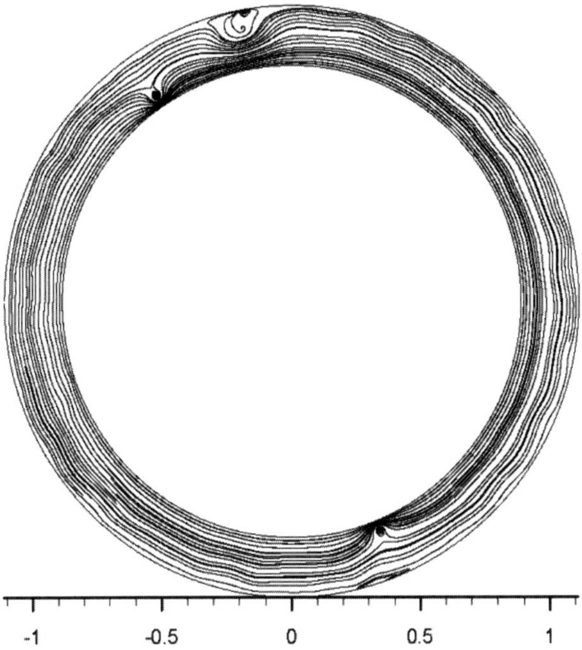


**Fig. B.9.**

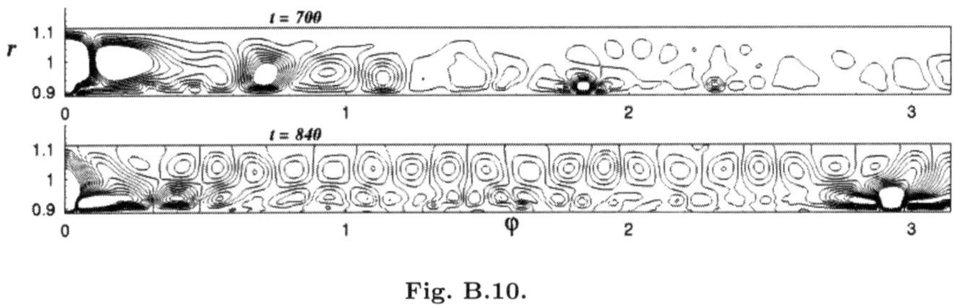


**Fig. B.10.**

heat decreases and this is additionally manifested by the weak variation of entropy in the course of the computation. This trend is even more pronounced in the first two variants computed in this study.

Figure B.12 shows the field of vorticity (curl of velocity) at $t = 840$. The insert at the center shows a fragment of the flow structure around a large-scale vortex. Gray areas correspond to low magnitudes of vorticity, while lighter and darker areas represent regions of clockwise and

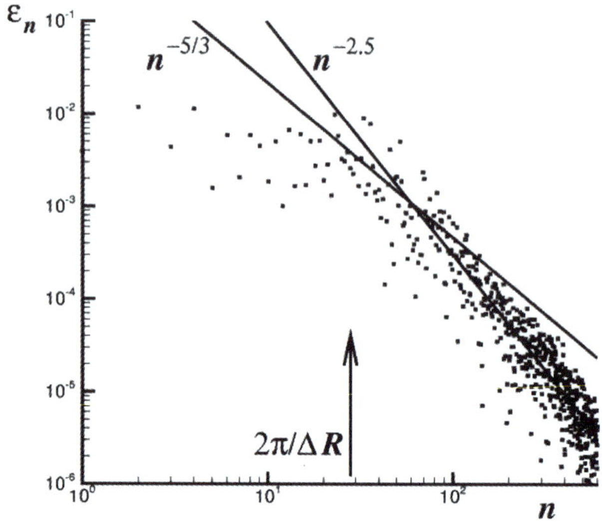


Fig. B.11.

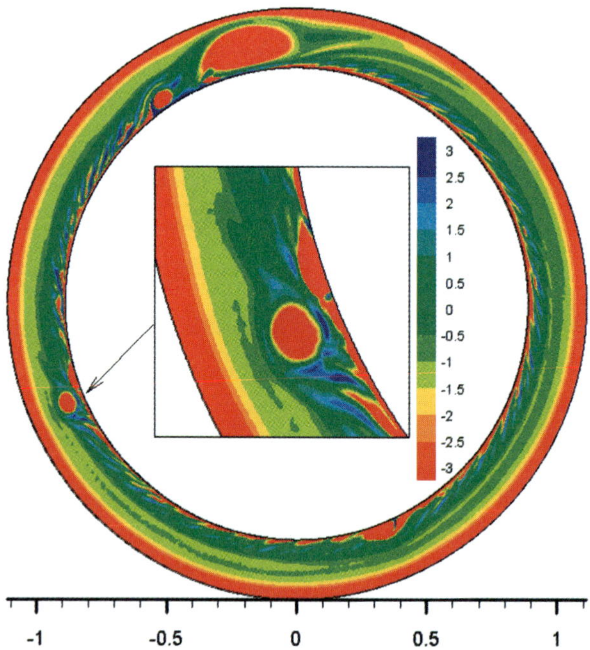


Fig. B.12.

counterclockwise vorticities, respectively. Note that small-scale structures (vortices) are localized around the large-scale ones and near the inner cylinder.

## B.5. Conclusions

The computations described here were performed to expose the evolution of the spectrum of developing turbulence. The results are particularly illustrative in the third case, when stability criterion (B.1) or (B.3) is violated. However, the size of the major large-scale structures identified in all computations was comparable to the gap between the cylinders, as illustrated by Figs. B.4, B.7, and B.10. High-frequency turbulence was also observed in our computations, but it was mainly localized between the large-scale vortices and the cylinder walls. The rest of the flow was almost laminar, with weak vorticity induced by the inner cylinder's rotation. In particular, a small-scale vortex pattern (corresponding to high-frequency turbulent spectrum) can be seen at the inner cylinder in the third variant.

The kinetic energy of turbulent motion is associated with large-scale vortex structures. We believe that the high-frequency portion of the spectrum is generated in the course of nonlinear interactions of large-scale structures with one another and with the cylinder walls. The effect of nonlinear interactions is not restricted to the drastic increase in effective viscosity over molecular viscosity (more precisely, the former should be called dynamic, rather than turbulent, viscosity, as suggested in Ref. 6). Unlike molecular viscosity (which is scale-independent), the effective (dynamic) viscosity is sensitive to the size of flow structures (vortices). Larger structure is associated with higher viscosity. The dynamic viscosity characterizes the intensity of nonlinear interactions. The computations performed in this study show that nonlinear interactions taking place in different regions of the computational domain are characterized by different intensities. This observation suggests that dynamic viscosity has a local nature.

As noted by Zeldovich (and confirmed by our numerical results), the structure of the flow between cylinders provides an example of intermittent turbulence, i.e. regions of almost laminar flow alternate with turbulent regions (large-scale vortices). The relative volumes of these regions may depend on the Reynolds number, but isotropic turbulence is almost never observed.

Further development in our research should rely on a higher grid resolution. Moreover, the present analysis can be extended to examine the evolution of a large-scale vortex on fine grids.

## References


1. O. M. Belotserkovskii and A. M. Oparin, *Chislennyi eksperiment v turbulentnosti: ot poryadka k khaosu* (*Numerical Experiment in Turbulence: From Order to Chaos*) (Nauka, Moscow, 2000).
2. O. M. Belotserkovskii, *Turbulence and Instabilities* (Mellen, Lewinston, New York, 2000).
3. S. A. Colgate and J. R. Buchler, Coherent transport of angular momentum — the Ranque — Hilsch tube as a paradigm. Astrophysical turbulence and convection, *Ann. NY Acad. Sci.* **898**, 105–111 (2000).
4. D. V. Bisikalo, A. A. Boyarchuk, O. A. Kuznetsov and V. M. Chechetkin, The impact of viscosity on the morphology of gaseous flows in semidetached binary systems, *Astron. Zh.* **77**(1), 31–41 (2000).
5. V. M. Canuto, Turbulence and laminar structures: can they co-exist? *Mon. Not. R. Astron. Soc.* **317**, 985–988 (2000).
6. Ya. B. Zel'dovich, On viscosity in flows between rotating cylinders, Preprint, Institute of Applied Mathematics, USSR Academy of Sciences Moscow (1979).
7. M. J. Molemaker and J. C. McWilliams, Instability and equilibration of centrifugally stable stratified Taylor–Couette flow, *Phys. Rev. Lett.* **86**(23), 5270–5273 (2000).
8. L. D. Landau and E. M. Lifshits, *Gidrodinamika* (*Fluid Mechanics*) (Nauka, Moscow, 1986).
9. H. I. Dryden, Recent advances in the mechanics of boundary layer flows, *Adv. Appl. Mech.* **1**, 1–40 (1948).
10. A. A. Townsend, *The Structure of Turbulent Shear Flow* (Cambridge University Press, London, 1956).
11. A. M. Oparin, Numerical simulation of problems associated with rapid development of hydrodynamic instabilities, *Novoe v Chislennom Modelirovanii: Algoritmy, Vychislitel'nye Eksperimenty, Rezul'taty* (*Advances in Numerical Simulation: Algorithms, Numerical Experiments, Results*) (Nauka, Moscow, 2000), pp. 63–98.

# Appendix C

# Universal Technology of Parallel Computations for the Problems Described by Systems of the Equations of Hyperbolic Type: A Step to Supersolver*

Universal technology for numerical solution of quasilinear sets of hyperbolic equations using parallel computations is presented. Unified formalized methodics, which includes a wide range of numerical methods as a special case, is proposed. A new method for the development of finite-difference schemes in nonconservative variables is offered. The architectural, algorithmic, and technical aspects of the implementation framework are studied. The results of the modeling of atmosphere flows in several two- and three-dimensional problems are shown.

## C.1. Introduction

We are involved in the numeric modeling of the problems described by hyperbolic systems of equations in partial derivatives. The process of modeling could be described as finding out the point in three-dimensional space, given by three axes, which we will briefly discuss.

Hyperbolic equations reside in many application domains: aerodynamics, hydrodynamics, acoustics, elastics, magnetohydrodynamics, shallow water, etc. In some domains the governing equations are of hyperbolic type, in other — the splitting of the system includes a hyperbolic step. But many methods for numerical solution of these equations could be used independently of the particular system of equations. So, the first direction, in which we are moving at, is that of the application domains and the particular physical problems. Here, the physical model, mass

*O. M. Belotserkovskii, M. N. Antonenko, A. V. Konyukhov, L. M. Kraginsky, A. M. Oparin and S. V. Fortova, *Comput. Fluid Dynam. J.* **11**(4), 456–466 (2003).

forces, particular form of the system, main calculation components, and Jacobi matrices vary.

In numerical solution of the hyperbolic system of equations in partial derivatives various methods are usually combined to work together. Some of these methods are orthogonal in the sense that they are independent and could be substituted with the method of the same type independently of the other components of the resulting method. This variety of the methods constitutes the second direction of numeric methods, where we find out the resulting method to solve our problem.

The last thing to choose on the last is the algorithm and implementation of chosen method. To obtain the most accurate results one should use parallel calculations on multiprocessor computers. It is worth mention that parallel and sequential algorithms for the same method differ much. There exists a diverse range of software packages and libraries of methods but most of researchers write their own programs, partially because most of these libraries are not as flexible, as researchers would like them to be, they are not multiplatform, they support only limited number of methods and they are hard to integrate and many of them do not provide parallel versions.

To simplify the development of the modeling programs in different areas with different methods and reuse existing software we propose the unified framework, which could be seen as the common kernel for all modeling problems, lying in the described space models. It should be universal, flexible, extensible, open, standard-oriented solver framework for parallel computers. Each coordinate should be described by a particular building block to be plugged into the framework. The most useful building blocks should be provided with the framework, but it should have the open interface for other users to plug in their own physics, methods, solvers, visualizers, etc. To build this framework, we need to base on the unified numeric methodic that should be also flexible and extensible. Below, we propose such methodics, present its reference implementation and show results of its application to numerical modeling of several actual problems in atmosphere flows modeling.

## C.2. Unified methodics

Consider the hyperbolic system of equations in partial derivatives

$$\frac{\partial \mathbf{Q}}{\partial t} + \mathbf{A}\frac{\partial \mathbf{Q}}{\partial x} + \mathbf{B}\frac{\partial \mathbf{Q}}{\partial y} + \mathbf{C}\frac{\partial \mathbf{Q}}{\partial z} = \mathbf{G}, \tag{C.1}$$

where $\mathbf{Q}$ is the vector of the main calculation components, $x, y, z$ — some independent coordinates in the modeled volume, $\mathbf{A}$, $\mathbf{B}$, $\mathbf{C}$ — Jacobi matrices (with dimension $N \times N$, where $N$ is the dimension of $\mathbf{Q}$). The model could also contain $K$ additional components $\mathbf{P}$, which describe the physical properties of the medium. In general case, matrices $\mathbf{A}$, $\mathbf{B}$, $\mathbf{C}$ can depend not only on coordinates $x$, $y$, $z$, but also on the field $\mathbf{P}$ (perhaps, variable in time) and solution vector $\mathbf{Q}$.

To reduce the multi-dimensional problem to a number of one-dimensional problems we should use a kind of splitting method. Those methods are called splitting by physical processes and spatial variables. Thus, we should solve three hyperbolic equations and consider right-hand terms: mass forces, viscosity, other differential terms, random terms, etc., which depend on the model:

$$\frac{\partial \mathbf{Q}'}{\partial t} + \mathbf{A}\frac{\partial \mathbf{Q}'}{\partial x} = 0, \tag{C.2}$$

$$\frac{\partial \mathbf{Q}''}{\partial t} + \mathbf{B}\frac{\partial \mathbf{Q}''}{\partial y} = 0, \tag{C.3}$$

$$\frac{\partial \mathbf{Q}'''}{\partial t} + \mathbf{C}\frac{\partial \mathbf{Q}'''}{\partial z} = 0, \tag{C.4}$$

$$\frac{\partial \mathbf{Q}''''}{\partial t} = \mathbf{G}. \tag{C.5}$$

The choice of initial values for each of equations: Eqs. (C.2), (C.3), (C.4), and (C.5) together with the method for calculating the resulting component for the $(n+1)$th time step $\mathbf{Q}^{n+1}$ on the base of $\mathbf{Q}'$, $\mathbf{Q}'$, $\mathbf{Q}'''$, $\mathbf{Q}''''$, and $\mathbf{Q}^n$ depends on the particular splitting method and is parameterized. Here we will mention the most popular splitting methods:

- "No" splitting ($\mathbf{Q}', \mathbf{Q}'', \mathbf{Q}'''$, and $\mathbf{Q}''''$ are calculated on the base of $\mathbf{Q}^n$ from the previous time step).
- Ordinary splitting (in this case $\mathbf{Q}'$ is calculated on the base of $\mathbf{Q}^n$; $\mathbf{Q}''$, on the base of $\mathbf{Q}'$; $\mathbf{Q}'''$, on the base of $\mathbf{Q}''$; $\mathbf{Q}''''$, on the base of $\mathbf{Q}'''$; $\mathbf{Q}^{n+1} = \mathbf{Q}''''$).
- Strang-type splitting[1] with alternation in the order of steps (so as to save the second-order approximation of the main equation).
- LeVeque spatial splitting.[2]

Sometimes Runge–Kutta method of increasing the order of approximation in time is used which is parameterized by the order $p$ and the $RK$ function:

$$\mathbf{Q}^{(p+1)} = F(\mathbf{Q}^{(p)}), \quad \mathbf{Q}^{(0)} = \mathbf{Q}^n, \quad \mathbf{Q}^{n+1} = RK(\mathbf{Q}^{(p)}, \ldots, \mathbf{Q}^{(0)}). \tag{C.6}$$

The system of the hyperbolic type is reduced to a number of transport equations while converting to characteristic variables. Instead of

$$\frac{\partial \mathbf{Q}}{\partial t} + \frac{\partial \mathbf{F}}{\partial x} = \frac{\partial \mathbf{Q}}{\partial t} + \mathbf{A}\frac{\partial \mathbf{Q}}{\partial x} = 0, \quad \mathbf{A} = \mathbf{\Omega}^{-1}\mathbf{\Lambda}\mathbf{\Omega}, \quad \alpha = \mathbf{\Omega}\mathbf{Q},$$
$$\mathbf{F} = F(\mathbf{F}^{+}, \mathbf{F}^{-}), \tag{C.7}$$

which could be rewritten as

$$\frac{\partial \mathbf{Q}}{\partial t} + \frac{\partial \mathbf{F}}{\partial \mathbf{F}^{+}}\frac{\partial \mathbf{F}^{+}}{\partial x} + \frac{\partial \mathbf{F}}{\partial \mathbf{F}^{-}}\frac{\partial \mathbf{F}^{-}}{\partial x} = \frac{\partial \mathbf{Q}}{\partial t} + \mathbf{A}^{+}\frac{\partial \mathbf{Q}}{\partial x} + \mathbf{A}^{-}\frac{\partial \mathbf{Q}}{\partial x} = 0,$$
$$\mathbf{A}^{\pm} = \mathbf{\Omega}^{\pm -1}\mathbf{\Lambda}^{\pm}\mathbf{\Omega}^{\pm}, \tag{C.8}$$

we are solving the following equation for the vector of characteristic variables $\alpha$:

$$\frac{\partial \alpha}{\partial t} + \mathbf{\Lambda}^{+}\frac{\partial \alpha}{\partial x} + \mathbf{\Lambda}^{-}\frac{\partial \alpha}{\partial x} = 0. \tag{C.9}$$

Possible alternatives for this step, called the splitting of the system, are:

- Steger–Warming flux–vector splitting[3]

$$\mathbf{F} = \mathbf{F}^{+} + \mathbf{F}^{-}, \quad \mathbf{\Omega}^{\pm} = \mathbf{\Omega}, \quad \mathbf{\Lambda}^{\pm} = \{\mathrm{diag}(\mathrm{l}_k^{\pm})\},$$
$$l_k^{\pm} = \frac{l_k \pm |l_k|}{2}. \tag{C.10}$$

- Van Leer flux–vector splitting, given by special formulae[4]

$$\mathbf{F} = \mathbf{F}^{+} + \mathbf{F}^{-}, \quad \text{for } |u| \geq |a| \quad \mathbf{\Omega}^{\pm} = \mathbf{\Omega},$$
$$\mathbf{\Lambda}^{\pm} = \mathbf{\Lambda}^{\pm}_{\mathrm{Steger-Warming}}. \tag{C.11}$$

- MUSCL-type schemes, spatial differencing[5] — no conversion to characteristic variables, calculate $\partial \mathbf{Q}/\partial x$ instead of $\partial \alpha/\partial x$, but $\mathbf{F}^{\pm} = \mathbf{F}^{\pm}_{\mathrm{Steger-Warming}}$.

The transport equation $f_t + \lambda f_x = 0$ could be solved with various methods, among which we mark out the group of approximate Riemann solvers:

- generalized Roe solver with different limiters,[6]
- Harten's TVD solver,[7]
- UNO schemes of second- and third-order,[8,9]
- Chakravarthy–Osher's scheme,[10]
- van Leer's MUSCL–TVD,[5]

- piece-wise parabolic method,[11]
- many other,
- hybrid schemes,[12]
- monotone hybrid second-order scheme McCormack–Beam–Warming,[13]
- hybrid scheme McCormack–Beam–Warming — third-order central differencing.[14]

Any conservative scheme for one-dimensional transport equation could be written as

$$f_i^{n+1} = f_i^n - \frac{\tau}{h}(\lambda f_{i+1/2} - \lambda f_{i-1/2}), \tag{C.12}$$

where the flux $F_{i+1/2} = \lambda f_{i+1/2}$ calculation parameterize the chosen method.

The curvilinear coordinates and nonuniform grid are parameterized with the help of generic view of the inner coordinates $\xi, \eta, \xi : x = x(\xi, \eta, \zeta), y = y(\xi, \eta, \zeta), z = z(\xi, \eta, \zeta)$. To obtain conservative behavior of the calculation, the finite-volume formalism should be used:

$$q_i^{n+1} = q_i^n - \frac{\tau(S_{i+1/2}^{\eta\zeta} F_{i+1/2}^n - S_{i-1/2}^{\eta\zeta} F_{i-1/2}^n)}{V_i}. \tag{C.13}$$

Sometimes the method uses approximation of the variables: $q_i \to \overline{q}_i$, so fluxes are calculated on the base of $\overline{q}$:

$$F_{i+1/2} = \text{Flux}\ (\bar{f}_i, \bar{f}_{i+1}, \ldots). \tag{C.14}$$

Some averaging is usually used to obtain the matrices at the cell boundaries (Roe average, simple mean value, etc.):

$$\mathbf{A}_{i+1/2} = \text{Average}\ (\mathbf{A}_i, \mathbf{A}_{i+1}). \tag{C.15}$$

Entropy correction consists of using slight modification of eigenvalues (Harten fix, Roe fix, etc.) and is parameterized with the set of functions $\psi^k$:

$$\lambda_{EF}^k = \psi^k(\lambda^k). \tag{C.16}$$

Artificial compression algorithm tries to sharpen the discontinuouties at distinguished regions and is parameterized with the functional

$$\text{Compress} : \Delta q \to \text{Compress}\ (\Delta q). \tag{C.17}$$

## C.3. A method for using non-conservative variables

The results presented in this paragraph were obtained by Konyukhov.[15] A new method for the construction of numerical schemes for hydro- and aerodynamical equations is proposed. Those schemes are equivalent to some other schemes in conservative variables, but are formulated in nonconservative ones. Let us denote the vector of conservative variables as $\mathbf{U}$. The governing equations are given by the formula:

$$\frac{\partial \mathbf{U}}{\partial t} = -\nabla_k \mathbf{F}^k = -\mathbf{A}^k \nabla_k \mathbf{U}, \tag{C.18}$$

where $\mathbf{A}^k = \frac{\partial \mathbf{F}^k}{\partial \mathbf{U}}$ – Jacobi matrix.

Let $\mathbf{V}$ be a vector of some other variables, perhaps, more convenient for the calculations, for example, $(\rho, v, P)$. The governing equations for $\mathbf{V}$ are rewritten as

$$\frac{\partial \mathbf{V}}{\partial t} = -\mathbf{B}^k \nabla_k \mathbf{V}, \tag{C.19}$$

where $\mathbf{B}^k = \mathbf{M}^{-1}\mathbf{A}^k\mathbf{M}$, and $\mathbf{M} = \frac{\partial \mathbf{U}}{\partial \mathbf{V}}$ is the conversion matrix.

The straightforward approximation of Eq. (C.19) tends to nonconservative schemes. The main idea of the proposed approach is to obtain the discrete approximation of Eq. (C.19) by taking the conservative approximation of Eq. (C.18) and then convert it to another expression, approximating Eq. (C.19).

From this point, we will deal with the mesh functions $\mathbf{U}_i^n$ and $\mathbf{V}_i^n$. Assume there exists an operator $\mathbf{M}_{ij}^{kn}$, such that

$$\forall k, n \quad \mathbf{M}_{ij}^{kn}(\mathbf{V}_j^n - \mathbf{V}_i^k) = \mathbf{U}_j^n - \mathbf{U}_i^k. \tag{C.20}$$

The main example of such an operator is given by Roe averaging.

Let us consider only explicit schemes, such that they use only data from the previous time step (implicit schemes and multilayer schemes are considered in a similar way). All such schemes could be written as

$$\mathbf{U}_l^{n+1} - \mathbf{U}_l^n = \mathbf{\Phi}(\mathbf{U}_l^n, \mathbf{U}_j^n - \mathbf{U}_i^n). \tag{C.21}$$

If substituting differences of $\mathbf{U}$ in Eq. (C.21) with the expression given by Eq. (C.20), we will have:

$$\mathbf{M}_{ll}^{n+1n}(\mathbf{V}_l^{n+1} - \mathbf{V}_l^n) = \mathbf{\Phi}(\mathbf{U}(\mathbf{V}_l^n), \mathbf{M}_{ji}^{nn}(\mathbf{V}_j^n - \mathbf{V}_i^n)), \tag{C.22}$$

which is the discrete analogue of Eq. (C.19), rewritten as

$$\mathbf{M}\frac{\partial \mathbf{V}}{\partial t} = -\mathbf{A}^k \mathbf{M} \nabla_k \mathbf{V}. \tag{C.23}$$

**Definition C.1.** Let us call two finite-difference schemes (a) and (b):

$$\mathbf{U}^{n+1} = \mathbf{\Phi}_a(\mathbf{U}^k), \quad k \leq n, \tag{a}$$

$$\mathbf{V}^{n+1} = \mathbf{\Phi}_b(\mathbf{V}^k), \quad k \leq n, \tag{b}$$

*equivalent*, if from

$$\mathbf{U}^k = \mathbf{U}(\mathbf{V}^k) \quad \forall k \leq n, \quad \mathbf{U}^{n+1} = \mathbf{\Phi}_a(\mathbf{U}^k), \quad \mathbf{V}^{n+1} = \mathbf{\Phi}_b(\mathbf{V}^k), \tag{C.24}$$

follows

$$\mathbf{U}^{n+1} = \mathbf{U}(\mathbf{V}^{n+1}).$$

Schemes (C.21) and (C.22) are equivalent in this sense. We will also call a scheme, written in nonconservative variables, *conservative*, if it is equivalent to any conservative scheme in conservative variables.

The inverse operator $\mathbf{M}^{-1}$ in Eq. (C.22) could be found with the help of one of the following approaches:

(1) Using inner iterations to obtain $\mathbf{V}^{n+1}$:

$$\Delta\mathbf{U}_j = -\frac{\tau}{\Delta x}(\mathbf{F}^n_{j+1/2} - \mathbf{F}^n_{j-1/2}), \tag{C.25}$$

$$\mathbf{V}^{(0)}_j = \mathbf{V}^n_j, \quad \text{then iterations in } k$$

$$\mathbf{V}^{(k+1)}_j = \mathbf{V}^{(k)}_j + \mathbf{M}^{-1}(\mathbf{V}^n_j, \mathbf{V}^{(k)}_j)\Delta\mathbf{U}_j, \tag{C.26}$$

until the desired accuracy is obtained on the step $K$

$$\mathbf{V}^{n+1}_j = \mathbf{V}^{(K)}_j.$$

(2) Using equivalence property:

$$\Delta\mathbf{U}_j = -\frac{\tau}{\Delta x}(\mathbf{F}^n_{j+1/2} - \mathbf{F}^n_{j-1/2}),$$

$$\mathbf{V}^{n+1}_j = \mathbf{V}^n_j + \mathbf{M}^{-1}(\mathbf{V}^n_j, \mathbf{U}^{-1}(\mathbf{U}(\mathbf{V}^n_j) + \Delta\mathbf{U}_j))\Delta\mathbf{U}_j. \tag{C.27}$$

Also it is possible to use explicit formulae for transformation from one type of variables to the other (third approach):

(3)

$$\Delta\mathbf{U}_j = -\frac{\tau}{\Delta x}(\mathbf{F}^n_{j+1/2} - \mathbf{F}^n_{j-1/2}),$$

$$\mathbf{U}^{n+1}_j = \mathbf{U}^{-1}(\mathbf{U}(\mathbf{V}^n_j) + \Delta\mathbf{U}_j). \tag{C.28}$$

In the last approach, the scheme ceases being linear, but the equivalence property is fulfilled exactly in this case.

Specific coefficients of the matrix $\mathbf{M}^{-1}$ should be obtained for each set of variables individually. In the case of Euler equations, matrix $\mathbf{M}^{-1}$ could be found using the properties of Roe averaging:

$$\Delta\mathbf{F} = \mathbf{C}(\bar{\mathbf{z}})\mathbf{D}^{-1}(\bar{\mathbf{z}})\Delta\mathbf{U} = \mathbf{C}(\bar{\mathbf{z}})\mathbf{E}^{-1}(\tilde{\mathbf{U}}, \bar{\mathbf{z}})\Delta\mathbf{V}, \tag{C.29}$$

where $\tilde{\mathbf{U}}$ is the mean (in the sense of Roe) of conservative variables $\mathbf{U}$

$$\Delta\mathbf{U} = \mathbf{U}_2 - \mathbf{U}_1, \quad \mathbf{z} = (\sqrt{\rho}, \sqrt{\rho}\mathrm{v}, \sqrt{\rho}H), \quad \bar{\mathbf{z}} = \frac{\mathbf{z}_2 + \mathbf{z}_1}{2}. \tag{C.30}$$

Then $\mathbf{M}^{-1} = \mathbf{E}(\tilde{\mathbf{U}}, \bar{\mathbf{z}})\mathbf{D}^{-1}(\bar{\mathbf{z}})$, where matrices $\mathbf{D}$ and $\mathbf{E}$ for variables $\mathbf{V} = (\rho, \mathrm{v}, P)$ look like as follows:

$$\mathbf{D}(\bar{\mathbf{z}}) = \begin{bmatrix} 2\bar{z}^1 & 0 & 0 \\ \bar{z}^2 & \bar{z}^1 & 0 \\ \dfrac{(\bar{z}^3 - 2\bar{z}^1 P_\rho - \bar{z}^2 P_{\rho\mathrm{v}})}{1 + P_{\rho E}} & \dfrac{-\bar{z}^1 P_{\rho\mathrm{v}}}{1 + P_{\rho E}} & \dfrac{\bar{z}^1}{1 + P_{\rho E}} \end{bmatrix}, \tag{C.31}$$

$$\mathbf{E}(\tilde{\mathbf{U}}, \bar{\mathbf{z}}) = \begin{bmatrix} 2\bar{z}^1 & 0 & 0 \\ -\bar{z}^2/\tilde{\rho} & \bar{z}^1/\tilde{\rho} & 0 \\ \dfrac{(2\bar{z}^1 P_\rho + \bar{z}^2 P_{\rho\mathrm{v}} + \bar{z}^3 P_{\rho E})}{1 + P_{\rho E}} & \dfrac{\bar{z}^1 P_{\rho\mathrm{v}}}{1 + P_{\rho E}} & \dfrac{\bar{z}^1 P_{\rho E}}{1 + P_{\rho E}} \end{bmatrix}, \tag{C.32}$$

$$P_\rho = \frac{\partial P}{\partial \rho}(\bar{\mathbf{z}}), \quad P_{\rho\mathrm{v}} = \frac{\partial P}{\partial(\rho\mathrm{v})}(\bar{\mathbf{z}}), \quad P_{\rho E} = \frac{\partial P}{\partial(\rho E)}(\bar{\mathbf{z}}).$$

And the matrix $\mathbf{M}^{-1}$ has the same form as the analytic expression for $\mathbf{M}^{-1}$, applied to the averaged by Roe variables:

$$\mathbf{M}^{-1} = \mathbf{E}(\tilde{\mathbf{U}}, \bar{\mathbf{z}})\mathbf{D}^{-1}(\bar{\mathbf{z}}) = \begin{bmatrix} 1 & 0 & 0 \\ -\tilde{\mathrm{v}}/\tilde{\rho} & 1/\tilde{\rho} & 0 \\ P_\rho & P_{\rho\mathrm{v}} & P_{\rho E} \end{bmatrix}. \tag{C.33}$$

It is worth mention, that the considered transformations do not affect computational stencil. Therefore this approach could also be applied to unstructured and adaptive grid.

The advantages of this method are:

- convenient form of state equation,
- convenient form of Jacobi Matrix,

- convenient form of boundary conditions,
- computational efficiency due to mentioned above factors,

smaller computational error (for example, in operations with energetic variables with different orders of magnitude).

## C.4. Parallel program implementation

The package for designed formalism has been developed and tested. Real three-dimensional problems in several application domains were simulated with its help. The program is designed for parallel computers as well as for uniprocessor ones. The package uses MPI standard interface[16] for parallel applications and thus can be run on any computer, independent of its hardware, even uniprocessor. SPMD model of calculation is implemented, which helps to support only one code, which is run on all processors. The program was tested on different platforms (Intel, Param, MVS-1000) with different OS (Linux, Solaris, Windows) and implementations of the MPI (MPICH, WMPI). Three-dimensional decomposition is implemented in the code, which results in a number of communications of the order $3\sqrt[3]{N}$ instead of $N$ in one-dimensional decomposition (still one- and two-dimensional decompositions are possible).

From the outer user's point of view, the program remains successive, as if it has no parallel behavior. Enhancements due to parallelism are needed only at a small number of points in a program:

- initialization and finalization of the parallel framework,
- scattering of the grid,
- broadcast of scalar parameters,
- scattering of the initial state,
- gathering data before its output,
- determining the minimum in time steps due to Courant condition,
- summing up the integral values of conservation laws for testing purposes,
- exchange of the data for extra outer cells (unified with setting boundary values), which are all provided by the framework but could be altered by the end user.

Data exchange is generalized and could be tuned to fit the method. Only two logical types of communications are usually used in our program: master–slave and communications between neighbours in communicator grid. These communications work with corresponding user-defined MPI datatypes: collective data (used only on master process), local data (used on slave processes) and three types for three spatial directions: intersections of local data arrays defined on adjacent nodes. The package uses standard blocking operations for master–slave communication and nonblocking optimized send–receive operations for data exchange between neighbors. Also support for nonstandard MPI implementations is provided:

- send–receive to/from self is replaced by simple copy,
- optimized send–receive could be replaced by interleaved send and receive,
- no exchange with out-of-the-range nodes in the communicator is done for nonperiodic boundary values.

Basic operation with the calculated components are implemented as vector operations, that helps to parameterize the number of components, and the flux calculation is parameterized by using the functions, calculating the matrices of eigenvectors and eigenvalues of corresponding Jacobi matrices. Initial and boundary conditions and right-hand side of the equation are also treated as parameters and are separated from the kernel. Users can easily modify existing conditions or set their own problem-specific functions, which implement simple and evident programming interface. New solvers could be easily plugged into the kernel, substituting existing ones, moreover, this could be done at several levels and if the new solver conforms to the flux calculation technique, then the modification is minimal and affects the small independent part of the code.

The parameters are divided into three groups: scalar, vector, and functional. Construction of the new application for solving new applied problem is accomplished by setting a number of scalar parameters and implementing a number of functions, describing the specific problem, which have the clear defined interface and then are compiled with the other parts of the program. Vector parameters, such as, initial values, grid data, etc. are set up in a function with certain interface and persist in the memory until the next calls. Among the scalar parameters user can vary the following ones:

- decomposition parameters,
- grid sizes,
- volume sizes,
- components number,
- start and finish time (in the problem),
- maximal Courant number, etc.

Functional and algorithmic parameters are implemented as functions with definite interfaces. The results of these function calls usually are intermediate calculations and do not persist for a long time. The user can provide his own algorithms, implementing the given interface. Any of these interchangeable algorithms accomplish only a strict number of clearly defined operations with the data. Among such algorithms we can mention:

- boundary values,
- Jacobi matrix eigenvalues and eigenvectors,
- curvilinear coordinates,
- mass forces and source flows,
- entropy correction function,
- artificial compression algorithm,
- limiter function (for approximate Riemann solvers),
- finite-difference method.

## C.5. Results of numerical simulation

The package was first tested on one-dimensional problems. Convergence, monotonicity, TVD property, entropy gaps were studied for the simple nonlinear problems — several configurations of Riemann problem. On Fig. C.1, several schemes are compared when applied to the Riemann problem with the following initial values $\rho = 1$; $v = 0$; $P = 1$ for $x < 0$, $P = 0.1$ for $x > 0$. It could be seen that the mentioned hybrid scheme,[13] Harten's TVD2 scheme,[7] and Chakravarthy–Osher's scheme[10] give similar results. This is due to the fact that all of them use procedures very similar to *MinMod* reconstruction. *SuperBee* limiter sharpens the discontinuities, but admits nonphysical nonmonotone behavior near the compression wave. The best results (from the tested schemes) are shown by the MUSCL–TVD3–ACM scheme,[17] in which the MUSCL-approach is applied to Chakravarthy–Osher's scheme with *MinMod* limiter and artificial compression with coefficient $b = 2$, in this case solution remains monotone. If using the same scheme with $b = 4$, near the rarefaction wave the nonphysical gap appears, which arises due to replacement of the rarefaction wave with the compression wave.

Different boundary values and splitting of mass forces calculation were studied to obtain stability in time for one-dimensional stratified atmosphere. Afterward, two- and three-dimensional tests were run. Here, double

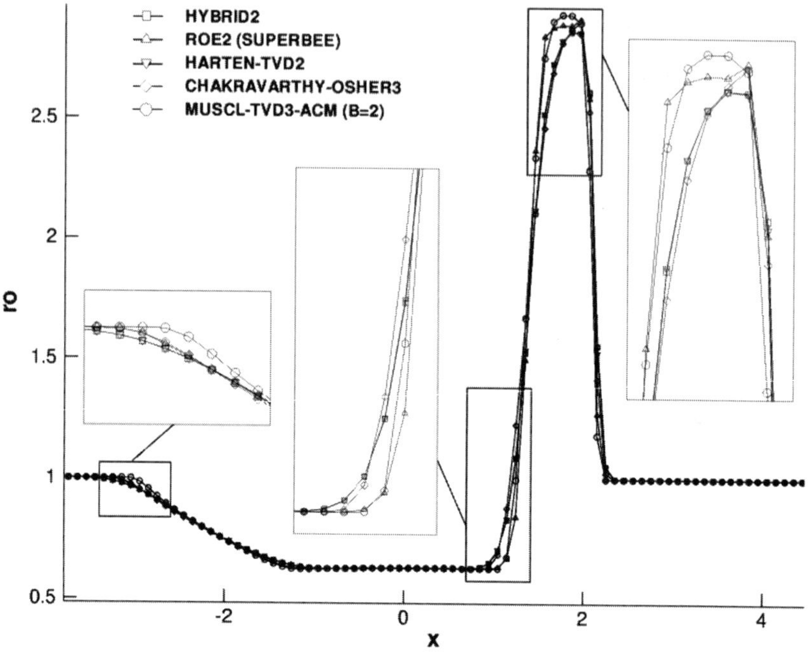


**Fig. C.1.**

Mach reflection and wave propagation in simple acoustic medium were modeled to test the approximation of cross-derivative terms, symmetry in all directions, dedicated directions and necessity for shear fix. For double Mach reflection problem, Mach number is 2.85. A shock wave is resolved quite good in all directions and it is orthogonal to boundary walls. Then, a few real modeling problems were simulated with the help of this package.

The first problem is the lifting of large-scale termic in standard stratified atmosphere. Initial state of the termic is the semi-spheric ($R = 1360\,\mathrm{m}$) cloud of hot gas (temperature in the center is 7085 K, and decreases exponentially till 2600 K on the boundary of the cloud), lying on the surface of the earth, without initial speed. This is the most simple model of the propagation of admixture cloud due to explosion of some products. In Fig. C.2, the evolution of the cloud after 0, 1, 2, 3, 4, 5 min is shown. The cloud turns into the torus curl, revolving in vertical plane. Points with maximum vorticity are shifted in the upper direction from the center of the torus equator. This determines the lift of the cloud. The height and the radius of the torus

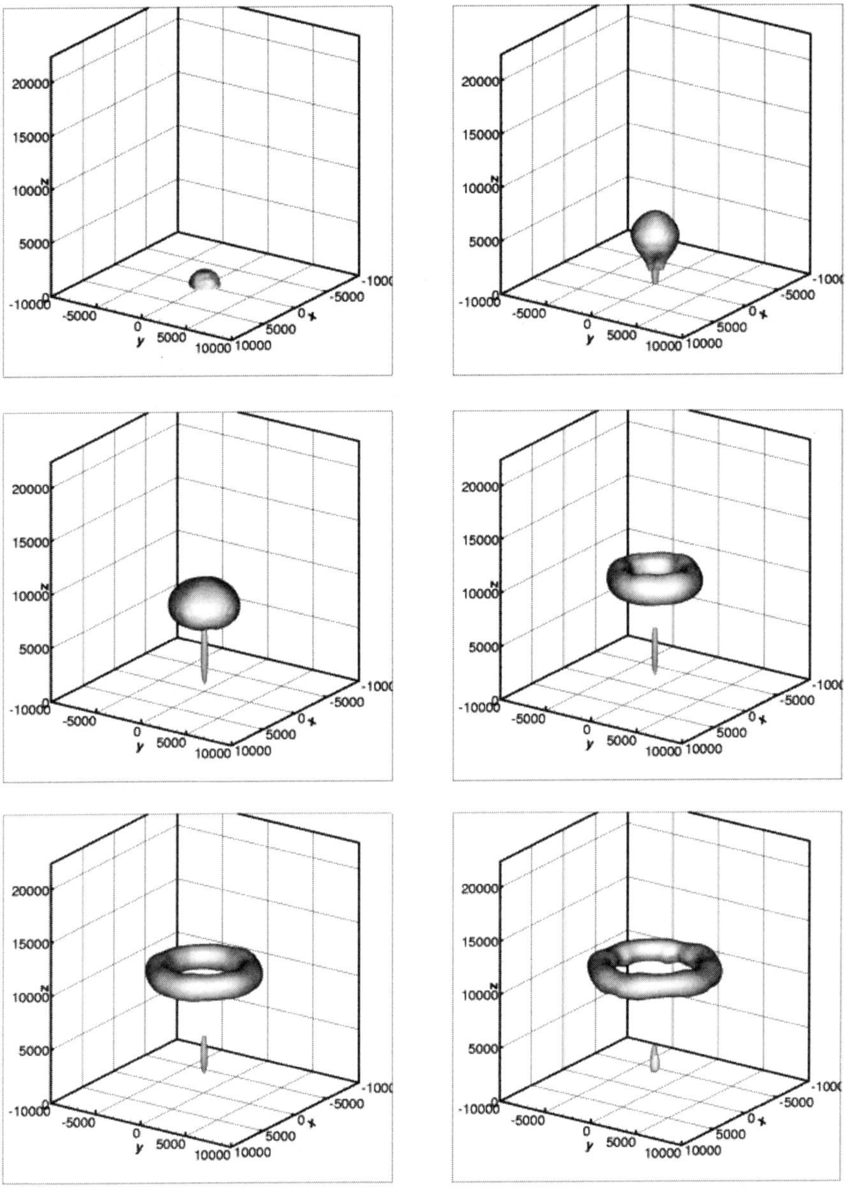


**Fig. C.2.**

cloud grow approximately as $C\sqrt{t}$ until the cloud reaches the stratosphere boundary at about 11–13 km.

The next series of simulations models the large-scale near-surface fire in stratified atmosphere. Two-dimensional as well as three-dimensional modeling were carried out. The fire is modeled as the near-surface source of energy- and light-weight admixture. The propagation of the fire cloud is visualized as the isosurface of admixture mass. Because the admixture is very light-weight it is almost fully carried with the fire. The source lies on the earth in the circle (with $R = 5\,\text{km}$ in two-dimensional simulation and 10 km in three-dimensional) and ejects energy with the power of $0.05\,\text{MWt/m}^2$. All the boundaries are impermeable walls. The upper boundary is 24 km high, and the side borders are 30 km long from the center of the source. In all directions, the grid is uniform near the source and then grows up exponentially. The grid sizes are $400 \times 150$ cells in two-dimensional and $128 \times 128 \times 64$ cells in three-dimensional model. In Fig. C.3, the distribution of admixture is shown for two-dimensional modeling at the moments $t = 18$, 27, 36, 45, 54, and 63 min. In Fig. C.4, some small isosurface of admixture mass is shown for three-dimensional modeling at the time $t = 10$, 20, 30, 50, 70, 90 min.

Two-dimensional fire cloud goes much more far than in three-dimensional case, while altitude of the spreading of the cloud is significantly less for two-dimensional case (about 5 km) then for three-dimensional, which reaches tropopause. It is worth mention the observed Rayleigh–Taylor instability,[18] which becomes apparent on the fine grids. Typical jets of hot gas go up where they are cooled and flowed down forming mushroom-like bubbles. The enlargement of scale of jets could be easily detected with the time. On the distant moments of time the cloud begins to oscillate near the average altitude with the Brunt–Vaissala frequency $\nu = \sqrt{-\frac{g}{\rho}\frac{\partial \rho}{\partial z}}$.

## C.6. Conclusion

- The technology for supersolver construction is proposed.
- The formalization for different methods for numerical solution of equations of the hyperbolic type is presented.
- A new method for constructing new schemes in nonconservable variables, having the conservative property, is proposed.
- The parallel code, implementing the mentioned technique, is overviewed.

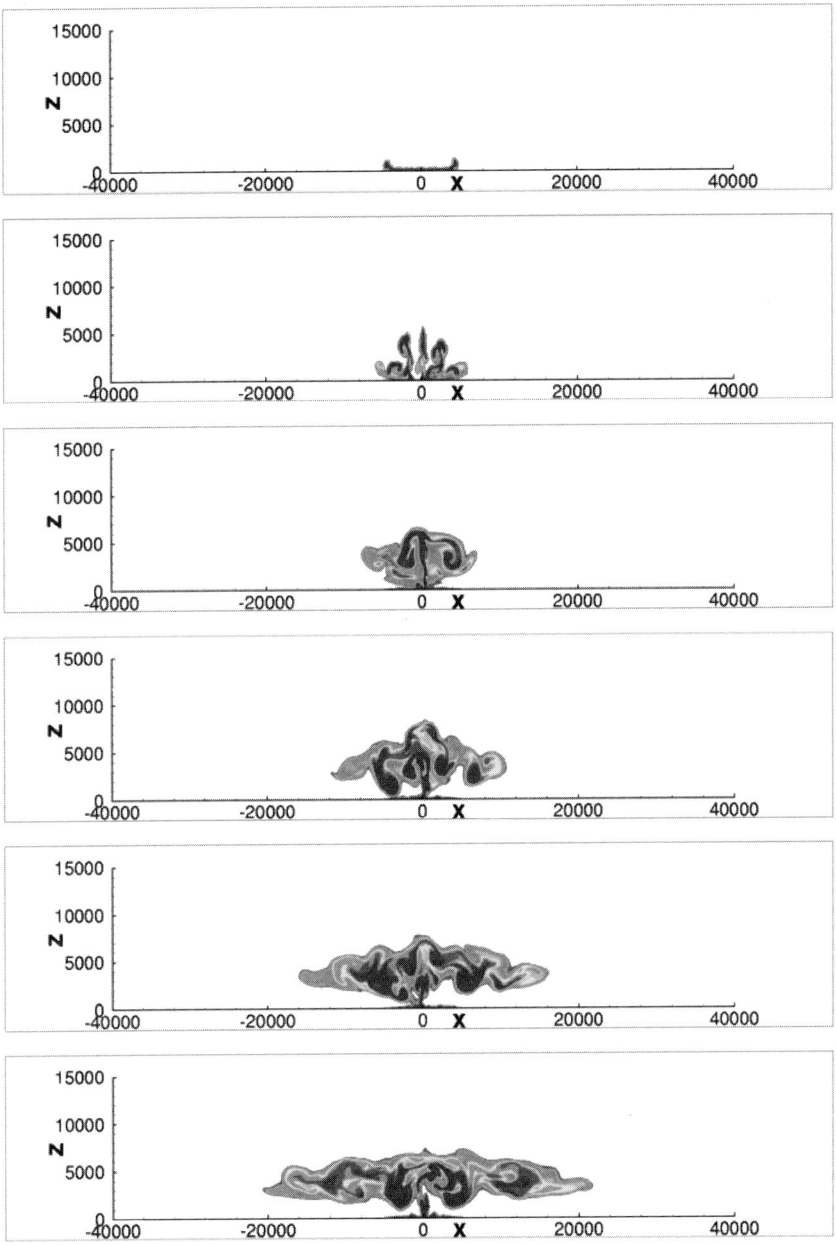


**Fig. C.3.**

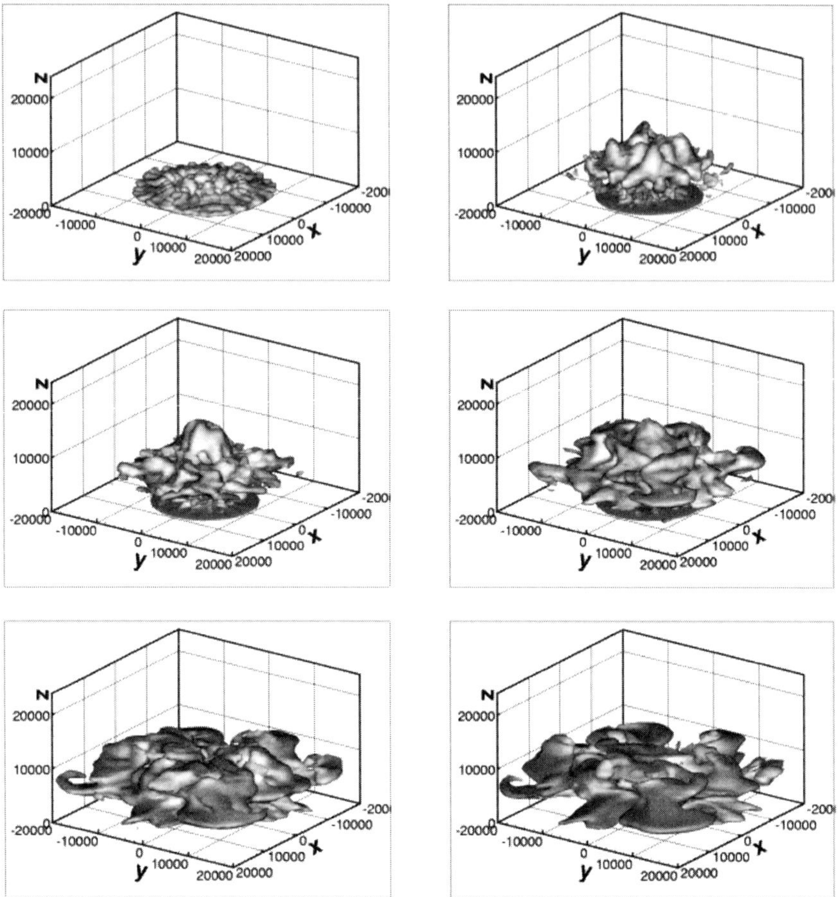


**Fig. C.4.**

- The programming techniques used for developing the parallel program are covered.
- Test and real three-dimensional simulations in aerodynamics, done with the help of the mentioned program, are presented.

## References

1. J. Strang, On the construction and comparison of difference schemes, *SIAM J. Num. Anal.* **5** (1968).

2. R. J. LeVeque, Wave propagation algorithms for multi-dimensional hyperbolic systems, *J. Comput. Phys.* **131** (1997).
3. J. L. Steger and R. F. Warming, Flux vector splitting of the inviscid gasdynamics equations with application to finite difference methods, *J. Comput. Phys.* **40** (1981).
4. B. van Leer, Flux-vector splitting for Euler equations, Lecture Notes in Physics, Vol. 170, pp. — (1982).
5. B. van Leer, Towards the ultimate conservative difference scheme. V. A second-order sequel to Godunov's method, *J. Comput. Phys.* **32** (1979).
6. P. L. Roe, Characteristic-based schemes for the Euler equations, *Ann. Rev. Fluid Mech.* **18** (1986).
7. A. Harten, High resolution schemes for hyperbolic conservation laws, *J. Comput. Phys.* **49**(2) (1983).
8. A. Harten and S. Osher, Uniformly high-order accurate nonoscillatory schemes, *SIAM J. Numer. Anal.* **24** (1987).
9. J. Y. Yang, Third-order nonoscillatory schemes for the Euler equations, *AIAA J.* **29**(10) (1991).
10. S. R. Chakravarthy and S. Osher, Computing with high-resolution upwind schemes for hyperbolic equations, Lecture Notes in Applied Mathematics, Vol. 22, pp. ——- (1985).
11. P. Colella and P. R. Woodward, The piecewise parabolic method (PPM) for gas dynamical simulations, *J. Comput. Phys.* **54** (1984).
12. K. M. Magomedov and A. S. Kholodov, *Grid-Characterstic Numerical Methods* (Nauka, Moscow, 1988).
13. O. M. Belotserkovskii, V. A. Gushchin and V. N. Konshin, Splitting method for the study of flows in stratified fluid with free boundary, *Zhournal Vychislitelnoy Matematiki i Matematicheskoy Fiziki* **27** (1987).
14. Y. A. Kholodov and J. V. Vorobjev, About one method of numerical integration of discontinuous solutions in gas dynamics, *J. Math. Mod.* **8** (1996).
15. O. M. Belotserkovskii and A. V. Konyukhov, Change of grid functions of dependent variables in finite-difference equations, *Comput. Math. Math. Phys.* **42**(2) (2000).
16. M. Snir, S. Otto, S. Huss-Lederman, D. Walker and J. Dongarra, *MPI: The Complete Reference* (MIT Press, 1996).
17. W. K. Anderson, J. L. Thomas and B. van Leer, Comparison of finite volume flux vector splittings for the Euler equations, *AIAA J.* **24**(9) (1986).
18. O. M. Belotserkovskii and A. M. Oparin, *Numerical Experiment on the Turbulence: From Order to Chaos* (Nauka, Moscow, 2000).

# Appendix D

# Supercomputers in Mathematical Modeling of the High Complexity Problems

In the paper the review of many works executed by our scientific school during last years is presented. The general principles of construction of numerical algorithms for high-performance computers are described. Several techniques are highlighted, which are based on the method of splitting with respect to physical processes and widely used in computing nonlinear multi-dimensional processes in fluid dynamics, in studies of turbulence and hydrodynamic instabilities, and in medicine and other natural sciences. We present new Russian developments here.

## D.1. Introduction

Numerical simulation plays a particularly important role when the physics of the phenomenon under study is not quite clear and the intrinsic mechanisms of interactions are not fully understood ("ill-posed problems"). *An numerical experiment* (where the statement of a problem, the solution method, and the implementation of an algorithm are treated as an integrated complex) essentially serves to refine the starting physical model. By computing various modifications of the model, facts and results are accumulated. Eventually, this leads to a possibility for selecting the most feasible variants.

At present mathematical modeling becomes more and more actual, and that is caused by new opportunities given by significant achievements in the development of high-performance computers with parallel architecture. As a rule, it is necessary to consider the complex nonlinear multidimensional unsteady equations with complicated internal structure and rheology. Literally, for the last 1–2 years, there has been a powerful jump in the development of the supercomputer engineering as itself (Japan, USA).

So the performance of new Japanese supercomputer NEC Earth Simulator (NEC ES) comes to 40 trillion mathematical operations per second. American specialists develop the supercomputers with yet better performance (hundred trillions per second). To this and other questions also, the *6th International Conference on "High Performance Computing in Asia Pacific Region" (HPC-6)*, Bangalore, India, December 2002, was devoted.

It is extremely important, in our opinion, for users of such computing machinery ("End Users") to comprehend the priorities of problems and methods (algorithms, solvers), allowing *rationally* using such high-performance tools for research. Now at the Russian Academy of Science, the project on development of effective parallel algorithms ("supersolver") for solution of *high complexity problems*, has been started.

The purpose of the project is to develop effective ("general") parallel algorithms to solve wide circle of the problems of high complexity, described by equation systems in partial derivatives. The basis of the development is application of some modern algorithms applicable for calculations by parallel computers, such as method of splitting, "large particles" method, FLUX-method of flows, statistical approaches, Riemann's schemes, method of finite elements, etc. The idea consists in creation of a certain technology of parallel computing as a set of original codes of the software documentary prototypes, parallel — dependent part of which will not be changed, and the applied part is created by the user's modification of the closest program prototype. Realization of the project will allow to reduce sharply the terms of development of new parallel applications and to use effectively available multiprocessing computing resources.

In this paper some stages of the development of such supersolvers are briefly formulated.

As the application the solution of high-complexity problems in such most actual directions of science and engineering as aerodynamics and fluid dynamics (on the basis of "large particles" method the package of applied problems "*Gas-Dynamics Tool*" is developed) are considered, problems of seismology (construction of three-dimensional seismograms is especially important), and dynamic fractures are investigated. Application of computer technologies in medicine (new and very perspective field of applications) seems to be very important.

Practically, these directions (and the problems connected with the developments in nanoelectronics also) were noted by the majority of contributors at the conference HPC-6. We present new Russian developments here.

## D.2. Turbulence and hydrodynamic instabilities

The approaches, which are mainly discussed, further on, are those, which use for a description of the *free-developed shear turbulent compressible flow* for extended temporal intervals the complete (and closed) set of the dynamical equations for true values of velocities and pressure, as well as the statistical methods. The combined application of both these approaches (based on the use of hydrodynamic equations and on the statistical Monte Carlo methodics) permits to understand in more detail the structures of turbulence, and to determine the rational ways of the construction of corresponding mathematical models. The cycle of works in this field was started from Karman Lecture given by O. M. Belotserkovskii in March 15–19, 1976.[a]

The problem of construction of turbulence theory has a long history beginning from classical works of Reynolds and Richardson. In the middle of 20 years of last age, Keller and Fridman have proposed the idea of stochastic description of turbulent processes that had invaluable principal significance. A number of important achievements was obtained (for instance, Kolmogorov–Obukhov spectrum, etc. see Refs. 1–9). Nevertheless, physical principles of turbulence development remain unclear (for instance, what is the source of energy of chaotic motion). Currently, experimental data point clearly on the existence of the large-scale "coherent" structures (especially, for fully developed turbulence), where the main part of energy transferred is disposed.[9] Let us note that, for high Reynolds number, the energetic part of spectrum is far from the dissipative one in terms of wave numbers (see Fig. D.1).

The last decades are marked by a new approach (Belotserkovskii, 1980–1985) in the study of turbulence, namely, by direct numerical simulation (DNS) of processes of hydrodynamic flow on the basis of solution of hydrodynamic and kinetic equations.[10,11] Preliminary results of such approach demonstrate an important difference from the traditional, statistical approaches. For instance, Belotserkovskii in his works[10–14] showed *significant* role of large-scale structures in the turbulence development.

[a] O. M. Belotserkovskii, Computational experiment: direct numerical simulation of complex gas-dynamics flows on the basis of Euler, Navier–Stokes and Boltzmann models, Karman Lecture, Von Karman Institute for Fluid Dynamics, March 15–19, 1976, Brussel, in *Numerical Methods in Fluid Dynamics*, eds. H. J. Wirz and J. J. Smolderen (Hemisphere, Washington, London, 1978), pp. 339–387. You can see here also the original description of FLUX-approaches (1973).

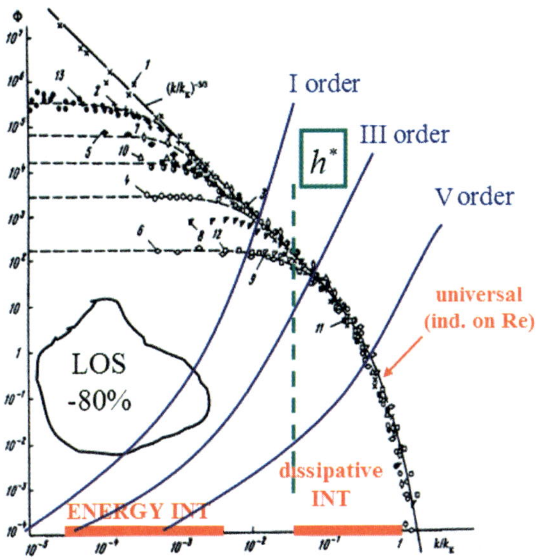


**Fig. D.1.** Spectra of energy density $E_1(k)$ of longitudinal velocity component fluctuations for various flows. $\Phi = E_1(Ev^3)^{-1/4}$, $k$ is a wave number. Choice of the "optimal" scale of resolution $h^*$.

Direct numerical simulation allows to understand the influence of nonlinear interactions on the development of turbulence and on the structure of flow.

In the future, because of the abovementioned, it would be expedient to carry out the numerical investigations of unsteady hydrodynamic phenomena on the basis of DNS by means of creation of theories of instability development and transition to turbulent stage.

The fundamental principles for constructing mathematical models for fully developed free turbulence and hydrodynamic instabilities are considered in the present section. Such a "rational" modeling is applied for a variety of unsteady multi-dimensional problems.[11–14]

The main ideology of "rational" approach for direct numerical simulation of the characteristics of fully developed shear turbulence by the high *Re* — investigation of large ordered structures (LOS) and small-scale stochastic turbulence (ST) is based on two hypotheses: "independence of LOS and ST" and "weak influence of molecular viscosity (or more generally, the mechanism of dissipation) by the study of LOS".[14–17] Eventually, Kolmogorov has received the form of a spectrum in an inertial interval assuming nothing in general about the kind of dissipative members.[1]

Really, for the wide class of phenomena, by the *high* Reynolds numbers within the low-frequency and inertial intervals of turbulent motion, the effects of molecular viscosity and of the small elements of flow in the largest part of perturbation domain are not practically essential neither for the general characteristics of macroscopic structures of the flow developed, nor for the flow pattern as a whole. This makes it possible not to take into consideration the effects of molecular viscosity when studying the dynamics of large vortices, and to implement the study of those on the basis of models of the ideal compressible gas (discrete Euler equations) using the methods of "rational" averaging ("upwind"-oriented monotone FLUX-schemes), but without special subgrid approximation and application of semi-empirical models of turbulence. The mentioned type of the oriented numerical schemes introduces *automatically* some scheme viscosity, which plays a role of implicit subgrid model.[11–14]

Figure D.1 demonstrates an experimental data[14] for spectra of energy density $E_1$ of longitudinal velocity component fluctuations for various turbulent flows ($k_k$ — Kolmogorov wave number; $E$ — local energy dissipation rate). For low-frequency and inertial parts of a spectrum (where LOS, which accumulate up to 80% of flow energy, are formed) turbulent flows strongly differ from each other depending on the kind of flow and external conditions. And it demands direct calculation of LOS. But, in dissipative interval (ST) practically universal behavior of a spectrum is observed. It once again confirms very weak dependence of molecular viscosity on the properties of LOS.

Using dissipative-stable and divergent-conservative oriented monotone FLUX-schemes,[12,14,20] having, for example, second- or third-order of accuracy, we choose the "optimal" scale (size) of resolution $h^*$ for direct numerical simulation of the discrete Euler equation. The insignificant contribution of energy from dissipative (high-frequency) part of a spectrum is suppressed by scheme viscosity or it is taken into account approximately. Curves in Fig. D.1 show the influence of scheme viscosity on the numerical solution for different orders of accuracy of the scheme. We can see that the schemes with second–third order of accuracy allow getting solutions for LOS without substantial influence of scheme dissipation (the latter plays the role of a filter).

By using the abovementioned approach (method of "large particles"[12,18]), the package of applied programs *Gas-Dynamics Tool* was developed for numerical solving of a wide range of gas-dynamical problems (Zibarov[19]).

At the same time, the properties of flows within the boundary layers and within the thin layers of mixing, at the viscous interval of turbulence, as well as those of flows by the moderate Reynolds numbers and in the domain of laminar–turbulent transition, are primarily determined by the molecular diffusion, and for these flows it is necessary to consider incompressible Navier–Stokes models (Gushshin, Belotserkovskii, and others[12,14,20–24]). The pulsation motions in turbulence are of chaotic type and have an unstable, irregular character, thus constituting a stochastic process. In view of that, one can speak here only on the obtaining of the mean characteristics of the motion of that type (like the moments of various orders) by way of a statistical processing of the results using, for example, kinetic approaches: direct Monte Carlo approaches (similar to the Crook equation, Yanitskii[12–14, 25–27]) and free turbulence based on the analogy of turbulent mixing with the Brownian motion of liquid particles (Ivanov and Yanitskii[27]). Numerical simulation of the rarefied hypersonic flow around bodies is of especial interest. Analysis of macroscopic fluctuations in such a flow around a cylinder was performed on the basis of Monte Carlo approach (Stefanov[28]).

The numerical studies of various kinds of *hydrodynamic instabilities* (Rayleigh–Taylor, Richtmyer–Meshkov, Kelvin–Helmholtz) are of an unquestionable interest, especially by the three-dimensional simulations extended to the large temporal intervals including turbulent stage. Multimode perturbations and spectral characteristics in 2D and 3D are studied numerically by various conditions. For example, a multimode problem, evolution of 2D perturbation into 3D structure, and growth of the thickness of a three-dimensional turbulent mixing zone are analyzed for Rayleigh–Taylor instability development (Oparin[13,14,29–31] and others).

Parallel software complex for simulation of spatial atmospheric flows induced by large-scale conflagration or explosion was developed. Both natural catastrophes (for instance, large-scale forest fire) and big industrial accidents can be considered as a source of such soiled admixtures. By the development of convective columns over the seat of fire, a large amount of dust is involved in these columns by rising fluxes of air. Then, soiled admixtures (dust, soot, etc.) ascend on big heights over the Earth surface, spread and gradually fall out, results in the soiling of regions of the order of 100 km and more in size. By industrial accidents the products of chemical trade or radioactivity can be spread by a similar way. A lot of numerical work was made in 2D geometry (both plane and axisymmetric) by different scientists. The experience shows that only numerical modeling in 3D geometry can answer different relevant questions. 3D modeling requires powerful computational resources and use of parallel computation technologies.[14,32–34,36]

The development of the pointed approach is contained in the abovementioned papers[12–14,18,27,31,43] and the new monograph,[34] where the problems of astrophysical turbulence, convection, and instability are also included.

We note again, that developed methods for investigations of coherent ordered structures are based on the consideration of "discrete" Euler equations for compressible flow with dissipative — stable and divergent — conservative finite-difference monotone FLUX-schemes, ("upwind" approximation[12,14,35] possessing scheme viscosity), and do not use special (explicit) subgrid approximations and semi-empirical turbulence models. Stochastic flows in the core of the wake (jet, mixing layer) are investigated at kinetic level by Monte Carlo methods.

Recently, the methods, which are rather close to the approach developed by us from 1976, began to appear (without references to our works, unfortunately) abroad also (for example, Refs. 37–40 and others).

## D.3. Supersolver

Finally, software for the solution of nonlinear problems of high complexity (*supersolver*) is considered.[43]

Using the experience of numerical simulation of various 3D problems, Universal Parallel Technology for numerical investigation of 3D problem (with mathematical model based on the set of partial differential equations of *hyperbolic type*) is under development for multiprocessor supercomputer. The usage of such technology will reduce significantly the time for probing new 3D applications.

This approach is based on selecting independent software blocks (where only parallelization is carried out), implementing various constituent parts of the combined numerical method. Software package for parallel computations of the problems within common approach is presented. The source code is a framework for the development of high-scale parallel software, mathematical model of which is based on the equations of the mentioned type.

Goals and advantages of using the parallel framework package are: acceleration of development of new parallel applications, software reuse, effective use of existing parallel computers, possibility to switch rapidly to a new

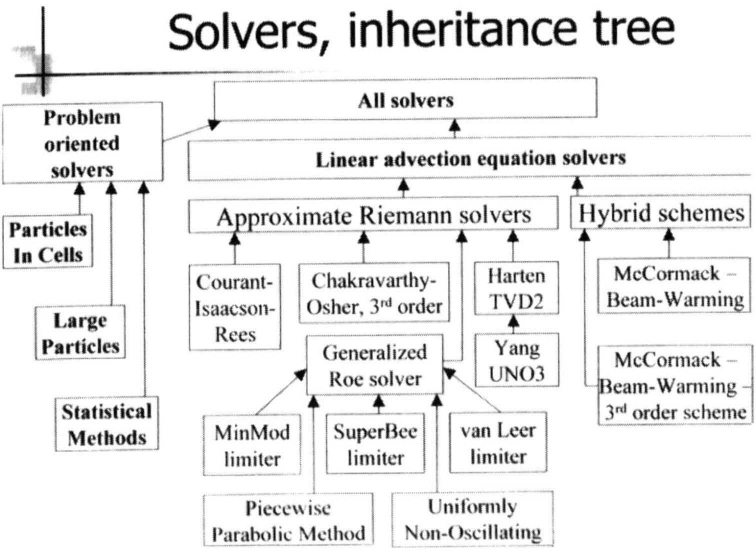


**Fig. D.2.** Supersolver: inheritance tree.

method, object-oriented and component-based programming, possibility to use IDE to create programs for standard models, etc.

Applications of the technology and the methods used are demonstrated on a few multi-dimensional problems in different areas (for example, simulation of the development of forest-fire, fire storm, tornado, etc.). Figure D.2 demonstrates three types of "inheritance tree", on the basis of which supersolver was created problem-oriented Riemann approximation and hybrid schemes (some details in Ref. 43).

## D.4. Applications

### *D.4.1. Gas-dynamics (CFD)*

The first calculations of LOS by FLUX-method[12] were made (Babakov) many years ago (1978). Turbulent wake (vortex sheet) downstream of a cylinder are shown in Figs. D.3(A) and D.3(B) for $M_\infty = 0.54$ (0.64) and $M_\infty = 0.9$. A lot of calculations for different approximation grids with the perturbations of the different kinds have clearly shown, that stable unsteady solution, which defines large structures and has good agreement

(a)

(b)

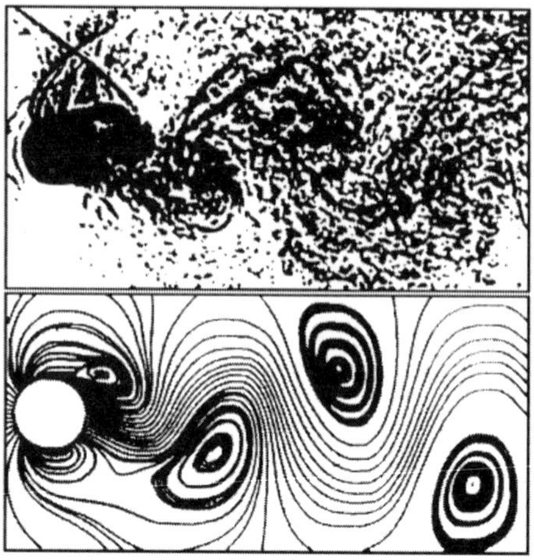

(a)

**Fig. D.3(A).** Turbulent wake downstream of a cylinder. Comparison of experimental results (a) ($M_\infty = 0.54, Sh = 0.18, C_x = 0.9$) with calculation (b) ($M_\infty = 0.64, Sh = 0.178, C_x = 0.89$).

(a)

(b)

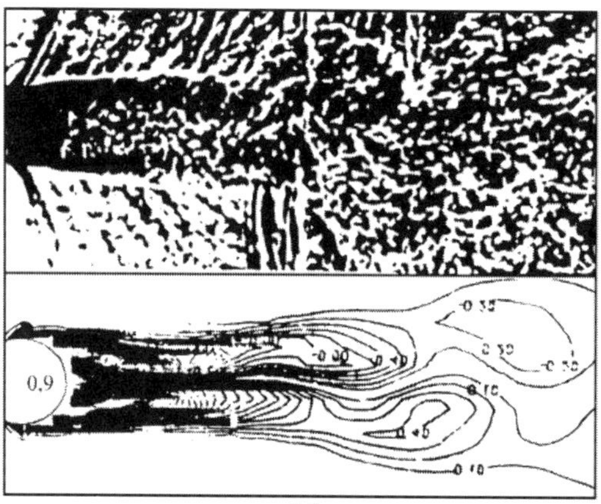


(b)

**Fig. D.3(B).** Comparison of calculated (b) and experimental results (a) for transonic flow past a circular cylinder. Constant-vorticity lines, $M_\infty = 0.9$.

with experimental data,[4] was obtained. Recent calculations (temperature fields) behind the cylinder are presented in Fig. D.4.

The flows near the landing modules in Mars's atmosphere at sub-, trans-, and super-sonic flight conditions for both zero and nonzero angles of

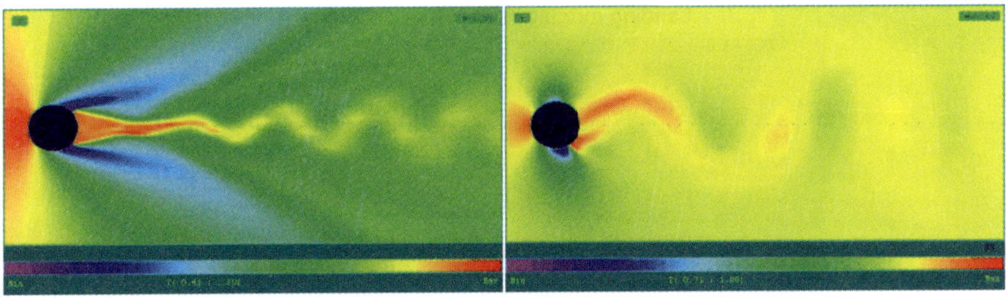

**Fig. D.4.** Unsteady regime (cylinder) for $M = 0.5$ (left) and $M = 1$. Temperature field.

attack are examined with the nonviscous perfect compressible gas model. For numerical simulation of flows and aerodynamics properties of axisymmetrical shaped vehicles at zero angle of attack a tailor-made FLUX program package is employed. This package includes a graphics module that works both in run-time mode and as a graphics post-processor. For the conical-shaped landing module the flow separation is governed by the presence of the edge. For the other form of landing module the separation at similar regimes is governed by the position of the shock wave closing a local supersonic region, not by the viscosity effects.

Simulation of the steady and unsteady 3D flows is carried out on multiprocessor computational systems using a tailor-made program package FLUX-MCS (FLUX for Multiprocessor Computer Systems).

For sub- and tran-sonic regimes the flow in a near wake is unsteady (Figs. D.5 and D.6). In Fig. D.6, the temperature fields are shown in two perpendicular sections.

For numerical simulation of gas-dynamical processes, *Gas-Dynamics-Tool*,[19,41] the multipurpose package, based on Large Particle Method, was developed. Package exploits numerical simulation of two- and three-dimensional unsteady processes in multicomponent systems of nonviscous/viscous compressible gases. The processes of thermoconductivity, diffusion, and chemical reactions might be taken into consideration. A universal method of resolving Euler and Navier–Stokes equations allows applying the package to wide range of gas-dynamical phenomena in science and industrial applications. The package is fully featured by visualization and pre-processor tools that allow analyzing complex 3D flow with different boundary geometries. The code has several types of solvers with numerous submodels describing different physical and chemical processes. Figures D.7(a), D.7(b), D.7(c) illustrate the possibilities of this approach.

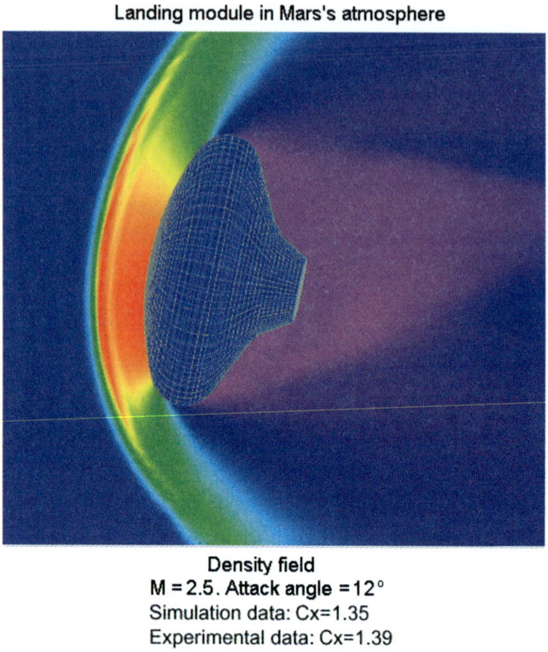


**Fig. D.5.** Density field in flow around landing module in Mars's atmosphere ($M = 2.5$).

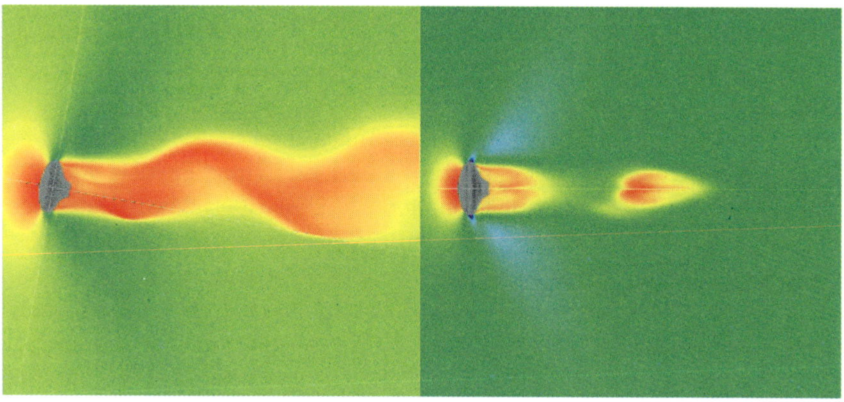

**Fig. D.6.** Unsteady regimes. $M = 1.0$. Temperature field is shown. At left — side view, at right — top view.

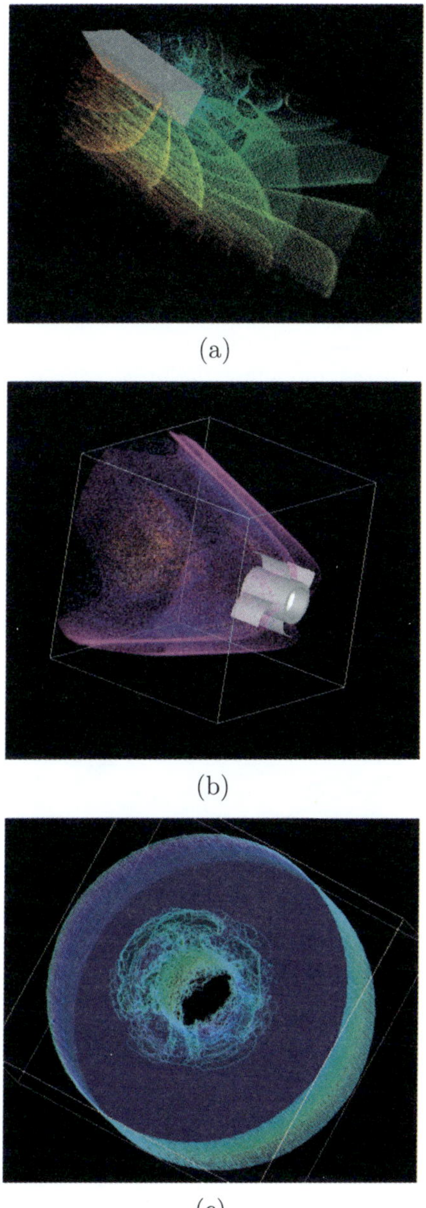

(a)

(b)

(c)

**Fig. D.7.** (a) 3D animations allow to achieve more comprehensive recognition of phenomenon or process under consideration. (b) Hypersonic flow over rocket stabilizer. (c) Impacted penetration in ceramic target 3D cloud image, voxel graphics.

### *D.4.2. Hydrodynamic instabilities*

For the specified class of the phenomena, calculation of 3D problems represents special interest. As an example, in Fig. D.8(a), the turbulent stage of the development of 3D Rayleigh–Taylor instability ($t = 12$) under random initial perturbations is presented. Results were obtained by Oparin. Figure D.8(c) illustrates the accent of upper boundary of bubbles in dependence on combination $A_t \cdot gt^2$ ($A_t$ — Atwood number) that characterizes turbulent mixing zone (TMZ) $Z = \alpha A_t \cdot gt^2$. We can see that in the 3D case, value of $\alpha$ coefficient is significantly different from that of 2D calculation, which represents great practical interest.

A number of problems, modeling *the large-scale high-energy convective flows* in the atmosphere, is solved numerically with the help of the software package based on the described parallel framework. This framework is also utilized (Kraginskij and Oparin[43]) in the other application fields.

One of the studied problems (with the help of Supersolver) is to investigate numerically the dynamics of pollution, distribution, and toxic dust precipitation in the Earth's atmosphere for large-scale fire. Natural

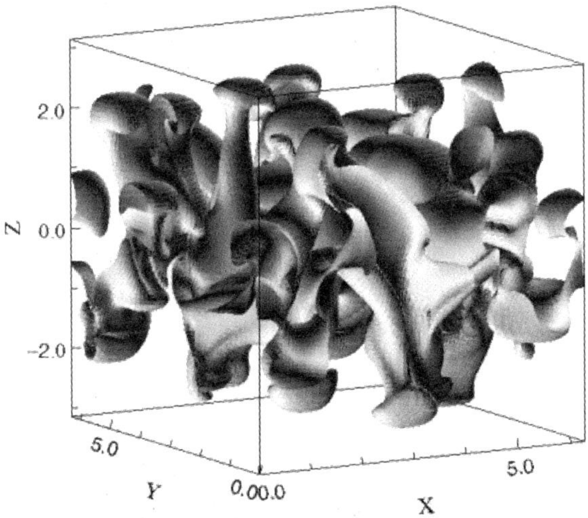


(a)

**Fig. D.8.** (a) Growth of the thickness of 3D turbulent mixing zone (TMZ). (b) The graph of ascent of the upper boundary of bubbles for 3D computation and for the corresponding 2D one. (c) 3D turbulent mixing zone. Simulated self-similar behavior of "light-to-heavy" penetration depth in 3D in comparison with 2D.

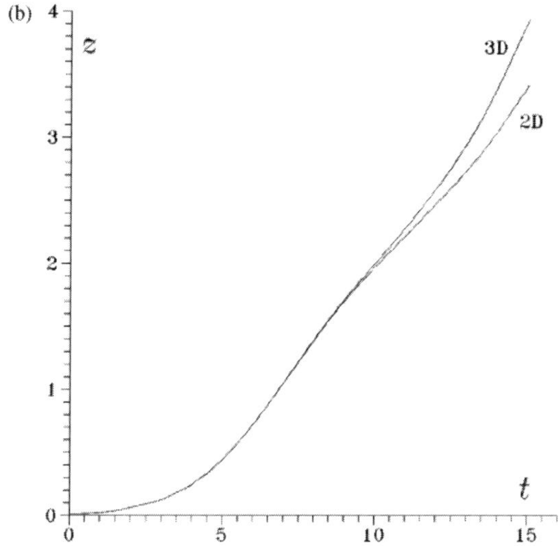


(b)

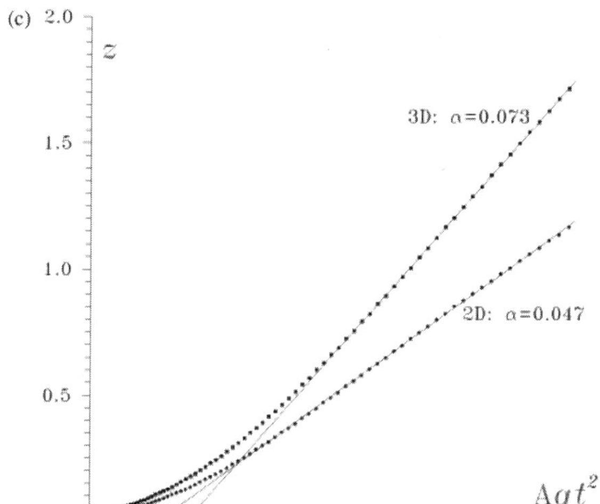


(c)

**Fig. D.8.** (*Continued*)

catastrophes and large industrial accidents are sources of such kinds of pollution. The results of the numerical experiment help us to explore the phenomenon better and to get a dynamical picture of precipitation of the dust on soil. The results obtained could be used as the basis for predicting high-energy accident subsequences and reducing damage and life losses.

Another series of numerical modeling is connected with the phenomena of firestorm, tornado, and similar flows with large whirls. We try to build the spatial model of the tornado, and show some progress in understanding its dynamics on different stages: from the very beginning to the developed flow with stable structures. Figures D.9–D.11 illustrate the results of these calculations.

### *D.4.3. Seismic data processing*

Solving of direct seismic problem is a matter of great importance for exploring the medium with complex structure as well as for exploring untraditional deposits of hydrocarbonates: oil, gas, and condensed gas. It was recently stated that oil deposits could not have organic nature. So, they

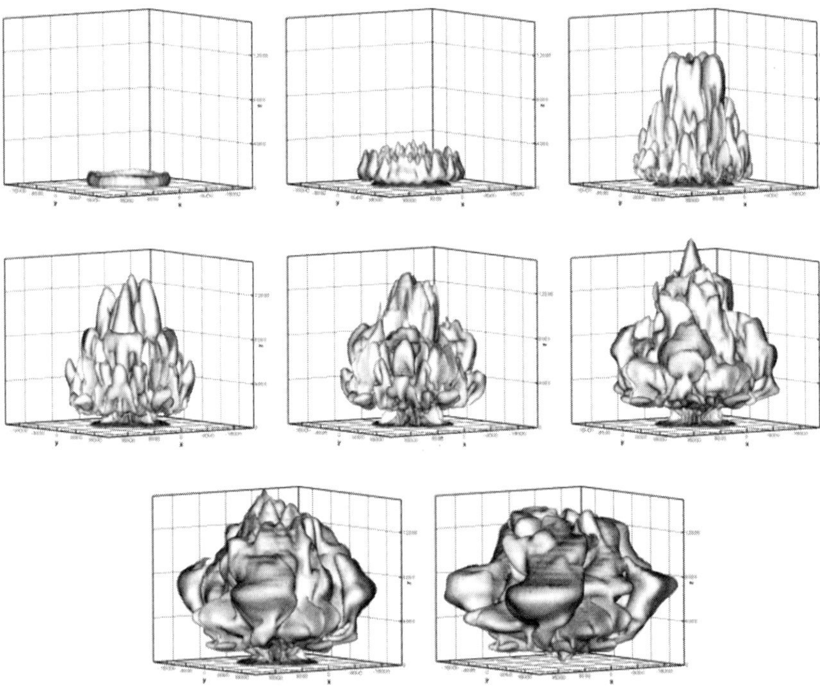

**Fig. D.9.** Large-scale fire ($R$ = 10 km, $Q$ = 0.05 MWt/sq. m., $t_{\text{off}}$ = 60 min). Time moments: 5, 10, 15, 20, 25, 40, 60, and 90 min are shown.

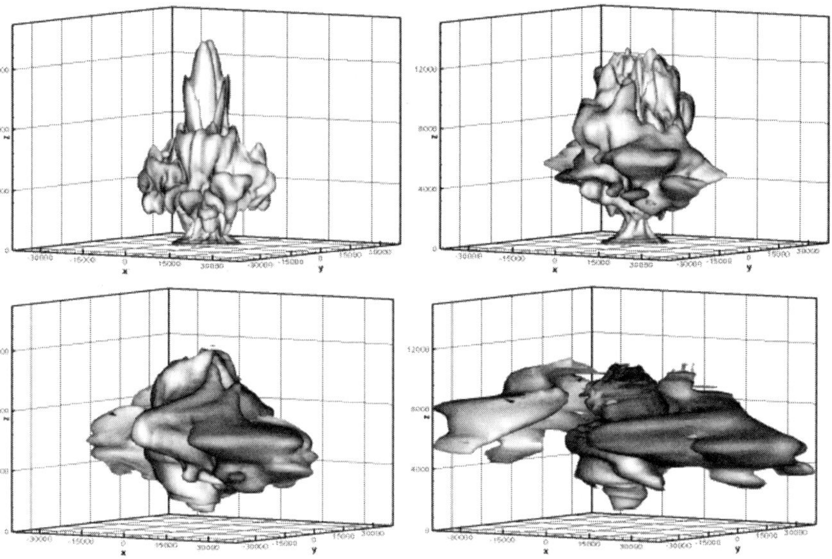

**Fig. D.10.** Large-scale fire storm ($R = 10\,\text{km}$, $Q = 0.05\,\text{MWt/sq. m.}$, $t_{\text{off}} = 60\,\text{min}$). Time moments: 25, 50, 100, and 200 min are shown.

could be found not only in the upper layers under the surface, but also in the crystalline base in porous collectors.

A numerical model is proposed to solve direct seismic 2D and 3D problems (wave propagation) in complex heterogeneous medium. The wave-field initiated by different sources of initial disturbance in elastic heterogeneous medium is treated. The effect of acoustic–elastic boundary is demonstrated in Fig. D.12 (Antonenko).

### *D.4.4. Safety of housing and industrial constructions under intensive dynamic loadings*

One of the most actual problems — safety of housing and industrial constructions under intensive dynamic loadings (industrial accidents, seismological activity, acts of terrorism, falling of planes and fragments of spacecraft, etc.) is considered. The numerical simulation (by Petrov) of such problems permits to formulate a quantitative estimation of stability and safety of constructions, i.e. to predict the placement and the size of the probable destruction area in dependence of the intensity and character of

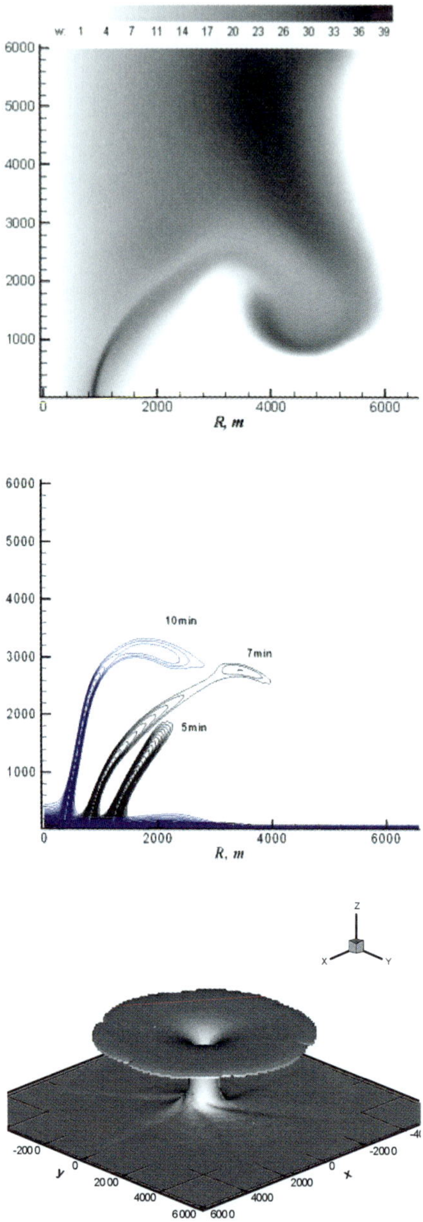


**Fig. D.11.** Evolution of vertical mesa-scale vortex in tornado-like structure: tangential velocity component (at top), isolines of passive substance at 5, 7, and 10 min (at center), isosurface of passive substance concentration (at bottom).

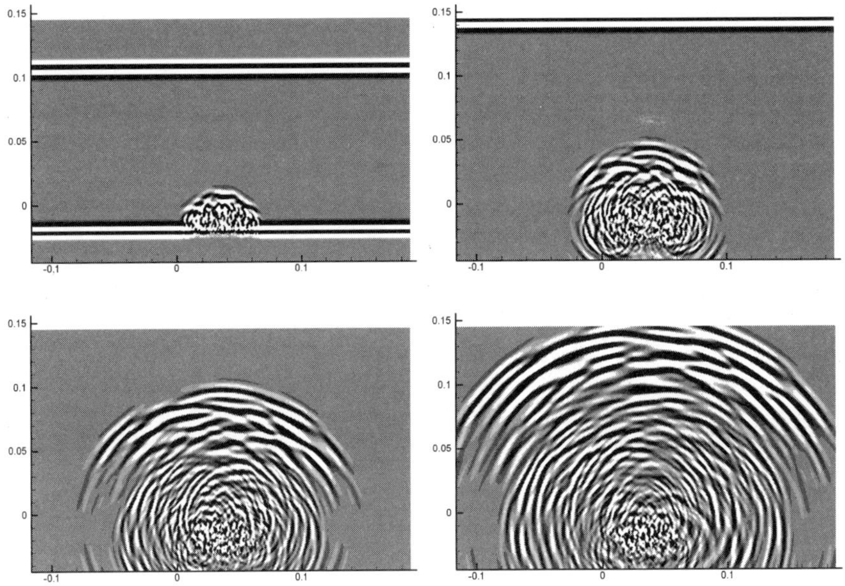


**Fig. D.12.** Results of numerical simulation. Interaction of plane wave with oil collector: dynamics of wave field. Velocity of vertical displacement shown for times 12, 18, 27, and 36 ms.

the impact, a place of loading, geometrical characteristics of a construction, and mechanical properties of materials of which it consists.

The behavior of a lattice concrete construction under a lateral impulse of loading, velocity fields, pressure fields, deformations, areas of destruction are simulated. For mathematical modeling of the behavior of considered constructions material, the dynamic system of the equations in partial derivatives of hyperbolic type of the rigid deformable body mechanics is used. In Fig. D.13 we can see the character of destruction of lattice building obtained by numerical simulation. It is interesting to note that the propagation of the disturbances has arrow-headed character, not spherical (Fig. D.13(b)).

### *D.4.5. Nonlinear contact shell dynamics*

The problems of deformable rigid body mechanics and simulation of nonlinear contact shell dynamics in incompressible fluid for 2D and 3D cases were also investigated (V. Yakushev). In Fig. D.14, we can see finite element model of a liquefied gas-carrier. The calculation of its deformations has great

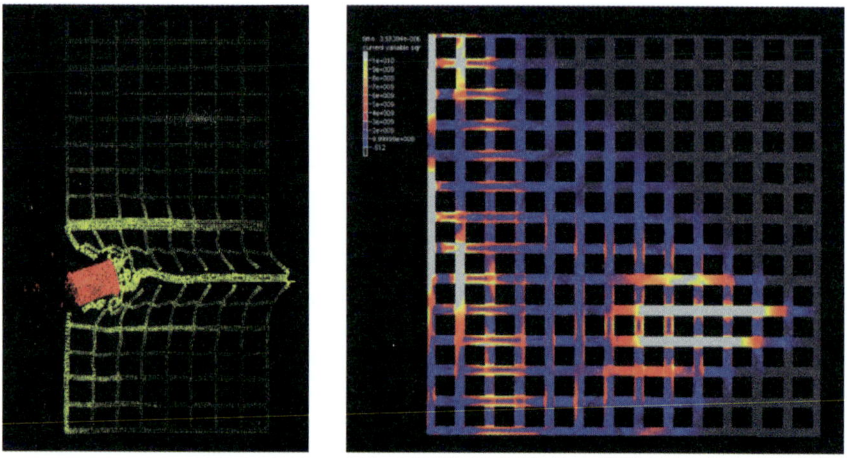

**Fig. D.13.** Framed construction destruction (at left). Squared stress deviator in lattice (at right). Impact at 45°.

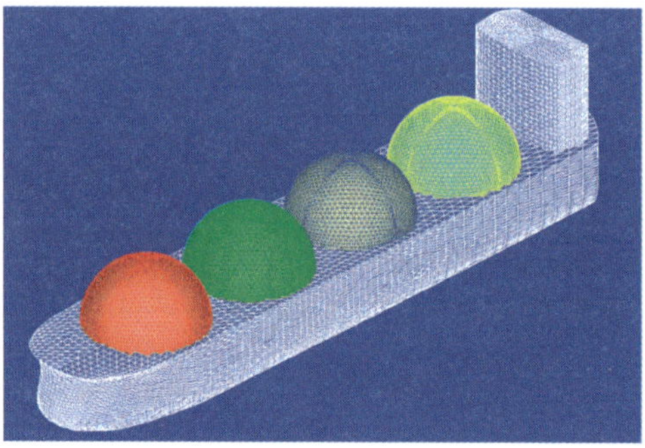

**Fig. D.14.** Finite-element model of liquefied gas-carrier.

practical importance. In Fig. D.15, numerical modeling results of collision of the tanker with quay are presented. On the character of deformations we can spot the scenario of the collision.

### *D.4.6. Computer models in medicine*

And finally, construction of computer models in medicine is the new perspective direction allowing here again to use rich saved-up experience. As examples, the results of mathematical modeling of circulatory

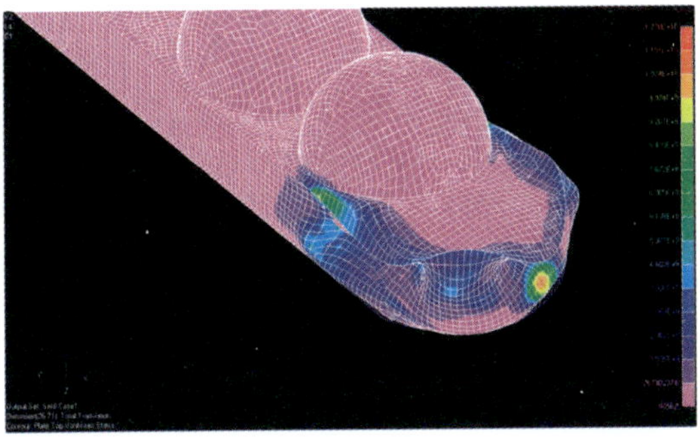

**Fig. D.15.** Numerical modeling of collision of the tanker with quay wall.

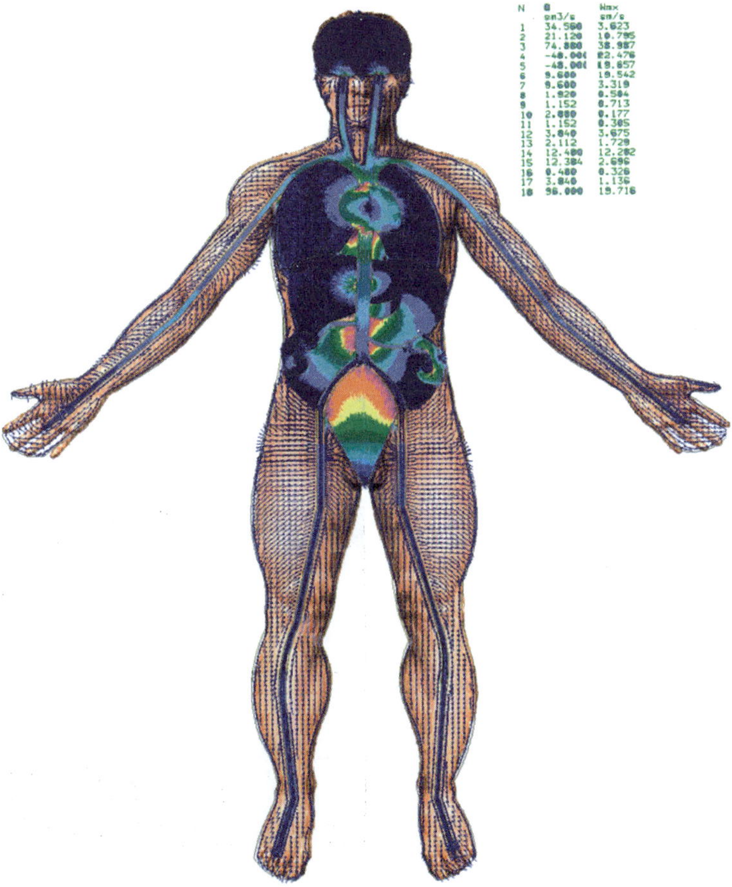


**Fig. D.16.** The example of calculated average velocity field in arterial system.

and respiratory systems of human organism and others are presented (Kholodov,[42] Petrov,[44] V. Kljuzhev, A. Talalaev and others).

In Fig. D.16, we can see the example of calculated average velocity field in the arterial system (A. Kholodov). For modeling of circulatory and respiratory systems of human organism, different systems of equations are used in dependence of the anatomic constitution of corresponding human organs: Navier–Stokes equations, equations of filtration type, etc. Special numerical methods (major schemes on unstructured grids) were developed for solving such problems. Complicated multiconnected domain of integration is considered.[42]

In Fig. D.17, the model of cranial trauma simulation is presented (Petrov). The results of calculation have shown that the propagation of the disturbances after impact is realized on the surface of the skull much faster, than through brain tissue. This fact is very important in clinical practice for the choice of the treatment scenario. In Fig. D.17(b), numerical results are presented. We can see clearly the tissue injure.

The unique medical and biologic stand was developed in Burdenko Main Clinical Hospital, specially for modeling and studying a collective brain of people (V. Kljuzhev, A. Talalaev, A. Lisitsky, 1997–2002). The results of brain condition modification are analyzed on the basis of methods not realized for acoustic suggestion with a position of consideration of a human brain as unique "decoder" that received meaningful and unconscious information.

The main purpose of the present work is the development of the medical and biologic stand created for studying the influence of consciousness

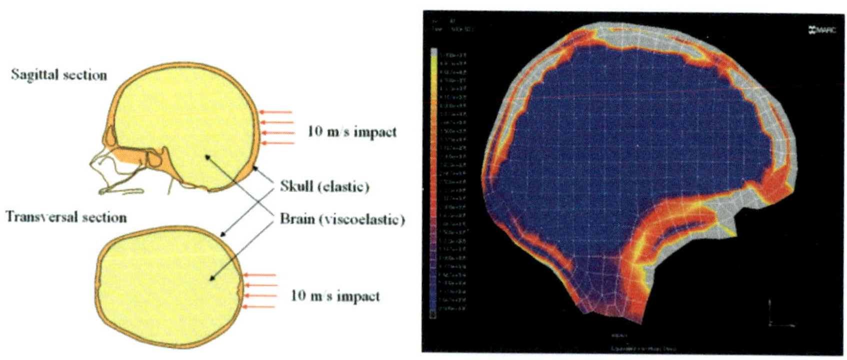


**Fig. D.17.** The problem of cranial trauma modeling (at left). Equivalent stresses are shown (at right).

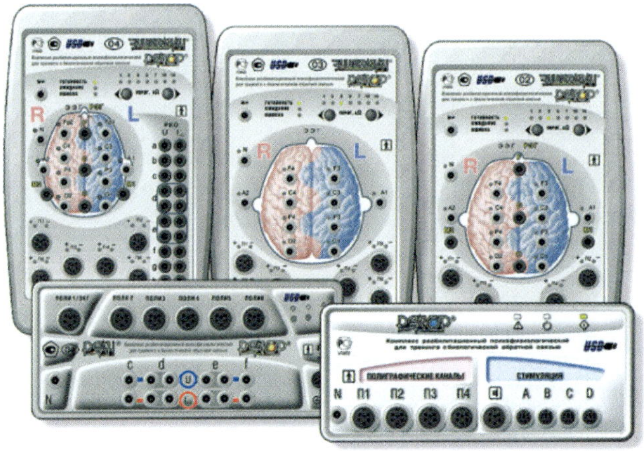


**Fig. D.18.** A line of Russian electroencephalographs with an opportunity of realization of biofeedback, used in the creation of the stand of "a virtual collective brain" modeling.

and mentality of people on a physical reality by modeling virtual "a collective brain" of the person. In Fig. D.18, the general view of the stand is shown. The stand unites eight gauges (polygraphs) and allows to register electric processes of human brain synchronously. In Fig. D.19, electroencephalograms (EEG) of four healthy persons are presented (all persons have received identical task). By special mathematical processing

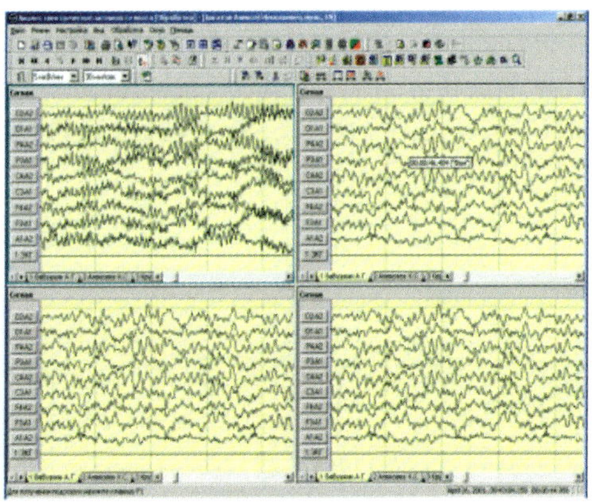

**Fig. D.19.** Synchronous records of electroencephalograms of four people in eight assignations in the making "a virtual collective brain".

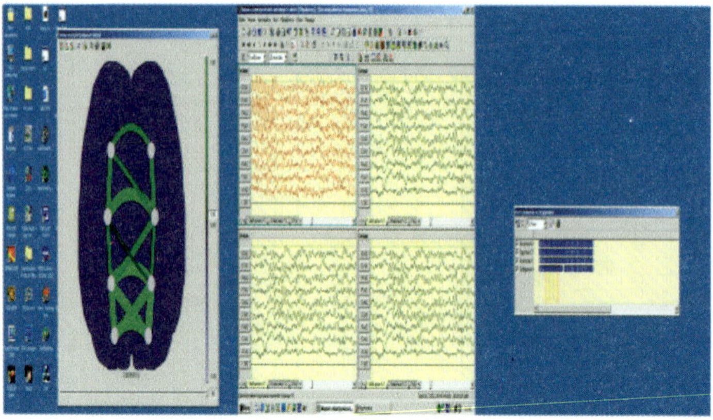

**Fig. D.20.** Schematic map of the intercentral communications "virtual collective brain" on an average level of correlation (A), based on EEG of four persons (B).

of EEG, generalized "virtual collective brain" is created (Fig. D.20(a)). In Fig. D.20(b), EEG of four patients are presented (at the left above — injured brain).

*All this allows modeling the major processes in the human brain!*

("...3D computer map of the human brain is composed. 20,000 neurosurgers and neuropathologists worked during several years in 6 countries. The data for this map were obtained as a result of brain scanning of 7000 of people. The sense of this work is extremely important — it was clear what regions of brain manage by it's different functions. For example, where is speech born, where is the appetite stimulated, where are aggressive thoughts placed and where is the passionate love pulled out?"... Newspaper "*Izvestiya*," August 13, 2003, Russia.)

## D.5. Conclusion

In the nearest future the development of rational models rather than high-speed computers will determine the effectiveness of the numerical experiment in various branches of science and technology.

$$\Downarrow \qquad\qquad \Downarrow$$

$$\boxed{\textbf{HARD} + \textbf{BRAIN} = \textbf{const!}}$$

## References

1. A. N. Kolmogorov, Equations of turbulent motion of incompressible fluid, *Izv. Acad. Nauk SSSR. Ser. Fiz.* **6**, 56–58 (1942) (in Russian).
2. G. K. Batchelor, *Theory of Homogeneous Turbulence* (Cambridge University Press, 1955).
3. A. A. Townsend, *The Structure of Turbulent Shear Flow* (Emmanuel College, Cambridge, 1956).
4. M. Van Dyke, *Album of Fluid Motion* (Parabolic Press, Palo Alto, CA, 1982).
5. J. O. Hinze, *Turbulence* (McGraw-Hill, New York, 1975).
6. A. S. Monin and A. M. Yaglom, *Statistical Fluid Mechanics: Mechanics of Turbulence* (MIT Press, Cambridge, MA, 1975).
7. S. A. Orszag, *Handbook of Turbulence. Fundamentals and Applications*, W. Frost and T. H. Moulden (eds.) (Plenum Press, New York, London, 1977), Vol. 1, pp. 311–347.
8. H. L. Swinney and J. P. Gollub (eds.), Hydrodynamic instabilities and the transition to turbulence, in Topics in Applied Physics, Vol. 45 (Springer-Verlag, Berlin, Heidelberg, New York, 1981).
9. B. J. Cantwell, Organized motions in turbulent flows, *Ann. Rev. Fluid Mech.* **13**, 457–515 (1981).
10. O. M. Belotserkovskii, Direct numerical modeling of transition fluid flows and turbulence problems, *Mechanics of Turbulent Flows* (Nauka, Moscow, 1980), pp. 70–109 (in Russian).
11. O. M. Belotserkovskii, Direct numerical modeling of free induced turbulence, *USSR Comput. Math. Math Phys.* **25**(12), 166–183 (1985).
12. O. M. Belotserkovskii, *Numerical Simulation in the Mechanics of Continuous Media*, 2nd edn. (Fizmatlit, Moscow, 1994) (in Russian); See also *Modern Solution Methods for Nonlinear Multidimensional Problems: Mathematics, Mechanics, Turbulence* (The Edwin Mellen Press, Lewinston, Queenston, Lamper, USA, 2000).
13. O. M. Belotserkovskii and A. M. Oparin, Numerical experiment on the turbulence: from order to chaos, 2nd edn. (Nauka, Moscow, 2000) (in Russian) (1st edn., translated by Bogell House Inc., Publishers, USA, *Int. J. Fluid Mech. Res.* **23**(5–6), 321–488 (1996).
14. O. M. Belotserkovskii, *Turbulence and Instabilities*, 1st edn. (Moscow Institute of Physics and Technology (MIPT), Moscow, 1999) (2nd edn., The Edwin Mellen Press, Lewinston, Queenston, Lamper, USA, 2000).
15. L. D. Landau and E. M. Lifshitz, *Fluid Mechanics* (Pergamon, London, 1956).
16. J. R. Herring, S. A. Orszag, R. H. Kraichnan and D. G. Fox, Decay of two-dimensional homogeneous turbulence, *J. Fluid Mech.* **66**, 417–444 (1974).
17. V. R. Kuznetsov and V. A. Sabel'nikov, *Turbulence and Combustion* (Nauka, Moscow, 1986) (in Russian).
18. O. M. Belotserkovskii and Yu. M. Davydov, *Method of Large Particles in Gas Dynamics: Numerical Experiment* (Nauka, Moscow, 1982) (in Russian).
19. A. V. Zibarov, Package of applied programs GAS DYNAMICS TOOL and its application in problems of numerical simulation of gasdynamic processes, Doctoral thesis, Moscow (2000) (in Russian).

20. O. M. Belotserkovskii, V. A. Gushchin and V. N. Kon'shin, A method of splitting for analyzing stratified free-surface flows, *USSR Comput. Math. Math. Phys.* **27**(4), 181–191 (1987).
21. S. O. Belotserkovskii, A. P. Mirabel and M. A. Chusov, On the construction of supercritical modes for a plane periodic flow, *Izv. Akad. Nauk SSSR Ser. Fiz. Atmosf. i Okeana.* **14**(1), 11–20 (1978) (in Russian).
22. S. O. Belotserkovskii, Simulation of viscous incompressible flows based on Navier–Stokes equations, Doctoral thesis, Moscow (1979) (in Russian).
23. V. A. Gushchin and P. V. Matyushin, Numerical simulation of separated flow past a sphere, *Comput. Math. Math. Phys.* **37**(9), 1086–1100 (1997).
24. V. A. Gushchin, Direct numerical simulation of transitional separated fluid flow around a circular cylinder, in *Proc. Colloquium Transitional Boundary Layers in Aeronautics*, Amsterdam, December 6–8, 1995 (Royal Netherlands Academy of Arts and Sciences, The Netherlands), pp. 113–121.
25. V. E. Yanitskii, A statistical method for solving some problems of the kinetic theory of gases and turbulence, Doctoral thesis, Moscow (1984) (in Russian).
26. O. M. Belotserkovskii and V. E. Yanitskii, The statistical method of particles-in-cells for solving problems of rarefied gas dynamics — 1, 2, *USSR Comput. Math. Math. Phys.* **15**(5) (Part 1); (6)(Part 2), 184–198.
27. O. M. Belotserkovskii, S. A. Ivanov and V. E. Yanitskii, Direct statistical modeling of some problems in turbulence theory, *Comput. Math. Math. Phys.* **38**(3), 474–487 (1998).
28. S. K. Stefanov, I. D. Boyd and C.-P. Cai, Monte Carlo analysis of macroscopic fluctuations in a rarefied hypersonic flow around a cylinder, *Phys. Fluids* **12**(5), 1226–1239 (2000).
29. N. A. Inogamov and A. M. Oparin, Three-dimensional array structures associated with Richtmyer–Meshkov and Rayleigh–Taylor instability, *J. Exp. Theor. Phys.* **89**(3), 481–499 (1999).
30. N. A. Inogamov, A. M. Oparin, A. Yu. Dem'yanov, L. N. Dembitskii and V. A. Khokhlov, On stochastic mixing caused by Rayleigh–Taylor instability, *J. Exp. Theor. Phys.* **92**(4), 715–743 (2001).
31. O. M. Belotserkovskii, A. M. Oparin, A numerical study of three-dimensional Rayleigh–Taylor instability development, *Comput. Math. Math. Phys.* **40**(7), 1054–1059 2000.
32. I. F. Muzafarov and S. V. Utyuzhnikov, Numerical simulation of convective columns over large conflagration into a atmosphere, *High Temp.* **33**(4), 588–595 (1995).
33. A. V. Konyukhov, M. V. Meshcheryakov and S. V. Utyuzhnikov, Numerical simulation of the processes of propagation of impurity from large-scale source in the atmosphere, *High Temp.* **37**(6), 873–879 (1999).
34. O. M. Belotserkovskii, A. M. Oparin and V. M. Chechetkin, *Turbulence: New Approaches* (Nauka, Moscow, 2003) (in Russian).
35. A. I. Tolstykh, *High Accuracy Non-Centered Compact Difference Schemes for Fluid Dynamics Applications*, Series of Advances in Mathematics for Appl. Sciences **21** (1994).

36. A. V. Konyukhov, M. V. Meshcheryakov, S. V. Utyuzhnikov and L. A. Chudov, Numerical modeling of turbulent large-scale thermic, *Izvestiya RAS, Series Fluid Mechanics* (3), 93 (1997).
37. D. Drikakis and P. Smolarkiewicz, On spurious vertical structures, *J. Comput. Phys.* **172**, 309–325 (2001).
38. L. Margolin and W. Rider, A rationale for implicit turbulence modeling, in *ECCOMAS Comput. Fluid Dynam. Conf.*, UK (2001), pp 1–20.
39. J. P. Boris, F. F. Grinstein, E. S. Oran and R. L. Kolbe, New insights into large-eddy simulations, *Fluid Dynam. Res.* **10**(4–6), 199–228 (1992).
40. C. C. Fereby and F. F. Grinstein, Large Eddy simulation of high-Reynolds-number free and wall-bounded flows, *J. Comput. Phys.* **181**(1), 68–97 (2002).
41. O. M. Belotserkovskii and A. V. Zibarov, Use Gas-dynamics tool 5.0 for three-dimensional shock wave processes simulation, *Abstracts of Russian–Indian International Workshop on High Performance Computing in Science and Engineering*, Moscow (June 2003).
42. O. M. Belotserkovskii and A. S. Kholodov, *Computer Models and Progress in Medicine* (Nauka, Moscow, 2001) (in Russian).
43. O. M. Belotserkovskii, M. N. Antonenko, A. V. Konyukhov, L. M. Kraginskij, A. M. Oparin and S. V. Fortova, Universal technology of parallel computations for the problems described by systems of the equations of hyperbolic type. A step to supersolver, *CFD J. (Japan)* **11**(4), 456–466 (2003).
44. I. B. Petrov, On the numerical modeling of biomechanical processes in medicine practice, *J. RAS: Inform. Technol. Comput. Syst.* **1–2**, 102–112 (2003) (in Russian).

# Appendix E

# On Nuts and Bolts of Structural Turbulence and Hydrodynamic Instabilities

As we have already seen constructively, the analysis of the phenomenon of turbulence must and can be performed through direct numerical simulations of a mechanics supposed to be inherent to secondary flows. This one reveals itself through such evidences as large vortices, structural instabilities, vortex cascades, and principal modes discussed below. Like fragments of puzzle, they say of a motion ordered with its own nuts and bolts, whatever chaotic it looks like at first sight. This opens an opportunity for a multi-oriented approach, of which prime ideology seems to be a rational combination of grid, spectral, and statistical methods. An attempt is made to unite the above evidences and produce an alternative point of view on the phenomenon in question when based on the main laws of conservation.

## E.1. Rational Constructivism

An approach suggested has in fact been initiated in Karman's Lecture of 1976, March 15–19.[29,31,33] Since then, a lot of work has been conducted in this direction by its author in collaboration with his former colleagues Belotserkovskii[30] and Yanitskii.[32] Some statements might seem to have only heuristic or intuitive characters; however, the results obtained as coming from nature phenomena, experimental evidences, and numerical simulations including those in physics and astronomy speak for themselves. There have been many papers and presentations in the seminars of P. L. Kapitza, N. N. Yanenko, V. V. Struminskii, A. M. Obukhov, A. S. Monin, G. Batchellor, H. Daiguji and many authors. A certain scrutinized verification had been held during 1994–1995 in Los Alamos (USA) in collaboration with Dr F. Harlow and his colleagues. As a result of almost 4 months of discussions, the concept proposed has been apprehended as quite adequate to what would be referred to as a *rational constructivism*.

Similar questions alternatively arise in connection with such notions, as the mixing zone, accretion discs, topology of an ideal fluid, and scaling analysis.[35,39,43] As a result of numerous discussions with A. M. Oparin, O. V. Troshkin, and V. M. Chechetkin, a viewpoint had been worked out according to which the treatment of turbulence in terms of chaos and statistics is required to be corrected with mechanics. The *reasons* for such a correction (to be detailed below) are the following:

(R1) pulsations are rather *definite* disturbances, than *random* fluctuations;
(R2) disturbances *produce* the stresses that balance other impulses;
(R3) instabilities *are produced* with the Euler's convection $\mathbf{u}\nabla\mathbf{u}$;
(R4) instabilities *are localized* in zones of large velocity gradients;
(R5) instabilities *are evidenced* with streams rolling up in vortices;
(R6) *vortices* give rise to pulsations;
(R7) pressure, density, and temperature fields are formed *first* within a vortex.

Let us turn to the main considerations.

## E.2. Back in Mechanics

Generally, the turbulence may be thought of as uniting two oppositions. When developed from initial data in direct numerical simulations, it proves to be predictable, while complex motion of a medium with a definite large-scale vortex structure of the vector field of velocities is $\mathbf{u}$.[37,39] Meanwhile, according to the known theoretical approaches attemping to describe this phenomenon it looks like an unpredictable process treatable for the most part through the small-scale statistical account.[26]

As this takes place, the turbulence reveals itself through its forces as produced with mentioned stresses, which makes it possible to take the turbulence as *a part of physics.* Such a consideration was forwarded by W. Thomson (Lord Kelvin) in his pioneering work on turbulent waves in an incompressible ideal fluid submitted to random pulsations.[3] After averaging the convection ($\langle \mathbf{u} \cdot \nabla u \rangle = \nabla \cdot \langle u\mathbf{u} \rangle$) he had found out cross waves of small disturbances due to the turbulent stresses (cf. with Ref. 36, Sec. 15). Later, Reynolds showed *vortices of turbulence* with colored jets,[4] and turbulent stresses had been named after him.

Be it as it may be, the main task of turbulence, as a part of mechanics, seemed to determine the sources of stresses for the velocity and pressure pulsations. While appreciating this difficult task, we state yet that with no relevant mechanics of turbulence, or with no required nuts and bolts, there will be no true understanding of the phenomenon in question. To this purpose we take the turbulence as *a regular* (*not chaotic*) *motion* submitted to the main laws of mass, impulse, and energy conservation and decomposed structurally in *vortices as elements*.

In other words, we proceed from the assumption that in order to get a visually chaotic picture, any flow develops deterministically from the initial state, rolling up its layers consecutively in definite directions through a series of instabilities, whose principal disturbant modes remain to be limited in number.

Some details of the mechanics suggested could be illustrated by the following concepts of large vortices, structural instabilities, time cascades, and boundness of noninertial part of the spectrum.

## E.3. Large Vortices

Many know why wires noise: when an incoming flow meets a wire, eddies originate one by one behind the rigid obstacle, develop in dimensions, break away from the surface of wire, and go with the wind down the flow to send a ray of reactive impulses to the rest mass. As a mechanical system with its own frequencies, the wire begins to oscillate, which produces the noise. When taken in the ocean, the long tubes (for transporting oil from the bottom to the surface), like the wires mentioned, give the same effect known to be *vortex-induced vibrations*.

Surprisingly as it might be at first sight, but some may guess the *birds accelerating* by the same reason: while moving up, a wing creates a vortex behind itself; while moving down, the wing cuts the vortex off the rest mass to take a reactive impulse pushing the bird ahead.

And practically, contrary to the orthodox aerodynamics, a bumblebee can fly. Its wings are too small and thin to keep its mass floating as with birds, while they are sufficiently mobile (frequently moving up and down) to create a reactive stream in the form of a street of the cut-off vortices and turn the reactive stream to the side opposite to the direction of the flight. Shortly speaking, a bee needs no lifting force: it simply shoots with swirling lumps of air to the *reactive vortices* pushing it ahead (see Fig. E.1).

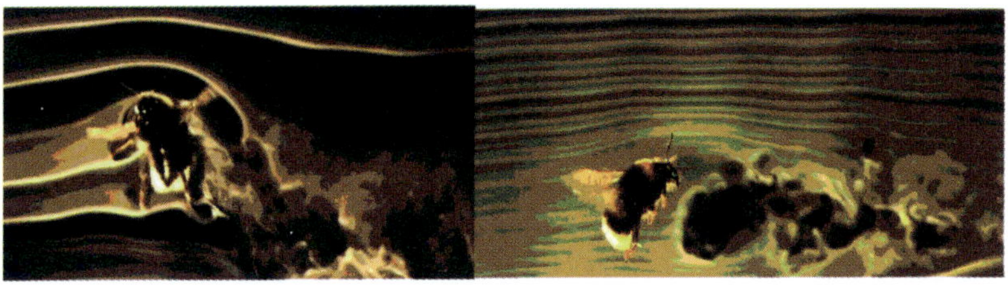

**Fig. E.1.** For its fat body and thin wings alone, a bumblebee might could not fly at all. Yet, it does fly, with only vortices. When flapping up, the wings produce the lumps of a swirling air behind themselves. Those are found to be increased in dimensions from tiny eddies to large curls in a short time. When flapping down, the wings cut vortices off the body. Like balloons, those are gone down with the upward flow. In so doing, they push a bee exactly in the opposite direction, no matter, with its head or back ahead at the moment.

More than 20 cents of payment for a flight by an airplane go to support large vortex tubes originated steadily between wings and fuselage.

Due to a large vortex, formed after stirring some tea, the tea-leaves floating at the bottom of the cup begin to cluster near the center of rotation instead of the expected motion to the walls of the cup (*Einstein's paradox*[23]). Namely, the mentioned convector mixing $\mathbf{u}\nabla\mathbf{u}$ produces the pressure that drops in direction to the central axis of any stable vortex tube.[36] The floating mixture moves in the direction of the lower pressure, which resolves the above paradox.

By the steady vortex tube, one can explain as well the decrement of temperature in the known *effect of Ranque–Hilsch*[10,12] that operates in mobile coolers used by American firemen. While coming through a hole, a gas inside the apparatus starts to rotate while producing the pressure drop in the direction of the axis of rotation. With a practically unchangeable density near the axis of rotation, the temperature as being proportional to the pressure, hence, drops, too. So, the rotation makes the incoming gas to be cooler to the center of rotation. As this takes place, the rest of the warm masses at the boundary of the tube go away through another hole.

By these outgoing warm masses from the center to the exterior of a separate vortex tube, a medium is capable of transfering the heat in its own way (and does it faster than the slow molecular diffusion). Hence, to heat water with mixing required a lot of vortices (after storming, a sea gets warmer). For this reason Joule used his trick with grids rotating inside a thermo-isolated volume.[1] Due the adiabatic *heating with vortices*, he found out the mechanical equivalent of heat, subsequently referred to as the specific heat at a constant volume.

## E.4. Structural Instabilities

Now, let us turn to pulsations as disturbances produced with two processes known to be the *parametric* and *structural* hydrodynamic instabilities. These are traditionally attributed to Reynolds and Lord Rayleigh,[4,5,9] respectively.

When related closely to small oscillations in mechanics, a theory of parametric instabilities is well developed and reduced to the so-called local (or linear) and nonlocal (or nonlinear) bifurcations of solutions of hydrodynamic equations.[14,17,18,21,22,27,28,34,38]

The idea of structural instability is much less quantitatively accomplished. Qualitatively, a structural instability is closely related to the topological one and excludes any inverse continuous transforms of a flow picture.[20,24,35,36] To be close to hydrodynamics, one may add that it assumes only the origin of a vortex. As with colored jets,[4] it takes place whenever layers are rolled up in vortices. It can also be caused by a difference in directions of density and pressure gradients[11] in tangential[2,8] or normal[16,25] velocity components, in densities,[5,13] temperatures,[6,7] eigenvalues,[19,20] inverse Rossby numbers[35] and other factors.[36,37,39]

On the other hand, the fact that the origin of vortices is due to the Euler's internal mixing can be illustrated by the *Saffman–Taylor instability*[15] as induced by the differences in viscosities. This one excludes the former mixing (here, the pressure gradient is proportional to the vector of velocity). As a result, fingers of instability have no centers of vortices, however curved they are,[42] as illustrated in Fig. E.2.

As to the viscous fluid taken in the presence of internal mixing, the large Reynolds numbers with no-slip condition lead to the large gradients of velocity, hence, to the structural instability near the rigid wall. Such structures as vortex horseshoes, rings, tubes, and protuberances are formed in the mixing zone between the laminar sub-layer and the turbulent kernel. And, it is the mixing zone where the logarithmic law is valid for the average velocity.

As this takes place, the above topological definition of the structural instability becomes too general to satisfy physics. So, one may put a question: *what makes the parallel layers roll up?* If the main reason for this is the large gradient of velocity, why the large gradient?

Be it some variational principle or something else, the only thing that remains to be affirmative is that the above internal mixing is basically responsible for this.

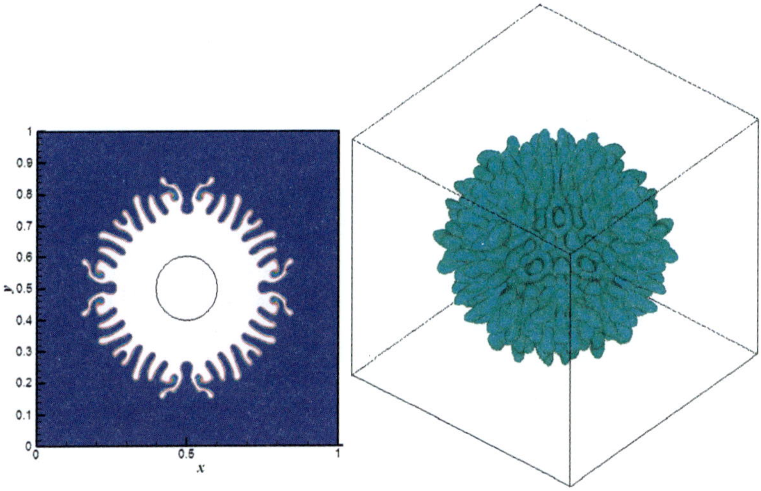


**Fig. E.2.** Viscous instabilities with fingers without vortices.[42]

## E.5. Vortex Cascades

Even though not knowing the answer to the question *why* instabilities arise, one can yet try to trace *how* they come into existence through the direct simulation of the main laws within a mixing zone. The Coette flow problem seems to be most suitable for the instability in question.

First, we have in this case the character Taylor vortex tube as if it is a cylinder passing through the flow (Fig. E.3).

Secondly, as time passed, we obtain a ray of cross-vortex tubes located on the first one whose axes are directed along the outer flow (Fig. E.4).

The process of origination of new rays of cross tubes is repeated, which leads finally to the picture of the developed turbulence (Fig. E.5).

## E.6. Principal Modes

Instabilities are ordinarily generated from basic Poiseule, Coette, or Kolmogorov flows that can be united in a boundary problem correctly stated for the proper linear equations.[40] In the case of the mixing zone producing vortices for the most part, we may restrict ourselves to the second of them to consider a layer bounded with two parallel planes, periodic in two orthogonal directions of planes. Then, in the related spectral method,

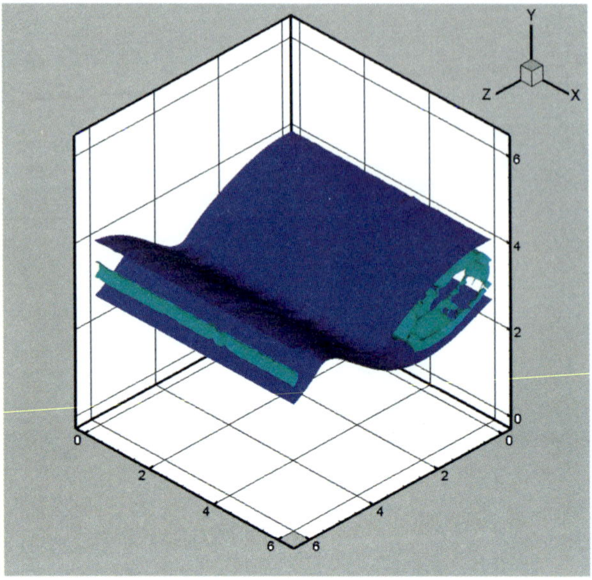


**Fig. E.3.** First instability of the Coette flow.

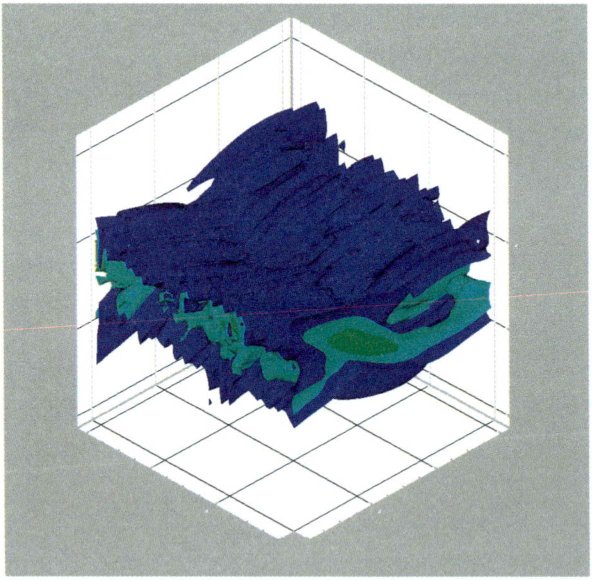

**Fig. E.4.** Secondary instabilities of the Coette flow.

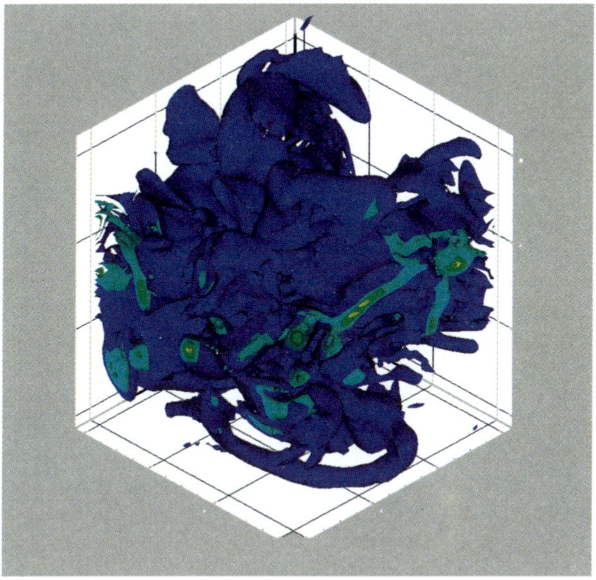

**Fig. E.5.** The developed turbulence in the Coette flow.

velocities of basic flow are found to be Fourier averages while the *Fourier pulsations* (with no averages) play part of disturbances. The latter counteract with the former to form a system equivalent to the initial one while dealing with four new variables: two horizontal averages of velocity and two vertical pulsations of velocity and vorticity.[41]

No matter, which form of equations are used, the Fourier decomposition has to satisfy the main principle that seems to justify the spectral method itself: the Fourier sum has to cover all the possible disturbances in a reasonable spectrum interval with a finite number of modes. Numerically, it will mean that only a finite part of spectrum will not be inertial. In other words, it is sufficient to have a finite number of main physical modes to cover secondary flows.

This hypothesis *on sufficiency of a finite number of modes* to cover the noninertial part of spectrum is in good agreement with experiments, where approximately only 2–5% of the total energy fall on the small-scale pulsations and at least 20–50% of it go to the large vortices.

## References

1. J. P. Joule, On the mechanical equivalent of heat, *Phil. Trans.* **140**(Part 1), 61–82 (1850).
2. H. L. F. Helmholtz, *Uber discontinuierlich Flussigkeitsbewegungen* (Monatsberichte konigl. Akad. Wissenschaften Berlin, 1868).
3. W. Thomson, On the propagation of laminar motion through a turbulently moving inviscid liquid, *Phil. Mag.* **4**(47), 342 (1887).
4. O. Reynolds, On the dynamical theory of an incompressible viscous fluid and the determination of the criterion, *Phil. Trans. Roy. Soc.* **A186**, 123 (1895).
5. Lord Rayleigh (J. W. Strutt), Investigation of the character of the equilibrium of an incompressible heavy fluid of variable density, Scientific papers. Cambridge, England, pp. 200–207 (2007).
6. H. Bénard, Les tourbillions sellulaires dans une nappe liquide, *Révue des Sciences* **11**, 1261–1271, 1309–1328 (1900).
7. H. Bénard, Les tourbillions sellulaires dans une nappe liquide transportante de la chaleur par convection on réqime permanente, *Ann. Chim. Phys.* **23**, 62–144 (1901).
8. Kelvin (W. Thomson), Mathematical and Physical Papers, Vol. 4 (Cambridge University Press, Cambridge, 1910).
9. Lord Rayleigh (J. W. Strutt), On the dynamics of revolving fluids, *Proc. Roy. Soc.* **A93**, 148 (1916).
10. G. Ranque, Expériences sur la déten giratoire avec productions simultanées d'un énchappement d'air chaud et d'un énchappement d'air froid, *J. de Physique et de la Radium* **4**, 1125–1150 (1933).
11. A. A. Friedman, *Experience on Hydromechanics of an Compressible Fluid*, (ONTI, Moscow, 1934) (in Russian).
12. R. Hilsch, The use of the expansion of gases in centrifuga field as a cooling process, *Rev. Sci. Instrum.* **18**(2), 108ff (1947).
13. G. I. Taylor, The instability of liquid surfaces when accelerated in a direction perpendicular to the planes, *J. Proc. Roy. Soc. London, Ser. A* **201**, 192–196 (1950).
14. C. C. Lin, *The Theory of Hydrodynamic Stability* (Cambridge University Press, Cambridge, 1955).
15. P. G. Saffman and G. I. Taylor, The penetration of a fluid into a porous medium of Hele-Shaw sell containing a more viscous liquid, *Proc. Roy. Soc. London, Ser. A* **245**, 312–329 (1958).
16. R. D. Richtmyer, Taylor instability in shock acceleration of compressible fluids, *Commun. Pure Appl. Math.* **13**, 297–319 (1960).
17. L. D. Meshalkin and Ya. G. Sinai, Investigations of the stability of a stationary solution of a system of equations for the plane movement of an incompressible viscous fluid, *J. Appl. Math. Mech.* **25**, 1140–1143 (1961).
18. O. A. Ladyzhenskaya, The mathematical theory of viscous incompressible fluid, *Math. Appl.* **2** (1963).

19. V. I. Arnold, Sur la topologie des écoulement stationnaires des fluids parfaites, *CRAS* **261**(1), 17–20 (1965).
20. V. I. Arnold, Sur la géometrie différentielle des groupes de Lie de dimension infinie et ses applications à l'hydrodynamique de fluides parfaits, *Ann. Inst. Fourier (Grenoble)* **16**, 319–361 (1966).
21. Yu. P. Ivanilov and G. N. Yakovlev, On the bifurcation of fluid flows between rotating cylinders, *J. Appl. Math. Mech.* **30**, 910–916 (1966).
22. V. I. Yudovich, On the bifurcation of rotational fluid flows, *Soviet Phys. Dokl.* **11**, 306–309 (1966).
23. A. Einstein, Reasons for formulation of curved riverbeds and the so-called Baire law, in Collected Scientific Works, Vol. 4 (Moscow, Nauka Publ., 1967), 74–77 (in Russian).
24. D. V. Anosov, Geodesic flows on closed Riemannian manifolds of negative curvature, *Proc. Steklov Inst. Math.* (1969).
25. E. E. Meshkov, Instability of an interface between two gases accelerated by a shock wave, *Mekh. Zhidk. I Gaza, Izv. Acad. Nauk SSSR* **5**, 151–157 (1969) (in Russian).
26. A. S. Monin and A. M. Yaglom, *Statistical Fluid Dynamics* (MIT Press, Cambridge, MA, 1971).
27. D. Ruelle and F. Takkens, On the nature of turbulence, *Commun. Math. Phys.* **20**, 167–192 (1971).
28. D. D. Joseph, Stability of fluid motions, I, *Springer Tracts Nat. Philos.* **27** (1976).
29. O. M. Belotserkovskii, Computational experiment: direct numerical simulation of complex gas-dynamics flows on the basis of Euler, Navier–Stokes and Boltsmann models, Karman's Lecture, Von Karman Institute for Fluid Dynamics, Numerical methods in fluid dynamics, H. J. Wirz and J. J. Smolderen (eds.) (March 15–19, 1976) (Hemisphere, Washington, London, 1978), pp. 339–387.
30. S. O. Belotserkovskii, Modelling of flows of a viscous incompressible fluid on the basis of the Navier–Stokes equations, Candidate Science thesis, Moscow Institute for Physics and Technology (1979) (in Russian).
31. O. M. Belotserkovskii, Direct numerical modeling of transition fluid flows and turbulence problems, in *Mechanics of Turbulent Flows* (Nauka Publ., Moscow, 1980), pp. 70–109.
32. V. E. Yanitskii, A statistical method for solving some problems of the kinetic theory of gases and turbulence, Doctoral thesis, Moscow (1984) (in Russian).
33. O. M. Belotserkovskii, Direct numerical modeling of free induced turbulence, *Comput. Math. Math. Phys.* **25**(6), 166–183 (1985).
34. D. N. Zubarev, V. G. Morozov and O. V. Troshkin, A bifurcational model of turbulent flow in a channel, *Soviet Phys. Dokl.* **31** (1986).
35. O. V. Troshkin, Topological analysis of the structure of hydrodynamic flows, *Russ. Math. Surv.* **43**, 153–182 (1988).
36. O. V. Troshkin, Nontraditional methods in mathematical hydrodynamics, *Trans. Math. Monographs* **144** (1995).
37. O. M. Belotserkovskii, *Turbulence and Instabilities* (MIPT, Moscow, 1999).

38. E. I. Oparina and O. V. Troshkin, Stability of Kolmogorov flow in a channel with rigid walls, *Dokl. Phys.* **49**, 583–587 (2004).
39. O. M. Belotserkovskii, A. M. Oparin and V. M. Chechetkin, *Turbulence: New Approaches* (CISP, Cambridge, 2005).
40. A. M. Oparin, E. I. Oparina and O. V. Troshkin, *Physical Modes in a Layer of a Viscous Incompressible Fluid* (Sputnik Company, Moscow, 2006) (in Russian).
41. O. V. Troshkin, On theory of a periodical layer of an incompressible fluid, *Comput. Math. Math. Phys.* **47**(4), 707–712 (2007).
42. M. S. Belotserkovskaya, Numerical simulation of filtration processes using the method of imbedding grids, Candidate Science thesis, Lomonosov Moscow State University (2007) (in Russian).
43. V. M. Canuto, Turbulence and laminar structures: can they co-exist? *Mon. Not. R. Astron. Soc.* **317**, 985–988 (2000).

# Appendix F

# List of the Main Publications of O. M. Belotserkovskii

### *Monographs*


1. O. M. Belotserkovskii, *Flow of Supersonic* Gas *around Blunt Bodies: Theoretical an Experimental Studies* [in Russian], (Vychisl. Tsentr Akad. Nauk SSSR, Moscow, 1967), 402 p.
2. O. M. Belotserkovskii, *Numerical Methods for Solving Problems of Mechanics of Continuous Media* (NASA, TTF-667, Washington, 1972), 212 p.
3. O. M. Belotserkovskii, Yu. P. Golovachev, V. G. Grudnitskii, Yu. M. Davydov *et al.*, *Numerical Investigation of Modern Gas Dynamics Problems* [in Russian], (Nauka, Moscow, 1974), 400 p.
4. O. M. Belotserkovskii and Yu. M. Davydov, *The Method of Large Particles: Computational Experiment* [in Russian], (Nauka, Moscow, 1982), 392 p.
5. O. M. Belotserkovskii, *Numerical Simulation in Mechanics of Continuous Media* [in Russian], 2nd ed. (Nauka, Moscow, 1994), 442 p.
6. O. M. Belotserkovskii, *Etude (sketch) about turbulence* [in Russian], (ed.) O. M. Belotserkovskiy (Nauka, Moscow, 1994), 286 p.
7. O. M. Belotserkovskii, *Information Science and Medicine, Ser.: Cybernetics, Unbounded Capabilities and Possible Restrictions* [in Russian], (eds.) O. M. Belotserkovskii and A. V. Vinogradov (Nauka, Moscow, 1997), 221 p.
8. O. M. Belotserkovskii, V. A. Andryushchenko and Yu. D. Shevelev, *Dynamics of Three-Dimensional Vortices Flows in Non-homogeneous Atmosphere: Computational Experiment* [in Russian], (Moscow Yanus-K Publ., 2000), 350 p.
9. O. M. Belotserkovskii, *Modern Solution Method for Nonlinear Multidimensional Problems: Mathematics, Mechanics, Turbulence* (Edwin Mellen Press, Lewiston Queenston–Lamper, USA, 2000), 411 p.
10. O. M. Belotserkovskii and A. M. Oparin, *Numerical Experiments in Turbulence: From Order to Chaos* [in Russian], (Nauka, Moscow, 2000), 224 p.

11. O. M. Belotserkovskii, *Computer Models and Advances in Medicine* [in Russian], (eds.) O. M. Belotserkovskii and A. S. Kholodov (Nauka, Moscow, 2001), 312 p.
12. O. M. Belotserkovskii, A. M. Oparin and V. M. Chechetkin, *Turbulence: New Approaches* [in Russian], (Nauka, Moscow, 2002), 286 p.
13. O. M. Belotserkovskii, *Computer and Brain* [in Russian], (ed.) O. M. Belotserkovskii (Nauka, Moscow, 2005), 320 p.
14. O. M. Belotserkovskii, A. M. Oparin and V. M. Chechetkin (eds.), *Turbulence: New Approaches*, Revised and Extended (Cambridge Int. Sci., UK, 2005), 285 p.
15. O. M. Belotserkovskii, V. M. Chechetkin, N. N. Fimin and A. M. Oparin, *The Physical Foundation of Fluid Dynamics: Macroscopic and Kinetic Approaches* (Russia, Publ. House "Ekonomika", 2007), 212 p.

## *Papers*

1. O. M. Belotserkovskii, Flow around á circular cylinder with á bow shock wave, *Dokl. Akad. Nàuk SSSR* **113**(3), 509–512 (1957).
2. O. M. Belotserkovskii, Extended abstract of candidate's dissertation in mathematics and physics, *Steklov Mat. Inst. AN SSSR*, Moscow, 1957 [in Russian].
3. O. M. Belotserkovskii, Flow around á symmetric airfoil with á Bow shock wave, *Prikl. Mat. Mekh.* **22**(2), 206–219 (1958).
4. O. M. Belotserkovskii, A. Bulekbayev and V. G. Grudnitskii, Algorithms for numerical schemes of the method of integral relations for calculating mixed gas flows, *USSR Comput. Maths. Math. Phys.* **6**(6), 162–184 (1966).
5. O. M. Belotserkovskii and E. G. Shifrin, Transonic flow behind a detached shock wave, *USSR Comput. Maths. Math. Phys.* **9**(4), 908–931 (1969).
6. O. M. Belotserkovskii and Yu. M. Davydov, The non-stationary large particle method for gas dynamics calculations, *USSR Comput. Maths. Math. Phys.* **11**, 182–207 (1971).
7. O. M. Belotserkovskii and L. I. Severinov, The conservative flux method and the calculation of the flow of viscous heat conducting gas past a body of á finite size, *USSR Comput. Maths., Matyhs. Phys.* **13**(2), 141–156 (1973).
8. O. M. Belotserkovskii, S. D. Osetrova, V. N. Fomin and A. S. Kholodov, Hypersonic flow of á radiative gas around blunt bodies, *USSR Comput. Maths. Math. Phys.* **14**(3), 992–1003 (1974).
9. O. M. Belotserkovskii, V. A. Gushchin and V. V. Shchennikov, A method of splitting for solving problems of viscous incompressible flows, *USSR Comput. Maths. Math. Phys.* **15**(1), 190–199 (1975).
10. O. M. Belotserkovskii and V. E. Yanitskii, The statistical method of particles-in-cells for the solution of problems of the dynamics of a rarefied gas — Computational aspects of the method, *USSR Comput. Maths. Math. Phys.* **15**(5 part 1), (1975).

11. O. M. Belotserkovskii and V. E. Yanitskii, The statistical method of particles-in-cells for the solution of problems of the dynamics of a rarefied gas — Computational aspects of the method, *USSR Comput. Maths. Math. Phys.* **15**(6), 184–198 (part 2) (1975).
12. O. M. Belotserkovskii, V. V. Demchenko, V. I. Kosarev and A. S. Kholodov, Numerical simulation of some problems of laser compression of shells, *USSR Comput. Maths. Math. Phys.* **18**, 420–444 (1978).
13. O. M. Belotserkovskii, The direct numerical simulation of complex flows in gas dynamics based on the Euler, Navier–Stokes and Boltzman Models, Karman's Lecture, von Karman Institute for Fluid Dynamics, 15–19 March 1976, in *Numerical Methods in Fluid Dynamics*, (eds.) H. J. Wirz and J. J. Smolderen (Hemisphere, Washington-London, 1978), pp. 339–387.
14. O. M. Belotserkovskii and A. S. Kholodov, Numerical investigation of some problems in gas dynamics using grid characteristic methods, in *Proc. Intl. Conf. on Numerical Methods in Hydrodynamics* (Tbilisi 1978, Moscow, 1978), pp. 3747 [in Russian].
15. O. M. Belotserkovskii and A. S. Kholodov, On the numerical solution of problems for multidimensional hyperbolic equations, *Gas and Wave Dynamics* (3), 6–17 (1978) [in Russian].
16. O. M. Belotserkovskii and V. Ya. Mitnitskii, Some exact solutions of the problem on the magnetic dipole squeezed by superconducting fluid, *USSR Comput. Maths. Math. Phys.* **19**, 464–473 (1979).
17. O. M. Belotserkovskii, New numerical models in mechanics of continua and plasma physics, *Akad. Wiss. DDR, Zentralinst. Math. und Mech.* (5) (1980).
18. O. M. Belotserkovskii, A. I. Erofeev and V. A. Yanitskii, On the non-stationary direct statistical modeling method for rarefied gas flows, *USSR Comput. Maths. Math. Phys.* **20**, 1174–1204 (1980).
19. O. M. Belotserkovskii and V. G. Grudnitskii, Investigation of nonstationary gas flows with a complex internal structure using the integral relation method, *USSR Comput. Maths. Math. Phys.* **20**, 1400–1415 (1980).
20. O. M. Belotserkovskii, On numerical models in plasma physics, *Modern Problems of Mathematical Physics and Computational Mathematics* (1982), pp. 48–63 [in Russian].
21. O. M. Belotserkovskii, A. P. Byrkin, A. P. Mazurov and A. I. Tolstykh, A difference method with improved accuracy for the calculation of viscous gas flows, *USSR Comput. Maths. Math. Phys.* **22**, 1480–1490 (1982).
22. O. M. Belotserkovskii, V. G. Grudnitskii and Yu. A. Prokhorchuk, A difference scheme of the second order of accuracy on the minimal stencil for hyperbolic equations, *USSR Comput. Maths. Math. Phys.* **23**, 119–126 (1983).
23. O. M. Belotserkovskii, Application of mathematical methods and computers in cardiology and surgery, *Vopr. Kibern.* (83), 3–17 (1983) [in Russian].
24. O. M. Belotserkovskii, S. M. Belotserkovskii and V. A. Gushchin, Numerical simulation of nonstationary periodic flow of a viscous fluid in the wake of a cylinder, *USSR Comput. Maths. Math. Phys.* **24**(12), 1204–1216 (1984).

25. O. M. Belotserkovskii, A. I. Panarin and V. V. Shchennikov, Generalized difference schemes and method of parametric correction of difference schemes, *Cybernetics and Computers* (1), 99–117 (1985) [in Russian].
26. O. M. Belotserkovskii, Direct numerical modeling of free induced turbulence, *USSR Comput. Maths. Math. Phys.* **25**(12), 166–183 (1985).
27. O. M. Belotserkovskii, A. I. Panarin and V. V. Shchennikov, Pipelining of algorithms for solving problems in mathematical physics on the basis of the parametric correction of difference schemes, *Vopr. Kibern.* (99), 10–35 (1986) [in Russian].
28. O. M. Belotserkovskii, A. S. Kholodov and L. I. Turchak, Grid-characteristics methods in multidimensional problems of gas dynamics, in *Current Problems in Computational Fluid Dynamics*, (eds.) O. M. Belotserkovskii and V. P. Shidlovsky (Mir, Moscow, 1986), pp. 125–189 [in Russian].
29. O. M. Belotserkovskii, S. M. Belotserkovskii and A. F. Pastushkov, Numerical simulation of internal waves in the flow of a stratified fluid around a semicircular obstacle, in *Problems in Applied Mathematics and Information Science* (Moscow, 1987), pp. 11–21 [in Russian].
30. O. M. Belotserkovskii, V. A. Gushchin and V. N. Kon'shin, A method of splitting for analyzing stratified free-surface flows, *USSR Comput. Maths. Math. Phys.* **27**, 181–191 (1987).
31. O. M. Belotserkovskii, Rational numerical simulation in nonlinear mechanics, (ed.) O. M. Belotserkovskii (Nauka, Moscow, 1990) [in Russian].
32. O. M. Belotserkovskii and V. V. Shchennikov, Principles of rational numerical simulation in aerohydrodynamics, in *Rational Numerical Simulation in Non-Linear Mechanics*, (ed.) O. M. Belotserkovskii (Nauka, Moscow, 1990), pp. 9–22 [in Russian].
33. O. M. Belotserkovskii, P. Yu. Skobelev and V. V. Shchennikov, New principles of difference discretization in mathematical physics, in *Rational Numerical Simulation in Non-Linear Mechanics*, (ed.) O. M. Belotserkovskii (Nauka, Moscow, 1990), pp. 9–22 [In Russian].
34. O. M. Belotserkovskii, V. V. Demchenko and A. M. Oparin, Sequential appearance of turbulence in the Richtmyer–Meshkov instability, *Dokl. Akad. Nauk* **334**(5), 581–583 (1994) [in Russian].
35. O. M. Belotserkovskii, V. V. Demchenko and A. M. Oparin, Nonstationary 3D numerical simulation of the Richtmyer–Meshkov instability, *Dokl. Akad. Nauk* **354**(2), 190–193 (1997) [in Russian].
36. O. M. Belotserkovskii, V. V. Demchenko and A. M. Oparin, 2D and 3D comparative study of the Richtmyer–Meshkov instability development, in *Proc. Intl. Workshop on Physics of Nomorassible Turbulent Mixing* (Imprimerie Carastere, Marseille, 1997), pp. 86–89.
37. O. M. Belotserkovskii, S. A. Ivanov and V. E. Yanitskii, Direct statistical simulation of some problems in turbulence theory, *Comput. Math. Math. Phys.* **38**(3), 474–487 (1998).
38. O. M. Belotserkovskii and A. S. Kholodov, Majorizing schemes on unstructured grids in the space of indeterminate coefficients, *Comput. Math. Math. Phys.* **39**, 1730–1747 (1999).

39. O. M. Belotserkovskii and A. M. Oparin, A numerical study of three-dimensional Rayleigh–Taylor instability development, *Comput. Math. Math. Phys.* **40**, 1054–1059 (2000).
40. O. M. Belotserkovskii, Mathematical modeling for supercomputers: Background and tendencies, *Comput. Math. Math. Phys.* **40**(8), 1173–1187 (2000).
41. O. M. Belotserkovskii, M. N. Antonenko, A. V. Konyukhov, L. M. Kraginskii and A. M. Oparin, Numerical simulation of 3D atmospheric flows caused by large-scale fires or explosions using parallel machine (Irkutsk, 2003) [in Russian].
42. O. M. Belotserkovskii and A. R. Pavlyukova, Direct numerical modeling of turbulence: Coherent structures, laminar–turbulent transition, Chaos, *Comput. Fluid Dynamics J.* **10**, 280–285 (2001).
43. O. M. Belotserkovskii, A. M. Oparin and V. M. Chechetkin, Formation of large-scale structures in the gap between rotating cylinders: The Rayleigh–Zeldovich problem, *Comput. Math. Math. Phys.* **42**(11), 1661–1670 (2002).
44. O. M. Belotserkovskii and A. V. Konyukhov, Change of grid functions of dependent variables in finite-difference equations, *Comput. Math. Math. Phys.* **42**(2), 224–237 (2002).
45. O. M. Belotserkovskii, M. N. Antonenko, A. V. Konyukhov, L. M. Kraginskii, S. V. Fortova and A. M. Oparin, Implementation of universal parallel computations technology to spatial problems described by hyperbolic set of equations, in *Abstracts of V-Int. Congress on Mathematical Modeling* (Dubna, Russia, 2002), Vol. 1, 60 p.
46. O. M. Belotserkovskii, M. N. Antonenko, A. V. Konyukhov, L. M. Kraginskii, S. V. Fortova and A. M. Oparin, Universal technology of parallel computations for the problems described by systems of the equations of hyperbolic type: A step to super solver, *Comput. Fluid Dynamics J.* **2**, 456–466 (2003).
47. O. M. Belotserkovskii, L. M. Kraginskii and A. M. Oparin, Modeling of large-scale atmospheric flow with the help of parallel framework for hyperbolic equations, in *Abstracts of 8th Japan-Russian Symp. on Computational Fluid Dynamic* (Sendai Japan, 2003), pp. 106–110.
48. O. M. Belotserkovskii, A. M. Oparin and V. M. Chechetkin, Physical processes underlying the development of shear turbulence, *J. Exp. Theor. Phys.* **99**, 504–509 (2004).
49. O. M. Belotserkovskii, Solving complicated problems using supercomputers: Experience and tendencies, in *Modern Problems in Applied Mathematics*, Issue 1 (MZPress Publ., Moscow, 2005), pp. 9–73 [in Russian].
50. O. M. Belotserkovskii and Yu. I. Khlopkov, Monte Carlo methods in applied mathematics and computational aerodynamics, *Comput. Math. Math. Phys.* **46**(8), 1418–1441 (2006).
51. O. M. Belotserkovskii, P. I. Agapov and I. B. Petrov, Numerical simulation of the consequences of a mechanical action on a human brain under a skull injury, *Comput. Math. Math. Phys.*, **46**(9), 1629–1638 (2006).
52. O. M. Belotserkovskii, V. M. Chechetkin, S. V. Fortova, A. M. Oparin, Yu. P. Popov, A. Yu. Lugovsky and S. I. Mukhin, The turbulence in free shear flows and in accretion disks, *Astron. Astrophys. Trans.* **25**, 419–434 (2006).

53. O. M. Belotserkovskii, On structural analysis of turbulence, in *Investigation of Hydrodynamical Instability and Turbulence in Fundamental and Technological Problems by Means of Mathematical Modeling with Supercomputers*, (eds.) O. Belotserkovskii, Y. Kaneda and I. Menshov (Nagoya University, Japan, 2007).